# PLANT BREEDING SYSTEMS

TITLES OF RELATED INTEREST

*Basic growth analysis*
R. Hunt

*Class experiments in plant physiology*
H. Meidner (editor)

*Comparative plant ecology*
J. P. Grime, J. G. Hodgson & R. Hunt

*Crop genetic resources*
J. T. Williams & J. H. W. Holden (editors)

*Introduction to vegetation analysis*
D. R. Causton

*Introduction to world vegetation (2nd edition)*
A. S. Collinson

*Light & plant growth*
J. W. Hart

*Lipids in plants and microbes*
J. R. Harwood & N. J. Russell

*Physiology & biochemistry of plant cell walls*
C. Brett & K. Waldron

*Plant development*
R. Lyndon

*Plants for arid lands*
G. E. Wickens, J. R. Goodin & D. V. Field (editors)

*Plant tropisms*
J. W. Hart

*Processes of vegetation change*
C. Burrows

# PLANT BREEDING SYSTEMS

A. J. RICHARDS

*Department of Plant Biology*
*University of Newcastle upon Tyne*

London
UNWIN HYMAN
Boston Sydney Wellington

Published by the Academic Division of
**Unwin Hyman Ltd**
15/17 Broadwick Street, London W1V 1FP, UK

Unwin Hyman Inc.
955 Massachusetts Avenue, Cambridge, MA 02139, USA

Allen & Unwin (Australia) Ltd
8 Napier Street, North Sydney, NSW 2060, Australia

Allen & Unwin (New Zealand) Ltd
in association with the Port Nicholson Press Ltd
Compusales Building, 75 Ghuznee Street, Wellington 1, New Zealand

First published in 1986
Second impression 1990

---

**British Library Cataloguing in Publication Data**

Richards, A. J.
Plant breeding systems
1. Plant genetics
I. Title
581.1′5 QK981
ISBN 0-04-581020-6
ISBN 0-04-581021-4 Pbk

---

**Library of Congress Cataloging in Publication Data**

Richards, A. J.
Plant breeding systems
Bibliography: p.
Includes index.
1. Plants—Reproduction. 2. Plant-breeding.
I. Title.
QK825.R53 1986 581.1′6 86-1148
ISBN 0-04-581020-6 (alk. paper)
ISBN 0-04-581021-4 (pbk.: alk. paper)

---

Printed in Great Britain at
the University Press, Cambridge

# *Preface*

Plant genetics had a distinguished start in the middle of the 19th century. A Bohemian monk laid the basis for the science of genetics through his crossing experiments with garden plants such as peas. He showed that the matter of inheritance was particulate, and obeyed certain laws, which are today named Mendelian in his honour. Simultaneously, but quite unaware of Mendel's work, the other founder of evolutionary genetics was also undertaking experiments and making observations in his garden in Kent, England. Charles Darwin can be considered to be the true father of breeding system studies, for in his two books *The effects of cross and self-fertilisation in the vegetable kingdom*, and *The different forms of flowers on plants of the same species* he pioneered work on self-incompatibility, inbreeding depression, heteromorphy, gynodioecy, pollination biology, and gene flow, although he himself had no concept of the gene.

The discovery of plant chromosomes and the important role played by chromosomal mechanisms in plant evolution, dominated plant evolutionary genetics in the first half of the 20th century. During the same period, Anderson aroused interest in the evolutionary importance of hybridisation, and Turesson, Gregor and others demonstrated how plant species undergo localised adaptation to form 'ecotypes'. However, despite the lead taken by Darwin, studies on plant breeding systems proceeded slowly, although work in the early years of the 20th century revealed the genetic basis of self-incompatibility mechanisms. After the Second World War, Bateman and others published pioneering work on gene-flow in crop-plants such as fruit trees. At about the same time, Grant was working on the effect of pollinator behaviour and flower syndromes on speciation in North American genera, while Crosby, Allard and others were interested in the population genetics of inbreeding plants. Another important pioneer of this period was Dan Lewis, who published influential papers on topics as diverse as self-incompatibility, heteromorphy and gynodioecy, while Herbert Baker was also working in this field.

However, the past two decades have witnessed an explosive expansion of work on plant breeding systems, much of it originating in the United States of America, although workers in New Zealand, Australia and the United Kingdom have also played significant roles. This expansion has been triggered by a number of theoretical and technical advances in the subject, the most important of which have been:

(a) developments in algebraic and computer-based modelling of populational and genetic systems;
(b) analysis of the genetic structure of populations by the gel electrophoresis of enzymes;
(c) realisation of the importance of resource allocation to components of sexual function in determining the nature of breeding systems;
(d) interest by zoologists in plant pollination as a system to which foraging theory can be usefully applied;
(e) development and application of biochemical techniques, notably to recognition phenomena associated with self-incompatibility.

Perhaps the most interesting trend in modern work has been a move towards synthesis. Mechanisms which influence patterns of plant mating and reproduction do not operate in isolation, but as part of a total breeding system syndrome. This syndrome includes the pattern and structure of the population in time and space, the structure and growth of the individual plant (genet), the structure and attributes of the individual flower, the distribution of and energetic expenditure by male and female organs, the pattern of pollen dispersal (which may be influenced by pollinator behaviour), the receptivity of the stigma and style to pollen from various donors, and the number, size and dispersal of the seeds. These factors may be studied by population ecologists, plant physiologists, pollination ecologists, animal behaviourists, theoretical geneticists, plant breeders, analytical geneticists and plant biochemists, not forgetting amateur horticulturists and natural historians, using a very wide armoury of techniques. However, such specialised information is misleading when viewed in isolation, and must always be interpreted in light of our knowledge of the total plant breeding system. Increasingly this is occurring, and teams who work on plant breeding systems often contain workers with different kinds of biological expertise.

There seems to be a tacit philosophical assumption amongst nearly all workers in the field that the synthetic goal, the rationale that underlies studies on plant breeding systems, is that of evolutionary genetics. It is taken for granted that the attributes observed are under genetic control, are heritable and of adaptive significance. Darwin and Mendel continue to influence the subject after 130 years.

The study of plant breeding systems is of more than theoretical importance or academic interest. Most of the techniques and ideas which support the vital and heavily supported applied science of plant breeding depend on the 'pure' botanical study of plant breeding systems. It is my view that the study of plant breeding systems will continue to grow in importance as the calamitous, and possibly terminal, acceleration in human population growth creates ever greater demands on our ability to produce food and raw materials from plants. As the human population increases, so the pressure on natural communities will become ever more severe. As habitats dwindle, rates of plant extinction will spiral and precious resources of new forms of medicine, food, structural material and fuel will be lost for ever. Even for species of known economic importance, most genotypes will become extinct, and the gene-base on which the development of new genotypes depends will be critically depleted. For self-incompatible crops, fertility will be impaired, and we will be unable to breed new disease-resistant cultivars to combat the threats imposed by disease in modern non-rotating monocultures.

We still know very little about how big, how spaced and how widespread reserves of natural communities need to be if we are to allow breeding systems to maintain genetic variabilities and fecundities of potentially useful plants at viable levels. Already for many plant species depletion has probably occurred to a point where extinction has occurred, is likely to occur, or at which much useful natural variation has been lost. Although the loss of habitat is already horrific and the future appears very dismal, the importance of habitat and species conservation is now becoming clear to many governments. It is essential that reserves, game parks and forests are preserved in a way that will also preserve their gene stocks. Not only is much plant breeding system work on wild populations required now, but it will be required for the foreseeable future and as governments become more conservationally enlightened, so such studies should grow in importance.

Much money is currently being spent in the 'developed world' on techniques of biotechnology, cell and gene cloning and genetic engineering. Doubtless these exciting new techniques hold out much promise for the future. However, in the 'Third World' such techniques will prove too complex and energy-expensive to have much impact in the next few decades. In these areas with rapidly increasing and often hungry human populations it will be vital that climates and soils are safeguarded, new crops and new strains of crops are developed, and natural habitats are protected. Plant breeding system studies and plant breeding will play a vital role as we face the worldwide environmental crisis of loss of habitat, and loss of gene stocks that is already with us. The next 'scientific revolution' will have to address these problems, and very much more money and resource should be directed towards these areas by governments and international bodies than is the case at present.

It is remarkable that at this time of growth in work on plant breeding systems when the need for such studies becomes ever more acute, that there is no textbook which attempts to cover the field. There are many excellent texts on experimental taxonomy, cytogenetics, population genetics and evolution, and some of these touch upon plant breeding systems. There are also books on plant breeding, such as Frankel and Galun (1977), and books on particular aspects of plant breeding system studies. Good examples of the latter include Proctor and Yeo (1973) on pollination, de Nettancourt (1977) on self-incompatibility and Charnov (1982) on sexual resource allocation. Other fairly recent influential books, peripheral to the field, have been on the evolution of sex by Maynard Smith (1978), and the population biology of plants by Harper (1977) and by Solbrig *et al.* (1979). Recently, the Princeton University Press have published several small books which are relevant to plant breeding systems as monographs in population biology, for instance Willson and Burley (1982).

It seemed to me that there was a need for a book which attempted to bring together the various aspects of plant breeding systems under one cover in a synthetic way, and this I have tried to write. I have carefully considered the audience to whom the book should be addressed. The audience with which I am most familiar is the advanced undergraduate student at my university who will graduate with an honours degree in Plant Biology. I have long felt the need of a single textbook to which I could direct such a student for further reading, an introduction to the literature, an account of modern facts and theories in plant breeding systems, and a synthetic approach to the subject.

I have been interested in plant breeding systems since I was a graduate student 20 years ago, working on the genetics of agamospermy in *Taraxacum*. Since then I have taken an interest in heteromorphic systems (*Primula*), dioecious and gynodioecious systems (*Potentilla fruticosa* and *Saxifraga granulata*), selfing species (*Senecio vulgaris* and *Epipactis*) and pollination biology. I also have a life-long interest in natural history and field botany. This book naturally reflects these particular interests, and the theme that runs through it is that of the plant in its habitat, its natural history and evolutionary genetics, rather than that of theory and mathematical modelling. Another book complementary to this one, with emphasis on the latter theme needs to be written, but by another author!

I have tried to combine an explanation of the basic features of plant breeding systems, and a review of the literature, with modern concepts and theories, and some of the ideas are frankly speculative. I have assumed a basic understanding of genetics, evolution and botany such as it would be reasonable to expect an advanced undergraduate student to have. However, I hope that the book will prove of interest to the undergraduate, postgraduate and research worker alike, for I have tried to combine basic concepts with advanced ideas. I even hope that parts of the book will prove to be of interest to natural historians and horticulturalists without a formal biological training. I have tried to draw on my experience as a natural historian and a gardener to make certain sections accessible and interesting to the layman.

One matter that has greatly concerned me is the scope of the book. I had little doubt what should be included under breeding systems, but what plants should I be discussing? There is a good deal of interesting and important work on the breeding systems of the fungi, algae, bryophytes and pteridophytes. Partly because I know little of these groups and partly because the book was already far too long, I have regretfully abandoned any attempt to cover these groups. The book might more accurately be entitled 'Flowering plant breeding systems', for it is largely concerned with the Angiosperms. I have also included some material on the Gymnosperms, but this is relatively scanty.

A. J. Richards

# Acknowledgements

I owe an especial debt of gratitude to the research students who have worked under my supervision. They have provided me with ideas, discussion, companionship and a stimulating environment, as well as very many interesting results, some of which are used in this book. My thanks go to John Faulkner, Neil Mitchell, Trevor Booth, Halijah Ibrahim, Liz Gynn (née Culwick), Mike Mogie, David Stevens, Guido Braem, Chris Haworth, Jane Hughes and Fran Wedderburn. My thanks are also due to A. D. (Henry) Ford who worked with me as a postdoctoral research fellow for five years, and contributed many ideas and discussions during the writing of this book. He has read part of the manuscript, and has produced many useful comments as have David Stevens, Jane Hughes and Fran Wedderburn.

I should also like to thank Terry Crawford of the University of York, who read all the manuscript, and produced many invaluable criticisms. Miles Jackson, from George Allen and Unwin has been a constant source of encouragement and useful comment, and I am grateful to him. Brian Waters produced the line illustrations, and Mark Wilson, Dan Mitchell, Geoffrey Chaytor and Michael Proctor took many of the photographs. To these, my friends, I give my thanks.

I should also like to thank our charming and helpful secretaries Sandra and Susan Clothier, and Lynn Wilson for typing the manuscript.

Many botanical friends over the years have produced discussions, ideas or data which have been a help in the writing of this book. I cannot list all of them, but I would especially like to mention John Lawton, Richard Abbott, Richard Wilson, Clive Stace, Quentin Kay and Stan Woodell.

My thanks are due to David Valentine and Jack Crosby, my teachers at the University of Durham many years ago, who first stimulated my interest in plant breeding systems.

I would finally like to thank the following organisations and individuals who have given permission for the reproduction of illustrative material (figure numbers in parentheses):

M. C. F. Proctor (2.4, 4.3–5, 4.7, 4.11, 4.15, 4.20, 4.31–2, 4.36, 4.43, 7.8); D. Mitchell (4.14, 4.33a); D. C. Lang (4.33b); Figure 4.10 reproduced from *Pollination and evolution* (J. A. Armstrong *et al.*, eds) by permission of L. W. Macior and the publisher; the Linnean Society of London and P. G. Kevan (4.22–5); M. Adey (4.28); G. Chaytor (4.30, 4.37, 9.3); Figure 4.35 reprinted by permission of Harvard

University Press from *The orchids, their natural history and classification* (R. L. Dressler); J. Schmitt and the Editor, *Evolution* (5.2); R. W. Cruden (5.3–4); D. C. Paton (5.5); Figure 5.6 reproduced from Y. B. Linhart, *American Naturalist* **107**, 516–17, by permission of the author and publisher; K. D. Waddington and the Editor, *Oikos* (5.9); Figure 5.10a reproduced from G. H. Pyke, *Oecologia* **36**, 281–93 by permission of the author and publisher; Figures 5.10b–c reproduced from B. Heinrich, *Oecologia* **40**, 240 by permission of the author and publisher; Figure 5.11a reprinted by permission from *Nature* **284**, 450–1, © 1980 Macmillan Journals; S. N. Handel and the Editor, *Evolution* (5.11b); M. Wilson (7.10, 8.3, 10.9); T. Booth (9.2); Figure 9.7 reproduced from *Principles of plant breeding* (R. W. Allard) by permission of the author and John Wiley & Sons; Figures 9.8–9 reproduced from R. W. Allard *et al.*, *Advances in Genetics* **14**, 55–131 by permission of R. W. Allard and Academic Press; the Botanical Society of the British Isles (9.10); A. Strid and the Botanical Society, Lund (9.11–12); J. Faulkner (10.6); S. Asker and the Editor, *Hereditas* (11.9); Table 2.1 reproduced from *The evolution of sex* (J. Maynard Smith) by permission of the author and Cambridge University Press.

# *Contents*

CHAPTER ONE

# Introduction

---

Most animals, and all vertebrates, behave. They are able to choose their mates, and they unconsciously influence the genetic make-up and evolutionary patterns of their populations through this behaviour. Breeding system studies in animals concentrate on display, song, territoriality, migration, dispersal, mate-choice, fidelity, clutch size, after-care and many other features that are mostly direct or indirect consequences of behaviour. Male and female function is often separated into different individuals (almost always so in vertebrates), and thus males can behave as males and females as females.

Plants do not behave. Unlike most animals, higher plants are essentially stationary within a generation, although some plants have considerable powers of vegetative dispersal (Ch. 10), and pollen and seed dispersal between generations can also be sizeable (Ch. 5). They are not conscious, and so they cannot choose a mate. In most seed plants, male and female function is not separated into different individuals; they are hermaphrodite. It would be easy to conclude that plants do not control their breeding systems, and that pollen is dispersed on to stigmas of the same or different flowers in a haphazard and uncontrolled way. This is far from the case. Plants have an even more diverse armoury of methods by which breeding systems manipulate and control the genetic structure of populations, and patterns of evolution, than animals have. Like animal behaviour, plant breeding systems are under genetic control and can themselves be selected for. They are rarely fixed and static, but are fluid and respond to selection pressures in an infinite variety of subtle and interrelated ways.

Because breeding systems control genetic variability, and are themselves controlled by components of genetic variability, they are liable to feedback. This can take three forms:

(a) **Positive feedback**; for instance, inbreeding reduces genetic variability, and so should reduce the variability in the breeding system, which may reinforce the inbreeding (Ch. 9).
(b) **Negative feedback**; for instance, agamospermy (Ch. 11) usually arises through the recombination of several different genes or groups of genes mediated by outcrossing events, often of an extreme kind, such as hybridisation. The genetic effect of agamospermy is to restrict or totally halt all genetic recombination and variability.

(c) Stabilising feedback; for instance, the frequency of self-incompatibility *S* alleles in a gametophytic system should be equal for all alleles in a population. This equality is maintained by the frequency-dependent selection inherent in the mating system, and is reproductively the most efficient distribution of *S* alleles (Ch. 6).

Various invertebrate animals have a wide choice of reproductive options, comparable to that in plants. For instance, certain tapeworms can be unisexual or hermaphrodite. If hermaphrodite, they can undergo self-fertilisation or cross fertilisation. They can produce eggs sexually or parthenogenetically, and they can reproduce 'vegetatively' as well as by eggs. It may be significant that animals such as these with a variety of reproductive strategies (corals are another example) are static, as are plants.

However, in nearly all vertebrates, and in most 'advanced' invertebrates such as arthropods, reproduction is limited to sexual cross fertilisation between unisexual individuals. This is a mechanism analogous to dioecy in seed plants; but dioecy is unusual, being found in only about 5% of species (Ch. 8).

In contrast, most seed plants have a wide range of reproductive options available for selection. These may arise constantly, through genetic variation, or intermittently, through mutation, in most plant populations. Broadly speaking, these are:

(a) **Hermaphrodity versus unisexuality**; hermaphrodites can self-fertilise whereas unisexuals cannot (Ch. 8). Self-fertilisation tends to reduce genetic variability (Ch. 9).
(b) **Self-pollination versus cross pollination**; hermaphrodites have many interrelated features which influence the transfer of pollen from anthers to stigmas in the same flower (autogamy), or between flowers (allogamy). Allogamous pollen transfer may occur between flowers of the same genet (geitonogamy), or between different genets (xenogamy or outcrossing) (Fig. 1.1). The amount of within-flower pollination (autogamy) depends on the degree of separation that occurs between pollen donation (from anthers) and pollen reception (on to stigmas) in time and space within the flower. Separation in time is called dichogamy, and separation in space is called herkogamy (refer to the glossary on p. 463 to interpret unfamiliar terminology!). The amount of between-flower pollination (allogamy) that results in selfing (geitonogamy) depends on patterns of pollen travel between flowers, on the size of the genets (number of ramets and area of ground covered), and on the number of donating and receiving flowers open together on each ramet (see Figs. 5.1 and 5.2, Ch. 5). [A genet is a genetically distinct individual, which has arisen from a single zygote (mating event). It may be composed of one to

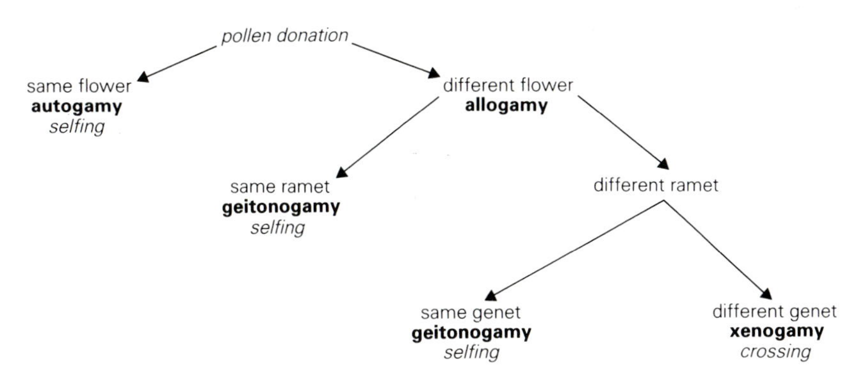


**Figure 1.1** Patterns of pollen transfer within and between flowers and plants.

many ramets, units that may be physically and physiologically dependent on one another, which has arisen from a genet by vegetative reproduction.]

(c) **Self-fertilisation versus cross fertilisation**; successful pollination does not necessarily imply successful fertilisation. Many plants reject selfed pollen through a mechanism known as self-incompatibility (Chs. 6 and 7). Self-incompatibility will lead to more outbreeding and greater genetic variability than will self-compatibility (Ch. 9) (see Fig. 1.3).

(d) **Sexuality versus asexuality**; there are two main mechanisms by which plants can reproduce asexually. These are vegetative reproduction (Ch. 10) and agamospermy (seeds without sex) (Ch. 11). Agamospermy is relatively unusual, but options for greater or lesser amounts of vegetative rather than sexual reproduction are available to most perennial plants. Asexuality produces little if any genetic variability, but maintains levels of heterozygosis.

The sexual organs of a seed-plant are contained within a flower. Genetic variation in distribution of anthers and gynoecia within and between flowers, and within and between plants, allows selection to occur for different patterns of distribution of sex organs. These are described in detail at the beginning of Chapter 8. Some of the commoner types are listed in Table 1.1.

In dioecious species, a genet can usually only express one sex. In hermaphrodite species (the majority of the Angiosperms, or flowering plants) every flower usually expresses both sexes. However, the majority of Gymnosperms, and a small proportion of Angiosperms, are monoecious, that is, the plant is hermaphrodite, but the flowers, the inflorescences, or even the ramets, are of one sex only. Monoecy is a little-studied phenomenon with respect to breeding systems, so it will receive

**Table 1.1** Some common types of sex distribution within and between flowers, and within and between genets of seed plants. Anthers are designated ♂, and gynoecia are designated ♀. Hermaphrodites are designated ⚥.

| Name | Distribution of sex organs within a flower | Distribution of sex organs within a plant | Breeding system | Angiosperm species (%) |
|---|---|---|---|---|
| dioecy | ♂ or ♀ | ♂ or ♀ | xenogamous (out-crossing) | 4 |
| gynodioecy | ⚥, ♂ or ♀ | ⚥ or ♀ | xenogamous, geitonogamous, autogamous | 7 |
| monoecy | ♂ or ♀ | ⚥ | allogamous, some selfing, some crossing | 5 |
| gynomonoecy | ⚥ or ♀ | ⚥ | allogamous and autogamous | 3 |
| hermaphrodity | ⚥ | ⚥ | allogamous and autogamous | 72 |
| (other) | | | | 9 |
| | | | | 100 |

little attention in this book. However, it usefully illustrates a number of features of general importance in breeding system evolution, and so it is discussed in some detail at this early stage.

Monoecy is widespread in large wind-pollinated plants such as trees and sedges and in water-plants, including those pollinated by water (Ch. 4). It is uncommon in plants that are visited by insects for pollen, presumably because female flowers, inflorescences, or ramets are unrewarding (Fig. 1.2).

One of the more interesting features of monoecious species is that they can change sex, or change their resource allocations to the sexes. Frankel and Galun (1977) showed that the proportions of male and female flowers are under the control of plant growth substances, and can be experimentally influenced by the quality and quantity of light, and levels of nutrition, water and mineral supply (see also Ch. 8). There is a tendency for male flowers to be produced earlier, and over a longer period of time, in relatively proximal positions on the plant. Female flowers tend to be produced later in the season, and in relatively distal parts of the plant. These trends are related to concentration gradients within shoots of plant growth substances such as auxin and gibberellins.

There is also a tendency for younger, smaller plants to be mostly or wholly male, and older, larger plants to be mostly or wholly female. A monoecious plant will change its sex expression during its lifetime, and this switch may well be reproductively beneficial. Larger, older females

**Figure 1.2**
The monoecious inflorescences of the wind-pollinated alder *Alnus glutinosa*. Female flowers on left, male flowers on right (×1).

will have more resources available to optimise expensive female functions (the production of fruits and seeds, etc.). Charnov (1984) showed that larger, more female plants tend to predominate in microenvironments within a habitat where the species performs more vigorously, whereas the smaller males predominate in less salubrious sites. Such optimisation of sexual function with respect to microenvironment is well illustrated by Policansky (1981) and Lovett Doust and Cavers (1982) for the North American woodland jack-in-the-pulpit, *Arisaema triphyllum*. The relationship between size and sex expression can vary between sites in this species. Males in one site were as large as females at another, but, in any one site, females were always much bigger than males.

Sex expression can be controlled by the environment irrespective of size and age. Dodson (1962) changed the sex of female *Catasetum* orchids from female to male by moving them from sun to shade; here, as in most instances, less salubrious niches tended to induce maleness, and more optimal sites tended to induce femaleness.

All monoecious species are potentially hermaphrodite, and most of them bear both male and female flowers. They are capable of self-fertilisation, at least potentially, through geitonogamy. However, most tend to be dichogamous, usually protandrous, so that males flower before females within a genet. As a result, most seed-set will be outcrossed (xenogamous). Even if they are homogamous (male and female function together), some seed is likely to be outcrossed. Because monoecious plants can separate male and female function in time and space within a genet, they are less likely than hermaphrodites to encounter sexual disharmony of a structural or resource allocation type (Ch. 2). Competition between male and female function should be minimised.

It may also be advantageous for a plant to be monoecious for only one sex, i.e. gynomonoecious (⚥♀) or andromonoecious (♂⚥). The unisexual sex will be that on which the plant must expend more to optimise fitness. If seed-set is restricted by inadequate pollination, selection will favour plants that produce extra male flowers (andromonoecious). If seed number, or seed and fruit size, is limited by male expenditure, it may pay the plant to produce some female flowers (gynomonoecious) (Webb 1981a,b). There is further consideration of these topics in Chapter 8. Gynomonoecy and andromonoecy are widespread in plants with complex coadapted inflorescences in which division of labour between florets occurs, as in the Umbelliferae (Apiaceae) and Compositae (Asteraceae), which are discussed further in Chapter 4.

Differential allocation of resources between male and female function is by no means restricted to monoecious conditions. It is maximised in gynodioecious and dioecious states, and may be a potent influence in the evolution of such diclinous conditions (Ch. 8). It may equally be found in plants with hermaphrodite flowers (Ch. 2) (see Charnov 1984 and Lloyd 1983 for reviews).

Likewise, the separation of male and female functions in space and in time is not a prerequisite of monoecy. Diclinous conditions separate sexual functions in space by definition, and separation in time is also common (Ch. 8).

Very many plants with hermaphrodite flowers also separate male and female function in space and time (Ch. 4). When anthers and stigmas are spatially separated within a flower, this is termed herkogamy. Within-flower pollination (autogamy) cannot occur, or can only occur as a result of an animal visit, and herkogamous flowers are mostly allogamous. Herkogamy may be especially common in self-incompatible species (Chs. 6 and 7) as it may encourage pollen transport between genets, and reduce the clogging of stigmas with non-functional selfed pollen. Even in self-compatible species, herkogamous allogamy will encourage mixed-strategy mating systems with some selfing and some crossing (Ch. 5).

When male and female functions of pollen donation and pollen receipt are separated by time, it is termed dichogamy. If pollen donation precedes stigmatic reception, it is called protandry. If stigmatic reception precedes pollen donation, it is termed protogyny. Like herkogamy, dichogamy will promote allogamy, increase the efficiency of outbreeding, and in a self-compatible species it will encourage mixed-strategy mating systems with some selfing and some crossing. This is well-illustrated in the onion, *Allium cepa*, by Currah and Ockendon (1978) and

Ockendon and Currah (1979). An onion flower cannot self-pollinate but, because there are many flowers in a head, there is much geitonogamous selfing between flowers. If the number of flowers is artificially reduced, the amount of outcrossing increases. Outcrossing will be an inverse function of:

(a) The number of flowers per head.
(b) The number of flowers visited per head by an insect.
(c) The likelihood of an insect flying to another head of the same plant.
(d) The amount of flowering-time overlap between flowers of the same head.

Dichogamy is also further discussed in Chapter 5.

Other features of floral timing are important components of the breeding system. These may be divided into the following categories:

(a) **Season**. Most species only flower during a certain season. Individuals that flower out of season will be reproductively isolated (Ch. 5). Flowering season is under genetic control in all species that have been investigated, and is subject to natural selection. Early flowering genotypes may be at a disadvantage with respect to cross pollination (but less so if protandrous than if protogynous). They may, however, be rendered fitter as seeds ripen before later-flowering genotypes. The latter will be reproductively fit, but their seedlings may be outcompeted by those from earlier flowering plants. There will be stabilising selection for flowering time.
(b) **Length of flowering per genet**. Individuals that produce few flowers at any one time over a long period ('steady state') adopt a different strategy from those that produce their flowers all together ('big bang'; Ch. 5). The latter may be more likely to self.
(c) **Length of flowering per flower**. If the life of a flower is expressed as a fraction of the flowering period of a genet, those with short lives are more likely to be outcrossed than those with long lives, but they may be reproductively less efficient.
(d) **Time of day of flowering or flower opening**. Some plants restrict the variety of their flower visitors, and increase their reproductive efficiency (and perhaps their outcrossing) by only opening at certain times of day. For instance, the bat-pollinated kapok tree (*Ceiba pentandra*) only flowers from 17.00 to 08.00 hours, and the moth-pollinated cactus *Selenicereus grandiflorus* from 20.00 to 02.00 hours. Flowering time can vary with the age of flower. According to Synge (1947), the broad bean (*Vicia faba*) flower opens at 16.00 hours on its first day, 13.00 hours on its second day, and at 11.00 hours on its third day.

Even the size and vigour of a plant may influence its breeding system, especially if it has hermaphrodite flowers, as most do. Ramets with large numbers of flowers, and genets with large numbers of ramets, are more

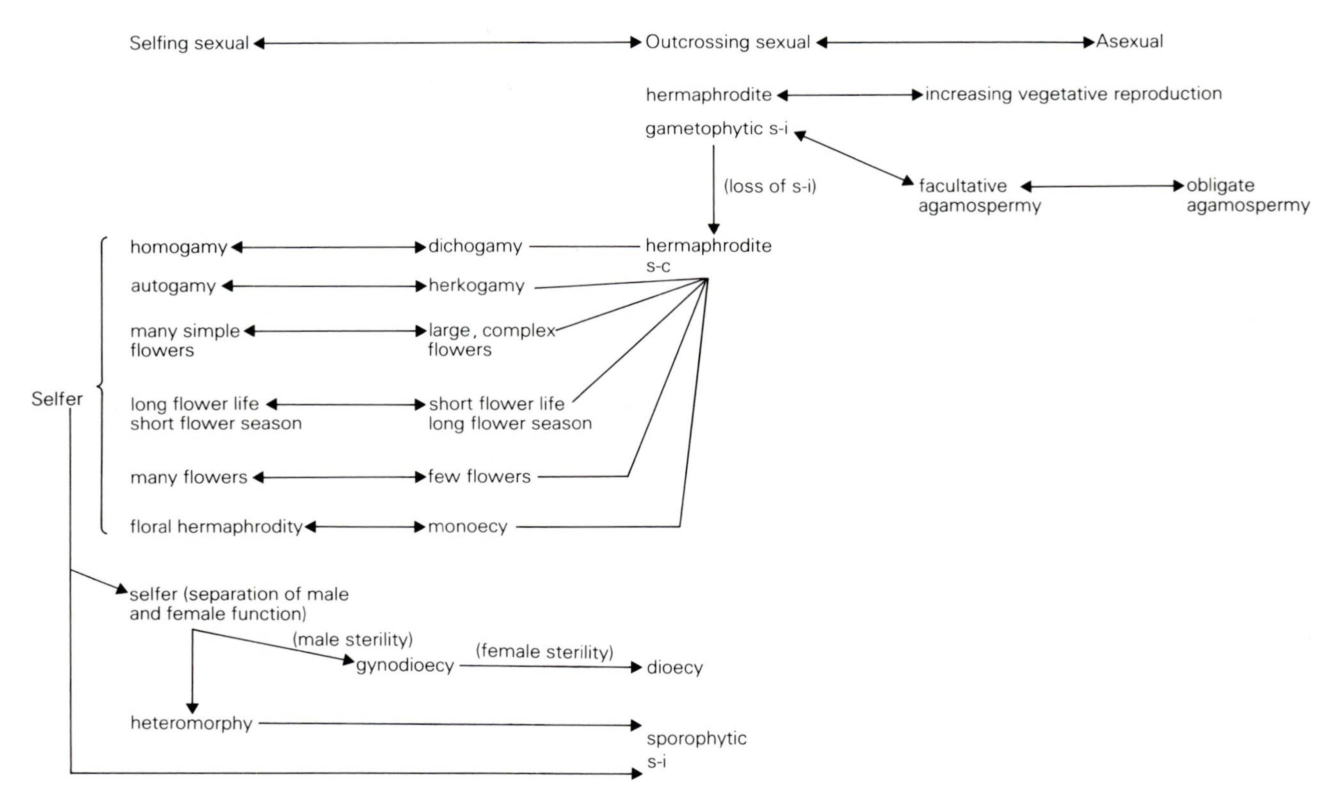


**Figure 1.3** Evolutionary trends in the breeding systems of seed plants.

likely to be geitonogamously selfed than smaller plants. If bee-pollinated, smaller genets and smaller ramets are likely to pollinate more individuals, and over a larger area, than are larger ramets, for pollinators will travel further in search of more rewarding 'patches' (Ch. 5). There will be stabilising selection for ramet size, and perhaps for genet size (number of ramets). If too small, they will compete poorly with surrounding vegetation, and will produce few seeds. If too large, their offspring may suffer a disadvantage from self-fertilisation.

As is discussed more thoroughly in Chapters 2, 4 and 6, it seems most likely that the original flowering plant was insect pollinated with hermaphrodite, self-incompatible flowers, and with limited powers of asexual reproduction. In short, it was a sexual outbreeder. Various levels of inbreeding, and various levels of asexuality, have since evolved secondarily, and in many cases these have once again given rise to outbreeders, through different mechanisms. A basic scheme of major trends in the evolution of breeding systems that I infer from modern-day distributions is presented in Figure 1.3.

The effects of different breeding systems on the genetic structure of populations, is a complex subject which is discussed in Chapters 2, 5 and 9–11. It is also summarised in the concluding section (Ch. 12).

The study of plant breeding systems, and their effect on the genetic structure of populations is of basic importance to students of evolution and population genetics. It is also of considerable applied importance.

Consider a tomato sandwich. All its constituent parts are made of the products of plant breeding systems, seeds and fruits. The bread is made from the seeds of the inbreeding annual cereal *Triticum aestivum* (wheat). The margarine may be made from the oily seeds of the monoecious, wind-pollinated maize, *Zea mays*, or the insect-pollinated hermaphrodite self-compatible sunflower *Helianthus annuus*.

The tomato itself is the fruit of an insect-pollinated annual, *Lycopersicon esculentum*, in which selection has occurred for self-compatibility and autogamy from an originally self-incompatible and herkogamous ancestor (Simmonds 1976). Pollen is only released from the poricidal anthers by vibration, so it is necessary to shake or hose the flowers in the absence of insect visits if fruit is to be set.

The pepper (*Piper nigrum*) is a perennial vine, the fruits of which are dried and ground to give the black pepper used to flavour the tomato sandwich. Originally dioecious, cultivated clones are monoecious and insect pollinated. Remarkably, little appears to be known about the pollination biology of this little-studied crop.

Most of the staple crops of the world, wheat, maize, rice, sorghum, millet, beans, bananas, coconut and many others, are fruits or seeds, the yield of which is a direct product of the breeding system of the plant and its efficiency. The better we understand this breeding system, the better shall we all be fed.

CHAPTER TWO

# *Sexual theory in seed plants*

---

## FUNDAMENTALS OF SEXUAL REPRODUCTION

### *Variability*

In eukaryotic organisms, such as seed plants, sexual reproduction has three major features that generate variability: recombination, segregation and sexual fusion.

**Recombination** Recombination constitutes the resorting of alleles of different loci linked on homologous or homoeologous chromosomes, by chiasmata during the heterotypic phase of meiosis (usually meiosis I).

**Segregation** Segregation is the inclusion into a gametophyte, usually at random, of any one of the four chromatids resulting from a bivalent association of two homologous chromosomes at the heterotypic phase of meiosis. Chromatids of different chromosomes enter gametes randomly with respect to each other, and alleles of heterozygous unlinked loci are randomly resorted with respect to each other between the parent and the gamete.

Segregation is a very powerful mechanism for generating variability. Irrespective of recombination, and assuming all pairs of homologous chromosomes in a diploid are heterozygous at at least one locus, the number of different gamete genotypes generated by a single diploid parent is $2^n$, where $n$ is the haploid chromosome number. In a plant where $n$ is 20 (a common and not untypical number), over $10^6$ (1 000 000) different gamete genotypes can be created by a single parent. In fact, for most plants, the number of potential gamete genotypes far exceeds the actual number of female gametes produced per parent; even for male gametes, this is likely to happen when $n$ is greater than about 12.

**Sexual fusion, or syngamy** Syngamy is the fusion of haploid, or functionally haploid, male and female gametes to form a diploid, or functionally diploid, zygote. As a result of the power of recombination and segregation in creating gametic variation, the number of potential zygote genotypes created even from matings between the same parents, or from selfing, is immense and can be estimated by $(2^n)^2$. In fact, the rare chance of exactly the same gamete genotypes meeting will render the actual figure slightly lower, and in largely homozygous habitual inbreeders it will be much lower. However, in the example quoted in the section on segregation, the potential number of zygote genotypes from a single diploid mother is about $10^{12}$. It must be realised that species with low chromosome base numbers will create much lower zygotic variability. Thus, where $n = 4$ (a condition not uncommon in higher plants) the potential number of zygote genotypes, ignoring recombination, is only 256, which may be fewer than the number of seeds produced per mother. Thus, an important consequence of low chromosome base number is a restriction on the release of variability (Ch. 9). However, for most plants, the chances of a single outcrossed parent giving rise to any two offspring with the same genotype will be very small indeed.

### *Gene migration*

A second important feature of sexual reproduction in nearly all plants is that it allows gene migration. Gene migration has two components, (a) **gene travel** (the motility of pollen and seeds (Ch. 5) allows genes to travel within, and to a lesser extent between, populations, and so a successful mutation, or linkage group can spread spatially) and (b) **gene incorporation** (cross fertilisation allows a successful mutation to spread between different mother–offspring lines).

The importance of gene migration is best illustrated by reference to asexual plants, for instance the agamospermous dandelions, *Taraxacum* (Ch. 11). In most *Taraxacum* species, seed is produced without fertilisation, and pollen is therefore non-functional. It might be supposed that a selective advantage would be gained by male-sterile plants without pollen (pollen and anthers comprise about 5% of the dry-matter production of reproductive effort in dandelions). Indeed, we find that many *Taraxacum* agamospecies do not produce pollen, and this is true of about 35 of the 200 or so agamospecies recorded in the British Isles (17%). All of the species in section Spectabilia are male sterile in contrast to only 2% of species in section Vulgaria. In addition, a number of species in sections Erythrosperma and Naevosa can vary in this respect. For instance, *T. rubicundum* usually produces pollen in the south of Europe, but is male sterile in the north. Many populations have male-fertile and male-sterile individuals coexisting. It is evident that continuing mutations produce further possibilities for male sterility. Occasional male-sterile individuals are found in otherwise polliniferous species, as I found for a single individual of *T. brachyglossum*, and there are many other instances on record (e.g. Richards 1972).

Assuming that male sterility is advantageous for agamospermous *Taraxacum*, and there is no direct evidence for this, the spread of this mutant is very seriously hampered by the asexuality of these plants. A new mutant gene is limited to the direct line of descent from the parent in which it arose, and is spatially limited to the seed-dispersal capabilities of that line. Its success will entirely depend on the total fitness of the line in which it occurs. It is useful, although perhaps not entirely accurate, to consider the genome of a line of descent (or seed clone) in *Taraxacum* as a single linkage group, and thus the new mutant as a 'hitch-hiker' (p. 17) on that linkage group. It will be subject to the selection pressures on that linkage group as a whole. Its success, unless it is extremely favourable, which in the present case is probably not so, will thus be subject to minority-type disadvantage. Male-sterile types can only become more common by repeated mutations, and by the relative success of male-sterile lines in contrast to male-fertile ones. Thus, it is not surprising to find the scattered and inconsistent pattern of male sterility in *Taraxacum* that we observe.

Yet, if a similarly successful mutant had arisen in a sexual population, it would be rapidly disseminated throughout the population, and would spread to other populations. It would not be limited to one linkage group, but could be associated with all the genotypes produced. By hitch-hiking on successful chromosomes, it might soon become fixed in the population.

## DISTRIBUTION OF SEXUAL REPRODUCTION

Sexuality is a striking and pervasive feature of nearly all eukaryotic organisms, and although many plants and animals have alternative asexual modes of reproduction, sexuality is absent from only a relatively very few parthenogenetic animals, agamospermous plants and sterile (usually hybrid) plant clones. In fact, recombination and a kind of sexual fusion substantially predate the eukaryotic cell, and may thus be nearly as old as life itself. Because all eukaryotic individuals have separate chromosomes, which undergo meiosis prior to gamete formation in a remarkably constant way, the principles of recombination, segregation and fusion are common to all higher organisms. This remarkable uniformity in the highly complicated procedures of sex among all eukaryotic organisms from algae to elephants, and protozoa to pine trees promotes two deductions. First, meiotic mechanisms, and consequently recombination, segregation and fusion, are primitive to the eukaryote cell, and are perhaps a vital part of its make-up and survival. Second, there are evolutionarily vital features of sexual reproduction, which have only allowed long-term survival of lines that maintain all the complex features of meiosis, and hence of sexual reproduction. This is particularly striking, for as we shall see later, some features of sexual reproduction appear to be disadvantageous in comparison with asexuality. Asexual mutants frequently arise in most sexual lines (Chs. 10 and 11), so the future of sexuality is periodically challenged by asexuality, which may convey short-term advantages.

## ANISOGAMY

Another remarkably constant feature of eukaryote sexuality is anisogamy. Although a few of the more primitive and mostly unicellular algae are isogamous, in which the fusing gametes appear identical, nearly all eukaryotes are anisogamous. These have gametes of very different sizes, invariably with one (which is called female) being much larger, sessile and less numerous, and the other (male) being smaller, actively or passively motile, and more numerous. It is generally accepted that anisogamy has evolved as a result of the vastly differing constraints operating on the gamete and the zygote. To be reproductively most efficient, gametes should usually be as small, numerous and as active as possible in order to create the maximum number of zygotes per unit of energy. Conversely, zygotes should be as large as possible, within physical and physiological constraints, so as to compete successfully for resource with other zygotes. Here is a classical example of disruptive selection at work, two directional selection pressures working in total opposition towards optimal size thresholds.

An early evolutionary solution to this problem was dimorphy in gamete size and number. One gamete type became smaller to a threshold in accordance with the requirements of gamete fitness, whereas the other became larger to a threshold in accordance with the requirements of zygote fitness. Being larger, the female gamete would be better fitted to receive post-zygotic parental (maternal) resource and protection, and would tend to be sessile in all bryophytes and vascular plants, although not most algae, and in most terrestrial animals.

The versatility of anisogamy in plants is clearly demonstrated by the replacement of the actively motile male gamete (antherozoid, which is sperm-like) in bryophytes and pteridophytes by the passively motile male gamete in seed plants. In seed plants, the male gamete, which is represented mostly by a nucleus, is transported by the male gametophyte (pollen grain and pollen tube; see Ch. 3.). It therefore becomes independent of water for sexual fusion. The requirement of water for sexual fusion is ecologically limiting for the bryophytes and pteridophytes. Motility in the male gametophyte may be favoured in plants which, unlike many animals, are themselves essentially sessile; pollen grains are much more mobile than are antherozoids. Seed plants have also developed specialised reception tissues (pollen chamber, stigma) for pollen-grain lodgement and pollen-tube growth. These tissues have been instrumental in the evolution of the ovule and ovary, and have been of great adaptive value to seed plants (Ch. 3).

## FUNCTIONS OF SEXUAL REPRODUCTION

The outstanding feature of sexual reproduction in outbreeders is the creation of an effectively infinite array of genetic diversity, even within populations and between siblings. In contrast, asexuals will only vary within and between lines by *de novo* mutations. These mutations will tend to accumulate within an asexual line (so-called Müller's ratchet), for the only way they can be lost is through natural selection. If they are recessive, or if they are only slightly disadvantageous, they will tend to persist. Only a tiny fraction will be advantageous and, as we have seen with pollen sterility in *Taraxacum*, even these will be restricted to their line and linkage group.

It is commonly supposed that the great success of sexuality has resulted from continuing requirements for evolutionary potential. The genetic variability inherent in sexual lines allows evolutionary responses to changing or novel environmental conditions; these are impossible in asexual lines. However, it must never be forgotten that genes controlling breeding systems can only evolve with respect to the selection pressures they encounter. If these pressures favour asexuality and genes that promote asexuality are present, asexuals will result.

A plant cannot develop a breeding system that anticipates its evolutionary future. At the same time, viewing as we do the products of an evolutionary past, it is possible to argue that the preponderance of sexuality today shows that sexual lines have persisted more successfully than asexual lines (Ch. 11). What this does *not* tell us is why all lines have not become asexual, and thus ultimately extinct. There is ample evidence that many sexual lines encounter opportunities to become asexual, but have 'shunned these opportunities'. Sexual reproduction is successful in the short term as well as the long. Why this should be so is the subject of continuing debate, stimulatingly reviewed by Maynard Smith (1978). Fundamentally, there are two main schools of thought, which are often regarded as being in opposition, but which in my view are complementary; I don't see why both should not be right, for both are intuitively plausible.

The first view is that all environments are inherently highly heterogeneous in time and space (although some more than others). Therefore a population that is genetically variable will be able to fill far more niches within that environment than one that is genetically invariable. It is curious that Maynard Smith dismisses this argument so cursorily (e.g. Maynard Smith 1978, pp. 89–93, p. 95). It is self-evident to any enquiring natural historian or ecological geneticist that most natural environments are infinitely variable in many independently varying attributes. No two individuals will ever inhabit exactly the same niche, and high levels of genetic variability for many characters will be at an evolutionary premium. Detailed evidence for this vital argument is misplaced in this book on breeding systems, but some support for such concepts, which are fundamental to ecological genetics, will be found in Chapter 10. For the present, suffice it to note that clones of dutch clover (*Trifolium repens*) are apparently selected strongly for coexistence with the genotype of ryegrass (*Lolium perenne*) forming their nearest neighbour (Turkington & Harper 1979). Asexual clones of *Taraxacum* from a single habitat differ in their competitive ability with respect to different grasses occurring in that population (Ford 1981). Examples of such fine-grain environmental heterogeneity favouring genetic variability are legion, and increasing yearly.

Maynard Smith has been concerned that arguments favouring sexual reproduction in a heterogeneous environment are group-selectionist. I can see no such difficulty. If a sexual parent leaves variable offspring able to fill more environmental niches than an asexual parent, it will be fitter than the asexual parent, and the inherited sexual mode of reproduction will predominate in a Darwinian manner.

That environments differ in time as well as in space is self-evident in all seasonal climates, and in all climates with weather, which is, by definition, unpredictable. There are many examples of time-niche adaptation within as well as between species, for instance between coexisting dioecious males and females (Ch. 8).

Maynard Smith may also be influenced by the difficulty of modelling environmental heterogeneity in any realistic way. His model environments tend to be uniform, which is quite unreal, and when they change, they do so in time in an oversimplified and unrealistically predictable way. Maynard Smith considers (1978, p. 89) 'the unlikely case that the correlation between features of the environment changes sign from generation to generations! Surely, such occurrences are commonplace, most simply when wet/dry, warm/cold seasons are examined.

That all environments, even very simple ones such as deserts, tundra, arable or aquatic, are also extremely spatially heterogeneous for many characteristics such as light (woodland), water (dunes) and soil, has become an ecological maxim. That such fine-grain environmental heterogeneity, over and above that presented by competing organisms, is of very great selective importance has been shown many times (e.g. Snaydon 1970, McNeilly 1968).

Environments are not only heterogeneous in time and space within and between seasons, but they may also show directional changes. Climates may warm or cool, most dramatically in response to glacial cycles. Desert margins may shift to and fro, as we have seen in North Africa in recent years. Air pollution, use of insecticides, rodenticides or antibiotics, and other man-made pressures, may lead to rapid directional selection and adaptation. It is important to realise that environments consist importantly of other organisms as competitors, predators or prey. If these other organisms have evolved so as to fill the maximum number of environmental niches, and at the same time show continuing evolutionary development, this adds to the increase in variability needed for successful competition and evolutionary development in the study organism. This idea has been called the 'Red Queen' effect from the anthropomorphised chess piece in *Alice through the looking-glass* who had to run as fast as possible to stay in the same place. To compete successfully, an organism must diversify and evolve as fast as possible to keep up with other organisms which are also having to diversify and evolve as fast as possible.

The second school of thought, persuasively argued in Maynard Smith (1978), concerns linkage disequilibria. These occur when the frequency of combination of alleles at two linked loci *A*/*a* and *B*/*b* on chromosomes, or gametes, differs from that predicted by the product of the allele frequencies. Linkage disequilibrium is likely to be increased by:

(a) Close linkage of *A*/*a* and *B*/*b*.
(b) Low rates of recombination (chiasma formation).
(c) Selection for particular allelic combinations.
(d) Small population size, or selfing.

Some forms of linkage disequilibrium are strongly favoured by selection, and may be maintained by cytological features, e.g. inversions or reciprocal translocations. In the heterostyly supergene in *Primula* (Ch. 7), most populations have maintained linkage disequilibrium to the extent that recombinants between the *S* supergene and the *s* supergene rarely occur (they are often homostyles), and when they do occur they usually appear to be outselected. The linkage groups *GPA* (thrum) and *gpa* (pin) are common, and although recombinants such as *Gpa* and *gPA* do occur, they rarely become established. This is an adaptive linkage disequilibrium, a linkage group or supergene. In this case, the *S* alleles at different loci are coadapted, as are the *s* alleles. If the linked loci were not coadapted, linkage disequilibrium might be harmful. Let us consider a situation in which the linked loci *A*/*a* and *B*/*b* have the following fitnesses (*W*) in environments Ā and B̄:

| | | |
|---|---|---|
| environment $\bar{A}$ | *A* 1.2 | *B* 0.9 |
| | *a* 1.0 | *b* 1.0 |
| environment $\bar{B}$ | *A* 1.2 | *B* 1.1 |
| | *a* 1.0 | *b* 1.0 |

Then, in environment $\bar{A}$, the preferred chromosome linkage group would be *Ab*, assuming simply additive fitnesses. On migration to environment $\bar{B}$ (or a change in the same locality from $\bar{A}$ conditions to $\bar{B}$ conditions), the preferred linkage group would be *AB*. However, in any of the situations listed above which favour linkage *dis*equilibrium, linkage equilibrium would not occur quickly enough to favour *AB* recombinants. Thus, the chromosome *Ab* might predominate to fixation although *AB* was more successful; particularly if, as would probably be the case, allele *B* was originally scarce. Thus, the disadvantageous allele *b* would 'hitch-hike' to success in its new environment $\bar{B}$ by being closely linked to the successful allele *A*.

In such a way, disadvantageous genes will accumulate in linkage groups without recombination, and if the whole genome is effectively one linkage group, as in an asexual form, the whole genome will accumulate disadvantageous mutants ('Müllers ratchet').

This combination of linkage disequilibrium and mutant accumulation has been termed the 'Hill–Robertson Effect' by Felsenstein (1974), and according to Maynard Smith is the most potent current selectant for sexuality. Although some accumulation of disadvantageous disequilibrium 'hitch-hikers' is likely to occur through tight linkage in all panmictic populations, this will be minimised by large haploid chromosome numbers, high rates of chiasma formation, outbreeding and large effective population sizes. These factors will encourage linkage equilibrium by maximising recombination. It is important to note that although recombination in this sense refers to the breakage of linkage groups through chiasmata, it will be rendered effective by the other elements of sexuality, the random segregation of the recombined chromosomes, and the random fusion of the resultant gametes. O'Donnell and Lawrence (1984) suggest that unequal distributions of incompatibility *S* alleles in wild populations of *Papaver rhoeas* may be caused by tight linkage of incompatibility loci to 'fitness genes'.

A further point of fundamental importance, which is readily overlooked, is the rôle played by sex in *creating* linkage groups. The processes of sexual reproduction will help advantageously coadapted genes to come together in tight linkage on the same chromosome. Thus sex can both promote useful linkage, and also dissipate damaging linkage.

I thoroughly agree with Maynard Smith that the negation of the Hill-Robertson Effect is a most vital component of the function of sexuality, but I believe very strongly that it has no meaning, except when viewed in the context of the highly heterogeneous and unpredictable environment. This is made clear by the 'raffle' argument of Williams (1975), quoted in Maynard Smith (1978). An asexual parent, he says, is like a man who buys 100 tickets in a raffle with only one prize, and finds that they all have the same number – he only has a minute chance of winning. Even if there were only one prize, the sexual parent with 100 different numbers would stand a better chance of winning. In the more realistic situation of there being many prizes, some of which may, however, be worth more than others, the chance of winning with only one type of ticket increases, but it is still much lower than with 100 different tickets.

This is elegantly demonstrated by a computer-generated model illustrated by Maynard Smith (1978, p. 104) (Table 2.1). In this simulation of the effect of sibling competition, the proportion of sexual to asexual individuals resulting after ten generations is given, where the initial frequency was 0.5. There are five heterozygous loci with one allele dominant, each allele fit, or not fit, and five paired features of the environment. There are thus $2^5$ (i.e. 32) different environments, and 32 different phenotypes, and each phenotype will be the fittest for one environment. Asexuality will only predominate where sibling and total competition (*RN*) is low, and the reproductive advantage of asexuality (*K*) is high (encircled values in Table 2.1). With realistic levels of sibling and non-sibling competition, and asexual advantage, sexuality is more likely to win, because more of the offspring produced sexually will be better adapted to the different environmental niches.

In a similar experiment in Maynard Smith (1978, p. 106), he shows that an allele for high recombination will always be favoured over an allele for low recombination when reproductive advantages are equal. Thus, even in the unrealistically low levels of environmental heterogeneity allowed

**Table 2.1** Results of simulating a model of sibling competition. Proportion of sexual individuals after ten generations (initial value, 0.5) (from Maynard Smith, 1978).

| Population size | Intensity of selection: Between families | Within families | Total | Advantage of asexual reproduction (*K*) | | | | | |
|---|---|---|---|---|---|---|---|---|---|
| *L* | *R* | *N* | *RN* | 1.0 | 1.2 | 1.4 | 1.6 | 1.8 | 2.0 |
| 400 | 20 | 1 | 20 | 0.48 | — | — | — | — | — |
| 400 | 1 | 20 | 20 | 0.51 | — | — | — | — | — |
| 200 | 2 | 4 | 8 | 0.97 | 0.69 | 0.29 | 0.09 | 0.10 | 0 |
| 200 | 4 | 4 | 16 | 1.0 | 0.92 | 0.79 | 0.37 | 0.38 | 0.14 |
| 200 | 6 | 6 | 36 | — | — | 0.92 | 0.95 | 0.56 | 0.54 |
| 400 | 6 | 8 | 48 | — | — | — | — | 0.78 | 0.64 |

by computer models, sexual reproduction will succeed over asexual reproduction. In most cases, this will be true even for unisexual species with a disadvantage, compared with asexuals and selfing hermaphrodites, as high as 1:2.

## DISADVANTAGES OF SEXUAL REPRODUCTION

Maynard Smith (1971, 1978) has stated that in a unisexual, or dioecious, species, sexuals will suffer a twofold disadvantage against asexuals, because they produce males and will usually be reproductively fittest when half their offspring are male. Thus, in contrast to asexuals which produce only females, half the reproductive effort is wasted by producing males, and half the species niches are filled with unproductive males. This argument is irrefutable with respect to unisexual species, and is indeed a powerful argument in favour of some compensatory selective advantage of sexual reproduction in groups such as the vertebrates, in which unisexuality is dominant. These arguments are further commented on in Chapter 8.

However, for over 95% of the organisms that are the subject of this book, the argument is much less important because they are hermaphrodites. For hermaphrodites, there may still be disadvantages inherent in sexual reproduction, but they are much less persuasive (Ch. 11, Table 11.9).

(a) Sexual mothers spend maternal resource on offspring (seeds) which are likely to include some unfit individuals. The variability of offspring resulting from the sexual process will not only engender individuals that can be fit in one or more of a wide variety of niches, but will also give rise to some individuals that will not be fit in any. Female expenditure on such unfit offspring has been called the 'cost of sex' or 'the cost of meiosis'. The offspring of an asexual will have the same genotype as their mother, and should therefore be just as fit as their mother in her particular niche.
(b) A newly arising asexual mother may still be potentially a sexual father. The asexual will be able to donate asexual genes to sexuals, but will not receive any sexual genes itself. This is disadvantageous to coexisting sexuals.
(c) Male-sterile asexuals will save on male reproductive effort in comparison with sexual hermaphrodites; as discussed on pp. 11–12 this advantage will not be great, and cannot easily be spread through the population.
(d) Outbreeding sexuals may be less reproductively efficient than asexuals in conditions of limited cross pollination.

It is sometimes found in flowering plants that asexuality, expressed as vegetative reproduction (Ch. 10) or agamospermy (Ch. 11), becomes more common in species or species-complexes in marginal environments. These may be at the edge of the species range, or the environments being very harsh, may approach the ecological limits suitable for plant growth (e.g. mountain summits, desert fringes, etc.). It may be that such environments impose heavy stabilising selection pressures, so that very few genotypes are fit. In such circumstances, asexuality may have an advantage. However, asexuality may also be reproductively more efficient in harsh environments as flowers and reproductive organs such as anthers and stigmas are liable to be damaged by frost, drought or disease. Vegetative reproduction, often by special organs such as bulbils, pseudoviviparous plantlets or bulbs (Ch. 10), or non-pseudogamous agamospermy (Ch. 11) may be favoured in these circumstances.

## SEXUAL RESOURCE ALLOCATION AND POLLEN : OVULE RATIOS

Model systems anticipate that if the genetic contribution to the next generation is equal from both parents, as will happen in hermaphrodites, resource expenditure should be allocated approximately equally between male and female function for a given individual (Maynard Smith 1978, p. 132). In Figure 2.1, 'fitness curves' are illustrated (derived from Charnov *et al.* 1976 and Charnov 1984). In these, the fitness of the individual sexes are scaled on the *X* and *Y* axes. In a dioecious (unisexual) population of stable size, the mean fitness of each sex should be two, that is, each individual should donate to the next generation on average two offspring. For an hermaphrodite in a stable population, the mean fitness of each parent should be one. However, male fitness may differ from female fitness. Consider a monoecious species which bears 20 flowers, and male and female flowers are equally expensive to produce. If it produces 10 male flowers and 10 female flowers, male and female functions are equally fit, and if each plant donates one offspring to the next generation, the male fitness $W_m$ and the female fitness $W_f$ will both be one. However, if there are 15 male flowers and 5 female flowers, male fitness will be $1 \times 5/10 = 0.5$ and female fitness will be $1 \times 15/10 = 1.5$ compared with the population average (Fig. 2.1). The intersect of these relative fitnesses is on the straight line for the fitness of the hermaphrodite plant $W_h = 1$. If the intersects of the relative fitnesses are to the right of the $W_h = 1$ line, $W_h$ is greater than one, and hermaphrodity will be favoured (the $W_h$ line will be concave). If, however, the intersects of the relative fitness are to the left of the $W_h = 1$ line, $W_h$ will be less than one, and dicliny will be favoured (Ch. 8). That is, the individual sexual fitnesses are greater than the

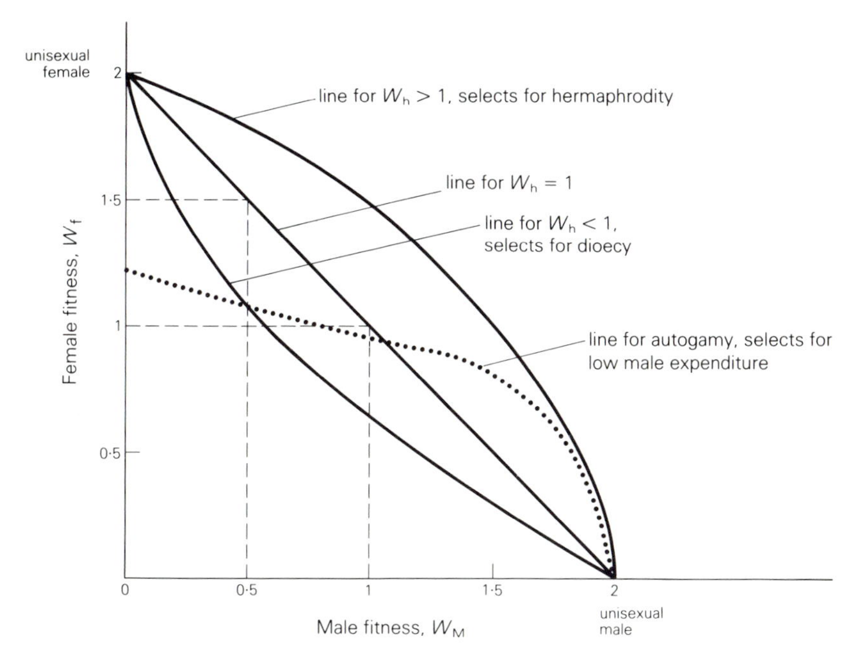


**Figure 2.1** Fitness curves for sexual function in hermaphrodite and dioecious plants. The fitness of each sex, in terms of contribution to the offspring relative to reproductive effort, are scaled on the *x* and *y* axes. See the text for further explanations (after Charnov 1984).

combined sexual fitness. Such conditions might arise if, for instance, a heavy female reproductive load resulted in insufficient allocation of resource to male function for total fitness $W_h$ to reach one (there might be too few male flowers in a monoecious species for pollination to be effective).

The smaller the successful contribution made by a sex to fitness, the greater will be the total fitness of that sex, and it will pay the plant to reduce expenditure on that sex to a threshold point below which the fitness of the plant $W_h$ will suffer. It is typical of an autogamous plant with high levels of selfing for male fitness to be optimised with very little expenditure compared with female expenditure (Ch. 9). That is, for all ovules to be self-fertilised, the flower need only produce a very small amount of pollen (dotted line on Fig. 2.1). It will pay an autogamous plant to only produce enough for all its ovules to be fertilised.

At times, authors have stated that the allocation of a resource to a sex should be a function of the genetic contribution made by that sex. I find this a confusing idea, because every zygote will receive an equal nuclear donation from a male gamete and a female gamete. It is the genetic contribution made to the next generation (or in other words fitness) *per individual* of that sex (if unisexual or dioecious) *or per unit of reproductive energy* of that sex if hermaphrodite (for instance a male flower or a female flower if monoecious) which is being referred to. I prefer to express these as fitnesses rather than 'genetic contributions'. Relative reproductive fitnesses will select for sexual resource allocation. The fitter a sex per individual, or per unit of reproductive energy, the lower the proportion of total reproductive resource that need be allocated to it.

A useful way of considering the partition of male and female resource is to examine pollen : ovule number ratios in different species. Cruden and Miller-Ward (1981) have also recorded stigmatic area, pollen-grain surface area (an attribute of pollen-grain size), pollen volume and pollen-bearing area on pollinators in 19 flowering plants in ten families occurring in a variety of habitats in North America. Highly significant trends were discovered which are summarised in Table 2.2.

From these trends, one would expect a species with inefficient pollination, for instance an outbreeder, to produce a lot of pollen per ovule to be fertilised, which allows small stigmatic areas with respect to the pollen-bearing area of the pollinator, and small, but expensive (in terms of pollen volume) pollen. In contrast, an efficient pollinator, for instance an autogam, would have few, large pollen grains of lesser total volume (expense), and large stigmatic areas. Because pollen number and pollen size are both so variable, the only approximation to an estimate of male cost is from pollen volume. The indication is that this is highly variable between species, and probably responds to selection resulting from different levels of male reproductive fitness.

Reproductive fitness may not be the only determinant of male and female resource allocation. Where pollination is inefficient, it would be expected that it should benefit the plant if there were a greater expenditure on male resource allocation. However, in three species Ter-Avanesian (1978) showed that when low pollen loads are artificially placed on stigmas, the genetic variability of offspring is higher than

**Table 2.2** Differences in pollen : ovule ratio, pollen surface area, pollen volume and number and stigmatic area: pollen-bearing areas in 19 flowering plants (from Cruden & Miller-Ward 1981).

| Characteristic studied | High values | Low values |
|---|---|---|
| pollen : ovule ratio | in outbreeders | in selfers |
| stigmatic area: pollen-bearing area | for low pollen : ovule ratio | for high pollen : ovule ratio |
| pollen surface area | for high stigmatic area: pollen-bearing area | for low stigmatic area: pollen-bearing area |
| pollen volume | for large pollen number | for small pollen number |
| pollen number | for large ovule number | for small ovule number |

when high pollen loads are used. It is considered that competition between pollen grains on the stigma and competition between pollen tubes in the style allow only those pollen grains that are particularly vigorous to succeed. If pollen-tube vigour is under genetic control, as seems likely, relatively few linkage groups carrying these vigour genes will pass to the next generation on the male side. In circumstances in which high levels of genetic variability are selectively advantageous, or in which selectively advantageous genes are linked to non-vigorous pollen-tube growth genes, it may pay the plant to produce inefficiently small quantities of pollen.

Currah (1981) also demonstrates competitive effects between different strains of pollen in the onion (*Allium cepa*), and in this instance selfed pollen often outcompetes crossed pollen at high density. Outcrossed seed is heavier, and probably fitter, than selfed seed, and thus it might pay an onion to produce inefficiently small amounts of pollen in order that outcrossed pollen can successfully compete with selfed pollen, and produce more of the large seed. These are cases in which the quality and the quantity of the reproductive output (seed) may be in conflict. Sexual selection often involves conflicts between different attributes, as is examined in the next section.

## SEXUAL SELECTION FOR DISADVANTAGEOUS OR DISHARMONIOUS ATTRIBUTES

Sexual selection was first identified by Darwin (1871), and has since received a good deal of attention from zoologists. Many animals show examples of directional selection from attributes that increase mating success, but may otherwise be disadvantageous. Such attributes usually arise from specialised features of the mating display, and will most often be found in the sex that assumes an active rôle in display and a passive rôle in mate-choice (usually the male). The mating display may originally have selected for features of general fitness (size, strength, vigour), but latterly selects for other features which have taken a symbolic rôle in the ritual. The large antlers of many male deer are frequently quoted in this context, and it is sometimes suggested that the giant Irish deer became extinct through inefficiency resulting from its vast antlers.

Study of plumage dimorphism in birds leads one to the interesting conclusion that pressures resulting from sexual selection for attractive, showy plumage may influence the pattern of mating choice, its original cause. Thus in some arctic waders such as the dotterel (*Charadrius morinellus*) and the phalaropes (*Phalaropus* spp.), the males are much less brightly coloured than the females, which is unusual in birds. These species have a short breeding and feeding season in the high arctic. After

**Figure 2.2** A stone plant *Lithops julii* (Aizoaceae) from the deserts of south-west Africa, an example of cryptic mimicry in plants (×0.7).

egg laying, the females, with the higher reproductive load, take little further part in reproductive activities, devoting their time to restoring lost body weight. The males have presumably become more cryptic in order to protect the nest and young from predators. In the mating ritual, the brighter female has assumed the dominant rôle in display, which is also very unusual in birds, leaving mate choice to the males.

It might seem that sexual selection will only occur in animals with behaviour-regulated choice of sexual partner, but this is far from the case. Consider, for instance, plants growing in largely unvegetated areas (e.g. deserts or mountain screes) which often show very efficient cryptic colouration as a protection against herbivores (stoneplants, *Lithops* (Fig. 2.2) or certain New Zealand alpines are good examples). These plants are cross-pollinated by insects, and in their hostile habitats require very showy flowers to achieve insect visits. Thus the sexual (conspicuousness) and vegetative (inconspicuousness) requirements are in direct conflict, and sexual selection for conspicuous flowers renders the plants liable to predation at flowering time.

Directional selection for features that encourage specialised flower visitors (Ch. 4) may also militate against fitness in other features. Bat flowers (i.e. flowers that attract bats) are generally very large, and produce copious nectar, so they must be very expensive to produce (Ch. 4, Fig. 4.19). It is not very surprising to discover that they are often rather few, and usually borne singly, so not many fruits are produced. Bat pollination may lead to efficient pollen transfer at the expense of heavy fruit production.

Pyke (1982b) showed that increases in conspicuousness of flowers and in potential reward to the pollinating birds of large inflorescences on tall stems in the waratah (*Telopea speciosissima*) may provide directional sexual selection for much larger inflorescences than the fruit-carrying capacity warrants. Fruit-set is therefore always low, even after efficient pollination (Ch. 9, Fig. 9.5). Wyatt (1981) similarly demonstrated that in five tropical leguminous shrubs, seed-set per plant is increased by producing more flowers and fruits rather than by increasing the seed-set per fruit. Sexual selection may favour the production of more flowers to increase fitness if this increases the attractiveness of the bush to pollinators in conditions of poor cross pollination.

Many flowers produce nectar to attract and reward pollinators, but although many pollinators (Lepidoptera, birds) feed on nectar alone, others (bees) also collect large quantities of pollen which becomes unavailable for cross pollination. Nectar may also attract general scavengers (ants, molluscs) which will also eat the sexually functional parts of flowers. A few plants have taken advantage of this problem, and have become specialist ant or slug flowers (Ch. 4, Fig. 4.40). Some plants have evolved features that discourage or trap ants, or provide other, more easily accessible extrafloral nectaries. Nectar may also encourage the establishment of fungal or bacterial facultative pathogens which lead to disease or rot.

Bawa (1980) has suggested that selection in tropical forest trees for large, energy-rich animal-dispersed fruits (another form of sexual selection) has encouraged the separation of male and female function, leading to dioecy.

Van der Pijl (1978) is a rich source of examples where sexual selection has led to disharmony of male and female function. To briefly mention a few examples, he suggests that monoecy in *Acer* species with winged fruits results from the development of the samara wing at an early stage in the flower. This interferes with stamen development spatially, leading to gynoecy in some flowers, and, by implication, to androecy in others (see Fig. 2.3, for *Ulmus*).

In the Asteraceae (Compositae) he suggests that the development of the capitulum with many small, tubular florets favoured pollen presentation by the gynoecium, usually on the outside of the style and stigma

**Figure 2.3** Male and female inflorescences in the wych elm, *Ulmus glabra* (×1.0). The female flowers are receptive to pollen, and ovaries are already developing wings which will allow the fruit to be wind-dispersed as a samara. It has been suggested that such ovary wings would interfere with male function in the same flower, and have thus encouraged the development of monoecy.

arms. Such a system results in abundant self-pollination, which may be disadvantageous to an outbreeder. Although some genera have avoided self-pollination by making stigmatic surfaces relatively inaccessible (e.g. the 'lantern-slits' in the style of *Solidago* (Fig. 2.4) which exhibits apical fusion of the stigma branches), these systems may be reproductively inefficient. Van der Pijl (1978) argues that the functional monoecy found within capitula of many Asteraceae has resulted from the disharmony of the presentation of a male function by a female organ.

He also suggests that pressures which have led to marked dichogamy (protandry or protogyny), result in either the male or the female function becoming inefficient when the time-spread of sexual function exceeds the useful life of the flower in terms of nectar production, odour production, or visual attractivity. In these cases, dicliny may be favoured, in which the remnants of the original dichogamy plainly persist (*Aesculus*). Pressures favouring dicliny may also lead to reproductive inefficiency, for instance in nectar flowers with carpellary nectaries. In *Rhamnus*, novel nectaries of a different kind are produced in male flowers.

**Figure 2.4** Flower of *Solidago virgaurea* (×6.0). Stigma branches initially fused at the apex and bearing pollen; the branches part latterly, exposing the receptive inner surface of the stigmas, and allow geitonogamous self-pollination to occur. Photo by M. C. F. Proctor.

These are a few of very many examples where directional selection for a specialised reproductive attribute in flowers has led to inefficiency in other (often also reproductive) features. Others are discussed with respect to heteromorphy in Chapter 7. It is increasingly conjectured that dicliny has evolved in many plants in response to such problems, especially when the maximisation of one or both of the sexual functions (pollen presentation and pollen receipt) is favoured (Bawa 1980) (Ch. 8).

## SEXUAL SYSTEMS AND GENOTYPE FREQUENCIES

Theoretical panmixis requires a sexual population of infinite size and random distribution, in which all individuals are equally as likely to exchange genes with each other and all are equally viable. Such constraints are considerable, and will be true for very few populations of plants. As is discussed below and in Chapter 10, plant population sizes are often very small, often indeed much smaller than is immediately apparent. As Chapter 5 shows, pollen flow is often far from random, and in Chapter 9 it becomes clear that many plants are far from being wholly outbreeding. Further, the distribution of plant populations is rarely random, but shows infinitely variable gradations of clumping, regular spacing and density, from dense groups to sub-isolated islands. However, mathematically, many populations may approximate to panmictic ideals sufficiently closely to give genotype frequencies that do not statistically depart from those predicted. It is when predicted frequencies vary from those observed, and particularly when heterozygote frequencies are lower, or higher, than those expected, that one may suspect that non-panmictic breeding systems are in operation.

As is well known, panmictic genotype frequencies are predicted by the formula independently conceived by G. H. Hardy, the Professor of Mathematics at the University of Cambridge and friend of the geneticist, R. C. Punnett, and by the German medic Weinberg, in 1908 (Punnett 1950). This principle, equation, law or equilibrium (for it is called all four) is the basic model of population genetics from which all others are derived, or which they assume, and is named after its inventors 'the Hardy–Weinberg Law'. It is not the function of this book to derive, or even expound, this law to any depth, for this is ably achieved by many primers of population genetics. Rather, I need to briefly recount its application at a simple level in breeding system theory, to which it is also basic.

The Hardy–Weinberg Law refers to genotype frequencies of individuals in a panmictic population for a locus polymorphic for two alleles which we will call $A$ and $a$. It requires the determination of allele frequencies $p(A)$ and $q(a)$, where $p + q = 1$. These frequencies are readily determined when there is incomplete dominance and the heterozygote $Aa$ is phenotypically distinct from either homozygote $AA$ or $aa$. In these cases:

$$p = \frac{2AA + Aa}{2n}$$

and $q = 1 - p$, where $n$ is the sample size. Where there is dominance, it will be necessary to assume Hardy–Weinberg genotype frequencies to establish allele frequencies, and if the object is to investigate departures from Hardy–Weinberg ratios, circularity is involved. The frequency of the homozygous recessive $aa = q^2$ (as will be seen below), from which $q$, and hence $p$, can be calculated, if panmictic assumptions are made.

The law predicts that genotype (and hence in the absence of dominance, phenotype) frequencies will occur at:

$$p^2(AA) + 2pq(Aa) + q^2(aa) = 1$$

This equation thus presents two very important, although perhaps rather obvious, conclusions:

(a) Genotype (and phenotype) frequencies are reliant on allele frequencies, *and nothing else*, in a panmictic population in which $A$ and $a$ are neutral with respect to each other.
(b) Allele, and hence genotype, frequencies will not vary significantly from one generation to another in the absence of differential selection, migration or mutation, in panmictic conditions.

To these, we may add another conclusion which is perhaps more important to the present topic:

(c) Significant excesses, or deficiencies, in the frequency of heterozygotes from those predicted by $2pq$ will most probably occur as a result of non-panmixis. However, excesses may also occur as a result of heterozygote advantage (heterozygote disadvantage is unusual). Differential selection of $A$ as against $a$ phenotypes may cause excesses or deficiencies for homozygote frequencies $p^2$ and $q^2$, but should not seriously bias $2pq$.

Non-panmictic excesses of heterozygote frequency in the absence of heterozygote advantage are caused by: (a) disassortive mating and (b) fixation of hybridity by agamospermy (Ch. 11) or vegetative apoximis (Ch. 10).

Non-panmictic deficiencies in heterozygote frequency will result from (a) assortive mating; (b) selfing; (c) small populations; and (d) sibling or near-relative mating.

Before we go on to examine those interesting abnormalities of breeding systems which lead to departures from expected heterozygote frequencies, it is worth briefly examining modifications of the formula for multi-allelic loci, and for polysomic loci.

### *Multi-allelic loci*

Hardy–Weinberg genotype ratios can be easily expanded to account for loci with more than two alleles, thus for three alleles:

$$p^2(AA) + 2pq(Aa) + 2pr(Aa') + q^2(aa) + 2qr(aa') + r^2(a'a') = 1$$

where $r$ is the frequency of the third allele $a'$ (and so on).

### *Polysomic loci*

These will occur in polyploids, where autopolyploidy, or repetition of loci in different, but homoeologous chromosomes, leads to a locus being present more than twice, and thus polysomic inheritance occurs. This leads to quite complicated frequencies; thus even at a tetrasomic locus:

$$p^4(AAAA) + 4p^3q(AAAa) + 6p^2q^2(AAaa) + 4pq^3(Aaaa) + q^4(aaaa) = 1$$

always assuming that you can estimate allele frequencies, which is likely to be very difficult.

### *Disassortive mating*

Disassortive mating, in which mating between two phenotypic expressions of different alleles at the same locus occurs more frequently than expected, is not common, even among animals. However, there are many examples of frequency-dependent dissortive mating amongst insects such as *Drosophila* and certain Lepidoptera, where distinctive phenotypes are particularly sought after as mating partners when rare in a population.

The only examples that I am aware of in plants that could be construed as disassortive mating are where the population is formed of two distinctive phenotypes which mate with each other, but cannot mate among themselves. Dioecy (unisexuality) is the most obvious example (Ch. 8), and heteromorphy (Ch. 7) also falls into this category. In these, loci linked to the chromosome on which control of the sexual dimorphism lies will be much more heterozygous in the heterogametic phase (XY in the case of dioecy) than predicted by Hardy–Weinberg equilibria, which is merely another way of describing the effects of sex linkage.

### *Assortive mating*

Assortive mating is a common phenomenon in both animals and plants, and is very important as a vital trigger to the inception of sympatric speciation. Clearly, if gene flow is restricted between two or more phenotypes in the populations, this population subdivision may allow genecological differentiation of the two subdemes which may in due course lead to speciation. Assortive mating in outbreeding plants can be due to one of two causes:

(a) Temporal separation, due to two phenotypes in a population which have different flowering times.
(b) Ethological separation, due to two phenotypes in a population which attract different animal visitors, or create different spatial patterns of pollen donation and reception.

There are numerous examples of both classes in the literature, and ethologically induced assortive mating is further discussed in Chapter 5, especially with regard to flower-colour polymorphisms.

Assortive mating in otherwise outbred plants will lead to an excess of homozygotes at those loci that control the assorting phenotypes, and at other loci linked to them. However, it will not lead to general homozygosis.

### *Selfing*

A special case of assortive mating is selfing, which is very common in plants, and occurs in nearly all self-compatible hermaphrodite seed plants, which may be a quarter or more of all species. In highly autogamous species it is the dominant breeding system (Ch. 9). Selfing will result in an excess of homozygotes, and obligate selfing may lead to total homozygosis, except perhaps at a few loci with marked heterozygote advantage. The lack of genetic variability, and inbreeding depression commonly associated with high levels of homozygosis may present evolutionary obstacles to this kind of breeding system; high levels of selfing are usually restricted to opportunist annuals for which fitness is maximised by reproductive efficiency.

### *Small populations*

Theoretical panmixis requires infinitely large populations of interbreeding individuals. In effect, most populations in which random gene exchange occurs between more than 100 genets will not show significant departures from panmictic expectancies. The effect of small population size on genetic structure is best shown by considering the sampling error of allele frequencies in populations of different sizes. We assume that these populations have been derived from a single founder at generation $t$ with allele frequencies of $p$ and $(1 - p) = q$ and have a size (number of interbreeding genets) of $n$.

For any generation $T$, the frequency of an allele will be distributed with a mean value of $p$ (assuming no selection) and variance of:

$$pq\left[1 - \left(1 - \frac{1}{2n}\right)^{T-t}\right]$$

The variance exhibited between populations will be a function of $n$. If $n$ is relatively large, variance between populations will be low. If $n$ is very small, the variance between populations in allele distribution will be high. This chance, or stochastic, fluctuation between populations, and for one population between generations, in allele distribution is known as genetic drift, or the Sewall Wright effect after the American geneticist who derived its theory.

The effect of genetic drift in small populations will be that chance fluctuations may cause a polymorphic allele to become extinct; that is a population originally polymorphic for alleles *A* and *a* at frequencies *p* and *q* will lose one allele by chance, and the population will become homozygous at that locus. Such genetic fixation will result in evolution in the absence of differential selection (we have assumed the alleles are selectively neutral with respect to one another).

Another effect is that small populations arising from originally panmictic founders will become homozygous at increasing numbers of loci as generations progress and different polymorphisms fix by chance. Thus, small populations will show a reduction in heterozygosis and genetic variability in comparison with larger, panmictic populations. In this they will resemble selfing populations, but the reason is quite different. Selfing populations will show a reduction in the frequency of heterozygotes, so that either or both of *A* and *a* may fix as homozygotes. Small, outbred populations will lose alleles by chance, so that they fix as homozygotes *AA* or *aa* in the absence of the other allele.

If small populations are completely isolated genetically, such homozygosity and loss of variability may be permanent. However, semi-isolated populations may show some genetic migration between themselves, and if the plants are cross fertilising, differentially fixed subpopulations may hybridise to restore heterozygosis and genetic variability at those loci again.

Earlier generations of British population geneticists, for example R. A. Fisher and E. B. Ford, did not consider genetic drift to be an important feature of evolution. Their studies of insect populations suggested that small populations (for instance less than 100) rarely if ever occurred. In fact, some isolated animal populations (for instance many birds) can be that small. Plant populations are very frequently hazardously small. Small population sizes in plants can be caused not only by local rarity, but also by clonality. As is more thoroughly discussed in Chapter 10, many apparently huge plant populations may in fact be composed of only a few genets or even of only one.

Most field botanists will not need convincing as to the very small populations in which the majority (although of course by no means all) species of plants occur. These small population sizes are augmented by very restricted pollen and seed dispersal in many cases (Ch. 5). There is remarkably little factual as opposed to anecdotal evidence on plant population sizes, and thus the large amount of data collected by the amateur botanist R. W. David on population sizes in the British Isles of seven of the less widespread species of *Carex* is of particular interest (David 1977–1982). David divides population sizes into four classes, 1–20, 21–100, 101–1000 and 1001+. For 484 populations, he finds the following distribution of size classes:

| | Size class | | | | |
|---|---|---|---|---|---|
| | 1–20 | 21–100 | 101–1000 | 1001+ | Total |
| number of populations | 156 | 167 | 105 | 56 | 484 |
| % populations per size class | 32.2 | 34.5 | 21.7 | 11.5 | |

*Carex* is a very large and important genus, particularly in temperate systems, and in many ways can be considered a rather representative temperate genus, consisting of largely outbred, sexual, perennial herbs with (in the species examined) some, but not extensive, vegetative reproduction. Although the species examined were not common, all existed in some large populations, and occurred in a variety of habitats from dry grassland to mires, and from woodland to mountain and seashore. It will be seen that about one-third of all populations can be considered very small (20 or less in size), a population size that will inevitably lead to major effects of genetic drift. About two-thirds of all populations are of less than 100 individuals, in which some genetic drift will be likely to occur. In most groups of plants, such critically small populations are not the exception, but the norm. I believe that the habitual small size of many plant populations is a phenomenon of major evolutionary significance which has crucially influenced their genetic structure and patterns of speciation. I also believe that this is a factor in plant evolution that has been almost entirely overlooked.

As yet, there has been very little work on wild populations of plants to assess the genetic effects experienced by small populations. This may be due, at least in part, to a 'Catch 22' situation. For populations to be sufficiently small for genetic drift to occur, so few individuals present themselves for sampling that sample sizes are too small for statistically significant departures from panmixis to be apparent. This becomes clear when allele frequencies are examined for the esterase locus in the cowslip *Primula veris* for one large and 13 small isolated populations in north-east England (Ibrahim 1979) (Table 2.3). Although two small populations were apparently monomorphic, and may have become genetically fixed for this locus, in no case did the other 11 show significant departures from panmictic expectations. It is just possible that the slight, but not significant, excess of heterozygotes in the large population indicates heterozygote advantage at this locus, although this is by no means clear. Such heterozygote advantage would tend to offset the effects of population size with respect to allele fixation.

Many plants, and especially those with many small and isolated populations of relict status, show marked between-population differentiation in characters that appear intuitively to be selectively neutral, for

**Table 2.3** Observed and panmictically expected numbers of heterozygotes at the esterase locus in the cowslip, *Primula veris* in 14 populations in north-east England (unpublished data in Ibrahim 1979).

| Population | Number of plants in population | Number of genotypes: *AA* | *Aa* observed | *Aa* expected | *aa* | *n* |
|---|---|---|---|---|---|---|
| Whittle Dene 2a | 500 | 3 | 24 | 21.3 | 25 | 52 |
| Gunnerton | 50 | 3 | 10 | 11.1 | 13 | 26 |
| Druridge | 40 | 5 | 6 | 6.2 | 2 | 13 |
| Lynemouth | 30 | 2 | 5 | 5.9 | 6 | 13 |
| High Shield, Hexham | 25 | 2 | 2 | 4.0 | 5 | 9 |
| Cassop 1 | 25 | 1 | 8 | 7.0 | 8 | 17 |
| Cassop 2 | 33 | 0 | 4 | 3.5 | 14 | 18 |
| Cassop 3 | 40 | 2 | 2 | 4.5 | 8 | 12 |
| Cassop 4 | 30 | 0 | 2 | 1.8 | 4 | 6 |
| Cassop 5 | 35 | 0 | 0 | 0 | 8 | 8 |
| Stannington 2 | 19 | 0 | 0 | 0 | 9 | 9 |
| Stannington 1 | 17 | 0 | 2 | 1.7 | 6 | 8 |
| Ashington | 33 | 2 | 7 | 6.7 | 5 | 14 |
| Acomb | 15 | 0 | 2 | 1.6 | 3 | 5 |

instance leaf shape or distribution of indumentum. These varieties are often given specific rank, although at least some are potentially interfertile. The genus *Euphrasia* (eyebrights) in north-west Europe is a good example. It seems very likely that, in many cases, the genetic differentiation that has led to such clear-cut morphological differences, and even to between-populational breeding barriers, may have occurred as a result of non-selective genetic drift in small, isolated populations. Such a thesis, although shown to be algebraically and biologically likely, is very hard to prove in any given case. It is also very dangerous to assign neutrality to character differences. Just because plausible selective differentials do not occur to the investigator does not mean that they do not exist. For example, Eisikowitch (1978) has shown that the relatively minor taxonomic differences that are employed to separate two subspecies of *Nigella arvensis* in the eastern Mediterranean (posture, branching, pedicle thickness) are highly adaptive with respect to pollination efficiency in contrasting habitat types.

Many examples have appeared in recent years of between-populational differences in genetic structure as exemplified by isozyme analysis. It is sometimes assumed that different alleles at a locus are neutral with respect to each other, population differences having arisen by genetic drift. Yet potential selective differences between alleles are rarely tested experimentally.

Since 1968, Kimura and co-workers have gathered together a considerable body of evidence concerning the differences in large molecules (e.g. proteins, haemoglobins, messenger RNA, cytochrome *c*) between different organisms (Kimura 1979 provides a popular review). From this, they suggest that the evolution of large molecules proceeds independently of selection, and is influenced merely by mutation rates, generation time and population size. This 'neutral theory of molecular evolution' therefore claims that if these parameters are known, the length of time that any two present-day taxa have been evolutionarily distinct can be calculated by reference to changes in their large molecules. They suggest that the time taken for a large-molecule polymorphism to proceed to fixation (and thus 'evolve') is dependent only on population size, and is calculated by $4N_e$ generations, where $N_e$ is the effective population number.

Such sweeping generalisations are clearly open to severe criticism and they have indeed been criticised, for instance by Johnson and Selander (1976) on the grounds of both premise and technique. They suggest that much of the earlier electrophoretic work on which many of Kimura's hypotheses are based is suspect, it being very easy to create apparent banding differences by slight changes in temperature or pH during the electrophoretic run. In relatively few instances are apparent allelic differences tested genetically. (Such criticisms do not, however, apply to results obtained from other techniques which lead to sequencing of large molecules; nevertheless these techniques may also be open to other criticisms.) Johnson and Selander also queried the basic premise of Kimura's work that differences between morphs of big molecules are necessarily neutral. They showed that levels of heterozygosity vary considerably, depending on the type of function of an enzyme. Regulatory enzymes, with very low variability, are apparently subject to much higher levels of stabilising selection than are enzymes that act on external substrates.

This 'selectionist' versus 'neutralist' controversy which has raged for more than a decade is dying as workers realise that large molecules have diverse evolutionary mechanisms, and that generalisations are unrealistic. It has, however, become clear that for apparently neutral genetic differences, rates of evolution may well be strongly influenced by effective population size. The Sewall Wright effect has come of age.

## *Near-relative mating*

Perhaps the most important difference between the mating systems of plants and animals is the ability of most animals to perceive a potential mating partner, and thus show behavioural choice in mate selection. This can lead to sexual selection, dissortive mating and assortive mating, and will usually lead to relatively large effective population sizes, as has already been discussed.

In contrast, plants are passive and sessile with respect to mate choice. One very important result of this distinction concerns near-relative mating (e.g. sibling-mating, or parent–offspring mating). In organisms with large populations and good dispersal, near-relative mating will be rare and unlikely to upset panmictic expectations. However, if populations are small, or dispersal is poor even in large populations, near-relative mating becomes more likely. Near-relative mating exacerbates the effects of small population size, and approaches the effects of selfing. As it is rarely a continuous process, but only occurs occasionally, its effects are often short-term. Although the level of heterozygosity may be somewhat decreased, and homozygous fixation increased, the offspring of a near-relative mating are more likely to suffer than are the offspring of other matings by the exposure of rare deleterious recessive genes in the homozygous condition. The occurrence of genetic diseases in Royal families, isolated communities, etc. in human beings provides familiar examples of this effect.

Selection pressures will favour attributes that discourage near-relative mating, as these will tend to promote fitter offspring. In the majority of animals, behaviour patterns that discourage near-relative mating have evolved. These include incest taboo in Primates, territoriality, juvenile dispersal, large mating assemblies, monogamy and seasonal migration. None of these mechanisms are available to plants, which are therefore more likely to suffer near-relative mating effects, especially in small populations. It also follows, however, that competition between families with different innate levels of near-relative mating will not occur in plants, and thus selective differentials are absent, except insofar as they promote offspring dispersal.

## CONCLUSION

This chapter has discussed the distribution of sexuality amongst living things, its advantages and disadvantages, and its genetic consequences. It has predicted the genetic structure of panmictic populations, and has discussed breeding systems that depart from panmictic expectations, whether inbreeding or asexual. Much of the rest of this book is devoted to a closer examination of the genetic control of different breeding systems, and their genetic consequences in terms of variation and evolutionary potential.

CHAPTER THREE

# Sexual reproduction in seed plants

This relatively brief chapter is concerned with the anatomy and physiology of sexual reproduction in seed plants, and in addition touches on the evolutionary history and adaptive success of spermatophyte reproduction. It is largely descriptive, and unlike other chapters does not dwell on modern advances and their implications. It merely attempts to provide a framework around which later chapters (especially Chs. 6, 7 and 11) can build.

## GENERAL

The development of sexual reproduction in seed plants is best conceptualised as a succession of enclosures of the sexual process within concentric containers. In order to understand how this has developed during the evolutionary history of plants, it is necessary to describe briefly the major features of reproduction in the Pteridophytes and the Gymnosperms, and to expound the alternation of generations.

## ALTERNATION OF GENERATIONS

In most animals, only the gametes are haploid, with one set of chromosomes each. These are produced as a result of the reduction process of meiosis from a diploid parent, and fuse to give a diploid zygote. In contrast, all plants that have sexual reproduction undergo an alternation of two generations. The gametophyte, which is haploid, produces haploid gametes which fuse to produce the diploid zygote. This develops to give the sporophyte generation, which undergoes meiosis, resulting in haploid spores (Fig. 3.1).

Some form of alternation of generations is common to all plants, although in some algae the diploid generation is limited to a single cell. From this statement two deductions may be made. First, alternation of generations evolved at an early stage in the development of sexuality among primitive photosynthetic eukaryotes (algae) and second, all more 'advanced' plants have evolved from such early plants, and in that sense are monophyletic.

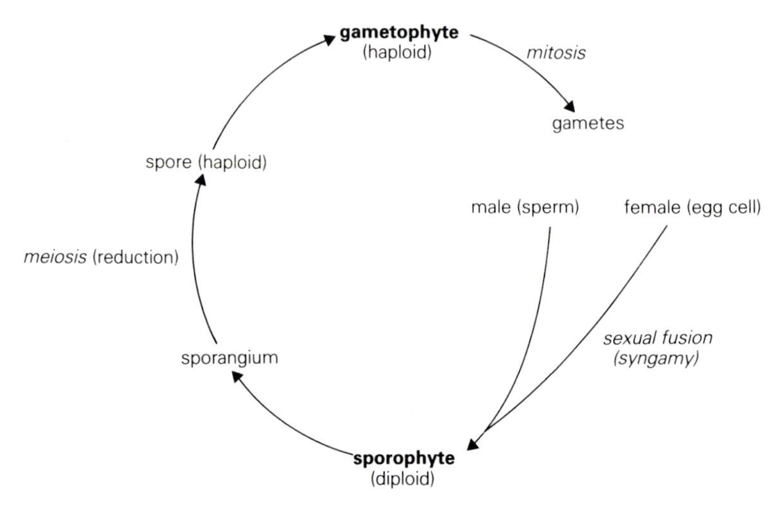


**Figure 3.1** Basic features of alternation of generation in plants.

## PTERIDOPHYTES

Both the Bryophytes and the Pteridophytes have a marked alternation of generations, which, however, differs conspicuously between these two groups. It is probable that both groups arose independently from the algae and developed in parallel. It is very likely that the seed plants arose from the Pteridophytes, and equally likely that they did not arise from the Bryophytes. Thus, we will pass over consideration of the Bryophytes, which have a dominant gametophyte generation and a dependent sporophyte, and briefly examine the Pteridophytes. In all extant members, these have a dominant sporophyte generation and a free-living, but generally much smaller, gametophyte (Fig. 3.2).

In particular, we should note three features of Pteridophyte reproduction:

(a) The self-motile male gamete (antherozoid) requires external water to achieve fertilisation.
(b) The sessile archegonium, containing the female gamete, limits the distribution of the dominant sporophyte generation to niches in which the female gametophyte can survive and reproduce.
(c) The sporangium, in which meiosis occurs in the diploid sporophyte to form haploid spores.

In nearly all living Pteridophytes, we find only one type of sporangium, producing only one type of spore. In four quite distinct and evolutionarily unrelated groups, however, we find heterospory, with megaspores (giving rise to female gametophytes) and microspores (giving rise to

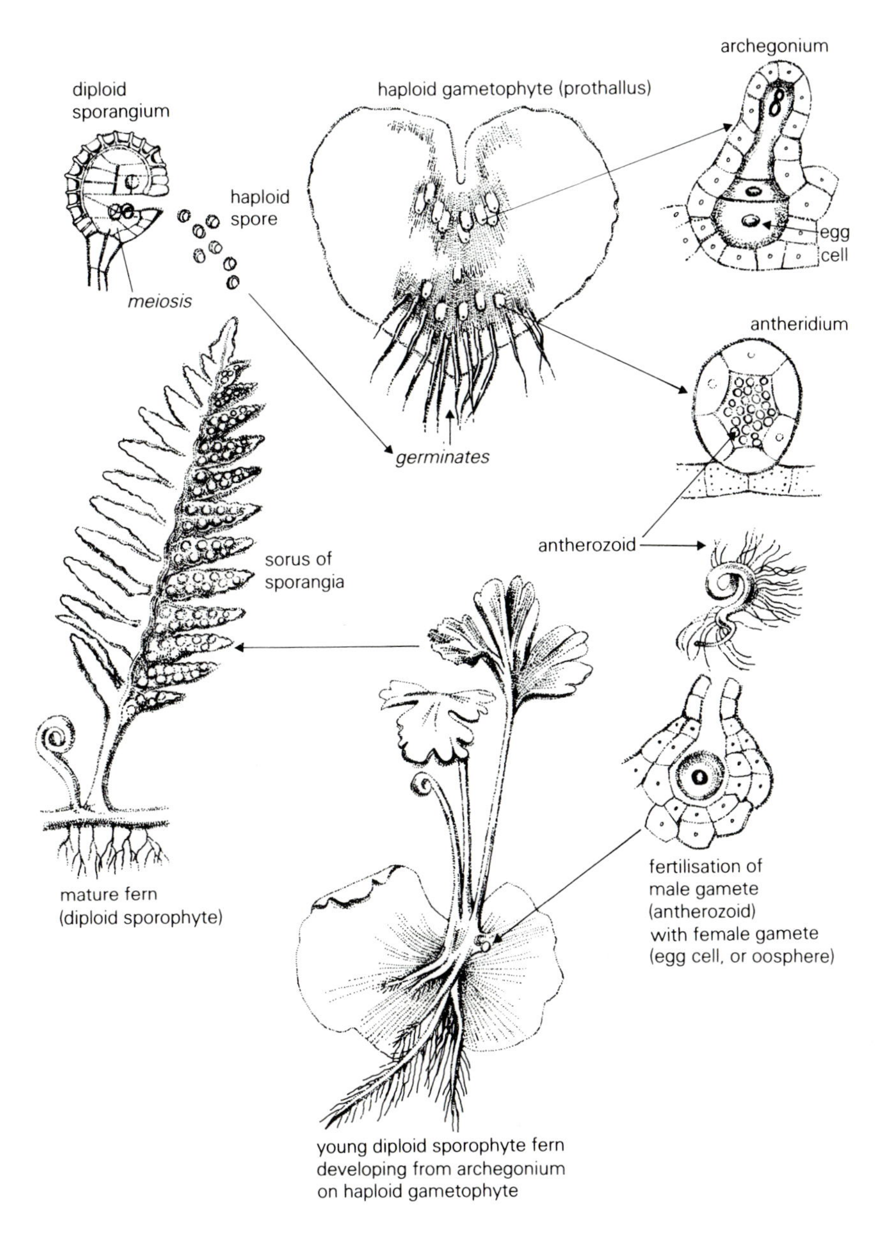


**Figure 3.2** Alternation of generations in a fern (Pteridophyta).

male gametophytes) being borne in different sporangia. These groups are the Selaginaceae, the Isoetaceae, the Marsileaceae and the Salviniaceae. This mechanism is particularly significant as the seed plants are also heterosporous, bearing microspores (pollen) in microsporangia (anthers), and megaspores in megasporangia (ovules). It is also noteworthy that in all these four families, the female gametophyte is contained within the megaspore, and in the Salviniaceae it is borne within the modified remnants of the megasporangia.

There is little dispute that among the ancestors of the seed plants can be numbered at least some of the Pteridosperms (seed ferns) which are extinct and disappeared from the fossil record as long ago as the Cretaceous. These were heterosporous and further resembled the seed plants by retaining the megaspore and the female gametophyte within the megasporangium. This therefore became an ovule, within which the sexual processes of female meiosis and fusion took place. This enormously significant development permitted four major changes in the sexual process:

(a) Sexual fusion no longer depended on external water, but took place internally.
(b) The male gamete was no longer externally motile, and thus male motility was restricted to the tough microspore (pollen).
(c) Growth of the male gametophyte, as well as the female gametophyte, was restricted to the megasporangium (ovule); thus a special site for microspore reception and germination was required on the megasporangium or megasporophyll.
(d) The zygote originated within the megasporangium of the mother, where it could be protected by the megasporangium wall from excesses of temperature or water stress, and from disease. It could be dispersed from its lofty position on the megasporophyll to 'safe sites', and could 'choose' the moment to undergo further development into a mature sporophyte ('germination'). In fact, it became a seed, and it is useful to consider the Pteridosperms as the first seed plants.

These changes in the sexual process would have allowed colonisation of a much wider range of habitats, which were unavailable to the Pteridophytes with their free-living, water-reliant gametophyte. In particular, the preadaptation by these changes to more arid environments would encourage the development of non-reproductive features (secondary thickening, bark, true roots, complex and often xeromorphic leaf anatomy) which would be successful in these conditions, and which are typical of most present-day Gymnosperms and primitive Angiosperms. It is probably no accident that the rise of the seed plants coincided with a period of increasing aridity in the early Cretaceous.

## GYMNOSPERMS

The Gymnosperms are represented today by five diverse groups, each of which is only distantly related to the others, and they all may have separate pre-Gymnosperm origins. At least one group (the Gnetopsida) is itself heterogeneous, and may be only distantly monophyletic. In reproductive terms, only the ovule/seed borne externally (hence gymnosperm, naked seed) is common to all, and there is great diversity in the form of the male gametophyte, the female gametophyte, the male gamete, the microsporangium, and its position, the megasporangium (ovule), and its position and arrangement, and the embryo and endosperm. Vegetatively, the Gymnosperms are equally diverse, although it is notable that all are in some sense woody, with secondary thickening.

It is not considered appropriate in this book to compare the different forms of the sexual process encountered in the Gymnosperms. As far as is known, they do not fundamentally alter the breeding system characteristics of the plants concerned in the same way as dioecy/hermaphrodity or dichogamy/homogamy. At the same time, it must be admitted that little seems to be known about the physiology of pollen germination, growth of the male gametophyte, or control of fertilisation in the Gymnosperms. Thus any rôle played by these features in controlling gene flow must remain for the present purely speculative. It is useful, however, to summarise some general features of Gymnosperm reproduction:

(a) Hermaphrodite cones do not occur and thus plants are always monoecious or dioecious. In the former case, dichogamy is common, and thus outbreeding is frequently obligate, being under temporal or spatial control.
(b) Pollen is almost always transported by wind. This is helped in the Pinaceae by the structure of the pollen grain, which has two lateral air-filled sacs (Fig. 3.3). Such pollen grains are remarkably mobile;

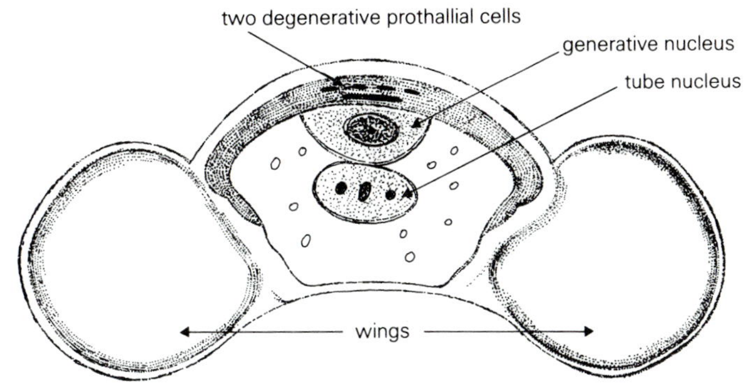


**Figure 3.3** Structure of a pollen grain in *Pinus* (Gymnospermae, Pinaceae). Note the gametophyte of four cells, and the two air-filled sacs.

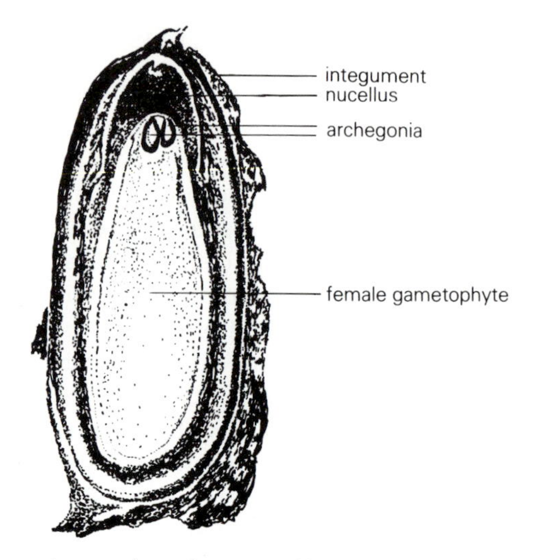


**Figure 3.4** Anatomy of an ovule at the time of fertilisation in *Pinus* (Gymnospermae, Pinaceae).

most pollen trapped in mid-Atlantic is of the Pinaceae (Ch. 5, Fig. 5.12). Only in *Ephedra* (Bino & Meeuse 1981) is it suggested that some entomophily occurs.

(c) Pollen is received at the micropile of the ovule, usually in a sticky drop of liquid containing both sugars and amino acids in solution. It passes through the micropile to the pollen chamber, adjacent to the nucellus, where it germinates to form the male gametophyte. In most species, the pollen grain has four cells, the generative nucleus, tube nucleus and two generative prothallial cells. After germination, the pollen tube (extended male gametophyte) undergoes two further divisions to yield six cells, a stalk cell and two sperm cells being added. In the Taxodiaceae and Cupressaceae, prothallial cells are missing, as they are in all members of the Taxopsida. In the Araucariaceae and Podocarpaceae there is considerable prothallial development leading to a substantial male gametophyte (pollen tube). This condition seems to be primitive, there being a progressive evolutionary reduction in the size of the pollen tube.

The pollen tube then grows through the nucellus to the archegonial chamber, where it releases both sperm cells. However, in the Cycadopsida, and *Ginkgo*, the pollen tube is short and haustorial, merely acting as an anchor for the pollen grain, which hangs partly free in the archegonial chamber.

(d) In most Gymnosperms, two sperm cells are released in the archegonial chamber, where one will fuse with an egg in one of the archegonia (from two to five archegonia occur, each containing an

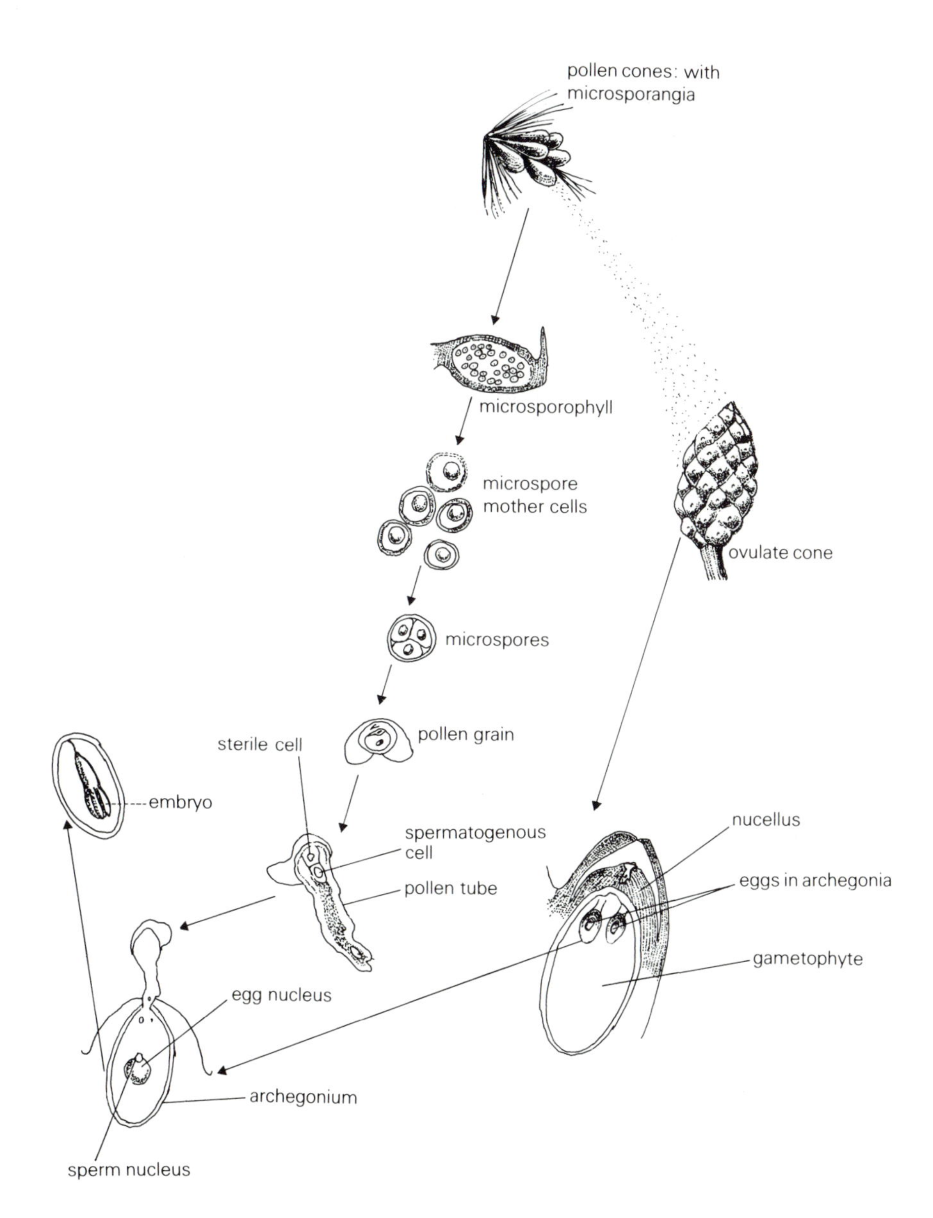


**Figure 3.5** Sexual reproduction in *Pinus* (Gymnospermae, Pinaceae).

egg, which often ripen sequentially). In the rare cases of both sperm cells fertilising an egg, one of the resulting zygotes will subsequently be eliminated by competition within the ovule. Sperm cells (male gametes) are usually nearly equal in size. If one is larger, as is markedly the case in the Taxopsida, the larger only will effect fertilisation. In some members of the Cupressaceae, multiple sperm cells are produced, although only one usually effects fertilisation. This attribute has been considered primitive, there being a progressive evolutionary reduction in the number of sperms. It has also been considered that the two ciliated sperms, of considerable size, which characterise the Cycadopsida and *Ginkgo* also represent a primitive condition. However, the evolutionary connection between these groups and other Gymnosperms is very remote, and these distinctive male gametes are best thought of as products of evolutionary isolation. They may aid fertilisation in these genera which have such degenerate male gametophytes and correspondingly further distances for the male gamete to travel.

A remarkable feature of Gymnosperm reproduction is the loss of some or all organelles from the egg cell prior to fertilisation. Thus all plastids, and in the case of *Cephalotaxus* mitochondria, inherited by the zygote originate from the male, not the female, parent, giving rise to patterns of extranuclear male inheritance (Gianordoli 1974, Willemse 1974).

In all Gymnosperms, male gametes are motile, and it must be presumed that pollen grain travel through the micropile into the pollen chamber, pollen germination, pollen-tube growth, release of male gametes, male gamete travel and fertilisation all occur in response to chemical stimuli, or chemical gradients. However, little appears to be known about the physiology of sexual reproduction in this group, and consequently we can infer little about the genetic control of mating partners within or between species. It is interesting, however, that in such an anemophilous group of plants, which lack the potential for prepollination isolation provided by floral mechanisms, interspecific hybrids are apparently very rare. It must therefore be inferred that strong post-pollination isolation mechanisms between related species have arisen, and these may well involve chemical signals employed by the sexual process.

## ANGIOSPERMS

Viewed comparatively, and from an anthropomorphic standpoint, seed plants such as Gymnosperms and Angiosperms have acquired two major advantages over Pteridophytes: removal of the dependance of the sexual process on external water; and protection and dispersal of the resultant zygote in a seed. Yet only 700 species of Gymnosperm survive today, in contrast with about 10 000 Pteridophytes and about 300 000 species of Angiosperm, and if it can be argued that the total biomass of extant Gymnosperms exceeds that of Pteridophytes, neither compares with the dominance of the contemporary terrestrial flora exhibited by the Angiosperms or flowering plants.

The Angiosperms have adopted a much wider range of growth forms, and have been able to inhabit a much wider range of ecological niches than either the Pteridophytes or the Gymnosperms, and this undoubtedly accounts for their success. They have also been able to adopt a much wider range of means of achieving sexual reproduction in terms of spore travel (pollination), control of mating partner both before and after pollination and fertilisation, and seed protection and dispersal, than have the Gymnosperms or Pteridophytes. This variety of mating systems, so characteristic of the flowering plants, has been largely responsible for the structural variety of this group, and hence its ecological success. This structural variety has been achieved in two ways:

(a) by maximising the opportunities for the reproductive isolation of demes, and thus encouraging their speciation; and
(b) by utilising manifold features of the environment, very frequently other organisms, to transport pollen and seeds to specific sites, thus greatly multiplying the potential number of species-specific niches within an environment.

Both these features, and hence indirectly the whole diversification of the flowering plants, have arisen as a result of the development of the gynoecium, which I consider to have been the single most significant advance in the evolution of land plants.

### *The gynoecium (pistil)*

The word 'angiosperm' translates as 'enclosed seed'. Just as the seed plants enclosed the process of sexual fusion and protected the zygote by allowing the male and female gametophytes to remain within the megasporangium (ovule = seed), so the Angiosperms have further enclosed the ovule within a modified sporangium-bearing leaf (megasporophyll), which is termed the carpel. The gynoecium (pistil) consists of the total megasporophylls of a flower and their contents; the carpels, whether free or fused into an ovary, and their ovules, styles and stigmas (Fig. 3.6). Thus, as stated earlier, the sexual process has been progressively enclosed in two sets of concentric containers. The second enclosure, of the ovules in the carpel, has had the following major consequences:

First, following the enclosure of the micropilar area of the ovule, it has been necessary for a pollen reception area to develop outside the carpel;

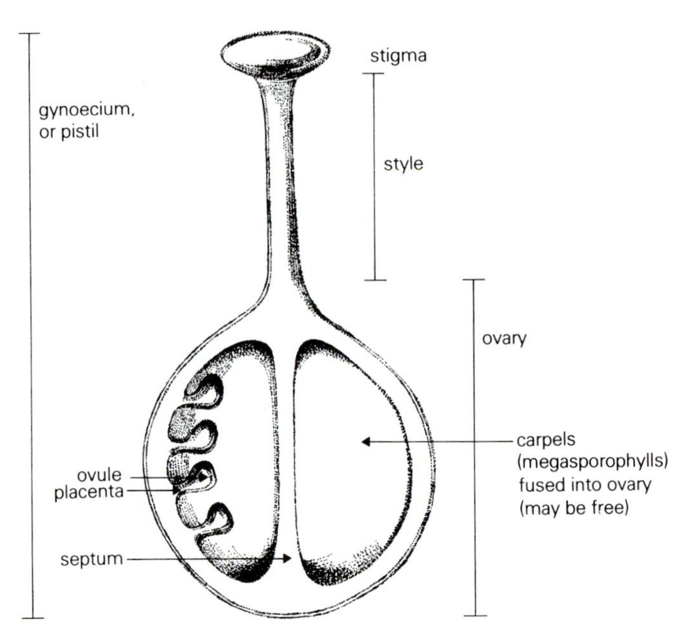


**Figure 3.6** Conventionalised diagram of a gynoecium in an Angiosperm in the family Cruciferae (Brassicaceae).

this is termed the stigma. The development of the stigma would subsequently have freed the function of pollen reception from the very small area on the ovule surface provided by the micropile in non-Angiosperm seed plants. It would have allowed the evolution of larger areas for pollen reception in optimal positions. In particular, it would have removed the potential antagonism between the functions of ovule protection (by the megasporophylls) and pollen reception which appears to be a feature of many modern Gymnosperms.

Second, the freeing of some of the foliar organs of the floral axis from the functions of ovule protection, which had been taken over by the carpel, allowed them to assume other functions such as attraction of animals (tepals, petals), reward of animals (nectaries), or protection of the floral axis (tepals, sepals, etc.). In this way the Angiosperm flower evolved. It also allowed the functions of pollen issue and reception to be combined within one floral axis; in other words, for the flower to be hermaphrodite, which is the usual and probably the ancestral condition in the Angiosperms (Ch. 4). This in turn allowed a greater diversity in breeding systems to evolve, including self-pollination. Animal pollination is greatly encouraged by the hermaphrodite condition, for all flowers produce pollen which many animals collect. Although some animal pollination may pre-date the Angiosperms, the development of the flower, and floral organs which attract, receive and reward visitors is unique to the Angiosperms. The great variety of flower types, and of pollinators, has been a very important influence in the diversification of the Angiosperms, providing both manifold niches and abundant opportunities for reproductive isolation.

Third, the continued enclosure of the ovule (seed) after fertilisation allowed great diversification to develop in the processes of protection, transportation and release of the seed, and in the structure of the seed itself. In the Angiosperms, ovaries (fruits) have developed colours, odours and food to encourage ingestion by animals; hooks, glue and external secretions to encourage external transport by animals; wings and hairs for transport by wind; bladders and tough exteriors for water transport; and many other devices. Neither could the minute, wind-borne seeds of the orchids and other families successfully reach maturity without external protection during their development. Much of the major niche diversification exhibited by the Angiosperms is a product of their highly specialised fruits, or of highly specialised seeds that develop in fruits. One may mention epiphytes, partial or obligate saprophytes, partial or obligate parasites, hydrophytes, halophytes and maritime psammophytes as being amongst those Angiosperm life-styles which are absent, or virtually so, from the Gymnosperms, and which rely heavily on a specialised fruit or seed.

Thus, the gynoecium has spectacularly influenced Angiosperm reproduction and breeding systems at all stages from pollination to seed germination, allowing the development of great diversity of function and ecological niches. This is well shown by a comparison with fossil and extant Gymnosperms, the other surviving group of seed plants. The breeding systems of Gymnosperms are limited to monoecy and dioecy, the pollination is limited to wind, and the large, thick-walled seeds have relatively unspecialised means of dispersal and establishment, through the agencies of wind or animals. Consequently, nearly all Gymnosperms are trees, occurring in limited niches in limited habitat types. Only rarely do more than two or three species coexist, and the low diversity of modern Gymnosperms is scarcely surprising.

## *Angiosperm reproduction*

The remainder of this chapter is concerned with the process of sexual reproduction in Angiosperms. It is necessarily brief, and is not intended as a thorough review of the subject. Neither does it attempt to portray the diversity of structure and function of Angiosperm reproduction which has been admirably fulfilled by books such as Maheshwari (1950). In a book on breeding systems, I am not concerned with embryology, or the structure and function of the seed or fruit, and I will limit myself to

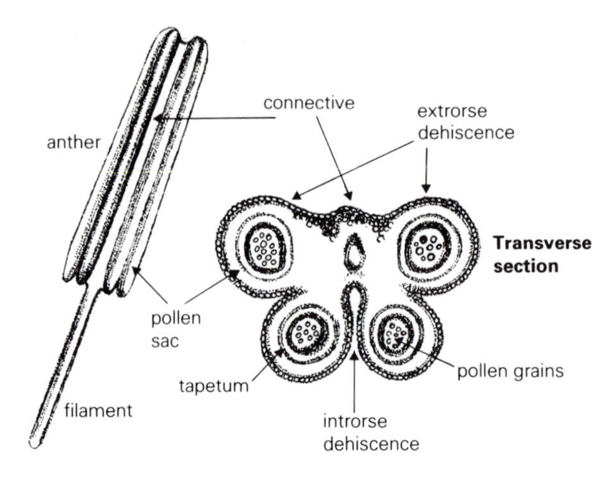


**Figure 3.7** Conventionalised diagram of a stamen in an Angiosperm.

considerations of pollen receipt and germination, pollen-tube growth, double fertilisation, the stigma, style, ovule and embryo-sac. The understanding of these basics of Angiosperm reproduction is fundamental to the study of breeding systems, yet modern school and university syllabi increasingly ignore these features, concentrating on more fashionable topics like ecology and molecular biology.

**Stamens, pollen and the male gametophyte** Most Angiosperm stamens are composed of a narrow stalk (filament), homologous with the microsporophyll of some Pteridophytes, and an anther, which is a microsporangium. The anther is usually composed of two segments, longitudinally joined by the connective, which each contain two pollen sacs (locules), although these may be fused within each segment (Fig. 3.7). Within each locule, an archesporial cell divides mitotically to yield many pollen grain mother cells (PMC), which undergo meiosis to give haploid tetrads of microspores. In most cases, these separate and develop as pollen grains within the anther. They are surrounded by a single layer of endothelial cells, the tapetum, which are secretory nurse cells, and are irregularly and highly polyploid. The mature anther most often dehisces longitudinally, towards the inside (introrse) or the outside (extrorse) of the flower, but may dehisce by apical or basal pores. In a few families, notably the Orchidaceae and Asclepiadaceae, there is a single indehiscent anther, the locules of which have formed two (or one) pollinia containing tetrads; the pollinia are dispersed whole.

Stamens vary from one to hundreds per flower and are borne laterally around the floral axis, proximally to (below) the gynoecium, but distally to (above) the corolla and calyx. Although most often borne free on the receptacle, the filaments and even the anthers may at times be fused to the petals, or more rarely to the nectaries or gynoecium. In the latter case, the anthers may be apparently fused to the gynoecium, often very close to the stigma, forming a gynostegium. This is once again typical of the unrelated dicotyledonous Asclepiadaceae, and monocotyledonous Orchidaceae, a clear case of parallel evolution (Ch. 4).

The pollen grain, a haploid microspore, is not dissimilar in structure to Gymnosperm pollen, or even Pteridophyte spores, but is consistently diagnosed by the columnar structure of the outer wall (exine), between the continuous inner layer (nexine) and the perforated outer layer (tectum) (Hickey & Doyle 1977). The exine is constructed primarily of a lipoprotein, sporopollenin, whereas the inner wall (intine) is made of cellulose (Heslop-Harrison 1975a). The exine may be variously decorated with spines, furrows, etc., and usually has from one to four pores, which may or may not be associated with furrows (Fig. 3.8). The characteristic size, shape and decoration of a pollen grain persists into the fossil record.

Within the grain, one or two mitoses have taken place to yield two or three haploid nuclei. The timing of the first pollen grain mitosis varies widely, occurring rapidly in those species with fast pollen grain development (e.g. most tropical species). However, in boreal plants in which floral initiation and meiosis occur the previous autumn (for instance

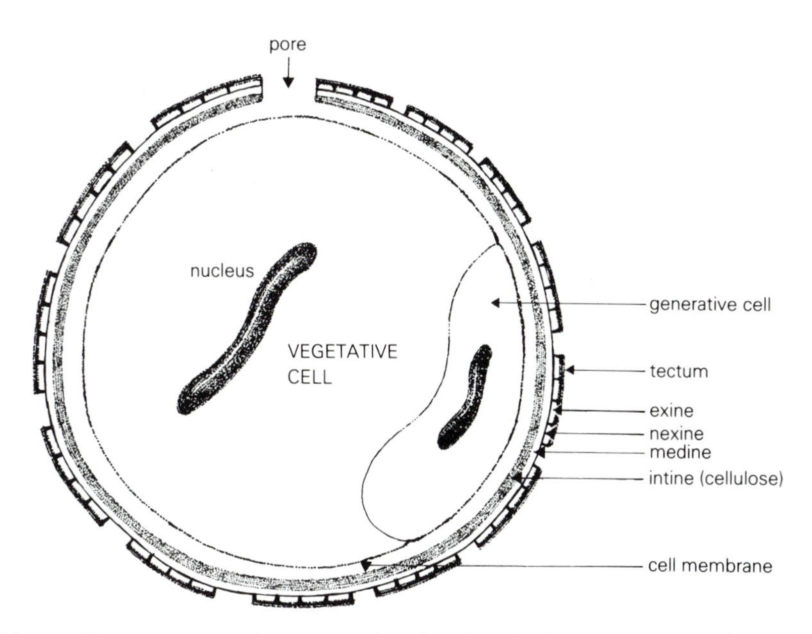


**Figure 3.8** Anatomy of a conventionalised typical Angiosperm pollen grain. The gametophyte may contain two or three cells.

many species of *Rhododendron*), pollen grains may over-winter in a uninucleate phase. The pollen grain mitosis is usually unsynchronised, except when the pollen grains are in intimate contact in pollinia, as in the orchids. The mitosis is abnormal, being asymmetrical, the pole next to the cell wall being blunt and the free pole acute in shape. After the first division, two unequally sized cells, without a cellulose partition, arise. The larger central cell (vegetative), and the smaller peripheral cell (generative) together form the highly reduced male gametophyte. The nucleus of the vegetative cell frequently assumes curious forms, sometimes becoming very thin and filamentous. The generative cell loosens itself from its peripheral position latterly, and often becomes spindle shaped. It can divide into the two male gametes at an early stage, or this second mitosis may wait until after the germination of the pollen grain, although in the latter case it may have entered prophase before germination. In some cases, the second mitosis occurs after the pollen grain alights on the stigma, but before germination, during the hydration phase (Maheshwari 1949). Thus some pollen grains are bi-nucleate, with a vegetative and a generative nucleus, whereas others are tri-nucleate, with a vegetative nucleus and two male gametes (sperm cells). There is a very strong correlation between these conditions, the nature of the cuticle of the stigmatic papillae (wet or dry), and the control of incompatibility, as follows (see Chs. 6 and 7):

| Pollen grain | Cuticle of stigmatic papilla | Control of incompatibility |
|---|---|---|
| bi-nucleate | 'wet' | gametophytic |
| tri-nucleate | 'dry' | sporophytic |

The second mitosis giving rise to the male gametes (sperm cells) is also usually abnormal in configuration, apparently because of limitations in space caused by the constriction of the generative cell or pollen tube. In wheat (*Triticum*), sperm cells show the following four developmental stages (Chu & Hu 1981):

(a) Naked cell stage in which the sperm cell is round and naked, enclosed only by a discontinuous plasma membrane.
(b) Walled cell stage in which the sperm cell is completely surrounded by membranes of its own and of the vegetative cell; the space is filled with callose.
(c) Stage of cytoplasm increase during which the size of the cell and the number of organelles increase and the cell changes in shape, becoming more elongated.
(d) Mature sperm in which the cytoplasm and organelles are concentrated at one end to form a tail-like structure.

Usually, both male gametes (sperm cells) are indistinguishable in shape and size, but Russell (1981) describes how those of *Plumbago*, which abnormally lacks synergids in the mature embryo-sac, has very distinct sperm cells. One only has a long slender projection which is associated with the vegetative nucleus; this sperm, which eventually fuses with the primary endosperm nucleus, contains the greater proportion of mitochondria. The other, which fuses with the egg cell to form the embryo, receives nearly all the plastids. There are a number of other cases cited in Maheshwari (1949) in which the sperm cell fusing with the egg cell is smaller than that which helps to form the endosperm.

**Pollen germination** When mature pollen is released from the anther it normally has a very short life. In some grasses this may be as short as 30 minutes, and even in insect-pollinated species with sticky pollen it rarely exceeds one day, although much longer periods of natural viability have been recorded in some fruit trees and Gymnosperms (Maheshwari & Rangaswamy 1965). Pollen which has been freeze-dried under a partial vacuum and stored at temperatures of about −5°C and less than 50% relative humidity may survive for much longer periods, certainly in excess of ten years. Heslop-Harrison (1979a,b) and Shivanna and Heslop-Harrison (1981) have assigned pollen viability to the condition of the plasmalemma of the vegetative cell. If this is able to reassume a continuous lamellar structure of lipid bilayer on rehydration, its osmotic properties are recovered, and the turgescence which is a vital preliminary to pollen germination can be attained. However, if the partially disordered plasmalemma encountered in the dry grain is further dissociated by heat or desiccation, it will become unable to reorder, and thus to act osmotically, and the grain will be non-viable. The state of the plasmalemma of the vegetative cell, which clearly plays a key rôle in the germination of the pollen grain, is readily assessed by the use of the dye fluorescein diacetate, which is cleaved by esterase activity to fluorescein and accumulates within the intact membrane in a viable cell.

The only essential requirements for the germination of many types of pollen grain seem to be water and oxygen. Pollen may germinate on the moist surface or a corolla (Maheshwari 1949) or in a water-filled corolla tube (Eisikowitch & Woodell 1975), and I have seen selfed pollen of a *Primula* germinate in nectaries. However, pollen tubes that germinate in water usually burst from turgor pressure, and a solution of the correct osmotic potential is necessary for pollen-tube survival, and a carbohydrate source is required for continued pollen-tube growth. Rather unexpectedly, boron is necessary for the successful germination of pollen in many species. Boron is deficient in pollen, but occurs at comparatively high concentrations in stigmas and styles. It has been shown to aid sugar uptake, and to be involved in pectin synthesis in the developing pollen tubes, perhaps by its coenzymic function in the synthesis of the D-galacturonosyl units of pectin (Stanley & Loewus 1964). Small quantities of calcium are also necessary for successful pollen germination and pollen-tube growth. The requirement seems to be complex, calcium being involved in pectate synthesis, suppression of inhibitory cations and osmotic regulation (Brewbaker & Kwack 1964). The population effect, or mass stimulation effect, observed when many pollen grains germinate much better *in vitro* than isolated grains, is also apparently due to calcium borne on the outside of the pollen, although other substances such as organic acids may also be implicated (Maheshwari & Rangaswamy 1965). Thus the rationale behind the use of mixtures of old and new date palm pollen by Arab cultivators is unclear.

Calcium gradients are also unambiguously implicated in the chemotropism of the pollen tubes within the gynoecium in some, but not all, plants. A very wide range of substances has been shown to have a stimulatory effect on pollen-tube growth *in vitro*, ranging from amino acids to auxins, and purines to sugars. Few, however, have been implicated in tropism, although it is often easy to show experimentally that stylar fragments exert a chemotropic effect.

Stigmas are usually receptive to pollen over a relatively short period while the flower is open, which is generally marked by the turgidity of the stigmatic papillae, and concluded by the necrosis of these cells. Legitimate pollen arriving on a ripe stigma will adhere to it. However, it is often possible for fertilisation to be achieved by artificial pollination of bud stigmas, particularly if the stigmatic surface is damaged.

There are two main types of stigmas recognised, termed 'wet' and 'dry'. These can be differentiated as follows:

| | Stigmatic exudate | Cuticle of papilla | Pollen hydration | Pollen entry | Incompatibility |
|---|---|---|---|---|---|
| wet | present | gappy | external | intercellular | gametophytic |
| dry | absent | continuous | internal | intracellular | sporophytic |

A wet stigma produces a stigmatic exudate containing free sugars, lipids, phenolic compounds and traces of enzymes. This exudate appears to perform four distinct functions:

(a) Attaching the pollen to the stigma surface.
(b) Hydrating the pollen via the discontinuous exine to the absorbent medine, the pectin- and protein-containing layer situated between the exine and the intine.
(c) Providing a suitable medium for the initial growth of the pollen tube.
(d) Preventing the stigmatic papillae with their discontinuous cuticles from drying out.

In contrast, a dry stigma, typical of most sporophytically controlled incompatibility systems, has no stigmatic exudate. The pollen adheres by virtue of its external cover of lipoprotein 'pollenkitt', or 'tryphine' which is secreted by the tapetum on to the developing grain. Pollen grains break down the cuticle of the stigma papilla enzymically. They are then able to absorb water from the turgid cells of the stigma papillae and become hydrated.

**Pollen-tube growth** On germination, after hydration, pollen tubes emerge through the pores of the pollen grain. In most plants, only one tube is produced per grain, but in a few cases (e.g. the Malvaceae), grains produce more than one tube (polysiphonous), although only one contains the pollen-tube nuclei. The other tubes may have haustorial or absorbative functions. Pollen tubes (microgametophytes) have a very simple structure, being composed of a pectin sheath and three cells lacking cellulose walls, and are filled largely with water and solutes. Organelles concentrate largely at the apex, which is rich in endoplasmic reticulum. The main amino acid in pollen grains is proline, which is rapidly metabolised and converted into glutamic acid in pollen tubes. RNA is synthesised by the developing pollen tube, and both RNA and ribosomes are released into interstylar tissue by the tube apex. Pectinase and $\beta$-1,4, glucanase are associated with the pollen tube, and probably mediate the plasticity of the growing tube apex, as well as assisting entry into the stigma.

In gametophytically controlled systems, with a wet stigma, pollen tubes grow down the outside of the stigma papilla, and penetrate the stigmatic tissues at the base of the papillae by dissolving the cell-wall middle layer of pectate. In systems with dry stigmas, such as the Cruciferae, the pollen tubes penetrate the cuticle (of lipoprotein), and the thin intermediate pectic layer, and grow in between the cellulose lamellae of the cell wall of the papilla. Thus both cutinase (esterase) and pectinase activity by the pollen-tube apex are required for successful stigmatic penetration in sporophytic systems.

Having reached the style, the pollen tube continues to grow intercellularly in the spaces between the elongated stylar transmitting tissue; these spaces are also pectic in nature. There are two main types of style. In open styles, typical of the monocotyledons, a central stylar canal is present, lined with a glandular canal that functions as transmitting tissue. In closed styles, as in most dicotyledons, the centre of the style is filled with transmitting tissue, between which the pollen tubes grow towards the ovules.

In most plants, growth of the pollen tube lasts between 12 and 48 hours from pollen germination to fertilisation. In exceptional cases, as for instance *Taraxacum kok-saghyz*, the Russian rubber dandelion, fertilisation occurs between 15 and 45 minutes after pollination. In contrast, in some trees such as oaks (*Quercus*), hazel (*Corylus*), alder (*Alnus*), *Garrya*, *Hicoria*, *Hamamelis* and *Ostraya*, pollen-tube growth may be arrested for weeks or months, with pollination taking place in the winter or early spring, but fertilisation occurring in the summer. In *Quercus velutina*, fertilisation is achieved only in the season following pollination, 13 months later. Similar situations prevail in some orchids, where the ovules only develop after pollination has occurred.

The distance travelled by the pollen tube varies widely, being less than 1 mm in some species with short or no styles and small ovaries, and as much as 20 cm or more in species with exceptionally long styles, which are frequently bird or bat pollinated. It is abundantly clear that the pollen tube is self-sustaining on resources of water, sugars and amino acids provided by the style, and that pollen size or pollen resource is not related to tube growth. [However, Plitmann and Levin (1983) showed that there is a close relationship between pollen size, pollen-tube diameter, stigma–papilla width and style length amongst a number of species in the family Polemoniaceae.] Growth rates also vary, and readings from 1.75 mm an hour to 35 mm an hour have been obtained in different species. Growth rates are temperature dependent, and are usually optimal between 25 and 30°C, being totally inhibited above 40°C and below 10°C. In cold climates, temperatures may thus be limiting to successful fertilisation, and various features which aid heat conservation, or promote 'solar furnaces' (Ch. 4) may be favoured.

As the growth of the pollen tube progresses, the vegetative cell, which had originally assumed an apical position, becomes diffuse and is left behind, often disappearing entirely by the time of fertilisation. In contrast, the two male gametes or sperm cells, which arose by the second pollen-grain mitosis from the generative cell and were initially proximal, actively move into an apical position. Some controversy still surrounds the mode of movement of these cells. By this time they are often elongated with a tail-like structure, and despite the absence of a well-marked flagellum or cilia, may still in some sense 'swim'. However, active cytoplasmic streaming is also observed, and their transport may be passive. Proximally, the mature pollen tube is characteristically filled with plugs of the complex carbohydrate 'callose' ($\beta$-1,3, glucan). This fluoresces under ultraviolet radiation when stained with, for instance, aniline blue and this is the basis of microscopic techniques for tracing the path of the pollen tube. Callose is a substance that can be a conspicuous feature of several different aspects of Angiosperm pollination and fertilisation, and appears to occur principally as a 'wounding response', not only to physical damage, but also as a result of chemical recognition or antagonism. Thus on the style prior to germination, callose appears outside the pollen grain in sporophytically controlled species, and inside the pollen grain in gametophytically controlled species, and it may be especially marked in incompatible reactions. It also occurs within the stigma papilla in incompatible reactions under sporophytic control and may even play a rôle in causing the incompatibility through blockage. In incompatible reactions under gametophytic control, it becomes increasingly evident around the progressing pollen tube before it halts in the style, although it is not apparently directly concerned with the cessation of growth. Thus its occurrence in the normal pollen tubes of compatible pollinations is something of an enigma.

On arrival at the ovary (carpel), the pollen tubes travel round intercellular spaces on the inside of the ovary wall until they reach an ovule placenta, up which they travel. At some stage during the progress of the tube from the ovary wall to the ovule, it emerges from intercellular spaces into free space, where it travels in extracellular secretion to the micropile of the ovule (porogamy). More rarely, the nucellus, or even the embryo-sac, protrude from the micropile, so that micropilar penetration is unnecessary. Occasionally, the chalazal end of the ovule is penetrated. Once one pollen tube has entered a micropile, growth of other tubes is stopped or redirected, and polyspermy caused by the entry of more than

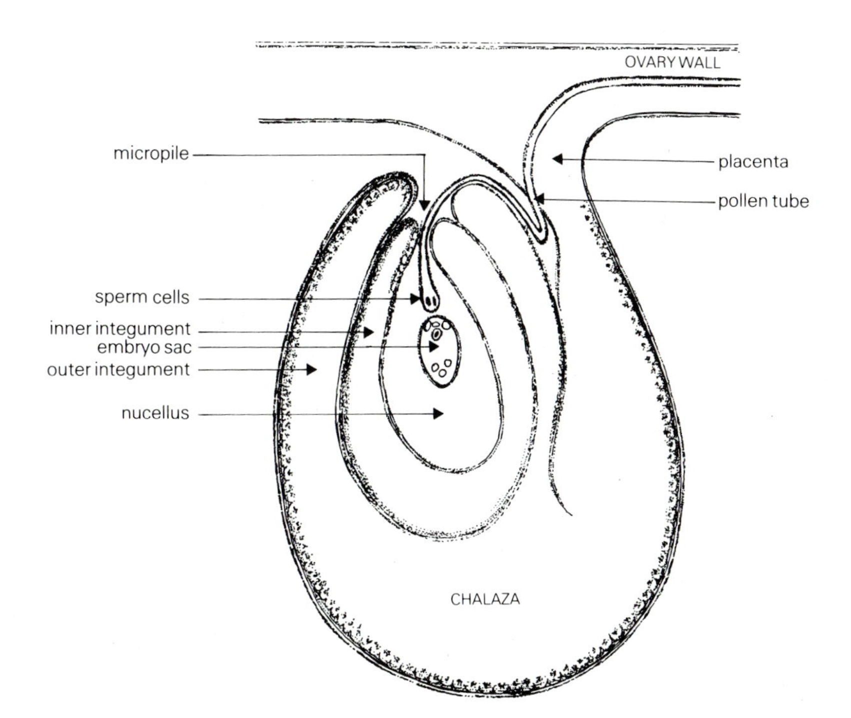


**Figure 3.9** Diagram of the path of the pollen tube in the ovary and ovule of an Angiosperm.

one tube into an ovule is rare, although it has been reported. At this stage, tubes occasionally branch, sometimes spectacularly so, branches lacking nuclei entering the integuments and the nucellus. It has been suggested that such anastomisation may be functional, allowing metabolite transfer from the nucellus and integuments to the embryo-sac via the pollen tube. In any case, the tube transverses the nucellus to the polar end of the embryo-sac, again in intercellular spaces in the ovule. The nucellar cells between which it passes frequently appear different from the remainder. On arrival at the embryo-sac the tube usually broadens before sperm release, and in some cases bifurcates in the directions of the egg cell and the primary endosperm nucleus, one sperm travelling into each branch.

In its later stages, the direction of travel of the pollen tube is remarkably changeable, 90° turns occurring at the style/ovary junction, the ovary/placental junction, at the micropile, and at the embryo-sac (Fig. 3.9). Strong chemotropic forces are clearly implicated, but these have not been satisfactorily identified. Promising results involving calcium gradients have been at least partially negated by the discovery that calcium levels in stylar transfer tissue are lower than in surrounding tissue.

**Double fusion** In the embryo-sac (female gametophyte), four unwalled cells lie at the polar (micropilar) end. Of the two synergids, the larger lies against the exterior of the embryo-sac, with the egg cell in close conjunction. The other synergid is close by, and the dikaryotic primary endosperm nucleus is distal to the egg cell (Fig. 3.10). The proximal (larger) synergid has produced outgrowths (filiform apparatus), and after pollination, but before the arrival of the pollen tube at the embryo-sac, it starts to degenerate. In the absence of pollination, similar effects can be produced by the application of gibberellic acid to ovules *in vitro*, and it is supposed that synergid degeneration is triggered by gibberellic acid carried on pollen grains. The vacuole of the synergid has high levels of calcium which are released on degeneration, and probably act as a close-range chemotrophic attractant to the pollen tube. This penetrates the synergid via the filiform apparatus, and ruptures, releasing the two sperm cells, starch grains and other organelles into the degenerating synergid. The synergid remains envelop the egg cell and the primary endosperm nucleus, carrying the two sperm cells to lie in conjunction with them.

At this point, double fusion takes place, a phenomenon that is unique to the Angiosperms among all living things. One haploid sperm cell (male gamete) fuses with the haploid egg cell to form the diploid zygote, which will form the embryo in the developing seed, and, ultimately, the new sporophyte generation. The second haploid sperm cell fuses with the primary endosperm nucleus. This has arisen from the fusion of two haploid cells of the embryo-sac (female gametophyte), and is at this stage

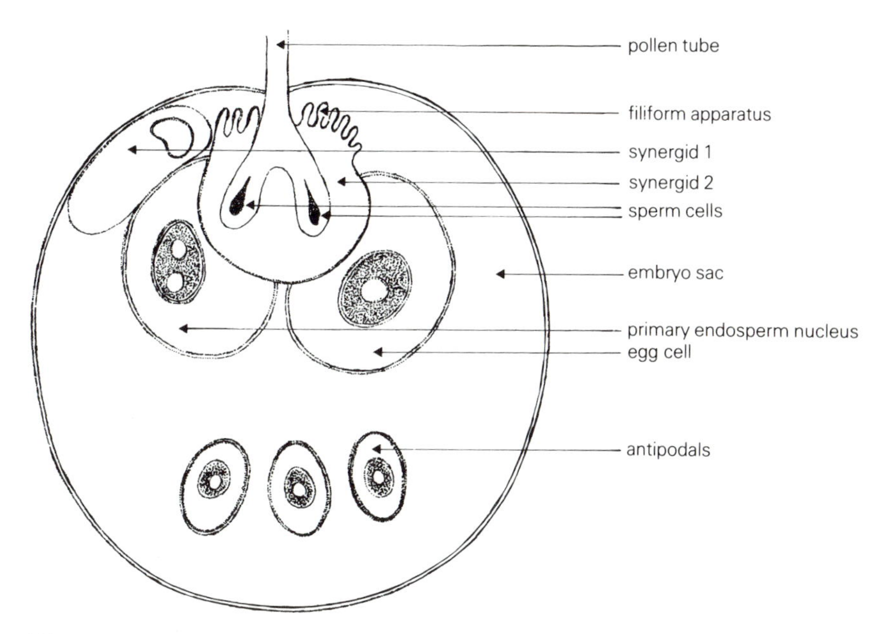


**Figure 3.10** Diagrammatic representation of the process of double fertilisation in an Angiosperm embryo-sac. The pollen tube has entered the embryo-sac by means of one of the two synergids that now engulfs it. The pollen tube has bifurcated, with one sperm cell in each branch, and the synergid leads one pollen tube branch and sperm cell towards the egg cell and the other towards the primary endosperm nucleus (fused polar nuclei). The nuclear membranes of the primary endosperm nucleus and the egg cell will shortly break down, admitting the degenerating synergid and the respective sperm cell.

either diploid, or dikaryotic. In either case, the result is a triploid primary endosperm cell, which develops into the endosperm.

Fusion occurs through the evagination of the female egg cell and enclosure of the sperm cell, apparently including its cytoplasmic contents. Once enclosed, the sperm cell membrane breaks down to release the nucleus, which lies against the female nucleus. The intervening nuclear membranes then disintegrate, but the respective nucleoli remain distinct, two in the zygote and three in the primary endosperm cell.

The complex mechanism of sexual fusion in the Angiosperms, just described, is remarkably constant throughout its many diverse orders, and has led many authors to suggest that the Angiosperms have a single origin, i.e. that they are monophyletic. Exceptions to the general pattern can be found. Thus in some plants, the synergids are absent. In *Plumbago*, the egg cell itself acquires a filiform apparatus and functions as a synergid, the synergid phase having been lost (Johri 1981). In other cases, both synergids are involved in sperm transport, often one sperm in each, or the synergids may survive but are not involved in sperm transport. In rare cases, multiple sperms may be produced by the pollen tube and multiple eggs by the embryo-sac (Favre-Duchatre 1974). In this last case, zygotes may also divide to produce multiple pro-embryos. In certain families with highly specialised seeds, there is no endosperm, and thus double fertilisation is not required. Thus in the Orchidaceae, there may be conventional double fusion, but no further development of the endosperm; the process may progress as far as double fusion, but no fusion with the primary endosperm nucleus occurs; or the embryo-sac may be incompletely developed, with four, five or six cells instead of eight (Savina 1974). In these latter cases, some or all of the antipodals may be missing, and the diploid primary endosperm nucleus is not formed. These are merely a few of the many variants observed to the sexual process in Angiosperms, but it is generally considered that they are secondary, later developments which do not counter the monophylesis argument.

**The ovule and female gametophyte** Being heterosporous, Angiosperms produce two types of sporangium on two types of sporophyll, giving rise to two types of spore, and eventually to two quite different gametophytes, one male and the other female:

| | Sporophyll | Sporangium | Spore | Gametophyte | Gamete |
|---|---|---|---|---|---|
| male | stamen | anther | pollen grain | pollen tube | sperm cell |
| female | carpel | ovule | megaspore | embryo-sac | egg cell |

At each of these stages, the various cells or organs are thus homologous with each other, but very different in structure and function. As in all seed plants, the female gametophyte generation has become extremely

**Figure 3.11** Evolutionary trends in Angiosperm ovaries and ovules. 1, Free multilocular carpels with parietal placentation; 2, fused multilocular carpels with axile placentation; 3, fused unilocular carpels with axile placentation; 4, fused unilocular carpels with basal placentation; 5, single multilocular carpel (derived from fused carpels) with axile placentation; 6, single multilocular carpel (derived from fused carpels) with basal placentation; 7, single unilocular carpels with basal placentation. A trend is observed in carpel fusion followed by loss of carpel walls (septae); in the reduction of number of ovules per carpel, and per ovary to one, and from axile placentation to basal placentation. 1 is typical of Magnoliaceae, Paeonaceae etc, 2 is typical of Cruciferae (Brassicaceae), 3 is found in Rutaceae etc., 4 is found in some Primulaceae, 5 is typical of the Leguminosae (Fabaceae), 6 is found in Papaveraceae etc., and 7 in Compositae (Asteraceae). ▶

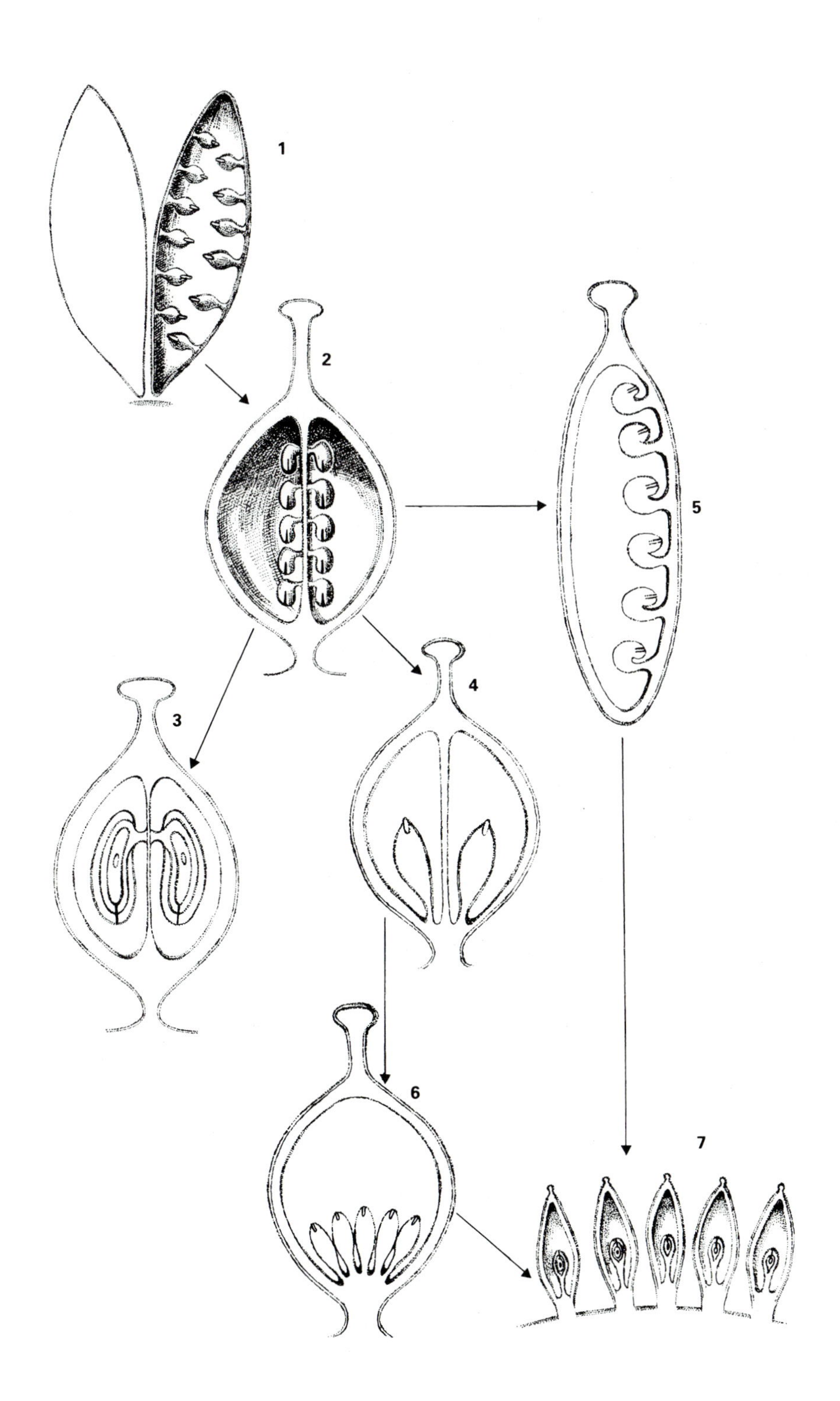
1
2
3
4
5
6
7

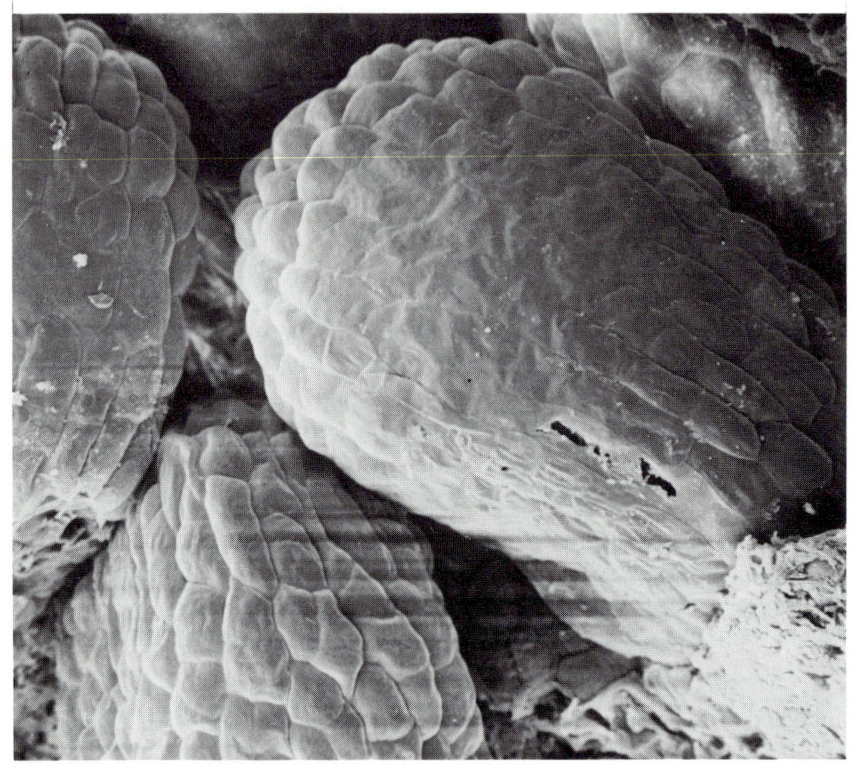

**Figure 3.12** Scanning electron micrograph of an ovule of *Primula modesta* (×160). The micropile is apical, and at the apex the outer and inner integuments are revealed.

reduced, and is parasitic on the sporophyte megasporophyll (carpel) and megasporangium (ovule).

In primitive Angiosperms, many ovules are borne in each carpel (part ovary), parietally and orthotropously, being attached to the ovary wall by a placenta at the chalazal end, distally to the micropile (Fig. 3.11). The ovule is composed of three layers; starting from the exterior (Fig. 3.12) these are:

| | | |
|---|---|---|
| outer integument | protective | form seed coat (testa) |
| inner integument | protective | |
| nucellus | nutritive | disappears after fertilisation |

The nucellus differentiates to form an archesporium with a central archespore. This enlarges to form an embryo-sac mother cell (EMC) which undergoes female meiosis. The resultant four megaspores form a row along the ovule axis, and the three at the micropilar end dwindle and eventually disappear. The fourth chalazal megaspore enlarges and undergoes three mitoses to give eight haploid cells. These differentiate to form the mature embryo-sac (female gametophyte) (see Fig. 3.10).

The embryo-sac is thus enclosed within the innermost tissue of the ovule, the nucellus; it lies centrally or towards the micropilar end and usually takes up about half the nucellar volume. The embryo-sac is readily distinguished from the parenchymatous, secretory nucellus, the cells of which bear cellulose walls, because it forms a rather empty looking coenocytic sac with a pectic wall and unwalled internal cells. These comprise, at the chalazal end, three antipodal cells of uniform size. At the micropilar end, as already discussed, lie two synergids and the egg cell. One synergid lies at the apex, in close conjunction with the egg cell, and the other, which may be somewhat smaller, is nearby. Distal to these cells, but still at the micropilar end, are the two remaining cells which fuse, usually prior to pollination, although sometimes remaining as a dikaryon. This is known as the primary endosperm nucleus (fused polar nuclei).

The fate of the eight cells of the embryo-sac after double fusion can be summarised as follows:

(a) The antipodals disintegrate, or undergo endomitoses, to form a block of antipodal tissue of unknown function in the seed; or they occasionally connect to the zygote by branches of pollen tube, apparently feeding it (rare).
(b) The primary endosperm nucleus forms triploid endosperm which acts as nurse tissue (protection and feeding) to the embryo in the seed, thus displacing the ovular (parent sporophyte) nucellus in nurse functions; the nucellus degenerates in the developing seed.
(c) The egg cell forms the zygote, which develops into the embryo in the seed, and thus into the new sporophyte generation.
(d) The synergids, one or both, start to disintegrate before accepting the pollen tube and sperms; if only one is involved in sperm transport, the other also soon disappears.

There are exceptions to this structure and function of the embryo-sac, the more important of which have been covered earlier in this chapter.

After double fusion, no less than four independent and genetically different generations coexist in the seed:

(a) The parent sporophyte (diploid) – integuments of the ovule, forming the testa.
(b) The female gametophyte (haploid) – the embryo-sac wall and antipodals.
(c) The offspring sporophyte (diploid) – the embryo.
(d) The endosperm (triploid) – the product of an independent fusion between a sperm cell and diploid primary endosperm nucleus.

The relationship between the endosperm and the embryo is of great interest. The male genome that enters the embryo should be exactly the same as that entering the endosperm, for the two sperm cells are genetically identical, resulting from the mitosis of a single spore (meiotic product). Similarly, the egg cell should be genetically identical to both of the polar nuclei. Thus the embryo should contain exactly the same nuclear, but not necessarily cytoplasmic, genes as the endosperm. Therefore, there is only one very important nuclear difference between the embryo and the endosperm. Whereas the embryo is diploid, with one genome from each parent, the endosperm is triploid, with two maternal contributions to one paternal. This also renders the diploid primary endosperm nucleus before gametic fusion entirely homozygous. It may thus potentially fail through the possession of homozygous lethals. This may account for the fact that it is frequently dikaryotic until fusion with the male gamete occurs.

Although it may be assumed that those genes controlling the independent developments of the endosperm and embryo have co-evolved harmoniously within a population, this may not be the case in hybrids between demes that have evolved different rates of seed development. As the embryo and endosperm receive different gene doses from the male and female parents, they may develop at different rates in the hybrid seed, resulting in embryo starvation, or in the smothering of the embryo by the endosperm, both of which lead to death of the seed, or poor viability. Such a case has been well described by Woodell (1960a,b) in hybrid *Primula*.

The independence of the embryo and the endosperm after double fusion is well illustrated by the work of Batygina (1974) on several species of cereals. In these, the fusion of the male gamete with the primary endosperm nucleus is completed in less than two hours, and is followed by a dormancy of less than one hour. In contrast, the fusion of the other sperm cell with the egg cell lasts from four to five hours, and is followed by a dormancy of about 18 hours, at which stage the endosperm may have four or eight cells.

The stimulus involved in the initiation of development in the embryo and endosperm has received some attention, and often seems to be dependent on the process of cytoplasmic fusion of the gametes. In this respect, observations on the aberrant phenomenon of hemigamy are of interest (Solnetzeva 1974). Here, cytoplasmic fusion of the gametes occurs, but nuclear fusion is delayed. Within the fused cytoplasm, nuclear division of the sperm cell and the egg cell proceed independently, the sperm cells forming some haploid tissue. In most of these instances, the embryo-sac is unreduced, and the diploid egg cell goes on to form an embryo, agamospermously, but only after cytoplasmic fusion with the sperm cell. In other agamosperms (Ch. 11), pollination but not fertilisation is required for embryony (embryo development) to proceed, and this can be substituted experimentally by an application of auxin to the stigma, auxin being carried externally on pollen (pseudogamy). [In all pseudogamous species, fusion of a sperm cell with the polar nuclei and 'conventional' endosperm development proceed as in a sexual plant.] Yet, embryony does not universally depend on an auxin message to the stigma, as is demonstrated by the experimental technique of placental pollination *in vitro* (Rangaswamy & Shivanna 1972). In fact, successful fertilisation and embryony can be achieved experimentally by the application of germinated or ungerminated pollen to a variety of maternal tissues, from the cut style to the ovule micropile. In all cases except non-pseudogamous agamosperms, however, some hormonal message from the pollen grain to the egg cell may be required for embryony to proceed, although this may not always involve actual gamete fusion. There is also some evidence that development of the endosperm always precedes that of the embryo, and that embryony rarely proceeds without prior division of the endosperm. This is clearly not the case in seeds without endosperm.

Once the harmonious development of the embryo and endosperm has started in the fertilised ovule (seed), the further embryological conditions of the seed, and its physiology and ecology do not concern us in this book. I will give some account of the rôle played by fruit dispersal in breeding systems at a later stage (Ch. 5). It is, however, worth briefly recounting some trends in the structure of the ovary and ovule at this stage.

## EVOLUTIONARY TRENDS IN THE GYNOECIUM

As stated earlier, the primitive condition in Angiosperms is probably for orthotropous ovules to occur parietally in free carpels, each with many ovules. Many later developments are observed in the orientation of the ovule (and its structure), in ovule placentation, and in carpellary development. These developmental trends may have been either primarily adaptive, resulting in specialised fruits or secondarily adaptive, resulting from constraints imposed by floral form and pollination biology. Although such trends in the form of the ovary and ovule in the Angiosperms are doubtless adaptive in origin, there is a good deal of evidence to suggest that those changes which have occurred have done so rarely, and that a good deal of homoeostasis is observed in these features. Such characters are traditionally employed in higher order systematics (delimitations of families, orders etc), and their high weighting in modern numerically based systematics can be justified statistically. It is probable that this evolutionary homoeostasis results from features of vascularisation, which impose severe constraints on the structure and function of the gynoecium. Preadaptation, parallel evolution and neoteny are unlikely to occur, and when they do so, can usually be detected by study of the vascular system:

**ovule orientation**: orthotropous → camplyotropous → anatropous

There appears to have been a trend for the micropile to change in orientation from a position away from the placenta, to a position in which it approaches the placenta (anatropous), the most frequently found form. Conceivably, this may facilitate the passage of the pollen tube to the micropile from the ovary wall:

**ovule and ovule wall**: ovule large, thick-walled with two integuments → ovule small, thin walled, often with only one integument, exceptionally only one cell thick

This trend is undoubtedly associated with specialised modes of fruit and seed dispersal:

**ovule placentation**: parietal → axile → free-central → apical and or basal → unilocular

In the primitive carpel, numerous ovules are arranged around the walls. As carpels fused to form segmented ovaries, ovules became restricted to the central septal axis (axile), where they would be better protected from predation. As the septae of the fused ovary were lost, ovules were restricted to the remains of these septae, apically and basally (free-central position), and latterly to apical and basal positions of the ovary wall (see Fig. 3.11). Reduction in ovary size accompanied complex many-flowered or compound inflorescences, to a point where ovule number was reduced, finally to one (unilocular), giving rise to one-seeded fruits (achenes) which are today a very common condition.

Two main trends are observed in carpel evolution, fusion of free carpels into segmented and latterly into non-segmented ovaries, and reduction of ovule number per carpel and per ovary to one (see Fig. 3.11):

free, multilocular carpels → free unilocular carpels
↘
fused multiseptate ovaries: carpels multilocular → carpels unilocular
↘
fused biseptate multilocular ovaries
↘
fused aseptate multilocular ovaries
↘
unilocular ovaries

These are associated with floral specialisation, and fruit specialisation, particularly with respect to reduction in size in compound inflorescences.

## CONCLUSION

Thus we see an overall pattern of the development of the sexual process in the flowering plants. The enclosure of ovules in a gynoecium has enabled great diversification to occur in the structure and function of the inflorescence and infructescence. However, the remarkable complexity and unique function of the sexual process between the male and the female gametophyte remains essentially unchanged throughout this vast class of plants.

CHAPTER FOUR

# *Floral diversity and pollination*

---

An outstanding feature of the Angiosperms is the amazing diversity in form and colour that has been adopted by the inflorescence, sufficient to inspire great art, fuel a major industry and serve as a solace for suffering mankind.

Yet the flower is merely a sex organ, and never has any function except to promote reproduction by seed, usually sexually. The beautiful, weird, sinister, astounding forms that flowers have acquired are strictly pragmatic, and have encouraged the ecological diversification, and dominance, of the flowering plants, as argued earlier (Ch. 3).

It is not the function of this book, neither is there the space, to catalogue all the manifold adaptations of the inflorescence to pollination by animals or abiotic forces, and to fruit dispersal. In recent years this has perhaps been most fully achieved, and certainly has been best illustrated, by Proctor and Yeo (1973). An interesting, but briefer, account is by Faegri and van der Pijl (1979). Of the earlier accounts, Knuth (1906–1909) and Müller (1883) are outstanding. In the present chapter, a brief survey of the adaptive radiation of floral organs in relation to pollination systems is attempted, with some fuller accounts of exceptionally interesting or complex syndromes. I shall reserve for the next chapter (Ch. 5) information on the control exercised by the inflorescence on patterns of gene exchange, and hence an evolutionary potential, which is more central to the theme of this book.

## PATTERNS OF FLORAL EVOLUTION

It is generally, although not universally, accepted that it is most useful to consider the flower of the genus *Magnolia* as the starting point for the adaptive radiation of floral diversity. In all features it exhibits what are thought to be primitive character states, and in this it may not be too distant in structure from the flower of a hypothetical archae-angiosperm. Thus, it is instructive to examine the structure and function of a *Magnolia* flower in detail (Fig. 4.1).

Flowers are very large (up to 30 cm in diameter in *M. grandiflora*) and borne individually on trees or large shrubs. They may flower sporadically

**Figure 4.1** The primitive flower of *Magnolia wilsonii* (×0.5).

over much of the season. Floral organs are borne spirally on an elongated floral axis (receptacle) in a manner reminiscent of a Gymnosperm cone. Proximally (on the outside) are borne the tepals (together these form the perianth), undifferentiated into petals and sepals. These are many (i.e. more than six which is the largest number of a type of floral organ which is usually constant), large, thick, with leaf-like veining, usually white (occasionally pink or purple, e.g. *M. campbellii*), and they are unscented, lacking markings, hairs or nectaries.

Distal to (inside) the tepals are borne the stamens (androecium), also spirally. These are also very numerous, and although smaller and narrower, also share the leaf-like structure of the tepal, the pollen sacs being borne in linear rows on the stamen surface. There is no clear differentiation into anther and filament; rather the organ as a whole is very recognisably a sporophyll. Anther dehiscence is longitudinal and extrorse (Ch. 3). Pollen grains are rather large, with a single pore.

Distal to the stamens, and apically (inside), are borne the carpels. These are quite free from one another, are relatively large, usually many, and they bear several large orthotropous ovules parietally. There is no style, the large, somewhat linear, stigmatic area being borne directly onto the carpel, usually subapically. In some species, the carpel is not fully closed.

The fruits form dry follicles, which dehisce longitudinally to release large, hard and thick-walled seeds which may lie for months or even years before the testa rots sufficiently to allow germination. There is no specialised mode of transport of the seed or fruit, although they may be eaten by birds and monkeys.

Observation of the flowers of modern Magnolias shows that a wide variety of insects, chiefly composed of the insect orders Coleoptera and Diptera, visits the large white bowls. Many are apparently attracted by 'rendezvous' stimuli. That is, the flower is a pleasant place to be for an insect, being sheltered and warmer than the ambient temperature. Most *Magnolia* flowers will probably act as a solar furnace (p. 475). However, many beetles (Coleoptera) and primitive flies with biting mouth parts (Diptera) actively feed on pollen, which is a food source rich in protein and fat, although with little if any 'instant energy' in the form of soluble carbohydrate.

Of the winged insects, the Coleoptera and Diptera were amongst the earliest orders to appear in the fossil record, being conspicuous by the Carboniferous period. At this early stage, long before the origin of the Angiosperms, spore feeding may have evolved in these groups of insects, using the cones of Pteridophytes and Gymnosperms. Other early groups of mobile winged insects, such as the Neuroptera and Odonata, were carnivorous or leaf eating (Orthoptera). Those groups of insects that evolved tubular mouthparts to suck up nectar (Hymenoptera, Lepidoptera and Syrphidae) appear much later in the fossil record, in the mid-Cretaceous, coincidentally with the main diversification of the Angiosperms. It is axiomatic that these groups co-evolved with nectar-producing flowers symbiotically. It is possible that nectar production predated the Angiosperms, and may have occurred for instance in the Pteridosperms or the Gnetopsida (leaf-borne nectaries are found in some present-day ferns). It is striking, however, that plants such as *Magnolia*, which show an apparently primitive state in every floral character, are only visited by the earliest types of insect, for pollen feeding alone. It is probable that pollen feeding predated nectar feeding among the visitors of early Angiosperms.

Modern Magnolias produce a lot of pollen (i.e. have a high pollen : ovule ratio; Ch. 2), and are therefore rather inefficient reproductively. They are homogamous, or weakly protandrous, but set little if any viable seed as isolated individuals in collections. It is almost certain that they are self-incompatible, and reference to other relatively primitive families of plants (Ranunculaceae, Papaveraceae, Theaceae) renders it very likely that they have a multilocus gametophytic incompatibility (Ch. 6). They have no means of vegetative reproduction (Ch. 10) and thus they rely exclusively on pollen-eating insects for reproduction (Fig. 4.2).

**Figure 4.2** A museid fly feeding on pollen in the 'solar furnace' primitive flower of the Mount Cook lily, *Ranunculus lyallii* (×0.6).

It is not surprising that the flowers are so large and attractive, or that such an excess of pollen is produced.

Using *Magnolia* as a baseline, I shall now consider developments that have occurred with respect to diversification of pollination mechanisms, which I have argued (Ch. 3) have been largely instrumental in the success of the Angiosperms.

## DEVELOPMENT OF PERIANTH SHAPE

Three major structural changes occurred in the development of the perianth:

(a) Differentiation of the calyx (sepals, small and green) from the corolla (petals, usually larger and white or coloured). These undertook different functions. The calyx protects the flower bud, and in hypogynous flowers the gynoecium, and in some instances forms nectaries, attractant organs or organs for fruit dispersal; whereas the corolla attracts, receives, shelters and succours the pollinator.

**Figure 4.3** A worker hive bee (*Apis mellifera*) 'robbing' the nectar from the base of the tube of the bluebell (*Hyacinthoides non-scripta*). Photo by M. C. F. Proctor (×1.5).

(b) Fusion of the calyx and corolla to form bells, trumpets, tubes and traps; although the free (unfused) perianths of more primitive plants may form such shapes, they are more liable to robbing of protected food sources (pollen, nectar) by pollinators not able to visit the flower legitimately (Fig. 4.3).

(c) Loss of radial symmetry (actinomorphy), and adoption of bilateral symmetry (zygomorphy) thus forming 'gullet' and 'flag' flowers (see below), usually characterised by a landing platform and dorsal, often 'sprung' stamens and styles; these are typical of 'bee flowers', but may also be found in other syndromes (Figs. 4.4 and 4.5a,b).

### *Zoophily (animal pollination)*

Faegri and van der Pijl (1979) have characterised six (or seven) main perianth shapes, and these are listed here with their main types of flower visitor (Table 4.1). It will be seen that beetles and simple flies are associated with dish- or bowl-shaped flowers, whereas birds, butterflies or moths are associated with trumpet-, brush- or tube-shaped flowers. Bees are the most catholic visitors, being commonly associated with all flower types except narrow tubes, and even in this class, long-tongued bees may often be effective flower visitors. In most ecosystems, bees are the most dominant and most diverse flower visitors. Broadly speaking, however, it is possible to associate various visitor types with various flower shapes, and to suggest that those in each category near the top of the list are the least specialised, and those at the bottom of the list are the most specialised (Kevan 1984).

Nevertheless, there has been a dangerous tendency in the literature for nearly a century to overgeneralise with respect to the syndromes of flowers associated with particular visitors. Certain flowers can only be effectively visited by certain visitors (*Arum*, flies; most papilionid flow-

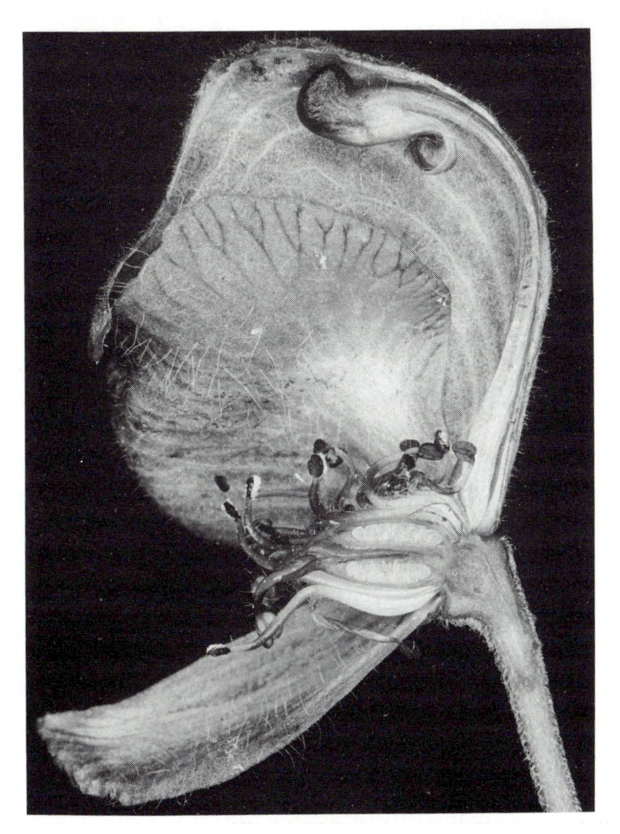

**Figure 4.4** Section of the zygomorphic bee flower of *Aconitum napellus* (Ranunculaceae), showing the large inflated posterior tepal containing two large tubular nectaries in the roof. *Bombus* bees land on the lower tepals, and climb upside down (sternotribically) into the roof, passing the stamens and stigmas as they do so. Photo by M. C. F. Proctor (×1.5).

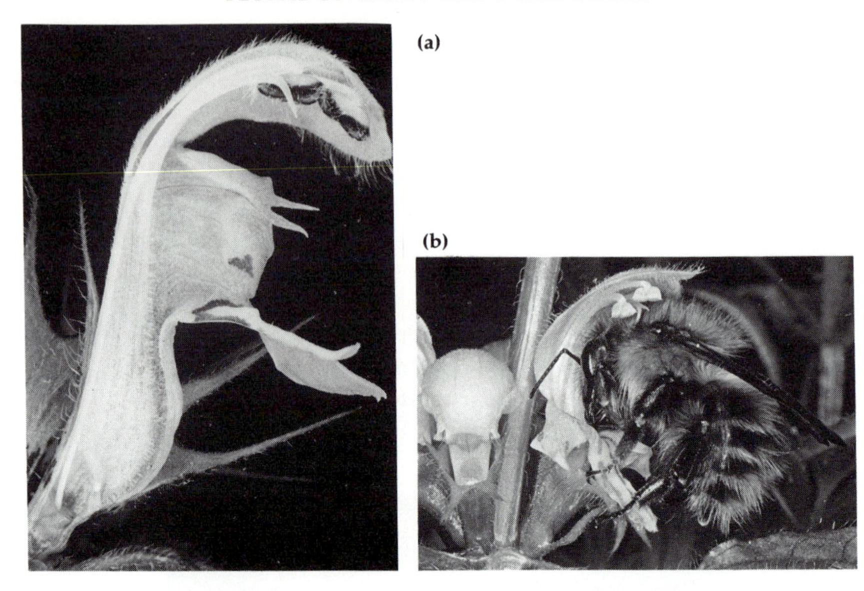


**Figure 4.5** Section of the zygmorphic bee flower of *Lamium album* (Labiatae) (a) which is nototribically visited by *Bombus* (b). The forked stigma, and the four stamens, two of which are shorter than the other two, each bearing divergent locules, are borne under the hood, and pollination takes place via the bee's thorax. Photos by M. C. F. Proctor (a ×3; b ×2).

ers, bees; figs, *Blastophaga*; *Yucca*, *Tegeticula* etc). However, in most cases such specificity that exists does so in the field in a certain season, where competing flowers, and competing pollinators, are diverse. Even where the nature of the reward, or the deceit, may be closely tailored to the morphology or behaviour of a single visitor, accidental pollination by other visitors may occur. Thus in many 'bird flowers', only birds may be able to reach the nectar, yet stamens and styles are far exerted, so that casual attraction of bees, for instance, may effect pollination. Flowers in gardens are frequently pollinated by species (very often the honey-bee) far removed in structure and behaviour from their native pollinator, and the introduction of either pollinators or flowers into foreign areas may break down pollination specificity. Thus many Australian banksias (bird flowers) are efficiently cross-pollinated by introduced honey-bees, and in Australia, African flame-trees (*Erythrina*) are pollinated by native birds such as honey-eaters, wattle-birds and even drongos. At times, any of the flower types listed in Table 4.1 may be visited by any of the pollinator types, and stable relationships may exist between the most unexpected pairs of species, as for instance, a beetle (primitive) and the orchid *Listera ovata* (in a very specialised group). Broad assumptions about coadaptation are thus only valid at a single time and in a single

**Table 4.1** 'Harmonic' relations between pollinators and perianth shape and colour (greatly simplified, adapted from Faegri & van der Pijl 1979).

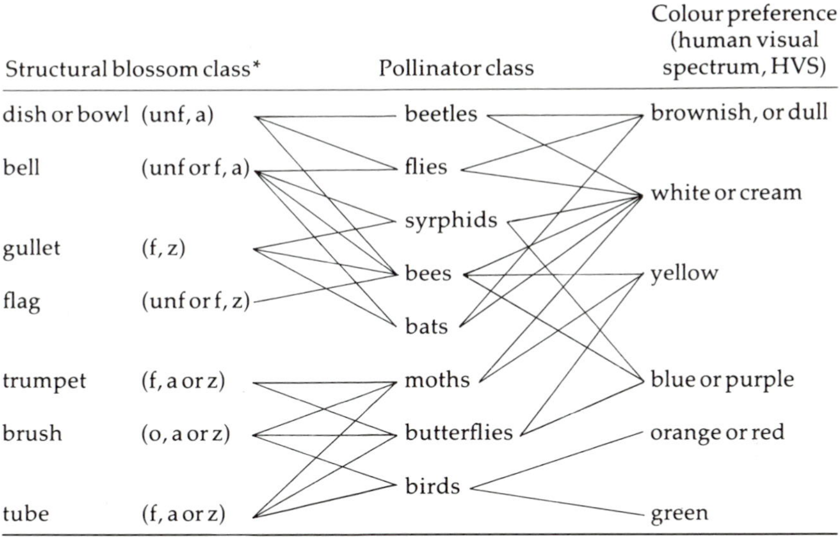

| Structural blossom class* | | Pollinator class | Colour preference (human visual spectrum, HVS) |
|---|---|---|---|
| dish or bowl | (unf, a) | beetles | brownish, or dull |
| bell | (unf or f, a) | flies | |
| | | | white or cream |
| | | syrphids | |
| gullet | (f, z) | | |
| | | bees | yellow |
| flag | (unf or f, z) | | |
| | | bats | |
| trumpet | (f, a or z) | moths | blue or purple |
| brush | (o, a or z) | butterflies | orange or red |
| | | birds | |
| tube | (f, a or z) | | green |

* Abbreviations: unf, corolla unfused; f, corolla fused; a, corolla actinomorphic; z, corolla zygomorphic.

place, and all the so-called 'adaptive syndromes' (Table 4.2) have many exceptions.

Nevertheless, it is also true to say that some flowers (and flower types) tend to be oligophilic, having very few species of visitors (often only one), and this tends to be more true of flowers with relatively specialised shapes (flags, trumpets, tubes) than those with unsophisticated shapes (cup, bell, brush). The latter may receive many different types of visitor

**Table 4.2** The major classes of pollination syndrome.

| | | | |
|---|---|---|---|
| beetles | cantharophily | entomophily | zoophily |
| flies | myophily, sapromyophily, bee-flies, syrphids | entomophily | zoophily |
| bees | melittophily | entomophily | zoophily |
| butterflies | psychophily | entomophily | zoophily |
| moths | phalaenophily | entomophily | zoophily |
| birds | ornithophily | | zoophily |
| bats | chireptenophily | | zoophily |
| also: small mammals, monkeys, slugs, ants, thrips, water skaters, others? | | | zoophily |
| wind | anemophily | | |
| water | hydrophily | | |

(polyphilic). Whether it is more advantageous to be oligophilic or polyphilic will depend on ecological constraints. Visitors to oligophilic flowers are likely to be obligate flower visitors, specialising on one or a few types of flower (oligotropic). Thus, pollen donated from a flower is likely to be received by a flower of the same species. As a result, less energy is required for pollen production per ovule fertilised, and more energy is available for the production of other floral attributes (e.g. more complex shapes, more or larger gynoecia, or more flowers). Oligophily is likely to be advantageous in ecosystems in which pollinators are numerous and diverse, and many species of animal-pollinated plants flower synchronously (Table 4.3).

In contrast, visitors to polyphilic flowers may not be obligate flower visitors, and in any case may visit many different types of non-specialist flower (polytropic; they have been called 'the insect riff-raff' small beetles, flies, thrips, bugs, etc). Thus in polyphilic flowers, pollen donated by a flower is much less likely to be received by a flower of the same species. More energy is therefore required for the production of pollen per ovule fertilised. Many such plants tend to be self-fertile, and most pollination will be selfed, geitonogamous or autogamous (Ch. 9). Flowers will tend to be numerous, often in large inflorescences, with unsophisticated shapes and simple pollen presentation and receipt. Such systems, in which accurate cross pollination is sacrificed for any pollination at all (but with the residual chance of cross pollination) are likely to be favoured in time/space niches in which potential pollinators are scarce and species-poor (i.e. early or late in the season, or in cold locations).

Although most zoophilous flowers can be classified in shape according to the types listed in Table 4.1, many defy such classification. Some flowers of great complexity have extraordinarily complex methods of pollination, and these lend much of the glamour to the 'story-telling' aspect of so-called 'pollination ecology', which has, perhaps rather unfortunately, tended to dominate texts on pollination biology, and latterly, television programmes. Some other flower shapes will be catalogued briefly here.

**Spurs** In a variety of families from the 'primitive' Ranunculaceae (*Aquilegia* (Fig. 4.6), *Delphinium)* to the Fumariaceae (*Corydalis*), Balsaminaceae (*Impatiens*), Scrophulariaceae (*Diascia, Linaria*), Lentiburiaceae (*Pinguicula, Utricularia*), Valerianaceae (*Kentranthus*), Violaceae (*Viola*), Tropaeolaceae (*Tropaeolus*), and the very specialised monocotyledonous Orchidaceae (*Platanthera, Gymnadenia*, and many others), corollas (and/or more rarely calyces) form spurs. These are broad and rounded to long and very narrow outgrowths near the base of the corolla, which usually contain nectar. Their function is twofold: to render

**Table 4.3** Patterns of flower visiting, and ecological constraints, on simple and complex flowers.

| Flower type | Flower visiting | Insect visiting | Diversity of potential visitors | Number of other synchronous species of flower | Pollination efficiency (reciprocal of pollen : ovule ratio) | Breeding system of flower |
|---|---|---|---|---|---|---|
| simple (bowl, bell, brush) | polyphilic | polytropic | low | small | low | mostly selfed |
| complex (gullet, flag, trumpet, tube, trap) | oligophilic | oligotropic | high | large | high | mostly crossed |

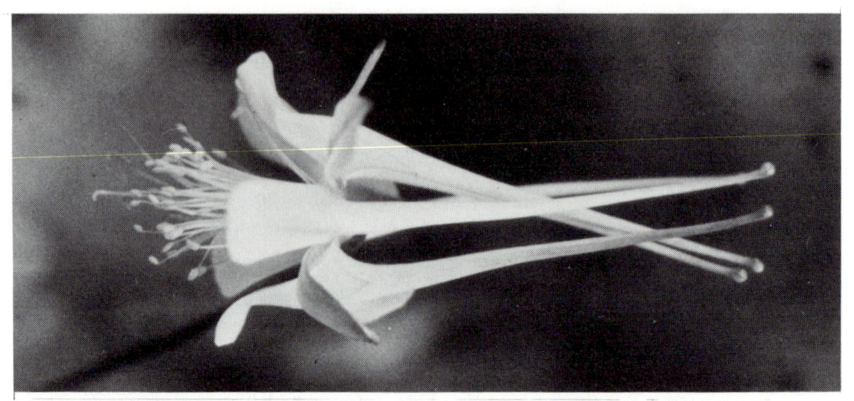

**Figure 4.6** Nectar-containing petal spurs in *Aquilegia longissima* (Ranunculaceae), a north American columbine visited by Lepidoptera which reach the nectar by means of a long proboscis (×1.5).

the nectar reward unavailable to all except a few structurally suited flower visitors; and to increase the distance from the reward to the anthers and stigma (pollen presentation and receipt) and/or the landing platform or hovering position. Thus, visitors that can 'work' spurred flowers will have long, narrow mouthparts, and include Lepidoptera (Fig. 4.7), long-tongued bees, and some birds. Spurs are very liable to illegitimate visiting, by 'nectar-robbing'.

**Constricted flowers** Some flowers are so constructed that no visitor can work the flower without the exertion of considerable force, or even without damaging the flower. Thus, once again, the visitors are limited to those with accurate structural specifications. Adey (1982) has recorded the ability of 45 species of bee to 'trip' 27 species of the tribe Genistinae, relatives of brooms and gorses. ['Tripping' is the depression of the keel and the forcing open of the wings of a papilionoid flag flower, thus releasing the spring-loaded stamens and style from a ventral position in the keel onto the pollinator (Fig. 4.8).] Using numerical methods she has grouped the bees into five classes based on body weight, and the flowers into seven classes based on structural and size characters. There is a strong relationship between bee types and flower types with respect to tripping ability. The papilionoid ('pea') type flower is a familiar example of constricted entry in a flag flower, but many mechanisms that constrict entry have been adopted. The corolla may be constricted at the apex (*Campanula zoysii*); the apices of the corolla lobes may even be fused, allowing only forced lateral entry (Fig. 4.9); the corolla tube may be filled with a plug of hairs allowing entry only to the hard, narrow mouthparts of some syrphids (*Primula primulina*); and there are many other examples.

**Figure 4.7** Lulworth Skipper butterfly (*Thymelicus acteon*) visiting the tubed and short-spurred flowers of red valerian, *Kentranthus ruber* which it is probing for nectar. Photo by M. C. F. Proctor (×2).

**Figure 4.8** Tripped flowers of the broom, *Cytisus scoparius*. The flowers have been visited by bees which have depressed the boat-shaped keel, so releasing the single style and the stamens, which hit the bee and then recoil, having been held under tension (×0.5).

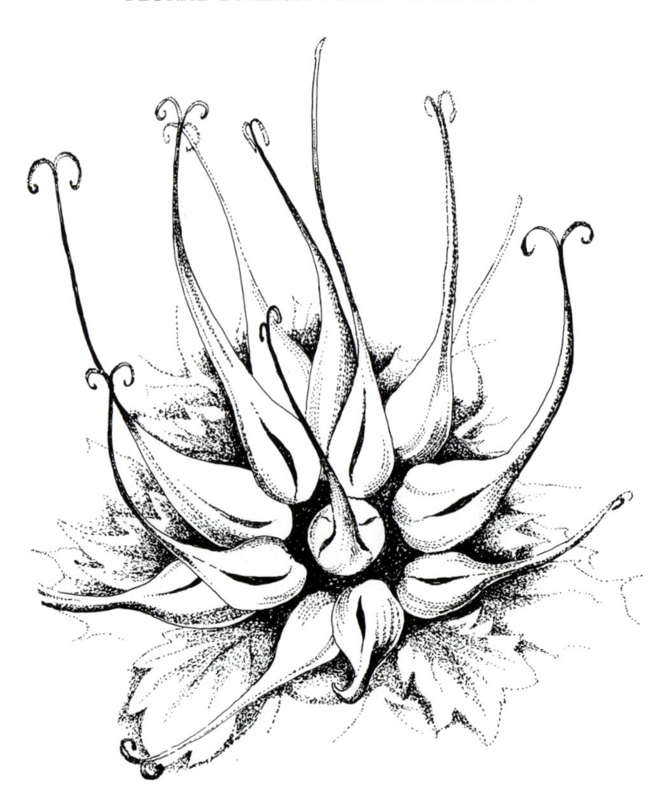

**Figure 4.9** The petal lobes in *Physoplexis comosa* (Campanulaceae), the devil's claw of the European Alps, are fused apically, allowing forced lateral entry by bees and syrphid flies which achieve pollination in the struggle to reach basally presented nectar (×2).

**Spring-release flowers** Often, although not invariably, associated with constrictions are spring-loaded mechanisms whereby the stamens and style(s) are forcibly released from a hidden location onto the pollinator as the flower is entered, often showering the pollinator with pollen. This has already been briefly described for papilionid flowers in the preceding section. In these cases, pollination is sternotribic, that is the underside of the visitor is dusted with pollen. In many other constricted flowers, for instance many Lamiaceae and Scrophulariaceae (the white dead-nettle, *Lamium album* is a familiar example; Fig. 4.5b) pollination is onto the back of the pollinator (nototribic). In a beautiful series of comparative observations made over many years on the many North American species of *Pedicularis* (Lousewort) Macior (1982) shows adaptation to both forms of pollen presentation in the same genus (Fig. 4.10). Pollen is often released explosively (as in the broom, *Cytisus scoparius*; see Fig. 4.8), even to the

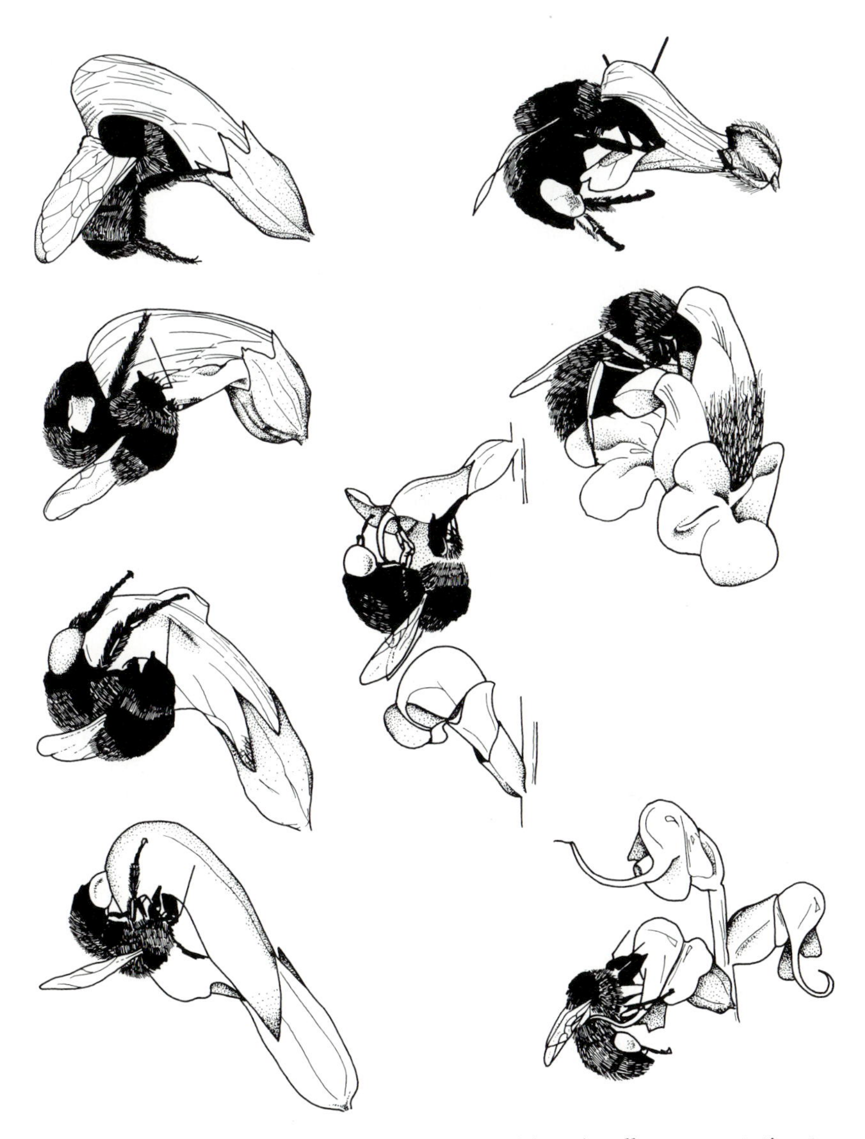

**Figure 4.10** Sternotribic (below) and nototribic (above) pollen presentation to *Bombus* species by the flowers of eight species of North American *Pedicularis* (Scrophulariaceae). Drawing reproduced by kind permission of L. W. Macior (all ×4).

**Figure 4.11** *Salvia glutinosa* being visited by *Bombus agrorum*. In the sages, each of the two stamens has an elongated connective which is pressed by the visiting bee, rocking the fertile anther lobe on to its back. Photo by M. C. F. Proctor (×3).

extent of upsetting and discouraging the flower visitor, but other forms of pollen release or presentation, such as through vibration, or a 'rocker' movement of the stamen (*Salvia glutinosa* ; Fig. 4.11) are also well known.

**One-way system trap flowers** Many kinds of trap flowers are recorded. Some, as in the Araceae are modifications of whole inflorescences, and are also dealt with later (p. 103). The incentive for visitors to enter trap flowers is very frequently sapromyophilous, the flower producing a decaying smell, and being coloured reddish, brownish or purplish in a manner reminiscent of dead flesh. This is true of Aristolochias, and some orchids, for instance Australian birdwing orchids (*Pterostylis*) and Paphilopediums. In other cases, visitors may be attracted by sweet scent and, to the human eye, bizarre shapes and colours. In the orchid *Coryanthes*, *Eulaema* bees attempt to collect scent droplets from the rim of the trap, but are intoxicated and fall in, only to escape eventually through a narrow exit passage, effecting pollination as they do so. In the Lady's slipper orchids (*Cypripedium*), so rare in Europe now but still very common in America, attraction into the trap is by nectar (Fig. 4.12). This is a rare exception to the general rule that trap flowers operate by deceit, not offering a reward. This deceit is not surprising, for an insect is unlikely to 'major' on a trap flower, in which it may be forced to spend hours at a time, as a food source. However, deceit renders the mechanism vulnerable to frequency-dependent selection. If deceit flowers are common in the environment, more so than their model, their strategy is likely to be less successful. In practice, they do tend to be uncommon.

Trap flowers usually form a pouch with a restricted, slippery entrance, often reinforced by unidirectional hairs. Exits may be separate, as in trap orchids, in which the visitor is forced to squeeze past the stigma and the pollinia (in that order). In *Aristolochia* and *Araceae* (see Fig. 4.30) the entrance and the exit are the same, so a female phase (pollen receipt) always precedes a male phase (pollen issue). The relaxation of stiff, unidirectional hairs mediates the operation of such a 'time-trap'. Modifications of the corolla wall into greasy, wax-secreting areas and into hairs enables traps to work. However, in their simplest forms, trap flowers may be merely formed by imbricated whorls of many stiff, inward-bending

**Figure 4.12** A slipper orchid *Paphilopedium* sp. The labellum forms a pouch containing nectar. Small bees enter the pouch distally, but can only exit proximally, having forced their way past the column and anthers (×1.5).

waxy tepals, which the insect can enter easily, but from which it cannot escape until the petals reflex during the later male phase (*Calycanthus*) (for obvious reasons one-entrance trap flowers are always protogynous). In extreme cases, early visitors during the female phase may be killed, either by drowning or by a toxin secreted into the stigmatic liquid (*Nymphaea*). As the male phase progresses, the stamens cover the toxic stigma, and visitors are able to escape, carrying pollen. In this primitive flower, pollen may adhere better to the stigma if the pollinator carrying it dies on the stigma, and thus pollen germination is enhanced.

**Flowers with food bodies** In most flowers, reward is provided solely by pollen and/or nectar. However, in some, nutritious food bodies may be produced on the perianth, stamens or staminodes. Such white, granular food bodies are the main reward for the visiting beetle *Colopterus* in *Calycanthus* (see above), and may also be provided by the labellum of certain orchids (*Maxillaria* and *Eria*). In other cases, oil bodies may be produced by special organs (elaiphores), which are collected for their broods by solitary anthophorid bees. This phenomenon seems largely restricted to South America, but is recorded for several different families of plants. It is well known in many species of *Calceolaria* (Scrophulariaceae).

Quite apart from true nectaries, which are located within the flower, usually on the receptacle, ovary or petal base, many plants also produce extra-floral nectaries, usually as stalked glands which secrete a sugar solution. These may be produced on many organs, and may have other physiological or protective functions unconnected with floral biology. But often they are located on the outside of the perianth or calyx (Fig. 4.13). They seem to have one or two functions; either distracting very small insects with poor capabilities of travel and pollen carry (thrips, small diptera) from taking true nectar from unprotected positions in bowl flowers; or encouraging ants to take up positions on the outside of the flower, thus discouraging flower-robbing by the pollinator at the base of the perianth (in constricted flowers such as *Thunbergia grandiflora*, visited by *Xylocopa* bees).

**Sexual mimic flowers** In these, the flower has come to resemble an individual of the pollinator species, and at least in the European 'insect orchids', *Ophrys* (Fig. 4.14) produces pheromone scents and tactile responses that also mimic the female insect. In the 20 or so species of *Ophrys*, it has been established that most are visited by the males of different species of solitary bee (in the genera *Andrena*, *Gorytes*, *Eucera*) and the wasp *Campsoscolia*. Usually, only one species of bee or wasp visits one species of *Ophrys*, although four or five species of *Ophrys* commonly coexist. In many cases, male bees are seen to attempt to

(a)

(b)

**Plate 4.1** Parallel evolution of flowers adapted to bird-visiting belonging to four unrelated plant families, native to four continents. All have tough red tube flowers and copious nectar. (a) *Columnea* sp. from the Caribbean (Gesneriaceae) – hummingbird flower. (b) *Aloe cameroni* from central Africa (Liliaceae) – sunbird flower. (c) *Rhododendron burtii* from south-east Asia (Ericaceae) – compare the bee-visited *Rh. hongkongense* (Fig. 4.18) and the moth-visited *Rh. hertzogii* (Fig. 4.21) in the same genus. (d) *Anigozanthus rubescens* (Haemodoraceae) from western Australia (honey-eater flower).

(c)

(d)

**Figure 4.13** Extrafloral nectaries on the calyx and pedicels of *Arbutus canariensis* (×1).

copulate with the flowers (pseudocopulation), and even ejaculate. Such visits are often brief, but I have observed a species of *Eucera* stay on a single flower of *Ophrys sphegodes* subsp. *helenae* for at least three hours, making copulatory movements throughout. It had pollinia on its head.

Sexual attraction is recorded for several other orchid genera, notably in Australia (*Caladenia*, *Drakaea*), and intermediate conditions between conventional attraction and sexual attraction are well demonstrated (Stoutamire 1983), illustrating that such bizarre syndromes can arise through Darwinian selection. In extreme cases, thynnid wasps, which abduct females, attempt to carry off mimic labella which are suspended on a long balanced lever. After depressing the labellum, it swings back upwards, throwing the wasp against the column and achieving pollination with remarkable accuracy. It is not correct to consider this as a form of pseudocopulation, however, as copulation seems never to be attempted.

Insect mimicry can also appeal to territorial instincts. The delicate brown and yellow striped hanging flowers of neotropical *Oncidiums* (Orchidaceae) give rise to aggressive behaviour in the territorial males of *Centris* bees, which try to fight the flowers, thus pollinating them.

**Figure 4.14** The 'pseudocopulation' flower of the insect-mimic fly orchid *Ophrys insectifera* of Europe. Males of the solitary bees *Gorytes campestris* and *G. mystaceus* mistake the flower for a female wasp, and attempt to mate with it, thus promoting cross pollination. The flowers produce a pheromone-like scent. Photo by D. Mitchell (×2).

### *Anemophily (wind pollination)*

Not all extreme floral adaptation is directed towards animal visitors. The wind is extremely important as an agent for pollen travel (anemophily), and many large families of plants, many of which habitually dominate communities, are wind pollinated. The disadvantage of wind as a pollen dispersant is its randomness. Whereas an animal may accurately transport a high proportion of the very little pollen produced large distances to a tiny stigmatic target, wind-transported pollen has no accuracy whatsoever. It is therefore not surprising to find this mechanism prevailing amongst grasses, sedges, rushes and trees in species-poor (temperate) communities, in which legitimate mating partners are readily reached. In species-rich communities, with low levels of individual ecological dominance, biotic pollen dispersal predominates (thus alpine grassland, Mediterranean maquis and tropical forest are full of pretty flowers).

The corresponding advantages of wind pollination are firstly, it is independent of weather, relatively speaking (and hence some wind-pollinated trees bloom very early in the year thus avoiding overmuch stigmatic contamination with foreign pollen) and secondly, that despite vast expenditures of energy on massive quantities of pollen, and to a lesser extent on large feathery styles, other floral parts can be reduced to a minimum. Thus, while rudimentary perianths persist in the Juncaceae (rushes), they have totally disappeared in the Poaceae (grasses), where they are replaced by derivatives of the bracts and bracteoles, and in most wind-pollinated trees.

It is a mistake to consider anemophily and zoophily as being non-overlapping categories. Work by Stelleman (e.g. 1979) and A. D. J. Meeuse (e.g. 1978) has clearly shown that in *Plantago* and *Salix* respectively, apparently wind-pollinated species may nevertheless receive appreciable quantities of animal-dispersed pollen, carried by pollinators as sophisticated as syrphid flies and bees respectively (Fig. 4.15).

No doubt many other flower types (for instance *Chenopodium* or *Urtica*)

**Figure 4.15** Hive bee, *Apis mellifera* visiting a female inflorescence of the sallow, *Salix cinerea*. Although willow flowers lack petals, and appear at first sight to be wind pollinated, they have nectaries and are visited by bees. Photo by M. C. F. Proctor (×2).

**Figure 4.16** The mostly anemophilous flower of the alpine meadow rue *Thalictrum alpinum*. Perianth segments are small and brownish, and the anthers relatively large and pendulous. Other species in the genus are entomophilous (×2).

undergo both wind and animal pollination, and in *Carex* there has been a recent account of syrphid visitation to at least the male spikes of *C. binervis* by a syrphid (Proctor 1978). In *Thalictrum*, a remarkable diversity of mechanisms are found, and whereas large-petalled, nectar-bearing species such as *T. tuberosum* are typical insect-visited bowl flowers, others such as *T. minus* and *T. alpinum* (Fig. 4.16) almost lack perianths having large hanging stamens, and are presumably mostly pollinated by wind. Here is a clear case of the evolution of anemophily within a genus.

### *Hydrophily (water pollination)*

Most water plants produce aerial inflorescences, and the flowers may be sophisticated, involving traps (*Nymphaea*), heteromorphy (*Hottonia*), or spurs (*Utricularia*). Occasionally, inflorescences are produced floating on the meniscus of the water surface, as in the South African water hyacinth (*Aponogeton*), which may be pollinated by water-skaters. However, in a few cases, flowers function under water. These may be entirely cleistogamous when submerged (*Subularia*, Ch. 9). Alternatively, pollen transport by water may occur (hydrophily). In epihydrophily, pollen grains, or in the famous case of *Vallisneria*, whole male flowers, are released to float to the water surface, where they encounter female flowers with stigmas at the meniscus. As has been pointed out by Faegri and van der Pijl (1979), these cases are unique in that pollination takes place in two dimensions only, which leads to a greater efficiency in pollen use. Other examples are *Ruppia, Callitriche, Hydrilla, Elodea* and the sea-grasses *Neptunia* and *Aeschynomene*.

In hyphydrophily, pollination takes place under water, in three dimensions (e.g. *Najas, Halophila, Ceratophyllum* and *Zostera*). The case of *Zostera* is especially interesting in that the tendency to a loss of exine exhibited in most cases of hydrophily reaches an extreme, and a pollen tube-like male gametophyte is dispersed through the sea, wrapping itself around any likely object. In almost all cases of hydrophily, perianths are extremely reduced, being small and green or absent. In *Vallisneria*, however, considerable modifications have occurred to the tepals of the male flowers, which rise to the surface by virtue of aerenchyma in the tepals, which are at first tightly closed but immediately open on the surface of the water to form a floating platform for the tiny male flower.

## THE PERIANTH AS AN ATTRACTANT

Up to now we have been considering the development and function of perianth shape and structure with respect to the control of pollination. However, there are many other features of the perianth which have

**Figure 4.17** Brownish spots in the bearded throat of the gullet flower of the monkey flower *Mimulus guttatus* act as nectar guides to visiting bees (×2).

shown remarkable amounts of diversification from a Magnoliid baseline, and these are concerned with attraction. They can be divided into colour attractants and scent (odour) attractants; attraction chiefly mediated by perianth shape (sapromyophilous or sexual deceit and the production of food bodies) have already been discussed.

The behaviour of a sophisticated flower visitor such as a bee, butterfly or bird with respect to an inflorescence is complex. The behaviour of bees, and especially honey-bees (*Apis*), is best known (Kevan & Baker 1983). Long-range (primary) attraction is generally by colour; bee vision is highly responsive to colour, but resolves poorly. Thus a suitably coloured object such as an article of clothing, or a motor car will frequently attract bees which, however, fly off at a distance of a metre or so. Occasionally, a strongly scented plant can also act as a long-range attractant to bees and they will react in experiments to hidden, scented flowers. However, most strongly scented flowers primarily attract other pollinators, especially moths (pale, night-scented flowers often with long tubes), or flies (brown or purple disgustingly scented flowers). Bats, which in common with moths fly principally at night, also usually visit scented flowers.

Once within a metre of a flower, a bee will respond to flower shape and pattern, which it is now able to resolve. Identification of the shape of an individual pollination unit (e.g. a flower) will allow the bee to orientate correctly, and colour guides on the flower will reinforce this orientation. A high proportion of bee flowers have differential markings (sometimes invisible to the human eye) which characteristically lead from the landing site to the reward (Figs. 4.17 and 4.18). Such 'nectar-guides' are unusual in flowers primarily visited by other classes of pollinators.

On alighting on the flower, the bee will disregard visual and olfactory cues, and its behaviour at the flower is determined principally by touch. Thus, hairs, rugosities, callouses, etc., which aid adhesion of the bee to the flower, may also act as tactile guides. In a flower with a 'bearded' lip (Fig. 4.17) (as in many Labiatae) the bee will walk up the beard into the flower towards the reward. Such features may also act as constrictants (see p. 76) and aid in pollen issue and receipt. The stamens may also be bearded at the base (Fig. 4.18).

Complex behavioural responses such as these in bees render it easy to oversimplify the rôle played by any one factor in attraction.

**Figure 4.18** Brownish-purple spots on the cream flower of *Rhododendron hongkongense* act as nectar guides to visiting bees. Stamen filaments are hairy, perhaps to help the bee to obtain purchase on landing (×1.5).

**Figure 4.19** The large terminal male flower of the bananas and plantains (*Musa*) is coloured an inconspicuous brownish-purple, and is visited by bats (×0.5).

## *Colour attraction*

Relatively little is known about the receptivity of many types of pollinator to flower colour. The broad correspondence of certain flower colours to certain classes of pollinator is well known (for instance Table 4.1), but is little understood. Most flower-visiting bats are nocturnal and probably have no visual clues, thus the sombre colours of most bat flowers may help to prevent predation of these flowers by other animals (Fig. 4.19). This may also be true of flowers visited by other pollinators which work principally by scent (e.g slugs, ants and flies). In contrast, many moths visit flowers at night by hovering and, in the absence of close-range tactile clues, they need to employ acute night vision for an accurate approach (Fig. 4.20). Thus, most moth flowers are pale or white, reflecting the small amount of available light (Fig. 4.21).

Perhaps the most widely remarked association between flower colour and pollinator class is the affinity of birds for orange or red flowers. It is most striking that this association has clearly originated on several

**Figure 4.20** The silver-Y moth, *Autographa gamma*, visiting the pale coloured, night-scented flowers of the honeysuckle, *Lonicera periclymenum*. Note extra-floral nectaries on the perianth tube. Photo by M. C. F. Proctor (×1).

**Figure 4.21** The white, night-scented tube flower of the moth-visited *Rhododendron hertzogii* from south-east Asia. The tubes are up to 6 cm in length, and the basal nectar can only be reached by moths with long probosces. Compare the bee-pollinated *Rh. hongkongense* (Fig. 4.18) (×0.6).

different occasions. Thus, the main avian flower visitors in the Americas are hummingbirds (Trochilidae), and they do not occur outside that vast continent. Many important families of plants (for instance the Cactaceae) are largely or entirely American, and have many bird-pollinated species, with scarlet flowers. Yet, in South Africa the entirely Old World sunbirds (Nectarinidae) similarly visit scarlet flowers of endemic genera such as *Kniphofia* and *Hacmanthus*. Similar associations are found in Australia (honey-eaters, Meliphagidae and lorikeets, Trichoglossidae with many endemic scarlet-flowered genera of the Myrtaceae, Proteaceae, Ericaceae, Goodeniaceae, Rutaceae, Loranthaceae, even Lentiburiaceae and many more); New Zealand (the endemic bellbird associates with the endemic *Metasideros*); and Hawaii (the endemic honey-creepers, Drepanididae, so many of which are now sadly extinct, with endemic scarlet lobelias etc.) (Plate 4.1).

The reason for this colour association is not hard to find. Birds have poor colour vision, and although they receive a spectral range similar to that of humans, they do not distinguish well between many yellows, blues and purples. Only at the long-wave limit of their vision do they distinguish colours well, and thus they are very responsive to red. As it happens, most insects receive a range of wavelengths of light shorter than that of vertebrates, and do not see red well; scarlet flowers in a garden bed are not visited by bees as was observed by Darwin (1876) on *Lobelia fulgens*. Thus an ecological niche with respect to flower colour is left vacant by insects, and is readily filled by birds. The ready adaptation to this response is observed even in birds which would have had little chance of encountering bird flowers in their breeding areas. Thus, I have observed Old World warblers such as the willow warbler (*Phylloscopus trochilus*) and the lesser whitethroat (*Sylvia curraca*) repeatedly feed on the scarlet bird flowers of *Kniphofia* (native to South Africa) while on migration on the north-east coast of England. Whether they were feeding on nectar or small insects is not clear, but they hovered for long periods while repeatedly probing the flowers, and their heads could be seen to be liberally dusted with yellow pollen.

Of course, birds do not only see red flowers, and not all bird flowers are red. Thus, many of the Australian bird flowers are yellow or white, as is the case for most *Eucalyptus*, and many flowers habitually visited by birds in Hawaii, South Africa and other areas are not red. Yellow forms of *Crocus* are savagely attacked in British gardens by sparrows (*Passer domesticus*), although purple crocuses are generally left unharmed. Although the cause of this aggression seems to be mysterious, I have observed that yellow-flowered plants may set better seed after such an attack. Q. O. N. Kay (pers. comm.) notes that the catkins of willows (*Salix* species) are visited in the UK by Blue Tits (*Parus caeruleus*) which can achieve cross-pollination. It appears that the birds are drinking nectar from the yellow catkins.

In contrast to birds, bees generally visit flowers with colours in the middle of the human visual spectrum (HVS), that is to say yellow to blue, or with mixed and wide spectral reflectances including some of these wavelengths (purple, pink, white). It has been long known that bees, at any rate, do not receive the same spectral range as vertebrates, but this has given rise to some confused thinking on the subject. There is, however, a splendid account of the relationship between the HVS and the 'insect' visual spectrum (IVS) in Kevan (1978), although in fact most experimental work has only been done with honey-bees, *Apis*.

Human colour vision extends from violet (380 nm) to deep red (780 nm), and bee vision from 300 nm to 700 nm. Thus, as already discussed, most HVS red tones are invisible to a bee, which ignores red flowers, but not to birds, whereas wavelengths of reflected light in the ultraviolet (300–380 nm) range which are visible to bees, cannot be seen by humans. Equally importantly, the peak sensitivities to colour vision also vary between vertebrates (humans) and insects (bees and butterflies at least; Table 4.4). Thus HVS sensitivity reaches peaks at 436 nm (blue), 546 nm (green) and 700 nm (red), and those of insects are at 360 nm, 440 nm and 588 nm. Kevan's arguments rest on two important premises.

First, that all perceived colours are best interpreted as mixtures in varying energies of the three sensitivity peaks (blue, green and red in HVS); they are thus best interpreted on a trichromatic colour scale (Figs. 4.22 and 4.23) with these three pure colours each represented at an apex of a triangle; equal emissions of each sensitivity peak wavelength are perceived as white (and this is in fact Kevan's definition of white), which is thus represented centrally on the diagram.

**Table 4.4** Some equivalences between trichromatic colours in the human visual spectrum (HVS) and the insect visual spectrum (IVS) (after Kevan 1978). Peak sensitivities are given in parentheses.

| HVS (380–780 nm) | IVS (300–700 nm) |
|---|---|
| red (700 nm) | |
| orange | |
| yellow | purple to red (588 nm) |
| green (546 nm) | mauve to red |
| white | yellow to white |
| blue (436 nm) | green (440 nm) |
| violet–purple | green |
| | blue (360 nm) |
| | violet–purple |

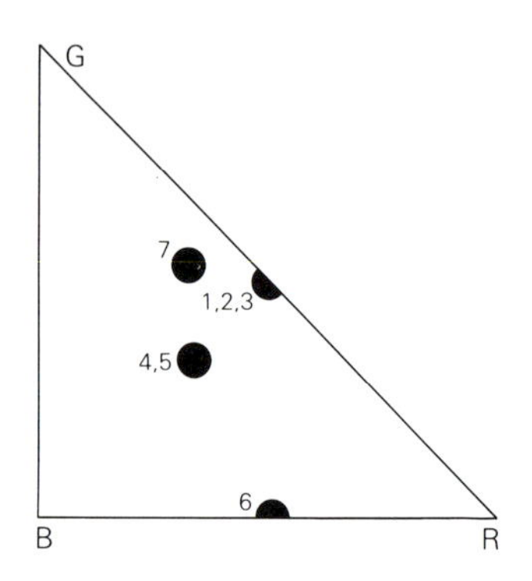


**Figure 4.22** Trichromatic plots for the HVS colour triangle. G, B and R are monospectral green, blue and red respectively. Points 1 to 7 correspond to those in Figure 4.23; 1, 2 and 3 are yellow to HVS, 4 and 5 are white, 6 is purple and 7 is greenish yellow (after Kevan 1978).

Second, that it is reasonable to equate the three sensitivity peaks in IVS with those in HVS in terms of perception (although this is not by its nature susceptible of proof), as follows:

| | HVS | IVS |
|---|---|---|
| red | 700 nm | 588 nm |
| green | 546 nm | 440 nm |
| blue | 436 nm | 360 nm |

As a result, to take a case of almost exact correspondence between HVS and IVS sensitivity peaks, human blue (436 nm) equals insect green (440 nm). But most colours perceived from reflecting surfaces are complex mixtures of the three peak sensitivities, and must be interpreted from trichromatic diagrams. Because peak sensitivities differ so much between human and insect perception, a single emitted source will take a very different position on HVS and IVS trichromatic diagrams (Figs. 4.22 and 4.23). In the seven colours represented here, only one (labelled 5) is by chance perceived the same by humans and insects (white). Thus, assuming the same colour names for their respective sensitivity peaks, some kind of 'translation' from human colours into insect colours can be achieved (Table 4.4).

From this argument stem three conclusions:

(a) There will be colour contrasts within the flower, and between the flower and its background, which will be visible to insects, but not to humans. 'Invisible' nectar guides which reflect in the ultraviolet region, and are thus visible to insects, but invisible to humans (except by the use of ultraviolet-sensitive camera film) are a case in

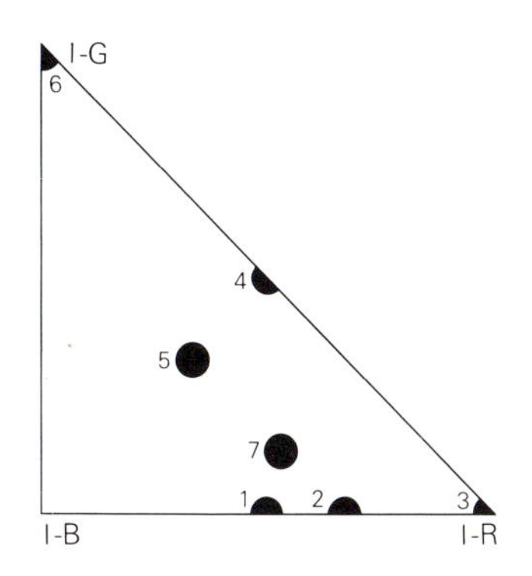


**Figure 4.23** Trichromatic plots in the IVS. I–G, I–B and I–R refer to the monospectral peaks of maximum sensitivity to insects (Table 4.4). Points 1 to 7 correspond to those in Figure 4.22. 1 is insect purple, 2 is insect red–purple, 3 is insect red (all yellow in HVS), 4 is insect yellow and 5 is insect white (both white in HVS), 6 is insect green (HVS purple), and 7 is insect mauve (greenish-yellow in HVS) (after Kevan 1978).

point. However, it is also worth reflecting that yellow flowers on a green background of foliage could be much less visible to bees than to humans. Galen and Kevan (1980) suggest that flower colour differences between shaded (pale blue) and alpine (purple) populations of *Polemonium viscosum* may relate to the visibility of flowers to insects against different backgrounds.

(b) That some flowers, and other floral organs, as well as guidemarks, which are at best dully coloured to the human eye, might be more attractive to an insect (in the ultraviolet region).

(c) That differences in flower colour within a flower or between flowers which cannot be perceived, or are poorly received by the human eye, may be much more vivid to insects. A good case is described by Kay (1982) for the yellow/white polymorphism of *Raphanus raphanistrum*, in which the yellow flowers visited by long-tongued bees and butterflies are insect purple, whereas the white flowers visited by *Bombus pascuorum* are insect cream.

It is important to emphasise the rôle played by spectral sensitivity. Thus bees do in fact perceive some red light, right up to 700 nm in wavelength, but their sensitivity at this wavelength is low, and they are not attracted by it.

The range of flower colours perceived by the human eye is large, and very nearly encompasses all the colours we can enjoy. When massed in a herbaceous border, or alpine meadow, such a range of colours can indeed make a fine spectacle. Yet Kevan has pointed out that the range of colours available to insects within a habitat is much greater than it is to the

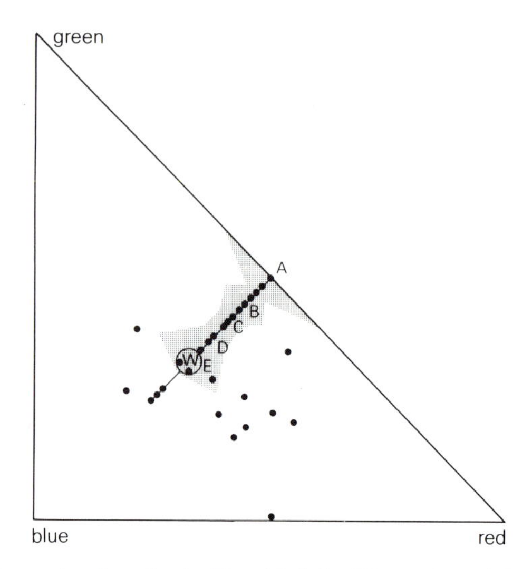


**Figure 4.24** Trichromatic plots (as in Fig. 4.22) for 53 species of Canadian weed in the HVS: most species appear white or yellow to the human eye (after Kevan 1978). W is the equal energy white point; A, B, C, D and E are points with multiple observations.

human eye. After all, the plants in the wild have adapted to pollinator perception, not to human perception, and in many areas pollinators are solely insects. It is no accident that the most dazzling flower colours to humans often come from vertebrate-flowers from tropical or subtropical areas, nor that the flamboyant colours of many of the birds of these areas (e.g. hummingbirds, sunbirds, lorikeets and parrots) are in fact cryptic. The birds are less visible while feeding and so less vulnerable if they are coloured like the flowers they visit.

Thus, Kevan has plotted the trichromatic receptivities in the HVS and the IVS of the colour reflectances of 53 species of Canadian weed (Figs. 4.24 and 4.25). It is clear that insects perceive a much wider range of colours amongst these flowers than we do. Adey (1982) has devised a technique by which insect colour vision can be conceptualised in human terms, using Kevan's theories. Monochrome photographs were taken of flowers using separately the following filters:

| Kodak filter no. | Transmission range (*nm*) | Human colour | Insect colour |
|---|---|---|---|
| 18A | 300–400 | ultraviolet | blue |
| 47 | 400–500 | blue | green |
| 61 | 500–600 | green | red |
| 29 | 600–750 | red | — |

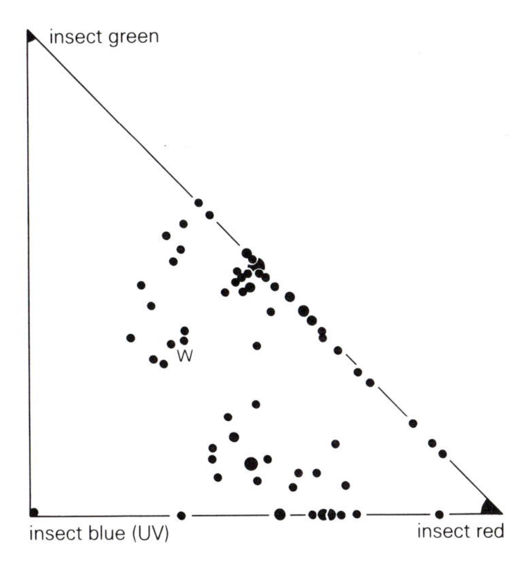


**Figure 4.25** Trichromatic plots (as in Fig. 4.23) for 53 species of Canadian weed in the IVS. There is a much greater variety of colour to insect vision than to human vision (Fig. 4.24). W is the equal energy white point for insects (after Kevan 1978).

These filters were used in conjunction with neutral density filters, and photographs taken against a uniformly reflecting 'grey scale' so that adjustments in density and contrast can be made (for the ultraviolet transmitting filter, the flash-gun output was adjusted). The three negatives obtained could then be sequentially printed on to colour-sensitive paper through colour filters appropriate to the IVS:

| Transmission range of negative filter (*nm*) | Insect colour | Printing filter |
|---|---|---|
| 300–400 | blue | 47 |
| 400–500 | green | 61 |
| 500–600 | red | 29 |

The resulting prints then contained colour mixes representative of the IVS conceptualised in HVS terms.

Petal colours depend on the wavelengths of light that are reflected from the petal surface, and these are influenced by two main factors:

(a) The pigments carried within the subepidermal cells of the petal, either in solution in the vacuole (e.g. anthocyanins, betacyanins, xantheines, flavonoids and phenolics) or on plastids (e.g. chlorophyll, caroteines and xanthophylls). The mixture of pigments

**Figure 4.26** The annulus or 'eye' of the flower of *Androsace villosa* (Primulaceae) changes colour from yellow to red with age, probably making this guide less attractive to insects (×2).

in a petal determines the wavelengths of incident light that are absorbed and thus those that are reflected.

(b) The structure of the epidermal and subepidermal layers of the petal. The epidermis itself has no pigments, and the structure of the epidermal cells in relation to the subepidermis strongly influences the nature of the reflected light. Thus, much light may be reflected at the epidermal surface (matt white), or reflected through lens-shaped epidermal cells from a reflecting subepidermis (brilliant white); clearly the shape of epidermal cells in particular can influence petal colour, by causing refraction effects, or interference effects.

Petal colours are not necessarily constant, and changes in petal colour are often adaptive, or apparently so. Thus in the bee-visited flowers of the Boraginaceae, there is frequently a marked change between the colour of the buds and recently opened flowers, and those that are donating or are receptive to pollen. This change is often from long wavelength reddish colours, not detected by bees, to short wavelength bluish colours to which bees respond (Müller 1883). It is a feature of many anthocyanins that they can change the wavelength of their reflected light markedly in response to minor physiological changes, for instance in pH value.

Floral guidemarks may often change colour as well, frequently as the flower ages, and the stigma cease to be receptive. Thus in *Androsace villosa* (Fig. 4.26), fresh flowers have a yellow annulus at the mouth of the tube, which turns red on ageing. This apparently occurs through an increase in, or change in the nature of, anthocyanins in this tissue, thereby masking the caroteine pigment, and rendering the guide invisible to insects. Similarly, the yellow spots on the whitish petals of the horse chestnut, *Aesculus hippocastanum* turn red after the stigmas wither. In the prophet flower, *Arnebia echioides* (Fig. 4.27), the fine black petal spots (supposedly representing the results of handling by Mohammed) fade with time, and older flowers do not therefore bear nectar guides. In such cases, the persistent corolla may protect the developing fruit after fertilisation, but the plant avoids unproductive, and perhaps discouraging, visits by the pollinator to overmature flowers. *Ranunculus glacialis* is

**Figure 4.27** In the prophet flower *Arnebia echioides* (Boraginaceae), the black petal spots on the yellow flowers fade with age, and may make the flower less attractive to insects (×0.5).

frequently quoted as a flower that starts white, but turns red after pollination. This is a much less certain case, for some populations are entirely reddish from anthesis, others stay white, and there is no evidence that those that change do so in response to pollination. As in other cases quoted, it is much more likely to be a response to ageing of the flower.

Petal colour may have many other functions, which have been surveyed by Kevan (1978). Thus, petals are frequently involved in deceit, as has been documented above (sapromyophily, pseudocopulation, etc.). It has not often been noted that flowers may show mimicry, not only to insects, but also to other species of flower. Clearly, it may be advantageous for a plant that is a minority in a patch to display signals to pollinators very similar to those displayed by the dominant zoophilous flower in the patch, particularly if pollinators are scarce or low in diversity. This advantage will be increased where the potential mimic is self-compatible, but reliant on insect visits for within-flower pollination, or geitonogamy (Ch. 9). Much of the incoming pollen on the shared pollinator is likely to come from the dominant model. A. P. Hamilton (personal communication) has shown that *Gladiolus illyricus* has flower colours very similar to foxgloves (*Digitalis purpurea*) where they coexist in north-west Europe, but further south it adopts the much paler shades of *Cistus albidus* in Mediterranean maquis. Other possible examples are quoted by Proctor and Yeo (1973), for instance between *Euphrasia micrantha* (an inconspicuous annual eyebright) and *Calluna vulgaris* (heather) on British heather moors. These authors also suggest that, in addition to such apparent examples of Batesian mimicry, Müllerian mimicry may occur between co-dominant species, particularly if they have unspecialised flowers. Thus, they suggest that the very similar shape, colour and size of the unrelated flowers of *Ranunculus* (buttercups), *Potentilla* (cinquefoils) and *Helianthemum* (rock roses) in temperate and alpine meadows in many parts of the world may be coadaptive. As yet, such suggestions seem not to have been the subject of experimental investigations, and this would seem to be a fruitful field for further research.

### *Odour attraction*

Despite the advent of the highly sophisticated and sensitive techniques of gas chromatography, relatively little is known of the chemistry of flower scent. Perhaps half of all animal-visited flowers appear scented to the human nose, and unlike many other features we have been discussing, floral odour may be primitive in the Angiosperms. Many primitive beetle flowers such as *Magnolia*, *Drimys*, *Nymphaea* etc. are strongly scented, especially at night, and odours with various functions are found in living representatives of still earlier groups such as the cycads, and certain Marchantiales (liverworts).

However, diversification in the nature of flower odour must have occurred during the adaptive radiation of the flowering plants, to give rise to odours as distinctive as the sweet scent of honeysuckle (*Lonicera*, a moth flower; see Fig. 4.20), the foetid stench of various aroids (sapromyophilous), the pheromone (sex hormone) imitating the scent of the pseudocopulatory *Ophrys* (Fig. 4.14) or the 'cabbagy' scent of many bat flowers (*Cobaea*). Variation in odour (to the human nose) can occur within a species. Galen and Kevan (1980, 1983) record 'sweet' and 'skunky' odours within populations of the *Polemonium viscosum*, and regret that *Bombus* bees preferentially visit flowers with sweet odours.

Flower odours are usually produced by the petals or tepals, from the epidermis. They are most usually produced by volatile oils, and, in these cases, the scent-producing areas can be stained by immersing flowers in 0.1% aqueous solutions of neutral red at room temperature for 2–10 hours (Vogel 1962, Adey 1982). Scent-producing areas in the Leguminosae are known as 'Duftmale', and Adey has shown that a wide variety of Duftmale patterns are produced by different species in tribe Genistinae, and that these are taxonomically distinctive (Fig. 4.28). Often, Duftmale patterns correspond to general morphology, thus they may be limited to the wings or the standard (but never to the keel), or they may sometimes correspond with guidemarks on the flower.

As has already been described, odour is most often employed as a primary long-range attractant to the flower. Experiments by Brantjes (1978) on various moths showed that they can respond to scent alone, and in some cases can show limited orientation to scent, although in others, scent merely elicits a feeding response. In practice, moths respond to stimuli of scent and vision simultaneously. The range at which scent can attract a pollinator is little known, but scents may be apparent to bees at a concentration of only one-hundredth of their apparency to man.

However, scent may have other functions. Bees carry scent back to the hive, where the scent of a productive source is detected by other workers and is used, together with other evidence from 'bee dances', to trace the productive patch. Some tropical euglossine bees collect perfume from a variety of types of flower (*Catasetum* and other orchids, Araceae such as *Anthurium*; *Gloxinia* and others). In scraping up the odorous cells, the bees appear to become drugged, but later they fill their leg baskets with the odour and perform apparently territorial flights (Vogel 1966). Whether the purpose of this behaviour is really territorial or sexual is not clear, but it may provide stimuli for flower visiting not concerned with energy budgets. Thus, bees may be encouraged to make much longer species-specific flights between widely dispersed flowers in tropical forests ('trap-lining'; Williams & Dodson 1972).

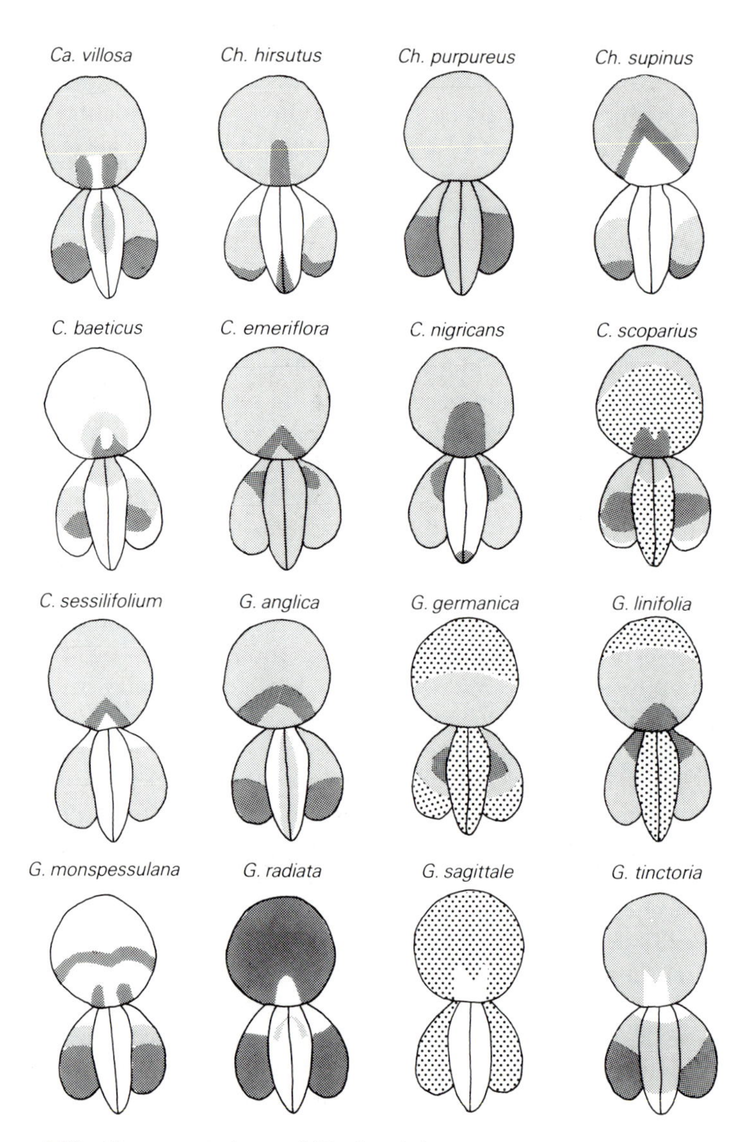


**Figure 4.28** Representations of 'Duftmale' patterns on the flowers of various species in the tribe Genistinae, family Leguminosae (Fabaceae). Duftmale secreting areas produce scents as volatile oils; heavily secreting areas are dark, moderately secreting areas are lightly shaded and lightly secreting areas are stippled. Ca, *Calicotome*; Ch, *Chamaecytisus*; C, *Cytisus*; G, *Genista*. Drawing reproduced by courtesy of Margaret Adey.

**Figure 4.29** The Australian orchid *Caladenia filamentosa* has very long, narrow perianth segments, the dark distal areas of which are 'osmosphores' and produce an unpleasant scent that attracts flies (×1.0).

In sapromyophilous flowers, scent may be limited to narrow hanging or erect appendages to the flower ('osmophores') (Vogel 1962), which may act as primary alighting sites for attracted flies (*Arisaema*, some *Caladenias* (Fig. 4.29), *Tacca*, etc.). More often, the foul scents emitted by sapromyophilous flowers, such as aroids (*Arum*, *Dracunculus*) are produced by the spadix (a modified end to the inflorescence axis) (Fig. 4.30). Amines, ammonia and skatoles are volatilised by a tremendous respiratory effort, in which the spadix may heat up to 22°C above ambient temperatures, rendering it hot to the touch (if the stench permits the experimenter to get so close) (B. D. J. Meeuse 1978). Such vile smells are usually preludes to a trap mechanism, but other aroids (*Amorphophallus*)

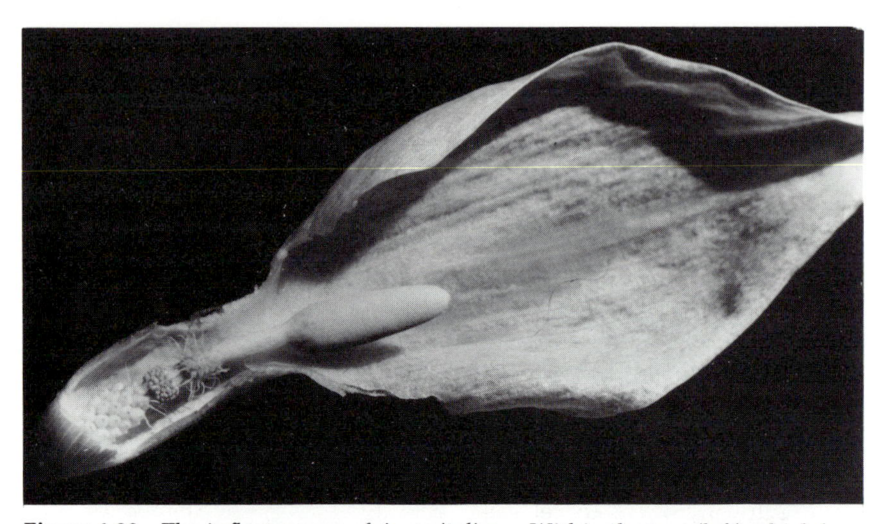

**Figure 4.30** The inflorescence of *Arum italicum*. Within the trap (left) which has been dissected open are the female flowers, male flowers and hairs (reading from left to right). During the initial female phase the hairs act as a unidirectional valve, allowing flies to enter the trap, but not to leave it. In the later male phase, the hairs collapse, releasing the flies which escape carrying pollen. The end of the inflorescence axis forms a spadix which heats up through respiratory activity to release foul-smelling skatoles which attract the flies. Photo by G. Chaytor (×0.5).

have no trap, yet flies are so attracted that they may spend several days in close proximity to the inflorescence.

Although Kullenberg (1956) has suggested that the pheromone-mimicking odours of *Ophrys* (Fig. 4.14), which are faintly discernible to the human nose, may in fact be pheromones themselves, later work renders this less likely. Gamma-cadineine is one of several substances produced by the *Ophrys* which elicit a mating response in the visitor. Flower odours are involved in other deceit mechanisms, as in *Arum conophalloides* which exclusively attracts blood-sucking midges (females only), although it does not trap them.

Of the more attractive odours (to the human nose), it is not easy to classify types of odour to types of pollinator. Moth flowers (Fig. 4.21) are usually very strongly and sweetly scented, principally by night. Bee flowers are poorly scented, but simple bowl flowers attracting beetles, syrphids and small flies (Fig. 4.2) are often sweetly scented. Wasp flowers may have a fruity scent, as do some bat flowers (Fig. 4.19), although others have disagreeable cabbagy scents. Bird flowers usually have no scent at all, and birds do not respond to flower odour (this negative evidence serves to suggest that most or all flower odours play a positive rôle in pollinator attraction, and odour production expends enough energy to render its loss evolutionarily successful where it has no function). As yet we know little about the chemistry of odours with respect to different attraction syndromes but it is likely to be complex. Analysis of human responses (25 individuals) to 12 different varieties of *Narcissus* which were hidden from the subject, suggested to me that considerable variation occurred not only with respect to odour detection (this is well documented in humans, for instance in the ability to smell *Freesia*), but also in the *Narcissus*. A study by Adey (1982) on the Genistinae produced similar results.

### *Nectar reward*

We have seen that the more primitive flower types are visited only for pollen, and that primitive flower visitors cannot utilise nectar. Forms of deceit attraction are by their nature likely to be frequency dependent, and will only work in relatively rare flower types. Alternative food sources (food bodies) are unusual. Thus, the population of specialised flower visitors (bees, wasps, butterflies, moths, birds and bats) will depend crucially on the production of nectar by many flower types.

Nectar is a solution of sugar, in which sucrose usually predominates, with fructose and glucose also being present, and maltose and mannose (which may be toxic) on occasions. It also contains amino-acids and lipids, not surprisingly as nectar is essentially phloem fluid, excreted from terminal phloem cells. It is secreted by localised densely packed groups of specialised cells (nectaries). These are most usually located on petals or sepals, often at the base, or hidden in a flap (*Ranunculus*; see Fig. 4.2), pouch (*Impatiens*), spur (*Aquilegia*; see Fig. 4.6) or tube (*Plumbago*), although the nectaries themselves may form special tubular organs derived from tepals (*Helleborus, Trollius*), or may be located on the receptacle (many Rosaceae etc). Some flower spurs merely act as a vessel to contain the nectar which is produced higher up (*Viola*).

In bird flowers and bat flowers, nectar production is often copious, and many millilitres may be shaken out of flowers. In many nectar flowers it may be difficult to locate, however, and a simple colorimetric test for sugar (for instance with diabetic test papers, e.g. 'diastix') is often useful. Once the nectary has been located, volumes of liquid nectar can be measured using microcapillary tubes, and the concentration of sugar estimated using a refractometer. In this way, the energetic reward to a visitor provided by a flower can be quantified in calories, and the diurnal output, and rhythm, of a flower producing nectar can be investigated (Corbett 1978, Best & Bierzychudek 1982). As monosaccharides have half the refractive index of disaccharides, it is possible to measure directly the nectar strength in sucrose equivalents without conversion factors; the calorific equivalent of sugar is 3.7 cals $mg^{-1}$ (15.5 J $mg^{-1}$).

A good deal of work has been published on the strength and composition of nectar (e.g. Percival 1961, 1965, Baker & Baker 1973a,b, 1975, 1977) but as Faegri and van der Pijl (1979) correctly point out, much work has unfortunately concerned itself with the concentration of nectar, and not its calorific value, which may be more significant to visitors. Maximum calorific rewards per flower per day have been observed as follows:

| | |
|---|---|
| *Sinapis alba* | 400 cal (1673 J) (Corbett 1978) |
| *Echium vulgare* | 100 cal (418 J) (Corbett 1978) |
| *Digitalis purpurea* | 50 cal (209 J) (Best & Bierzychudek 1982) |
| *Arctostaphylos otayensis* | 1.5 cal + (6.3 J) (Heinrich & Raven 1972) |

However, these papers also show that sugar content varies greatly with many flowers, for instance in *Digitalis*, having rewards of only 1 calorie (4.184 J). Many variables act on the amount of sugar available, including position of flower, age of flower, time of day, weather and vigour of the plant. Maximum rewards tend to be found in the first flower to open, in the middle of the flower's life, early in the morning (although some species show maximum secretion of nectar at other times of day, for instance mid-afternoon) on hot, but not too dry, days, on vigorous plants. Undoubtedly, rewards may be much higher than this; hummingbirds may acquire more than 60 cal $s^{-1}$ (250 J $s^{-1}$) while hovering (Wolf & Hainsworth 1971, Wolf 1975), and bird and bat flowers may contain well in excess of 1000 cal (4184 J).

The concentration of nectar also varies considerably, not only between species, but even within flowers. In most flowers, the concentration of nectar in sucrose equivalents lies between 25 and 75% (Percival 1961), but Corbett (1978) showed that in *Echium vulgare*, the normal daily fluctuation in concentration was between 20 and 58%, whereas in *Sinapis alba* it was between 20 and almost 100%. Both species secreted the daily supply of nectar early in the morning, and again in the evening; sugar concentration was highly linked with flower temperature, which reached a peak for *Echium* at 13.00 hours, and for *Sinapis* at 17.00 hours on experimental days. Examination by Percival (1965) of the range of nectar concentrations in flowers with different types of visitor, shows that nectar concentration can be highly adaptive in limiting pollinator visits (Table 4.5).

In contrast, Baker and Baker (1975), also using Percival's data, suggest that little difference in the mean concentration of nectar exists between flowers of different pollination types (all average between 20 and 25%), although within each group variation from 5 to 50% occurs. Insects are probably not responsive to sugar concentrations below 10%; Vansell *et al.* (1942) showed that only when orange blossom (*Citrus*) nectar was concentrated by daytime heat to 30% was it attractive to bees; undoub-

**Table 4.5** Sugar concentrations in the nectar of flowers of various pollination classes (Corbett 1978, after Percival 1965).

| Pollinator | Concentration of sugar (%) | Number of species of plant |
|---|---|---|
| moth | 8–18 | 2 |
| bat | 14–16 | 2 |
| bird | 13–40 | 7 |
| butterfly | 21–48 | 2 |
| bee | 10–74 | 24 |

tedly proboscis drinkers such as butterflies and moths will require more dilute nectar than this, as will 'lappers', such as bats, and suckers, such as many birds. At concentrations above 70%, nectar becomes very viscid, and at high concentrations it is crystalline. It will then become more suitable to feeding by insects with relatively unmodified mouthparts such as flies, and some beetles and bees; nectar of unspecialised umbel flowers may always be in this state. Corbett (1978) suggests that the pollinator suite of *Sinapis* changes through the day, long-tongued bees and butterflies being replaced by flies, and this broadens the chances of successful pollination, while limiting the pollinators at a given moment. This apparently successful strategy may be a very common one, and may explain why many flowers have a 'limb and tube' shape; the salver-shaped limb concentrates heat, and allows nectar to become more concentrated during the day; the tube limits visits earlier in the day to specialised long-tongued feeders. Much more work is needed on the rôle played by nectar in providing 'time-niches' in the pollinator suite.

For butterflies, larger moths and some birds, nectar provides the total food input (hence the presence, or at any rate function of, trace amounts of nitrogenous substances), but for bats, other birds and bees it acts as a liquid fuel. It may also act as a source of water; thus bees may visit more dilute nectar sources on hot days. Many pollen-collecting bees will collect pollen on one species, and fuel with nectar on another. Thus workers of *Bombus lucorum* in Cumbria, UK, on a June day in 1983 (ambient temperature 21°C) were pollen collecting exclusively from *Cerastium fontanum*, which has no nectar, and drinking exclusively from *Trifolium repens*. For many different individual bees, the ratio of pollen visits to nectar visits was 10 : 1. On a colder day, the proportion of nectar stops will increase. Bees rarely fly at under 10°C, below which temperature the thoracic–ambient temperature differential is such that any feeding flights will have a negative energy budget. Naturally, spring flowers tend to have higher nectar concentrations; some autumn flowers (*Solidago*, *Eupatorium*, *Campanula thyrsoides*) have dense heads so that pollen-collecting bees can crawl, thus saving energy.

**Figure 4.31** The flower of *Parnassia palustris* has five staminodes which bear about 10 filiform processes with shiny terminal knobs each. These have been regarded as dummy nectaries which attract the pollinator (here the hover-fly *Neoascia podagrica*) which on landing is able to discover the concealed true nectaries. Photo by M. C. F. Proctor (×3).

Nectar robbing may also save energy. I have observed the smaller-bodied (alien) honey-bee (*Apis*) visiting the flowers of the native *Iris pseudacorus* and the introduced *I. versicolor* legitimately in marshes on the shores of Windermere, UK. However, the larger bodied *Bombus lucorum* visited *I. pseudacorus* legitimately, but had difficulty with the smaller-bored flowers of *I. versicolor*, which it robbed by levering the base of the false tube. Legitimate visits lasted an average 25 seconds, but illegitimate visits only 5 seconds. A few species of flower visitors are habitual nectar thieves; thus *Bombus mastrucatus* usually bites holes at the base of corollas which it could readily visit legitimately (Fig. 4.3). Macior (1966) has described how the queens of *Bombus affinis* pierce the spurs of *Aquilegia* (bird and sphingid flowers) for nectar, while collecting pollen (and probably achieving pollination) legitimately. Examination of any population of, for instance, red clover, *Trifolium pratense*, will demonstrate that many flowers have been robbed, with perforated bases. Such behaviour is bound to affect gene flow by rendering robbed flowers unattractive to legitimate visitors.

Although many flowers have guide marks on petals which lead towards nectaries, nectaries themselves are usually inconspicuous, but they can be visible and glistening, especially in some beetle flowers. However, the glistening substance itself is often not nectar. *Parnassia* is among nectarless flowers that produce dummy nectaries (Fig. 4.31).

In general, despite great variation in the production of nectar even within a single flower, the chemical composition, concentration, calorific value and position of nectar produced by a flower will tend to be adaptive with respect to its habitual visitors. Larger visitors will receive greater energetic rewards, but may have to expend more energy in acquiring them (Heinrich 1975). Total flower feeders will receive a more balanced nectar diet (more amino acids and lipids) (Heinrich 1979a,b) than mixed feeders. Sucking feeders will receive more dilute nectar, in greater volumes, from narrower receptacles, at greater cost than dabbing feeders. Thus butterflies may expect to receive relatively large volumes of dilute nectar with amino-acids and spend long periods of time at each flower, in comparison with short-tongued bees, or flies, which will receive small volumes of concentrated pure-sugar nectar on short visits.

Baker and Fisher (1983) and Kevan and Baker (1983) give average concentrations of amino-acids in nectar of flowers with different visitor syndromes, and show that sapromyophilous carrion fly-visited flowers have about 12 times the concentration of amino-acid of any other flower type. They also list occurrences of individual amino-acids, and show that alamine is the most widespread amino-acid in nectar, but that 13 amino acids occur in more than half the sample of 395 species.

By such strategies, the plant tends to narrow the range of visitors to those to which it is best adapted for pollination; and to those most suitable to the time/space niche in which it flowers.

## DEVELOPMENT OF THE ANDROECIUM AND GYNOECIUM

In the Pteridosperm and Gymnosperm cone, a conflict is observed between the functions of ovule protection and receipt of pollen. The requirement for the protection of the ovule from predation or desiccation by the cone scale renders pollen receipt on to the micropile inefficient (p. 46). Neither function can act efficiently in the presence of the other. I have argued (Ch. 3) that the evolution of the carpel and stigma allowed these functions to be separated in the flowering plants, and created space for the coexistence, within the same floral axis, of stamens (a hermaphrodite flower). That this arrangement has been successful is demonstrated by the overwhelming preponderance of hermaphrodity (95% of Angiosperm species; Ch. 8). Yet it has been argued that hermaphrodity has created dissonance between the functions of pollen release and pollen reception, and that this is amongst the pressures favouring dioecy (Bawa 1980). Van der Pijl (1978) also lists cases of apparent dissonance in the male and female functions of hermaphrodite flowers (Ch. 2).

Despite being hermaphrodite, most Angiosperm flowers favour the transfer of pollen from one individual to another, as many floral features described in this chapter show, and may only function sexually when such transference has taken place (Ch. 6). Even where self-pollination occurs, this most often happens as a mixed strategy, in a 'fail-safe' rôle, for environmental heterogeneity and linkage disequilibrium favour cross fertilisation (Ch. 2), and habitual self-fertilisation is rare in most life forms (Ch. 9). Although geitonogamy may be common and unavoidable in most plants with multiple inflorescences, within-flower selfing is avoided in most allogamous species, by the separation of male and female functions in time (dichogamy, that is to say protandry or protogyny) or space (herkogamy). Complex flowers usually separate the stigma from the anthers spatially, as in heterostyly (Ch. 7), trap flowers (*Arum*, *Cypripedium*), flag flowers (Leguminosae), gullet flowers (Labiatae), tube flowers (*Lonicera*) and countless other examples. However, such spatial separation may vary significantly, even within a population, as in *Gilia achillaefolia* (Table 9.4, p. 340).

Dichogamy is achieved merely by differential timing, and requires no morphological adaptation. In contrast, herkogamy demands the localisation of the functions of pollen reception (small stigmata, on long styles, or secluded). It is difficult to imagine successful herkogamy in the unspecialised flowers of *Magnolia* (Fig. 4.1) or *Ranunculus* (Fig. 4.2); reduction in the number as well as the size of anthers and stigmata will also aid herkogamy, which is essential to the correct function of any specialised flower. In fact, any complex flower-visiting syndrome is initially reliant on herkogamy for its successful development, and subsequent function.

### *The androecium (stamens)*

Potentially, an antagonism may arise between the functions of pollen as food, and pollen transport. Primitive 'pollen flowers' which lack nectar may be visited entirely for the collection of pollen as a food. It is a complete food for many groups of insects, for instance bee broods, as it contains proteins (16–30%), lipids (3–10%) and some, rather unavailable, carbohydrates (0–15%). Thus, paradoxically, the reward must be transported to another flower, and inefficiently collected or ingested, if cross-pollination is to be achieved. In practice, most primitive flower visitors ('insect riff-raff') indulge in rather inaccurate flower-visiting procedures ('mess and soil pollination') and the niceties of efficient pollen collection and grooming are beyond them. Indeed, many are not in the flower for the pollen at all ('rendezvous attraction'). However, bees are quite a different matter, and bees that visit efficient flowers with low pollen : ovule ratios (Ch. 2) for the purposes of pollen collection, may transfer very little pollen to other flowers, especially if they groom efficiently. Various floral techniques overcome this problem, for instance the sequential opening of anthers (very common in long-flowering winter flowers such as *Viburnum farreri* and *Jasminum nudiflorum*), and the division of labour between stamens. This is most often achieved by conspicuous staminodes, which lack or have little pollen, or by conspicuous tufts of hairs on the filaments or connective; whereas polliniferous anthers are cryptic (*Commelina coelestis*). Alternatively, there may be provision of two types of stamen, one designed for feeding (conspicuous, extruded) and one for pollen carry (*Lagerstroemia indica*, and many Scrophulariaceae, i.e. *Verbascum*). In extreme cases, 'feeding stamens' may provide no pollen, but another type of food reward to the visitor (*Calycanthus occidentalis*, *Tibouchinia*). Vogel (1978) provides a useful review of changes in anther function.

Dichogamy may also cause antagonism of function, as in strongly protandrous pollen flowers. Thus, the familiar African violet, *Saintpaulia*, maintains attractive yellow anthers through the female phase, long after the pollen has been shed, thus attracting pollinators by deceit. Pollen flowers are almost never protogynous, as they would not be attractive to pollinators during the initial female phase.

In primitive pollen flowers, pollen is usually dry, and is often released simultaneously from many anthers, extrorsely, by the feeding action of the visitor. In more specialised nectar flowers, pollen is more often sticky and adheres to the visitor as it brushes past the anthers; in these cases, pollen dehiscence is more often introrse, towards the centre of the flower, thus mediating pollinator contact. Anther dehiscence may take place sequentially over a period of days, or even weeks. In some flowers with corolla tubes constricted distally to the anthers, anther dehiscence is apical and poricidal, onto the head of the long-tongued visitor (Ericaceae). Pollen release from such anthers (and some others) is mediated by vibration from the buzzing or hovering visitor, and vibration-released pollen is dry. Pollen may also be released explosively, as in the sprung-trap flowers of many Leguminosae, a puff of released pollen being clearly visible as the striving visitor suddenly releases the stamens from the keel (see Fig. 4.8).

In brush, trumpet and some tube flowers adapted to hovering visitors (hummingbirds, sphingids, some bees and syrphids), stamens are long-extruded from the flower on long slender, but strong, filaments (Figs. 4.20, 4.21 and Plate 4.1). In contrast, gullet and some bell flowers have included stamens (Figs. 4.4 and 4.5), and in these the filaments are frequently fused to the wall of the corolla, so that the anthers appear to arise from the corolla itself. In narrow bell flowers (Campanulaceae, Ericaeae, Dipsacaceae), the filaments may be fused to each other to form a staminal tube with an apical whorl of anthers between which the pollinator forces itself, or its mouthparts. This system reaches its evolutionary climax in the Compositae with its heads of reduced tube florets each containing a staminal tube surrounding the single style (Fig. 4.32).

Staminal filaments may also become fused to the gynoecium, so that the stamens are closely attached to the ovary or style, often adjacent to the stigma. Thus potential disharmony of pollen release and reception has been overcome by combining both on the same organ (gynostegium); the style replaces the filament as the anther-presenting organ. In many Compositae there is a further step, in that the style and stigma actually present the pollen which has been released onto the style by the anther tube. Initial self-pollination is avoided by protandry (the stigma lobes do not open and recoil to achieve self-pollination until a later stage; Fig. 4.32). In the gynostegial Asclepiadaceae and Orchidaceae, both dichogamy and herkogamy prevail. For instance, in many orchids pollinium release occurs before the stigmatic cavity is receptive (protandry), although this is not true of those with a trap mechanism (e.g. *Cypripedium, Coryanthes*). In addition, the stigmatic cavity is usually hidden with respect to the pollinia (herkogamy) in such a way that self-pollination is impossible. Only in rare cases can self-pollination occur in this vast group. This becomes possible either through the loss of the intervening viscidium in the primitive *Epipactis*, which lack a stigmatic cavity, stigma presentation being relatively open (Fig. 4.33), or through the possession of unusually long caudicles to the pollinia which droop onto the stigma (*Ophrys apifera*; Fig. 4.34) (Ch. 9). In any case, herkogamy prevails, despite the gynostegium. That is, although the anthers are borne on the column (= modified style), they are spatially separated from the stigma.

The Orchidaceae are indeed a remarkable case which demands further description. Replacement of the filament by the style as the anther-presenter has allowed a considerable reduction and specialisation to occur in the anther. In fact, only one anther remains in a flower, except in the primitive Lady's slipper orchids (*Cypripedium*; Fig. 4.12) and their relatives, which have two that still release particulate pollen. This single anther forms two, and in some cases only one, pollinia from its two locules. These pollinia have sticky stalks (caudicles) which may be originally joined together, but are more usually dispersed separately. These caudicles have been formed from an ancestral connective. The filament is entirely missing, and the pollinia merely consist of indehiscent membranous sacs full of blocks of pollen nuclei (massulae) still in the tetrad state, and without pollen grain walls. Each pollinium can contain between 500 and 5000 nuclei, and this becomes the pollination unit, being dispersed on the pollinator whole.

**Figure 4.32** Disk-florets of the golden rod, *Solidago virgaurea*, family Compositae (Asteraceae). Each capitulum bears many reduced flowers (florets) which are tube-shaped and have five small lobes, although the marginal ray-florets (Ch. 2, Fig. 2.4) form a strap-shaped ligule. The five stamens in each floret are fused into a tube, through which the style with closed pairs of stigma arms passes. The outside of the stigma arms forms the pollen-presenting organ. Later the stigma arms are recoiled and self-pollination can occur. Photo by M. C. F. Proctor (×7).

Pollinia may vary remarkably in form, even in a single tribe such as the Oncidineae (Fig. 4.35) and Dressler (1980) gives a splendid account of pollinium form and function. Pollinium form may depend on the shape and behaviour of the pollinator. As many as 18 different positions for normal pollinium attachment on an insect have been enumerated; these vary from a butterfly's proboscis (it may not be able to recoil the proboscis as a result) to the underside of the abdomen of a bee. Pollinium shape will depend on the position of its attachment on the pollinator, but also on the position and shape of the pollinium-receiving stigmatic

**(a)**

**(b)**

**Figure 4.33** (a) Inflorescence of the outbreeder *Epipactis helleborine*, a common European orchid. The dark shiny hypochile contains nectar and attracts wasps (*Vespa* spp.). Immediately above this is a white knob, the rostellum (arrowed), bearing a sticky viscidium on which rests the two pollinia which drop out of the anther. Within-flower pollination cannot occur, at least in the absence of a wasp visit, and pollinia are removed by means of the sticky viscidium. Photo by M. C. F. Proctor (×3). (b) Flowers of the inbreeder *Epipactis dunensis*, a rare British endemic orchid, in which the viscidium is absent, and the rostellum rapidly withers, allowing the pollinia to fall directly onto the stigma and self-pollinate, as can be seen in the upper flower. Photo by D. Lang (×3).

**Figure 4.34** In the European bee orchid, *Ophrys apifera*, pollinia drop from the anther on long sticky caudicles, allowing automatic self-pollination in the absence of an insect visit. The anther is within the hood at the top of the flower, and the caudicles and pollinia can be seen below this. Most *Ophrys* species can only be pollinated by pseudocopulatory solitary wasps (see Fig. 4.14) (×2).

cavity on the column, to which it is coadapted. In many orchids, as Darwin (1862) observed for *Orchis mascula*, the pollinia attach to the bee in an outward-pointing posture (Fig. 4.36), so that when the bee visits other flowers in the same inflorescence, or even colony, it cannot pollinate other flowers. Only when the bee undertakes a lengthy flight does the caudicle dry out, and the pollinium bends forward to the direction in which the bee is flying. It is then in a position to pollinate subsequently visited flowers, thus ensuring relatively long-range pollination.

All orchids are relatively specialised flowers. The invention of the gynostegium has encouraged the evolution of the pollinium, and this results in an 'all or nothing' approach to pollination. Roughly equal numbers of ovules occur in an ovary as pollen nuclei occur in a pollinium. Thus, on the relatively rare occasions that a successful pollination takes place, very large numbers of seeds are fertilised, but these are, by constraints of space, very small and reduced.

Although these seeds can travel long distances by wind (at least for

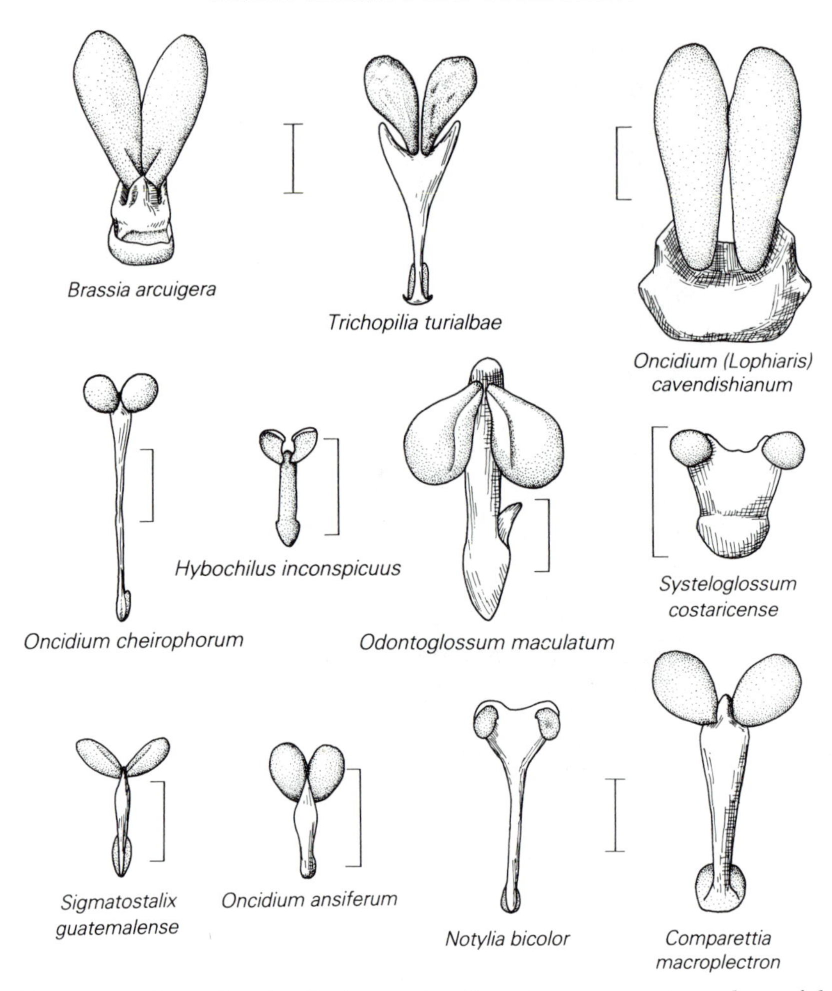


**Figure 4.35** Diversity of pollinium and anther shape amongst members of the orchid tribe Oncidineae. Scale 7 mm. Reproduced by kind permission of R. L. Dressler.

terrestrial orchids; sheltered epiphytic orchids show much more limited patterns of distribution and more local endemism), the safe sites for seed establishment are few, and most need to form a symbiotic association with a fungus before development can occur.

The gynostegium also requires very accurate patterns of behaviour by the pollinator if pollen donation and pollen receipt are to be effective. Target areas are very small, although they may be increased by columnar wings in the small-flowered, pendulous, wind-blown, *Centris*-attacked *Oncidiums* (Dodson & Frymire 1961). Thus orchids tend to have very specialised, complex, pollinator-specific flowers to aid this accuracy.

**Figure 4.36** Pollinia removed from the flower of the European common spotted orchid, *Dactylorhiza fuchsii*, are initially pointing in an outward direction and rarely achieve pollination. Only after the caudicle dries out, often after an insect flight, does the pollinium bend forward into a position in which pollination is likely. This mechanism will discourage geitonogamy and encourage xenogamy. Photo by M. C. F. Proctor (×3).

Curiously, relatively few orchids offer rewards; their complex modes of attraction usually involve deceit, as is the case in many highly oligophilic (pollinator-specific) species. Many of these remarkable features of the Orchidaceae owe their origin to the gynostegium.

There is an interesting parallel in the dicotyledonous Asclepiadaceae, which has also evolved a gynostegium, and this provides a test for the theory of the rôle played by the gynostegium in allowing the evolution of the other remarkable features of the orchids. Despite the very distant relationship of the two families, many features are indeed common to both. Both have pollinia, which can be remarkably similar in form in the two families, and both have groups, which have been divided into subfamilies, with one or two pollinia respectively. Both have dichogamous and herkogamous stigmatic cavities, and both produce, in most cases, many small seeds, although those in the orchids are more numerous and more reduced. The Asclepiadaceae do not, as far as it is known, have a symbiotic requirement for seedling establishment, although this is a matter that may deserve further investigation. Both also tend to have very specialised pollination mechanisms, although one such is unique to the Asclepiadaceae (tribe Ceropegieae, including the succulent stapeliads) in which the legs or proboscis of the visitor are trapped in elastic anther appendages. The struggle for release leads to pollinium attachment. However, although the variety of form of the plant (from lianes such as *Hoya* to succulents such as *Stapelia*) and of the flower found in the asclepiads almost rivals that of the orchids, the range of pollinators

**Figure 4.37** The common sedge, *Carex nigra*. This has monoecious inflorescences, and is wind pollinated. There are three male spikes (top), the upper-most of which has shed pollen, and the lower two have yet to do so. The two female spikes have receptive stigmas, so this species is homogamous. Photo by G. Chaytor (×1.5).

and pollination mechanisms is much more limited. Most asclepiads attract flies by deceit (sapromyophily), although they use a variety of mechanisms to do so. It is likely that the regular, actinomorphic flowers of the asclepiads preclude the attraction of many more sophisticated visitors such as bees. Such visitors prefer landing platforms as provided by zygomorphic flowers like the orchids. It is interesting, however, that deceit plays a major rôle in the pollination syndromes of both families.

Up to now, only the stamens of animal-pollinated plants have been discussed. The requirements for successful pollen release and dispersal in wind-pollinated plants naturally differ from those of insect-pollinated plants. However, Stelleman (for instance 1978) and others point out that pollen-eating insects such as beetles and some syrphids frequently visit flowers apparently adapted for anemophily, and that successful insect-mediated pollen transport can occur in these (for instance *Plantago* and *Carex* species). The stamens of wind-pollinated flowers are typically long-exerted, having long, slender filaments which may be pendulous (Fig. 4.37). They have large anthers containing much dry, rather smoothly ornamented, pollen, which dehisce readily, especially in warm, humid conditions (which therefore result in the greatest suffering to those with pollen allergies).

Pollination by water has already been discussed in this chapter; hydrophily has required the evolution of some distinctive male characteristics, such as anthers that will dehisce when submerged, floating pollen, pollen without walls, and even floating male flowers.

## *The gynoecium (carpels or ovary)*

The gynoecium shows a variety of adaptations to forms of pollination as does the androecium, but in addition gynoecial adaptations may be directed towards fruit and seed dispersal. The very different functions of pollen reception and fruit dispersal may in some instances be apparently antagonistic, as has been discussed in Chapter 2 and in van der Pijl (1978), or they may be coadaptive. Thus if we can once again use the orchids as an example, the very large number of highly reduced ovules has clearly originated in response to the evolution of the pollinium. But the very small wind-borne seeds have allowed the orchids to acquire several distinctive ecological functions (epiphyty, saprophyty) as well as an unusual capability for dispersal. However, the adaptive radiation of fruit and seed morphology in the Angiosperms scarcely merits full discussion in this book, and has been briefly reported on in Chapter 3.

Much more relevant is variation in the position of the ovary in the Angiosperm flower, the evolution and diversificiation of the style, and the stigma. Although the primitive, Magnoliid condition is for the carpels to be free, most Angiosperms (outside the Magnoliales and some Ranales) have carpels fused into an ovary. This development probably accompanied the evolution of a style, and localised stigmata. Flowers with free carpels, as in *Magnolia* (see Fig. 4.1), *Paeonia* or *Helleborus* generally have a diffuse stigma and a very short style, or no style, on each carpel. Although multistyled fused ovaries do occur, as in the Caryophyllaceae, they are not widespread, and are typical only of unspecialised flowers. Carpellary fusion was most usually followed by stylar fusion leading to localised stigmata typical of specialised flowers.

An evolutionary trend is also discernible in the position of the ovary with respect to the remainder of the flower. In the Magnoliales, Ranales, Nympheales, Liliales and other putatively primitive groups which tend to have unspecialised flowers (as well as anemophilous groups such as the Cyperales, Poales and Fagales), the ovary is superior (hypogynous) with the stamens and perianth inserted at its base. Such a syndrome may have some benefits to the plant, especially with respect to ovarian temperatures in solar-furnace bowl flowers in cold climates. However, a superior ovary may well be more liable to predation, being obviously presented, and it may interfere with the function of complex specialised flowers with tubes or traps. (However, the predominantly gullet-flowered, bee-pollinated Labiatae and Scrophulariaceae, and the flag-flowered Leguminosae have superior ovaries.) Nevertheless, it is significant that a parallel trend from superior ovaries to inferior ovaries occurs in both the monocotyledons (e.g. Liliaceae to Orchidaceae) and the dicotyledons (e.g. Ranunculaceae to Compositae), and in both classes it tends to be associated with increasing specialisation in flower form.

The number of stigmatic lobes to the single style in flowers with carpels fused into an ovary usually reflects the number of fused carpels, that is the number of segments in the ovary. Thus, the Geraniaceae have five stigmas, the Liliaceae have three stigmas, the Cruciferae have two stigmas (or sometimes one), and the Gramineae have a single stigma. Sometimes stigmas are more numerous, as in the Orchidaceae which have two stigmatic cavities, although an aseptate (but originally tripartite) ovary, the Cyperaceae have three or two stigmas, and the Compositae have two stigmas, both with unilocular ovaries. In other cases (e.g Ericaceae), the number of stigmas (one) may be fewer than the number of segments in the ovary (five).

Stigmas and styles show rather less development in form than stamens with respect to pollination biology. In anemophilous plants, stigmas are typically exerted, long and feathery (Fig. 4.37), being pinnate in shape. In entomophilous flowers they may vary very much in size, in relation to the rest of the flower, tending to be larger in unspecialised flowers, as has been discussed above and in Chapter 2. Flowers with efficient pollination will tend to have sparse pollen and small stigmas. The nature of the stigmatic papillae will depend on the incompatibility system, being 'wet' or 'dry' (Ch. 3). Even in very complex and specialised inflorescences such as the traps of *Arum*, or *Aristolochia*, the brood flowers of *Yucca* or *Silene*, or the enclosed inflorescences of *Ficus*, the form of the stigma does not differ a great deal. Only in the pollinium flowers of the Orchidaceae and Asclepiadaceae does it change radically to form a cavity in the column (style) to receive the pollinium, which, unlike free pollen, is too large to lodge successfully on an exposed stigma.

However, the number and position of styles and stigmas may be highly adaptive with respect to animal pollination. Flowers receiving large, hovering visitors with long tongues (e.g. hawk-moths and humming-birds) usually have long-exerted styles with smallish capitate stigmas; the styles may be up to 20 cm long (*Datura, Hibiscus*). Other bird flowers, and flowers pollinated by small mammals such as bats, mice and small marsupials, are usually brush flowers with large numbers of stamens, the filaments of which may, together with the tough, single styles, form the main and attractive part of the flower. This is particularly prevalent in Australian Myrtaceae (*Eucalyptus, Callistemon*) (Fig. 4.38) and Proteaceae (*Banksia, Grevillea, Hakea*). A parallel development occurs in the Mimosoideae, notably in *Acacia*, although these are more often visited by bees.

In sprung flowers, as has already been discussed, a rather conventional style is held in tension to hit the visitor when entrance to the flower is forced (Leguminosae, some Fumariaceae). In many more complex trap, gullet, trumpet, flag and tube flowers, the style and stigma are unconventionally positioned (herkogamy), thus maximising pollen receipt from other flowers, rather than from the same flower. This has resulted in an

**Figure 4.38** Inflorescences of the Australian gum *Eucalyptus calophylla* (Myrtaceae). Stamens are many and are coloured yellow, forming the main attractant to pollinators (×0.5).

(a)

(b)

interesting development in *Iris*. In the related, but apparently more primitive, *Crocus* the tepals are poorly differentiated, forming a bowl, and there is a tripartite stigma, which may be brilliantly coloured, for instance orange or red, as an attractant (Fig. 4.39a). In *Crocus banaticus*, however, the inner tepals are narrow and erect, forming a flag, as are the inner tepals of *Iris*. This tepal differentiation removes the unspecialised bowl shape of the flower (most *Crocus* are very early or late-flowering species, opening only in the sun to form solar furnaces to attract bees on cold days). The flag tepals in *Iris* render each of the three facets of the flower effectively zygomorphic to the pollinator, but do not localise the approach of the pollinator sufficiently for efficient operation as a zygomorphic flower. This problem has been solved brilliantly and uniquely by *Iris*, which has enlarged and broadened the style into a petaloid form, to become the upper half of a tube, the landing platform being formed by the lower, larger tepal (fall) (Fig. 4.39b). The flag tepal retains its function as a long-distance attractant, although in some species it has become secondarily reduced. Thus, the six-tepalled flower produces three pollination units (gullets) made of the fall and the style. Each contains a stamen and a stigma, the latter being on the inner side of the top (style) of each tube as a small flap. The visitor generally visits each of these units in turn.

A very remarkable stigma is found in a familiar and very aberrant member of the Liliaceae, *Aspidistra*, in which the flower, rarely seen in cultivation, produces a stigma which uniquely acts as the attractant and the reward. The flowers of *Aspidistra* are produced at the end of underground stems, away from the leaves, at ground level. The six brownish tepals form a bell, the mouth of which is entirely blocked by a very large, fleshy, disc-like stigma (Fig. 4.40), beneath which are hidden the stamens. The stigma is attractive to slugs, which partly eat the stigma, thus gaining access to the stamens below. When the slugs emerge from the flower they will pollinate the remains of the stigma on that or other flowers. Curiously, this remarkable pollination mechanism, which seems to have no parallel, has escaped mention in most standard texts on pollination biology.

---

◀ **Figure 4.39** (a) *Crocus angustifolius* (Iridaceae) from Turkey illustrating a simple early spring bowl flower with undifferentiated yellow tepals and a tripartite orange stigma (×1). (b) The more specialised bee-visited *Iris histrioides* (Iridaceae) from Turkey has six perianth segments differentiated into three falls (which are spotted and striped with nectar guides) and three narrow erect flags. The styles are petaloid and form the upper half of a gullet-shaped pollination unit; the fall forms the lower half (×1).

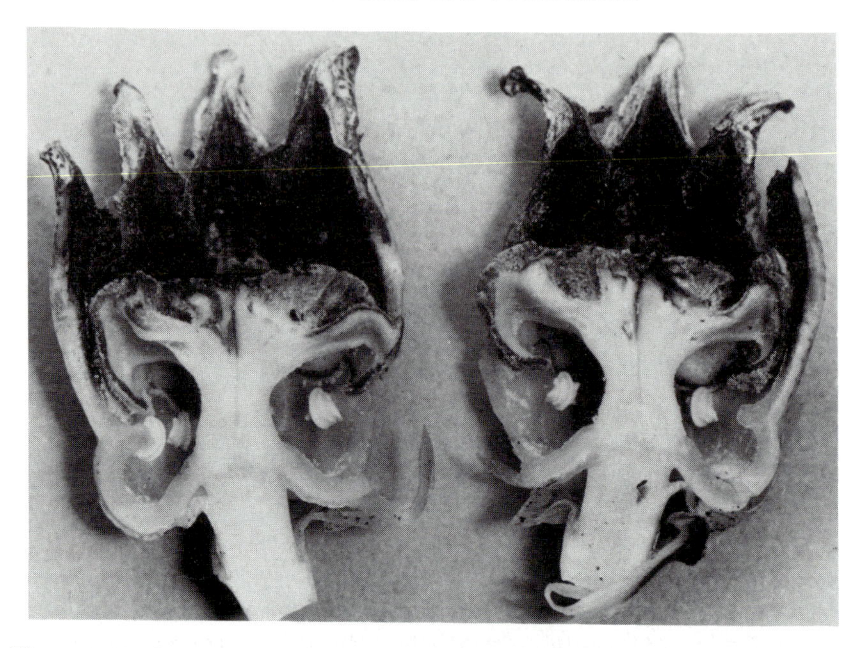

**Figure 4.40** Section of the flower of the tropical *Aspidistra lurida* (Liliaceae). Flowers are inconspicuous, brownish in colour, and borne on the ground. A large fleshy stigma which blocks the perianth is attractive to slugs which eat it, revealing the hidden stamens. Pollen sticks to the slug and, if part of the stigma is left uneaten, pollination takes place (×2).

## DEVELOPMENT OF THE INFLORESCENCE

In *Magnolia*, and other supposedly primitive woody Angiosperms, flowers are borne singly. With the evolution of the herbaceous habit in the Ranales came a tendency for flowers to be aggregated into inflorescences, probably originally as a panicle, as in *Ranunculus aconitifolius*, and later as a raceme, developed from the panicle, as in *Delphinium* and *Aconitum*. Aggregation of the inflorescence had the following consequences with respect to pollinator attraction and gene flow:

(a) The primary attractant (colour, scent) could increase in size, whereas the number of pollination units (i.e. flowers) could be increased within a single attraction unit.
(b) A greater reward per pollinator flight could be provided, thus tending to make specialisation by the pollinator on one plant species more worthwhile, and encouraging oligophily, oligotropy and pollination efficiency for both the pollinator and the plant.
(c) By separating receptivity of the different flowers in the inflorescence

in time, the inflorescence could act as an attraction unit for a longer period; coupled with dichogamy, sequential receptivity of flowers would encourage a mixture of allogamy and geitonogamy (Ch. 9) allowing both some outbreeding and acceptable levels of seed-set.

(d) Division of labour could evolve between different flowers in the inflorescence with respect to:

(i) attraction [the large, sterile marginal flowers in the umbel of *Viburnum opulus* or *Hydrangea villosa* (Fig. 4.41) or the terminal sterile florets in *Muscari comosum* (Fig. 4.42)];

(ii) sexual function (the female ray-florets contrasted with the hermaphrodite disk-florets in Compositae such as *Senecio* or *Doronicum*; Fig. 4.43);

(iii) reward (the nectarless marginal florets of *Scabiosa* in comparison with the nectar-bearing central florets); or

(iv) timing (the sequential opening of the flowers in most aggregate inflorescences).

**Figure 4.41** Inflorescences of *Hydrangea villosa* showing division of labour within a large umbel. Marginal florets are large, sterile and showy; central florets are relatively inconspicuous, but produce nectar and are fertile (×0.25).

**Figure 4.42** In *Muscari comosum* (Liliaceae) from southern Europe, showy sterile flowers at the top of the inflorescence act as an attractant (×0.5).

**Figure 4.43** Inconspicuous hermaphrodite tubular disc florets, and conspicuous female marginal ligulate ray florets dissected from the head of the Oxford ragwort, *Senecio squalidus*, illustrating division of labour within the capitulum of a member of the Compositae (Asteraceae). See also Figure 2.4. Photo by M. C. F. Proctor (×5).

(e) Other organs in the floral axis apart from flowers could be used as attractants, or more rarely as rewards or defence mechanisms. Most flowers are subtended by bracts or, in the case of some complex multiple inflorescences, the inflorescence is subtended by a bract, and the flowers by bracteoles. Such bracts may have a wide diversity of function some of which are listed in Table 4.6.

Stems may also act as agents of attraction or reward, by being brilliantly coloured (*Polygonum persicaria*), or by bearing extrafloral nectaries (e.g. *Arbutus*; Fig. 4.13). Even leaves can act as reproductive agents, especially in cases where bracts perform other specialised functions; thus in the familiar house plant 'poinsettia' (*Euphorbia pulcherrima*), the upper leaves are brilliantly red, presumably acting as a primary attractant, while the bracts act as landing platforms, and house the independent nectaries, in the curious division of labour between the floral units typical of that genus. Leaves can also aid in the dispersal of fruits, as in the grass *Sporobolus*, and other desert species which form balls of fruit which are blown along the ground by the wind. The extraordinary fibrous balls of the maritime grass, *Posidonia* discovered on many Mediterranean shores (for instance Mallorca) may also aid fruit dispersal, either by sea or, when stranded, along beaches by wind.

In some cases, flowers may move with respect to the inflorescence in a functional way. Some species of the epiphytic orchid genus *Oncidium* have small, solitary, pendant, yellow and brown spotted flowers which are thought to mimic territorial males of *Centris* bees, thus provoking aggression (Dodson & Frymire 1961) and pollination. G. J. Braem (personal communication) has observed in *O. henekenii* in the Dominican Republic that flowers move from a patent, apparent position to a reflexed, non-apparent position only after receiving pollinia. They do not so move after a visit during which pollinia are donated. Such non-apparency of pollinated flowers, often mediated by a colour change, as has been discussed above, may well be functional in limiting pollinator visits to receptive flowers. In a frequency-dependent function based on deceits such as this, a limitation of apparency would be particularly valuable.

**Table 4.6** Some functions of bracts in the reproduction of flowering plants.

| Species | Nature of bract(s) | Function of bract(s) |
|---|---|---|
| *Ajuga pyramidalis*, *A. orientalis* | large, showy, reddish or purplish | act as primary attractant to bees, flowers have become small and unshowy |
| *Saxifraga grisebachii* | showy, red, covered with glandular hairs | act as primary attractant and reward to small flies, etc. |
| *Dracunculus vulgaris*, *Arum muscivorus*, etc. | a spathe, enclosing inflorescence, mottled dark purple | primary attractant to flies by production of foul scent, secondary attractant by provision of visual guide marks, formation of trap by shape and provision of slippery surface |
| *Lamium maculatum*, *Colias* spp. | large, green with white blotches | provision of guide marks |
| *Freycinetia insignis* | coloured, fleshy, sugary | attractant and reward to fruit-eating bats |
| *Eryngium* spp. | stiff, spiny, sometimes coloured bluish or yellowish (Fig. 4.44) | protection of the inflorescence from grazing; primary attractant |
| Compositae | form an involucre around the capitulum, sometimes spiny | protection of young flowers from extremes of climate and from grazing |
| *Tilia* spp. | adnate to the peduncle, forming a wing | assists dispersal of the fruit by wind, using 'propeller' action |
| *Cornus* spp. | four white, yellow or red bracts closely subtend reduced inflorescences (Fig. 4.45) | form secondary bowl-shaped 'solar furnace' type floral unit |

**Figure 4.44** The prickly blue bracts of *Eryngium alpinum* (Umbelliferae) protect the inflorescence from grazing (Table 4.6) (×0.33).

**Figure 4.45** The white bracts of the pan-arctic dwarf cornel, *Chamaeperiсlymenum* (*Cornus*) *suecica* make an inflorescence of small flowers attractive to insects (Table 4.6) (×1).

## SPECIAL CASES OF POLLINATION SYNDROME

The co-evolution of the inflorescence, and the pollinator, in the flowering plants, is an immensely complex field of natural history which has engendered some of the most charismatic and visually appealing stories in biology. Many remarkable, and often symbiotic associations have developed between flowers and animals, and this chapter could be greatly extended by the description of such mechanisms in detail. However, such relationships are numerically unimportant with respect to pollen flow in most communities, and it is felt better to leave the telling of such stories to the specialist books on pollination biology, such as Proctor and Yeo (1973) and Faegri and van der Pijl (1979).

Some popular syndromes have been touched on earlier in this chapter, such as the sapromyophilous fly-trap flowers of *Arum*, *Aristolochia*, *Cypripedium* and *Nymphaea*, the sexual deceit flowers of *Ophrys*, *Caladenia* and *Oncidium*, perfume collection, water pollination, and provision of food bodies and oil-bodies. However, one interesting class of pollination syndrome has been scarcely mentioned until this point, that of the total symbiotic relationship seen in brood flowers.

A simple, non-obligate brood-flower syndrome has been recounted by Brantjes (1978) for moths in the subfamily Hadeninae, which lay eggs, usually singly, in the ovaries of flowers from which they have fed on nectar. The moth thus benefits at both the adult and larval stages (the larvae eat some but not all of the developing ovules), and the flower is pollinated and produces some seeds. This occurs in various members of the Liliaceae, Amaryllidaceae and Caryophyllaceae. In the familiar white campion, *Silene latifolia*, the moth *Hadena bicruris* visits flowers of both sexes, but only oviposits on female flowers, which it is apparently able to differentiate by a quantitative scent difference. Andrew Stewart and I (unpublished) have found that the frequency of oviposition by *H. bicruris* on red campion, *Silene dioica*, is dependent on both the size and density of campion plants. Most flowers are eaten (and pollinated) when the campion patch is biggest and densest, and the tallest plants, and those with most flowers, suffer the heaviest predation.

A more complex syndrome is observed in the American *Yuccas*, which are visited by a genus of moth (*Tegeticula*, formerly known as *Pronuba*), with only one species, *T. yuccasella* pollinating all the eastern species. The female moth gathers pollen which is shaped into a sticky ball and thrust into the stigmatic tube of another flower; at the same time the moth lays one egg in each cell of the ovary. By carefully ensuring pollination, the moth provides ovules for its larvae. One ovule in each *Yucca* carpel next to the egg grows abnormally large and feeds the larva; the remainder develop normally.

The ultimate brood inflorescence is *Ficus*, the figs, a very large genus of trees and shrubs, many of which are ecologically very important in the tropics. The *Ficus* mechanism is too complex to be dealt with in detail here, and the reader is referred to McLean and Cook (1956), Ramirez (1969), Galil and Eisikowitch (1968, 1969) or Storey (1975). The essence of the story is that the inflorescence is a closed sphere which may contain male, female and neuter flowers. The edible fig, *Ficus carica*, is gynodioecious and may have three types of inflorescence. One, produced by the hermaphrodite in the winter, produces neuter and a few male flowers ('caprifig'). Female chalcid wasps of the genus *Blastophaga* penetrate the sphere and lay eggs in neuter flowers. The resulting offspring hatch in the spring, the male wasps fertilise the females, and then die. The female wasps escape, and enter a second class of inflorescence, produced in the spring, which has female and neuter flowers. The female flowers are fertilised with pollen carried from the male inflorescence, and the wasps lay eggs in the neuter flowers. The next generation of female fertilised wasps seek out a third type of inflorescence with only neuter flowers where the annual cycle is completed. Although some primitive fig populations ('Smyrna figs') require cross pollination for fruit to set, and thus must contain some hermaphrodite 'caprifigs', most cultivated figs ('common type') set fruit without pollination (parthenocarpy), and are grown without males unless male caprifigs are required for breeding purposes. Thus, much fig culture is able to persist in the absence of the *Blastophaga* pollinator.*

## CONCLUSION: POLLINATION AND THE ECOSYSTEM

All flowers are adapted to mediate the transfer of pollen from anthers to stigmas, whether within a flower (autogamously), or between flowers of the same or different genets (allogamously). For seed to set in obligately outbred self-incompatible species, there must be adequate pollen transfer between flowers of different genets (xenogamously).

Flowers also harbour the processes of fertilisation, embryogenesis and seed production, and develop the fruit within which the seed is dispersed. The diversification of Angiosperm flowers has been paralleled by a diversification of flower visitors, thus allowing the following developments:

(a) Many species of plants with biotic and abiotic means of pollen transfer can coexist in a habitat.

* Although many exotic crops can produce fruit in the absence of their original pollinator, this is not always the case. The output of the oil palm, *Elais guineensis*, originally a native of west Africa, has been greatly improved in Malaysia after the introduction of its natural pollinator, the weevil *Elaidobius kamerunicus*.

(b) Energetically efficient means of pollen transfer have evolved.
(c) Pollen transfer can occur in environmentally adverse conditions.
(d) Speciation of related sympatric congeners into differential adaptive noda has occurred through prepollination reproductive isolation of demes.

All habitats are heterogeneous and with respect to seed reproduction, they are importantly heterogeneous in five variables: time, within a season; height above ground; number of species of potential flower visitors; number of individuals of flower visitors; sites for seedling establishment.

Thus, for any species of flowering plant a suite of flower characteristics has evolved which is coadaptive with respect to an equilibrium point of reproductive strategy for a niche based on at least one point amongst each of these five variables. Any one species will have its own characteristic flowering time, flower height above ground, means of donating and receiving pollen, and seed size, seed number and seed dispersal capability, which allows it to maintain this equilibrium point. Furthermore, the theory of competitive exclusion renders it unlikely that any other species in that habitat will inhabit the same equilibrium point on all five criteria.

For coexisting species with similar equilibrium points, there will also exist a threshold level of reproductive efficiency. This threshold will be a function of the proportion of the total energy budget of the plant spent in replacing that plant by seed reproduction. Coexisting species in a habitat may have very different equilibrium points of reproductive strategy, and will thus be able to coexist with very different levels of reproductive efficiency. A small annual on an anthill in a closed grassland is likely to flower early, be dwarf, have unspecialised, mostly autogamous flowers, and have large seeds with poor dispersal. This syndrome will favour an '*r*' type (Ch. 10) reproductive efficiency and more than 50% of its total synthesised energy may be expended on its sole reproductive effort.

In contrast, a neighbouring perennial orchid in closed grassland may flower in mid-summer, be taller, attract and reward only one species of pollinator, which is scarce in the habitat, and which alone can effectively mediate pollination. Thus it may rarely set seed. However, when seeds are set, they are very small and produced in very large quantities. Such a species may spend much less than 10% of its total annual energy budget on seed reproduction. It has a '*K*' type reproductive efficiency, in which much more of its energy is spent on perennation and vegetative multiplication.

However, we may expect other annuals on anthills, and other orchids with similar pollination systems, to show reproductive efficiencies similar to their respective counterparts in this example.

How many different types of reproductive strategy equilibria can coexist in a habitat will, to a great extent, depend on the total biomass productivity of the habitat. Thus, on an Arctic tundra, the flowering season is short and time niches few; exposure to wind and cold renders all plants dwarf, so height niches are almost absent; pollinators are species-poor and few in number, so most flower types are generalist; and requirements for seedling establishment before the winter, or seed survival, are rigorous, so seed and fruit types are limited. Conversely, in the very productive tropics, flowering can occur over 12 months of the year, trees may exceed 80 m in height, the variety and total quantity of potential flower visitors is vast, and seeds may establish on the forest floor, or 50 m up in the canopy. Clearly, there are very many more reproductive niches in the tropical forest, and it is no accident that tropical habitats harbour almost all the more spectacular and arcane pollination syndromes.

Unfortunately, we have very little data to support such energetically based models, and, in particular, our knowledge of the allocation of energy budgets in different plants, and in different habitats, is almost nil. Also, we have very few studies that compare different habitat types with respect to their pollination syndromes. That of Moldenke (1975) examines oligophily to polyphily in plants, and oligotropy to polytropy in pollinators in a wide range of Californian habitats over a considerable altitudinal transect. He found that the spectrum of flower visiting pattern and pollinator receipt pattern varied rather little over a wide range of habitats. For instance, most visitors were moderately oligotropic, visiting from two to five plant species, and extreme monotropy (one host only) and polytropy (more than ten hosts) were surprisingly rare. Extreme monotropy and extreme polytropy were surprisingly more frequent in high-altitude tundra vegetation.

Proctor (1978) divides plants into three broad types of pollination syndrome which he supposes evolved at an early stage in Angiosperm evolution, and between which types plants have been changing ever since. However, he points out that in some important plant families, adaptations have apparently reached a dead end, precluding further between-type evolution. This is of no apparent disadvantage, as almost every habitat will provide different niches (reproductive strategy equilibria) covering between them all three of Proctor's classes (Table 4.7).

Proctor's field survey is made of a number of phytosociological communities on the limestone region of the Burren, western Ireland. Like Moldenke, he finds a wide and not dissimilar spectrum of pollination types is typical of most communities: that is, each bears a diversity of reproductive strategy equilibria, and it is typical of no community of plants to specialise in only one reproductive strategy, or pollination

**Table 4.7** Pollination syndromes and the rôle of pollination in speciation.

| Pollination syndrome | Specialist successful family | Rôle of pollination in speciation | Number of ovules per flower |
|---|---|---|---|
| anemophily | Gramineae | none | 1 |
| generalist zoophily | Compositae | low | 1 |
| specialist zoophily | Orchidaceae | high | many |

syndrome. It is interesting that the more productive, more stable communities have slightly higher proportions of specialist flowers; this might well be predicted as they will have more reproductive niches available. One might also predict that a higher proportion of plant species in productive communities might have a low reproductive efficiency threshold ('*K*' strategy, Ch. 10) but, unfortunately, we do not have this information.

Neither Moldenke nor Proctor give absolute values for the number of species present in each community. One would predict that the more productive, more stable, communities would carry more species in each pollination syndrome and altogether. The diversity of reproductive niches or equilibrium points available in a habitat should be a major component in the floristic richness found in that habitat.

CHAPTER FIVE

# *Pollination biology and gene flow*

## INTRODUCTION

In the preceding chapter, emphasis was placed on the diversification of the Angiosperm flower from its Magnolian base. I attempted to show how various floral features have become adapted to different means of pollen transport, and how floral diversification has encouraged speciation, and floristic richness, within a habitat.

However, no attempt was made there to describe patterns of pollen flow within and between plants, nor the efficiency of various pollination mechanisms in dispersing pollen and achieving seed-set. This information is clearly of great importance, and is indeed more central to the theme of this book than mere descriptions of coadaptations of pollinating agents and flowers. The influence of the behaviour of pollinators on pollen flow, and the ways that plants have adapted to the behavioural characteristics of pollinators, can be said to play an overriding rôle with respect to patterns of gene dispersal, and the genetic structure of plant populations.

In contrast, abiotic pollination by wind and water is relatively unresponsive to subtle adaptation, and patterns of gene flow in wind-pollinated species depend to a great extent on chance. Some consideration will, however, be given to the distances over which pollen can disperse by wind.

In order to understand the coadaptation of zoophilous plants to pollinators with respect to gene flow, it is first necessary to give an account of pollinator behaviour and foraging strategies of flower visitors. Subsequent sections of this chapter will examine pollen travel within and between plants, the efficiency of pollen travel, neighbourhood, interruptions to pollen flow, and the rôle of ethological isolation in plant speciation.

## FORAGING THEORY

The foraging strategy of pollinators has been modelled by Pyke (1978a,b,c, 1979, 1980a,b, 1982a), Pyke *et al.* (1977) and Krebs (1978) and is based on food-gathering and predator–prey models devised by Royama (1971), Lawton (1973), Schoener (1969) and others. There is a useful review by Waddington (1983). Flower visitors gather food at the first trophic level, which is usually the most energetically efficient. It is normal to find that predators at higher trophic levels have a lower efficiency of food gathering. More time and energy is expended by these in holding territory, and in the locating, gathering, processing, eating or transporting of prey items. Data on energetic costs involved in flower foraging have been published by Wolf and Hainsworth (1971), Wolf *et al.* (1972), Stiles (1975), Heinrich (1972a,b, 1975, 1979b), Waddington (1981), Waddington and Heinrich (1981), Waddington *et al.* (1981), Best and Bierzychudek (1982) and Zimmerman (1979) among others.

At the most simple level, foraging theory predicts that animals will visit flowers in the most energetically efficient manner. Energy costs in foraging will be composed of:

(a) energy expenditure per distance travelled;
(b) mean distance travelled per flower visit;
(c) energy expenditure in maintaining body temperature, a function of the difference in temperature between ambient and body temperature, of wind speed, and of travel speed.

Energy gains or savings are considered as:

(a) energy reward per flower visited (from nectar for suckers, and from pollen or food bodies for chewers);
(b) energy savings made with respect to heat loss in solar furnace flowers, and in stasis, absence of predator evasion, reduced need to seek mates etc. provided by the protection of the flower.

Brood feeders (bees, birds) may also gain reward for non-feeding relatives, which is taken back to the hive or nest.

## FORAGING STRATEGIES

### *Social bees*

Foraging efficiency can thus be expressed in terms of profit, as a ratio of calories gained to calories expended. It can be predicted that flower visitors will cease to visit a patch of flowers when other more profitable sources exist nearby, and will cease to fly at all when all available sources are no longer profitable, as on cold days early or late in the season. Some important flower visitors, such as colonial bees ‘learn’ the position and characters of a rewarding patch, using information transmitted by experienced members of the hive or nest. Scent collection, and bee dances teach young foraging bees which major (a North American word for graduate) on currently rewarding patches (Heinrich 1976, 1979c). They are probably only able to remajor onto different sources on a few occasions, which may limit their foraging efficiency. Having majored on a flower type, bees increase their efficiency (rate of feeding) on that type with experience (Heinrich 1976, 1979c). Within a hive, different individual workers may have majored on different flower types, and thus relearning may involve other workers that have majored on other patches. It is expected that patches worked from a given hive would allow similar levels of net energy gain to the worker. Patches that drop below this threshold due to overworking or cessation of flowering would no longer be worked.

**Quality and quantity of reward** Thus, as has been suggested by Waddington and Heinrich (1981), colonial bees with workers have a flower-visiting strategy composed of the following actions:

(a) A novice bee will sample a number of different flower types, in addition to receiving messages about local rewarding flower patches.
(b) From this information, it will 'major' on a rewarding flower type, which will be the most rewarding one encountered in terms of energy received : energy spent. This flower will be common, clumped, in full flower, with good pollen and nectar production, and will be conspicuous and accessible to bees. It may not be the most potentially rewarding species present, which may already be heavily worked by other individuals, thus decreasing its attractiveness.
(c) Its complex behaviour pattern with respect to flower visiting will then depend on rewards subsequently received, as detailed in Table 5.1.

There is a good deal of evidence in the literature to show that bees do indeed respond in this complex way, and thus the quality of reward presentation, and the quantity of reward received, in an environmental context, is likely to be extremely important in determining patterns of pollen flow, and hence of gene flow (Waddington *et al.* 1981, Schmitt 1980, Zimmerman 1979, Thomson 1982). To take one example in detail, Waddington (1981) has investigated the behavioural responses of *Bombus americanus* to differences in nectar volume in wild populations of *Delphinium virescens*. He shows that nectar volume per flower decreases from the bottom flower of the raceme upwards, a common pattern also observed for instance in the foxglove, *Digitalis purpurea* (Best & Bierzychudek 1982). Visitors generally visit the bottom-most flower first, moving upwards, and leaving the spike when rewards per flower drop below the mean flower reward for the population. Racemes with large nectar volumes in the bottom-most flowers also have relatively large

**Table 5.1** Foraging strategies amongst workers of colonial bees.

| Components of foraging strategy | Strategy on major flower | | | |
|---|---|---|---|---|
| | Rewards high | Rewards intermittent | One reward poor or absent | Rewards poor |
| 1. number of flower species visited per trip, and frequency and distribution of visits to each | one species (oligotropic) | one (major) species to few (minor) species, majoring on one | two to few species, regular alternation between pollen flowers and nectar flowers | remajor to new species |
| 2. distance of hive or nest to foraging patch | near to distant | more distant | nearer | — |
| 3. number of inflorescences visited per foraging trip | few | more | more | many |
| 4. distance of flights between inflorescences | short | longer | longer | long |
| 5. directionality of flights between inflorescences | random | approaching 180° | variable | approaching 180° |
| 6. time spent at a flower | short | variable | short | long |
| 7. proportion of flowers in inflorescence visited | high | lower | high or low | low |
| 8. direction of movement within inflorescence | unidirectional from most rewarding flower, not skipping | more skipping | more skipping | random |
| 9. proportion of time spent on inflorescence on feeding | low | higher | variable | high |

**Table 5.2** Behaviour of bees with respect to nectar volume of the bottom-most flower of a raceme in *Delphinium* and *Digitalis*.

| Characteristic studied | Nectar volume Large | Small |
|---|---|---|
| quantity of pollen collected per flower visit | large | small |
| number of flowers visited per inflorescence | many | few |
| distance travelled between inflorescences | short | long |

volumes in other flowers, and there is a strong correlation between the nectar volume of a flower and the amount of available pollen in it. This probably depends on both the age of the flower, the number of visits it has received, and on the physiological vigour of the plant.

Behaviour of the bee with respect to nectar volume of the bottom-most flower is summarised in Table 5.2.

This shows that inflorescences producing a lot of nectar will receive a good deal of geitonogamous pollination due to pollen carryover between flowers of the same inflorescence (5% of pollen received from a flower is received by the sixth subsequent flower on the same inflorescence visited; the value is 44% for the first subsequent flower visited). At the same time, substantial quantities of pollen will be carried between inflorescences, but most will only travel short distances.

In contrast, inflorescences producing relatively little nectar will donate and receive relatively little pollen. Geitonogamy will be scarce and outcrossed pollen will travel further on average than for high nectar production.

It is very possible that within a flower-visitor syndrome (bee visiting in the present case) there is an optimum production of nectar per flower, on which act the constraints of energetic effort, self (geitonogamous) pollination, ovule fertilisation and restriction of pollen travel. The production of either too much or too little nectar per flower would be disadvantageous. It remains to be seen whether variation in nectar production is heritable. If so, co-evolution of differing pollination strategies based on different levels of nectar production may have occurred around this optimal point, perhaps in response to the varying demands of different seasons. Such broadening of the pollen-visiting niche might alleviate fluctuating seasonal crises in seed-set, on the one hand, and pollen travel (effective population number) on the other. Unfortunately, however, we have no information on the heritability of nectar production, and such mixed strategies may be only fortuitous.

In both *Delphinium virescens* and *Digitalis purpurea*, flowers are protandrous, and open sequentially from the bottom upwards (Fig. 5.1). Thus, the lower flowers are female and the upper ones male. Increasing nectar production during the life of a flower ensures that a bee moves upwards on an inflorescence, and will not usually leave an inflorescence until it reaches a flower in the male phase. It will therefore minimise geitonogamy, and carry pollen to the first flower of the next raceme visited. As it is advantageous for it to visit the most productive flower first, it will almost always visit a flower in the female phase first, thus

**Figure 5.1** The racemose inflorescence of the common foxglove, *Digitalis purpurea*. Flowers open sequentially upwards, and are protandrous. Usually about 10 are open together, the lower in a female phase, and the upper in a male phase. The calorific value of nectar reward increases downwards, so bees visit lower flowers first and work upwards. In doing so, they pass from flowers in a female phase to flowers in a male phase, and so maximise cross-pollination between inflorescences in this self-compatible biennial (×0.5).

promoting cross pollination (*Digitalis*, at least, is self-compatible, so that 'fail-safe' geitonogamy mediated between flowers of the same inflorescence will help seed-set in poor flying conditions, or for isolated plants; dichogamy is total, so that within-flower fertilisation is not possible).

Such detailed field studies can be complemented by laboratory work, as Waddington *et al.* (1981) have done for *Bombus edwardsii*, using artificial flowers with rewards of sugar solutions of known volume and concentration. This study reached the following conclusions:

(a) Better rewards encourage more rapid flower visitation (with more flowers visited, but less time spent at each), and more activity generally.
(b) More reliable rewards (high constancy of reward) are preferred (using artificial flower colours) to less reliable rewards, even if the total reward from the two types is equal. More reliable rewards are considered to reinforce a behaviour pattern more frequently.

**Directionality of flight** Directionality of between-inflorescence flights has often been observed. Mogford (1974), working with the marsh thistle, *Cirsium palustre*, distinguished between sequential and overall directionality, although these concepts seem to have been overlooked in later discussions. Woodell (1978), who examined the direction of bee flights on salt-marsh populations of thrift (*Armeria maritima*) and sea-lavender (*Limonium vulgare*), considered that strong sequential directionality in flights was merely a symptom of overall directionality imposed on the bee by almost unidirectional winds in this coastal locality (Scolt Head Island, Norfolk, UK). When the wind is strong bees prefer to fly, and especially to land, into the wind. Overall directionality might also be imposed by the topography of a patch, or habitat (e.g road verges, wood margins and cliff edges), by the position of the hive or nest with respect to the patch, or even by the position of the sun.

In contrast, Pyke (1978a) has considered sequential directionality to be a direct product of foraging strategy. He examined sequential flight directions of *Bombus flavifrons* working *Aconitum columbianum*, and *B. appositus* working *Delphinium nelsonii*, and in both cases found unimodal sequential directionality, with modal peaks very close to 0° (= 180° from direction of arrival) in each instance. He suggests that such patterns optimise foraging success, as they minimise the chance of a bee revisiting an inflorescence. However, overall directionality and environmental constraints were not considered.

Zimmerman (1979), also working with *Bombus flavifrons*, but visiting a Jacob's ladder, *Polemonium foliosissimum*, found no directionality in sequential inflorescence visiting, only totally random movements. This he ascribes to the very different pattern of flower presentation in this species. The mean number of flowers per plant was 90, and the mean number of plants per patch 193. Only an average of 3.3% of flowers per patch are visited on each foraging journey, and thus the chance of a bee encountering a flower it has already worked is much lower. Because the species grows in a highly aggregated manner, directional foraging is inefficient: the visitor would rapidly leave the patch.

Clearly, both sequential unidirectionality (optimal foraging) and overall directionality (environmental constraints) may occur, and in some localities both may operate together. Thus the bee may forage unidirectionally and the environment may determine in which direction this is. In either case, unidirectional between-inflorescence flights are likely to maximise pollen travel, and increase the number of individuals between which gene exchange occurs.

**Majoring patterns with time and patch 'switching'** Initial choice of a patch on which to 'major' will involve consideration of the profitability of a patch, as has already been discussed. As patches are only likely to be adopted as others become less profitable, they may receive few visits early in their flowering period, but remain popular right up to the time that flowering ceases. Thomson (1982) examined patterns of flower visitation to entomophilous flowers in sub-alpine meadows in the Rocky Mountains of Colorado, and described five patterns of visitation to a patch with time. The most frequent pattern showed low visitation early in the flowering of a patch, but high visitation right up to the end of flowering. He found no consistent relationship between the degree of overlap in flowering time of species with similar pollinators and visit frequency. For the monoecious cucumber, *Cucumis sativus*, Handel and Mishkin (1985) showed that, although pollinator visits remain high to the end of the flowering season, fruit-set declines considerably for late-season flowers. This was attributed to competition for resource by earlier set fruits causing fruit-set failure in later flowers.

It seems probable that, for dominant species, flowering time overlap can decrease insect visitation and thus be detrimental, but for less dominant or codominant species, flowering-time overlap can mutually enhance the attractiveness of similar patches through Müllerian mimicry. In such entomophilous species-rich communities, bees may thus major on cohorts of coadapted species. In some cases, similar species within the cohort may perform different functions for the bee, i.e. nectar fuelling and pollen gathering. Proctor (1978) and Proctor and Yeo (1973) have suggested that such cohorts of unrelated Müllerian mimics may be common especially among relatively unspecialised flowers of temperate grasslands. It is even conceivable that frequency dependence may act with respect to flowering-time overlap of mimics. Dominants might suffer from competition with minority species for pollinators and might eventually become sufficiently scarce, through inadequate legitimate pollination, that they would benefit from sharing pollinators with other codominants. In such circumstances they might be able to become dominant again. However, there is as yet little evidence to support such a tentative model.

Recent work by Macior (1983), also working in sub-alpine North American communities, in Montana, throws an interesting light on the effect of the majoring habit of bees on pollination efficiency. He concludes that the more frequent a bee species is in a habitat, the more different species of flower that species of bee will visit. That is, an abundant bee species will need to major on a greater diversity of patch types if it is to exploit the community successfully. However, oligophilic flowers, which receive visits from only one or a very few insect species, benefit from just as many visits in total as polyphilic (generalist) flowers that receive visits from many different species of insect. A consequence of the majoring habit is that floral specialisation does not disadvantage any plant species. Indeed, Thomson (1982) has suggested that increased flower-time overlap between similar flowers will increase pollinator visit frequency, but decrease seed-set, due to blockage of stigmatic sites with ineffective foreign pollen. There will be a pay-off threshold in specialisation between high-visit low-accuracy generalisers and low-visit high-accuracy specialisers. But in a pollinator-rich community both types of flower may be able to coexist, as Macior has shown.

**Flight height** One of the many features of foraging behaviour that enables a bee to remain faithful to its majoring patch is flight height, and foraging bees fly at a remarkably constant height. A species of plant may adopt an inflorescence height different from those of coexisting plants with similar syndromes. Waddington (1979) shows that species in Colorado meadows that share pollinators are less alike in inflorescence height than those that do not share pollinators. However, once again a 'pay-off' between too little and too much specialisation may operate in spatial niches such as this, as has been discussed for flowering-time niches. And once again, the greater the number of pollinators, the greater variety of generalist and specialist types that are likely to be able to coexist.

### *Butterflies and moths*

Up to this point, only the foraging strategies of colonial bees have been discussed, partly because these are most thoroughly known. However, generalist flowers may receive visits from other classes of pollinators as well as bees. Schmitt (1980) elegantly demonstrates the different flight patterns of foraging bees and butterflies, and the very different patterns

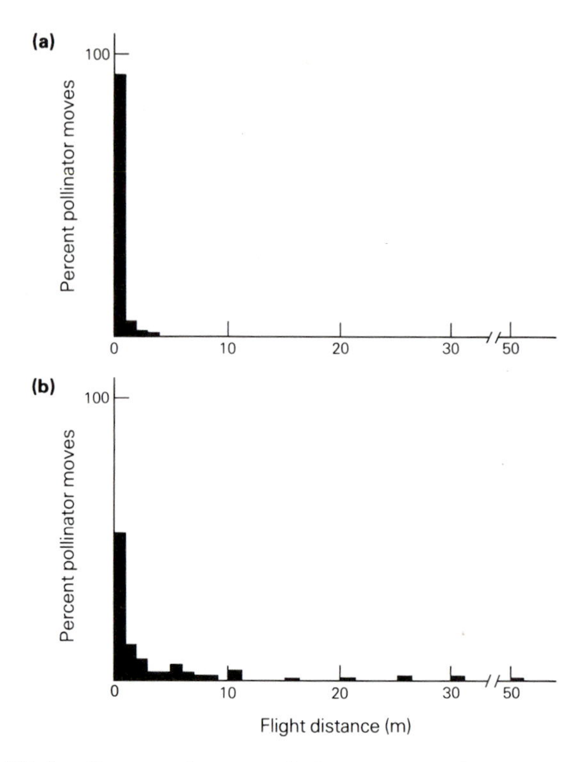


**Figure 5.2** Flight distances between inflorescences of three North American species of *Senecio* made by visiting bees (a) and butterflies (b). After Schmitt (1980).

of pollen flow that result, on three North American ragworts, *Senecio amplectens*, *S. crassulus* and *S. interrigimus* (Fig. 5.2). These are efficiently and effectively visited by both long-tongued bees and butterflies of many species. To quote Schmitt:

> 'the individual fitness of a butterfly depends not only on nectar uptake, but also on mating success, and for females, oviposition success. In a patchy environment, butterfly fitness may be maximised by flying longer interplant distances so as to maximise the probability of encounters with mates, or with larval foodplants. Levin and Kerster (1969) have shown that a strong relationship exists between plant spacing means and bee flight distance means. Beattie (1976) found a similar relationship for solitary bees, but not for lepidopterans, observed foraging on *Viola*. One can therefore expect differences in foraging behaviour on the same flower resource between bumblebees and butterflies that could be reflected in plant neighbourhood characteristics.'

Schmitt goes on to show that effective population sizes, expressed as neighbourhood sizes, differ by huge factors of between 50 and 500, depending on whether the visitor is a bee or a butterfly. It is also worth noting that butterflies thermoregulate by basking, whereas bees and syrphids require the expenditure of energy for thermal control. Thus butterflies are less dependent than bees on regular fuelling stops at flowers.

Anyone who, like me, has chased foraging butterflies for purposes of capture or photography, will indeed testify to the maddeningly indecisive, long-distance and essentially unidirectional flight patterns that they show (Baker 1969). Perhaps for this reason, butterflies have been very little studied as foragers and pollen dispersants. Yet they are undoubtedly very important visitors to the flowers of late summer with inflorescence heads of many narrow-tubed flowers (e.g. thistles and knapweeds, *Cirsium, Carduus, Carlina, Centaurea*, scabious, *Scabiosa* and *Knautia*, valerians, *Valeriana* and *Kentranthus*). Although these flower types are also visited by bees, syrphids and beetles, I estimated that ten times as many butterflies as bumblebees were visiting these flowers in a meadow in Auvergne, France in August 1983.

Relatively little is also known about foraging strategies of night-flying moths. Brantjes (1978) recounts how slow, aimless flights are replaced by much more urgent 'seeking flights' ('Nahrungflug') on the detection of odour. (In the dark, foraging strategies based on sight cannot be employed except at close range.) The moth then orientates along a scent gradient upwind until it makes visual contact with an inflorescence, from which it feeds. Flights between inflorescences are therefore likely to be relatively distant and random in direction. In common with other poikilothermic ('cold-blooded') foragers, moths are highly influenced by the ambient temperature, and as they fly at night will more frequently encounter low temperatures than day-flying congeners. Cruden *et al.* (1976) have shown that in Mexico and the western USA hawk-moth (sphingid) activity is severely limited by temperatures below 15°C at sundown (Fig. 5.3). They demonstrated that the proportion of flowers visited by moths in several moth-visited genera varies strikingly with altitude (Fig. 5.4), and that at high altitudes seed-set is severely impaired. They suggest that the primary determinant of altitudinal limit for moth-visited flowers in these genera is moth activity.

However, sphingid flowers such as *Aquilegia pubescens* and *Castilleja sessiliflora* occur at considerable elevations in the Californian Sierra Nevada. These produce nectar by day, in contrast with the Mexican moth flowers, and are visited by the hawk-moth, *Hyles lineata*, at elevations of over 3500 m during the (warm) day. It is suggested that this syndrome has evolved in the absence of hummingbirds which would otherwise compete for tube flowers during the day, but not the night. Being

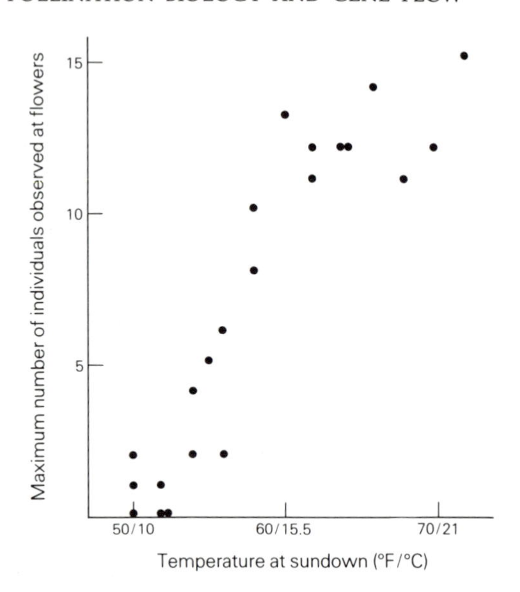


**Figure 5.3** Maximum numbers of hawk moths (sphingids) foraging at flowers of *Oenothera caespitosa* in the western USA at dusk at different temperatures. After Cruden *et al.* (1976).

homoiothermic ('warm-blooded') birds are more efficient flower visitors in the uncertain climates of high altitudes.

## *Birds*

A great deal has been learnt about hummingbird and other flower-bird foraging behaviour. In common with long-tongued bumblebees and lepidopterans, birds prefer to feed from long-tubed flowers (Ch. 4). A longer proboscis, or beak, renders nectar extraction from a single tubed or spurred flower more efficient, but also lengthens the time interval between probes of different flowers (Hainworth & Wolf 1972, Inouye 1980, Pyke 1982a). This relationship has been substantiated for *Bombus* foraging on *Delphinium* and *Aconitum*, and for hummingbirds feeding on artificial flowers. However, field data on the time taken at flowers by hummingbirds with different beak lengths is contradictory (Wolf *et al.* 1972, Stiles 1975). Nevertheless, it is likely that long-beaked birds would be best served by specialising on long-tubed flowers, which cannot be exploited by birds with shorter beaks, and therefore have relatively untapped resources. Short-beaked birds are perforce required to specialise on short-tubed flowers.

Although relationships between insect body sizes and flower features seem more often to be mechanical, as in the ability of bees to trip the flowers of brooms (Leguminosae tribe Genistinae, Ch. 4, p. 76), there is a well-established relationship between body weight and the calorific reward obtained per flower by hummingbirds. Although hummingbird body weight varies tenfold between different species (from 2.7 g to 20 g), stabilising selection for body weight seems to have occurred, based on the trade-off between energy requirements and aggressive defence of flower patches. Larger species are more likely to succeed in defence of a patch against smaller species of birds or insects, but will need more energy to do so. Plants adapted to large bird species may thus produce more flowers with greater rewards per flower in larger patches to make this defence worthwhile. Such strategies towards oligophily, which are expensive to the flower as well as the bird, may only succeed when increase in resource spent on flower quantity and quality is matched by superior reproductive performance. Oligotrophy by birds may, or may not, result in greater quantities of pollen carried more accurately, and over longer distances than is the case for more frugal flowers visited by smaller and less specialised species. Such expensive specialist strategies are also only likely to succeed when the diversity of pollinators is high. Feinsinger *et al.* (1982) and Feinsinger and Swarm (1982) show that birds

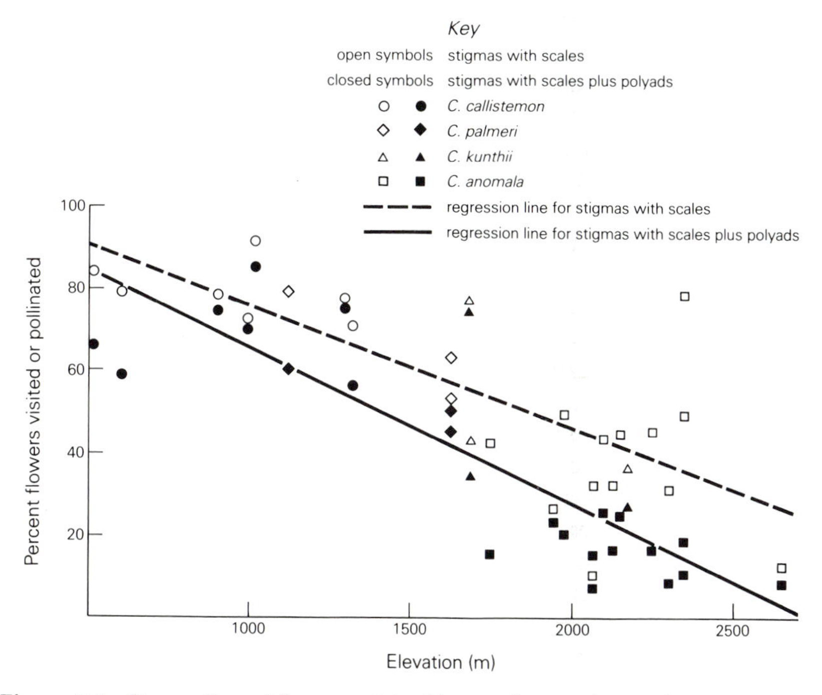


**Figure 5.4** Proportion of flowers visited by moths on plants of various species of *Calliandra* (Leguminosae) at different altitudes in Mexico. After Cruden *et al.* (1976).

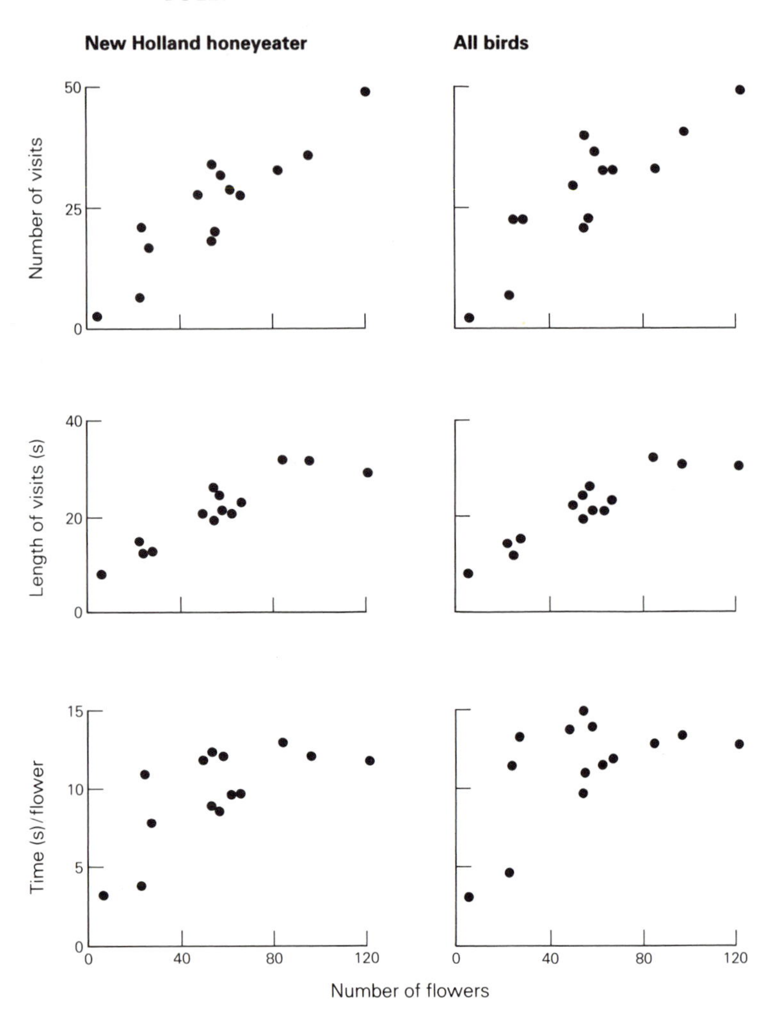


**Figure 5.5** Relationship between foraging behaviour of the New Holland honey-eater (left), and all visiting birds (right), and number of flowers open per plant of the gum *Eucalyptus cosmophylla* (Myrtaceae) in Cleland, Australia (after Paton 1982).

on small, species-poor Caribbean islands are more generalist (polytropic).

There may also be intraspecific competition between plants for visits, based on resource allocation. Pyke (1982b), for the Australian warahtah (*Telopea speciosissima*) and Paton (1982) for three other species of Australian bird flowers (Fig. 5.5) have shown that birds preferentially visit inflorescences with more flowers, and those in which the individual flowers have greater rewards. They also spend a longer time at these more productive inflorescences. However, there is no indication that such enhanced pollinator activity increases fruit-set; interestingly, Pyke (1982b) shows that better fruit-set is solely a function of inflorescence size, and hence presumably of available resource, in *Teleopea*. Artificially enhanced cross pollination did not increase fruit-set in open-pollinated plants. There is also no indication whether inflorescence size is heritable in these cases. If the size of the inflorescence is totally under environmental control, which might well be the case, it could not respond to selection pressures for optimal inflorescence size. In any case, there must be an equilibrium point beyond which increased inflorescence size leads to disadvantageously high reproductive loads (Ch. 2).

Although some hummingbirds may weigh as little as 2.5 g, these are amongst the smallest of all vertebrates. The relationship of body surface area to mass is such that very small homoiothermic individuals experience difficulty in stabilising body temperature. In contrast, invertebrates, which lack the lungs and efficient oxygen-carrying vascularisation of terrestrial vertebrates, are limited by gaseous diffusion to small body sizes, and rarely grow as large as 2.5 g. Vertebrate flower visitors, whether birds, bats, small rodents or marsupials, are almost always heavier than invertebrate flower visitors. Being homoiothermic, they require additional energy to maintain high body temperatures. Thus, vertebrate flower visiting is almost restricted to tropical and subtropical areas where high ambient temperatures, long flowering seasons, and highly productive plants render it feasible.

## TIME-NICHE STRATEGIES

By extending the feeding season of pollinators, it is of advantage to zoophilous flowers to fill as many time-niches in the community as ambient temperatures allow. Thus, in the non-seasonal tropics, different plants flower at different times, so that rewards for a class of pollinators are available at all times of year. Near-yearly time clocks (which may, however, vary from 9-month to 16-month cycles) for the flowering of each species ensure that this is so. Gentry (1974, 1976) has shown that of the 200 or more neotropical species of Bignoniaceae, most of which are lianes, not more than 20 usually coexist. In any one area there are usually about 50% red bird-visited species, and 50% yellowish butterfly- or moth-visited species. For each class, no two species has the same flowering time, there being about 10 time-niches in the year.

Even in tropical systems with a dry season, some species flower during the dry season, and may produce more copious and less concentrated nectar to provide water as well as fuel for visitors. Some may flower when they have shed their leaves (*Cochlospermum*, *Erythrina*) and quite often

**Table 5.3** The timing of flower production in tropical forest systems.

| Strategy | Flower production per plant per day | Flowering season | Chief pollinators | Species density | Distance of pollen travel | Position in forest | Species diversity |
|---|---|---|---|---|---|---|---|
| 'steady-state' | 1 or few | long | bee, butterfly | high | short, accurate | low (floor) | low |
| 'big-bang' | most together ('cornucopia') | short | bird, bat | low | long, accurate (mostly selfed) | high (canopy) | high |
| 'multiple-bang' | cornucopia simultaneously in several species (Müllerian mimicry) | short | bird, bat | low | variable, inaccurate (mostly selfed) | high (canopy) | high |

flowers are borne directly on to bare trunks or twigs (cauliflory) as in *Couroupita*. Specialist pollinators are more likely than generalist pollinators to need a succession of flower species in different time niches. Unlike the specialist hummingbirds, which are largely restricted to neotropical regions, the honey-eaters of seasonal Australia are insect eaters as well as flower feeders, and eat insects for about half the year. In colder seasonal climates, where vertebrate pollinators do not occur, specialist flower feeders may only have very short emergence periods to cope with the relative lack of time-niches within the season. Many solitary bees have quite short emergence periods as adult flower feeders, as do syrphids, wasps, butterflies and moths. Even social bees only produce workers for part of the year, and queens are more generalised feeders, specialising on different types of flower from the workers and hibernating during the winter.

In tropical systems, it is not surprising to discover a marked inverse correlation between the length of time that a species is in flower, and the number of flowers that are produced at any one time. Gentry (1974, 1976), working with neotropical Bignoniaceae has described markedly different strategies with respect to the timing of flower production, and Bawa (1981) has described how these may influence patterns of speciation and diversity in different levels of the forest (Table 5.3).

Foraging strategies of the pollinator will depend on spatial behaviour patterns. Linhart (1973) has shown that territorial hummingbirds in Costa Rica (*Amazalia* spp.) favour feeding on species of *Heliconia* which grow aggregated together in clumps, with a 'big-bang' phenology, on the forest edge. Non-territorial hummingbirds (*Glaucis hirsuta*, *Threnetes ruckeri*, *Phaethornis* spp.) range widely, and preferentially visit 'steady-state' *Heliconias*, with spaced individuals each producing one flower per day (Fig. 5.6). In these, pollen travel will at times be distant, although neighbourhood sizes might still be low. Other examples of distant travel between steady-state strategy plants are found in 'traplining' euglossine bees with specialised modes of scent collection (Dressler 1968, Williams and Dodson 1972). As a rule, however, pollinators can be expected to travel less far between steady-state plants with poorly rewarding patches, than between big-bang flowers with highly rewarding patches. This pattern contrasts markedly with that of temperate plants such as *Delphinium*, in which pollen travel usually varies inversely with quality of reward (Waddington 1981).

Thus, big-bang flowers will mediate distant pollen travel, and can grow at lower densities and higher diversities in the forest. However, they will suffer from very high levels of geitonogamous pollination, and in the case of 'multiple bang' strategies in which several similar, although often unrelated, species flower together, much crossed pollen may be wasted on a non-receptive species.

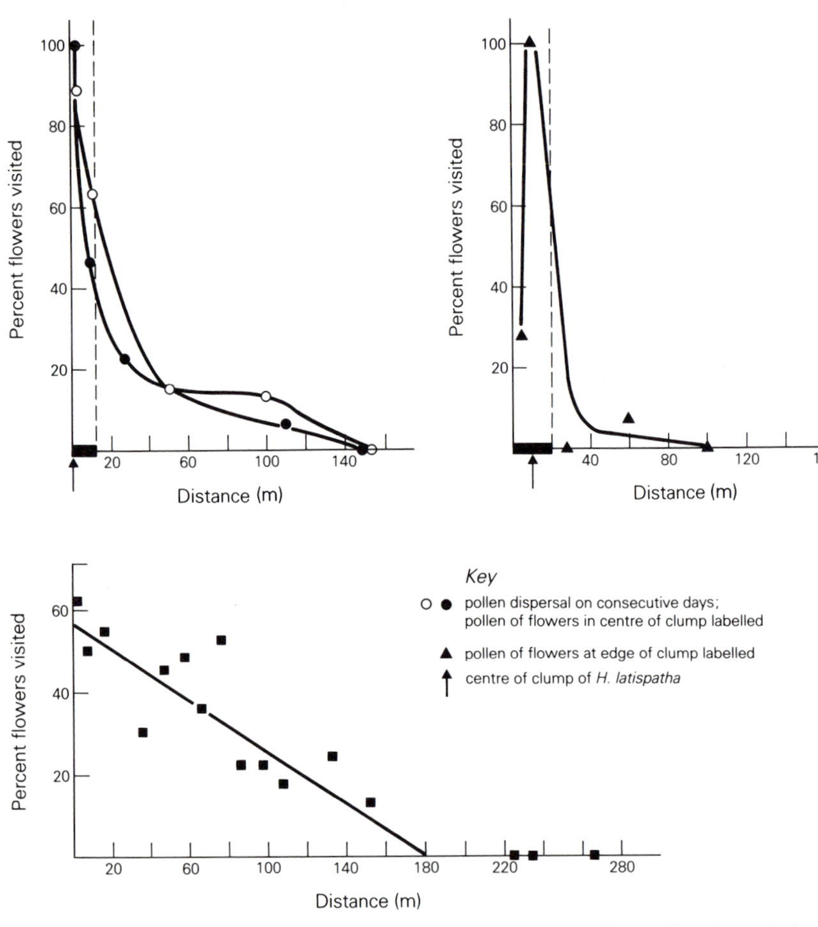


**Figure 5.6** Pollen movements in the hummingbird-pollinated *Heliconia latispatha* (top) (Musaceae) which has a clumped distribution, and *H. acuminata* (bottom) which has a scattered distribution, in Costa Rica (after Linhart 1973).

Many big-bang species are likely to be self-incompatible, to minimise unfavourable selfed fruit-set from geitonogamy. In a stable forest, opportunities for seedling establishment will be few, and a premium will be placed on the quality rather than the quantity of fruit-set. If self-compatible, crossed fruit may outcompete fruits of selfed origin.

## POLLEN TRAVEL WITHIN AND BETWEEN PLANTS

### *Techniques*

To understand microevolution, niche width, populational subdivision, evolutionary potential and the plant breeding potential of plant populations, it is necessary to investigate the genetic structure of those populations. This genetic structure is rigidly controlled by the breeding system of the plants, and a major factor in the nature of the breeding system is pollen travel. It is not possible to undertake serious field studies of plant populational genetics without an understanding of the patterns of pollen flow:

(a) What proportion of the pollen reaches a conspecific stigma.
(b) What proportion of the pollen reaching a conspecific stigma does so on another individual.
(c) How far does crossed pollen travel onto stigmas of other individuals (neighbourhood area), and how many individuals can it encounter (neighbourhood size, or effective population number).

For such estimates to be made, it is essential that pollen be marked, or be otherwise identifiable in some way. Handel (1983) gives a useful review of techniques used in studying pollen travel. Many studies of pollen travel are based largely or entirely on observations of the foraging animal, which can be a very inaccurate and misleading technique. Furthermore, no information can be gathered for wind-pollinated plants in this way; however, the travel of anemophilous pollen can be estimated by the use of pollen traps.

A simple but biased method of examining entomophilous pollen travel between plants is by restricting pollen sources to one plant, or one inflorescence, and subsequently examining quantities of pollen on stigmas, or seed-sets, on plants at various distances from the source (Table 5.4).

Unfortunately, all such techniques introduce biases, or artificial population structures, which give results of dubious value to investigations of pollen travel in natural populations. For instance, for dioecious species it is relatively easy to remove all males, or all male flowers from a population, except for the experimental pollen source. If, however, the pollinator is visiting the males of this species in order to collect pollen, it will forage far less thoroughly in an almost all-female population. Very important and scarce information can, however, be gleaned from such techniques concerning the proportions of between-flower pollen travel that are geitonogamous, and on amounts of pollen carry-over between flowers (Thomson & Plowright 1980). It is important to realise that results obtained may not accurately reflect pollen flow in unmanipulated situations.

More recently, techniques using markers have been introduced, and these fall into two distinct categories, which measure different, although related, phenomena:

(a) External marking of pollen grains, using dyes, stains or fine powders. These measure pollen travel onto stigmas or flowers.

**Table 5.4** Techniques for estimating pollen travel between entomophilous plants by examining foreign pollen on stigmas, or seed-set at various distances from a known pollen source.

| System | Manipulation | Recording | References |
|---|---|---|---|
| dioecy | remove male plants or male flowers, except for source | stigmas or seed | none? |
| interspecific pollination (pollen specifically recognisable) | remove flowers of one species, except for source | stigmas | Barons (1938), Crane and Mather (1943), Pope *et al.* (1944), Levin and Kerster (1968), Levin (1972) |
| heteromorphy, with heteromorphic pollen | remove flowers or plants of one morph, or use 'minimum distance' technique | stigmas or seed | Richards and Ibrahim (1978), Ibrahim (1979) |
| polymorphism in pollen colour | use source of single plant with aberrant pollen colour | stigmas | Thomson and Plowright (1980) |
| self-incompatibility | use single cross-compatible source with self-incompatible recipients (e.g. fruit crops) | seed | Wolfe *et al.* (1934), Roberts (1945), Bateman (1947), Free (1962) |
| protogyny | use single male-phase source with emasculated, clean-stigma female-phase recipients | stigmas | Thomson and Plowright (1980) |

**Table 5.5** Techniques for estimating pollen travel, or gene flow, between entomophilous plants using (a) artificial pollen marking or (b) dominant marker genes.

(a) Pollen marking

| Techniques | Species | References |
|---|---|---|
| methylene blue powder | *Gossympium arboreum, hirsutum* (cotton) | Stephens and Finker (1953) |
| basic fuchsin in 95% ethanol | *Delphinium virescens* (*Bombus americana*) | Waddington (1981) |
| Evans blue, neutral red, Bismarck brown | *Heliconia* species (hummingbirds) | Linhart (1973) |
| 'fluorescent pollen dye' | *Senecio* species | Schmitt (1980) |
| 'fluorescent powder' ? acriflavine, or quinacrine mustard | many species | Bawa (1981, 1983), Waser and Price (1982) |

(b) Genetic markers

| Species | Gene | References |
|---|---|---|
| *Cucumis sativus* | 'bitter' | Handel (1983) |
| *Cucumis melo* | green cotyledons (yellow cotyledons) | Handel (1982) |
| *Lupinus texensis* | allozyme *Pgi*-1 | Schaal (1980) |
| *Primula vulgaris, veris* | red flowers (yellow) | Ibrahim (1979) |
| *Gilia*? species | blue flowers (white) | Epling and Dobzhansky (1942) |
| *Gossypium* species | red cotton (white) | Stephens and Finker (1953) |
| *Raphanus sativus* and wind-pollinated species | ? | Williams (1964) |
| *Lolium perenne* (wind pollinated) | red tiller-base (white) | Griffiths (1950), Gleaves (1973) |

(b) Use of dominant genetic markers on a known pollen source, which can be detected in offspring of potential recipients, either readily (flower colour, leaf colour, taste) or by genetic analysis by isozymes. These measure patterns of gene flow between generations.

Neither of these techniques is ideal, and each involves certain types of potential bias, such as detection by pollinator of pollen marker, introduction into a population of abnormal genotype, or relaxation of selection upon offspring genotypes. The one used will depend on the exact question being asked by the investigator, and the availability of suitable floral types or suitable genetic markers (Table 5.5).

### *Pollen reaching a conspecific stigma (pollination efficiency)*

Pollination efficiency can be considered either as a function of the proportion of pollen grains produced per flower, which reach a stigma of the same species of plant, or as a function of the number of conspecific grains which reach a stigma in relation to the number of ovules to be fertilised via that stigma. As we have seen in Chapter 2, the number of pollen grains produced per ovule to be fertilised varies adaptively with the ease with which they can reach a stigma. Thus, autogamous species (Ch. 9) produce relatively small amounts of pollen, as self-pollination is readily achieved. In such autogamous species the ovules are almost always fully fertilised; pollination efficiency is high. In allogamous species in which, for some reason (e.g. dichogamy, herkogamy, dicliny or self-incompatibility), pollination efficiency, as measured by the proportion of seed-set, is lower, more pollen per ovule is usually produced (Cruden 1977).

In practice, very little information seems to be available concerning pollination efficiency. Levin and Berube (1977) have studied some features of the efficiency of the pollination system in *Phlox pilosa* and *P. glaberrima* visited by the sulphur butterfly *Colias eurytheme*. On probing a flower, approximately 10% of the available pollen is extracted on the proboscis of the butterfly, although some of this is lost when the proboscis recoils. Between 10% and 17% of the pollen on the proboscis was estimated to be deposited on the next flower visited, and thus a single visit between flowers only transports about 1% of the available pollen. Most other information relates to heteromorphic species (Ganders 1976, Ornduff 1970, 1971, 1976, 1979b, 1980a,b, Phillipp & Schou 1981, Schou 1983, Weller 1980). Heterostylous species with heteromorphic pollen (of two different sizes) are peculiarly suitable for assessments of the efficiency, per pollen grain produced, of within-flower, between-flower and between-plant pollination (Ch. 7, Fig. 7.1). Thus in *Primula*, pollen occurring on stigma will have the origins given in Table 5.6.

**Table 5.6** Origins of pollen found on stigmas of *Primula*.

| Source of pollen on stigmas | Stigma type Pin (P) | Thrum (T) | |
|---|---|---|---|
| within flower | none | some thrum pollen | not distinguishable after anthesis, unless one flower per plant only open, or after emasculation |
| between flower, within plant | most pin pollen | most thrum pollen | |
| between plant | all thrum pollen plus some pin pollen calculated by number of thrum grains × P:T plant ratio × P:T pollen production ratio | all pin pollen plus some thrum pollen calculated by number of pin grains × T:P plant ratio × T:P pollen production ratio | |

The total number of pollen grains produced per anther, and hence per flower has been calculated for each of the three western European species of *Primula* section *Primula* (Table 5.7). Rather similar patterns of pollen production and pollination efficiency are seen in each. The following features are outstanding:

(a) Many more pin grains are produced than thrum grains, apparently by virtue of the smaller diameter, and hence smaller volume, of pin pollen.
(b) More pin pollen is thus available for between-flower transport, both within and between plants. As the presentation of pin pollen is less apparent than that of thrum pollen, and as within-flower selfing should be rare in pins, this surplus should aid pin cross pollination.
(c) It can therefore be assumed that the greater number of illegitimate (pin) pollinations on to pin stigmas, by factors of between 3 and 15, is at least partly a function of the greater number of pin pollen grains produced.
(d) However, the number of legitimate pollinations by pin grains (on to thrum stigmas) only exceeds the number of legitimate pollinations by thrum grains on to pin stigmas in *P. veris*. The very considerable variation of samples, as expressed by the standard deviation calculated for *P. elatior*, renders differences between the amount of legitimate pollination experienced by the two morphs unimportant. Thus,

**Table 5.7** Pollen flow in three species of *Primula*.

| Species | Pollen production/flower | | illegitimate pollen/stigma (stigma type) | | illegitimate pollen/stigma (stigma type) | | legitimate pollen/ovule (stigma type) | |
|---|---|---|---|---|---|---|---|---|
| | T* | P | T | P | T | P | T | P |
| | | | (overall mean) | | (overall mean) | | | |
| *P. vulgaris* (Orndruff 1979) | 89 000 | 283 000 | 740 | 2427 | 255 | 403 | 4.5 | 6.5 |
| | | | (overall mean) | | (overall mean) | | | |
| *P. veris* (Orndruff 1980) | 87 000 | 211 000 | 298 | 4038 | 353 | 180 | 5.9 | 3.0 |
| | | | | | | | (estimates based on 60 ovules/flower) | |
| *P. elatior* (Schou 1983) | 86 645 ±16 680 | 138 710 ±29 649 | 348 ±252 | 3715 ±2492 | 692 ±584 | 889 ±433 | 11.0 | 14.0 |

| Species | % of all pollen-producing legitimate crosses (pollen type) | | % of all pollinations cross-pollinations (stigma type) | | % of all pollinations legitimate | |
|---|---|---|---|---|---|---|
| | T | P | T | P | T | P |
| *P. vulgaris* (Orndruff 1979) | 0.45 | 0.09 | 33.7 | 59.5 | 25.6 | 14.2 |
| *P. veris* (Orndruff 1980) | 0.21 | 0.17 | 76.6 | 14.6 | 54.2 | 4.2 |
| *P. elatior* (Schou 1983) | 1.02 | 0.50 | 84.0 | 50.2 | 66.5 | 19.3 |

* Abbreviation: T = thrum, P = pin.

we can see that the relative inaccessibility of the thrum stigma is as relevant to the success of legitimate pollination as the inaccessibility of the pin anther. There tends to be more pollination of both kinds of pollen on to the extruded pin stigma.

(e) There are, on average, between 3 and 14 legitimate pollinations per ovule to be fertilised; these rather surprising figures reveal that poor pollination is rarely a factor in limiting seed-set in these obligately outbred species, at least in the populations tested.

(f) Pollination efficiency (defined as the proportion of all pollen produced that results in legitimate pollination) is low, ranging from 1 : 1000 to 1 : 100 grains. The less numerous thrum pollen is from 1.5 to 5 times as efficient as pin pollen and this greater efficiency roughly balances the smaller output of thrum pollen, suggesting that the difference in number of pollen grains produced by the two morphs is adapted to the greater inaccessibility of the pin anther and the thrum stigma to the pollinator. It is thus possible to conclude that the difference in pollen size is a function of pollination efficiency, allowing more pin pollen to be produced.

(g) If it is assumed that the proportion of the two pollen types being carried from flower to flower on pollinators is a function of the relative production of the two pollen types (and the relative inaccessibility of pin pollen renders this assumption rather unlikely), it is possible to estimate the proportion of all pollinations made (which are known) that are cross pollinations because the number of legitimate cross pollinations is also known. These estimates of cross pollination range from 14.6 to 84.0%, and seem surprisingly high for a multiflowered species with a flower of rather simple construction. In *P. veris* and *P. elatior*, with scapose flowers, pin pollen travels better between plants than it does in the non-scapose *P. vulgaris*. This may be a function of the foraging time spent at an individual in each species. In the less rewarding *P. vulgaris*, flowers may be more thoroughly foraged, resulting in more efficient travel of thrum pollen. It is notable that relatively little within-flower illegitimate pollination of thrums occurs, although physically this seems likely to predominate; in the absence of rain on the upward-pointing primrose flower (Eisikowitch and Woodell 1975) the sticky pollen may fall on to its own stigma only rarely.

(h) Darwin (1877) hypothesised that the heteromorphy of the *Primula* flower should maximise between-morph pollination, as seems reasonable through the correspondence in the position within flowers of anthers and stigmas of cross-compatible morphs. In fact, within-flower selfing makes it difficult to decide whether this is in fact the case (Chs. 7 and 8). Although about two-thirds of cross pollinations on to the hidden thrum stigmas are legitimate, less than

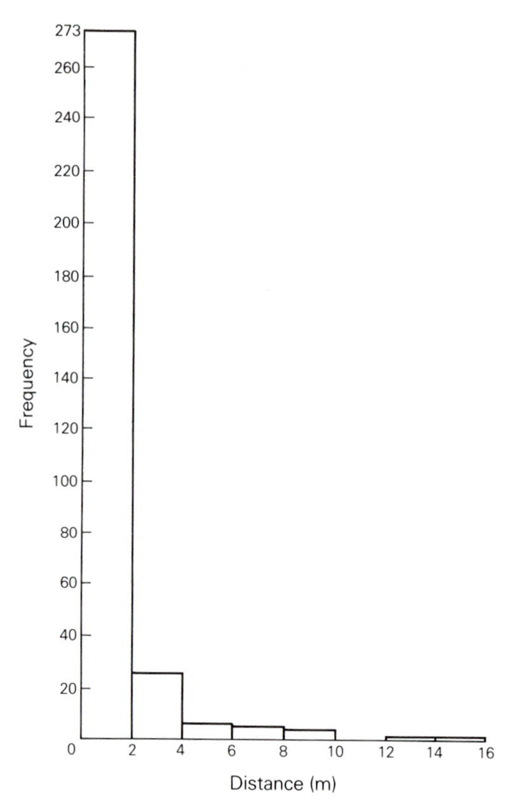


**Figure 5.7** Distances travelled by bees (*Bombus* spp.) between inflorescences of the cowslip, *Primula veris* in a population in Northumberland, UK.

one-third of cross pollinations on to the exposed pin stigmas are legitimate. This proportion is remarkably constant between species and it may reflect visits by hasty or non-specialist visitors to pin flowers.

## *Pollen carryover and geitonogamy*

In *Primula* we are in the privileged position of being able to monitor exactly the pollination efficiency and pollen travel within and between flowers in this particular and unusual dimorphism. Yet, although we are able to estimate what proportion of pollen travels between plants, both legitimately and illegitimately, we have little idea what proportion of within-plant pollen travel is within-flower travel, and what proportion of the pollen that adheres to a pollinator travels to the first, and to subsequent flowers. This last is called carryover, and although it is of little genetic consequence to differentiate between different types of selfing, it is of great importance to determine the carryover sequence of animal-carried pollen, as it will determine how much pollen is geitonogamously selfed, and how much is crossed, and over what distance, for non-heteromorphic plants.

We can get a rough estimate of carryover in *Primula veris* by a comparison of the distribution of bee flights (Fig. 5.7) with the travel of pollen in the same population (Fig. 5.8). It can be seen that while occasional bee flights between flowers were recorded up to 16 m in distance, and whereas individual pollen grains travelled minimum distances of at least 13 m, the vast majority of bee flights (over 95%) were of less than 2 m. Although the majority of pollen also travelled less than 2 m, a higher proportion travelled from 2 to 6 m minimum distance, and most of this further travel can be assigned to carryover from pollen not deposited on the first flower visited.

The patterns of carryover, and its consequences, have been modelled by Crawford (1984b), partially based on Bateman (1947), Levin and Kerster (1971) and Primack and Silander (1975). Crawford examines a situation in which the number of pollen grains on a pollinator is in

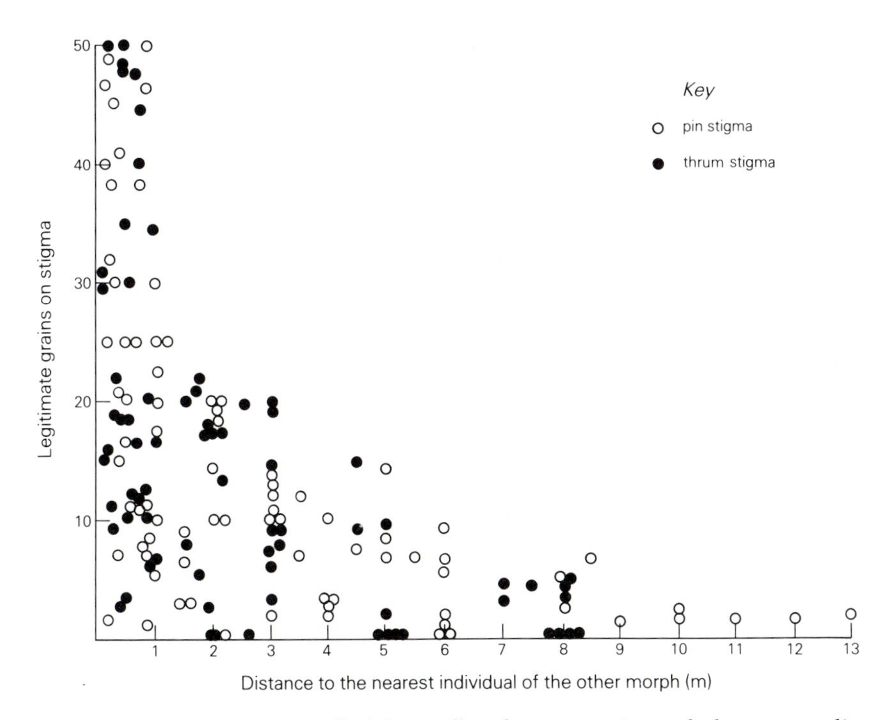


**Figure 5.8** Distances travelled by pollen between pin and thrum cowslip, *Primula veris*, as estimated by numbers of legitimate grains on stigmas and distance to nearest legitimate pollinator.

dynamic equilibrium so that each time it visits a flower, a proportion of the pollen grains on it, $p$, is exchanged. It is assumed that pollen is deposited on a stigma of a flower before fresh pollen is collected from that flower (which is probably unrealistic), and that the plant is fully self-compatible. It is also assumed that $n$ flowers are visited, in sequence, on a plant (genet). The proportion of all pollen collected from a plant (genet) geitonogamously selfed on that plant is described by:

$$1 - 1/np[1 - (1 - p)^n]$$

The proportion of pollen collected from a genet deposited on the $i$th plant genet subsequently visited is described by:

$$1/np[1 - (1 - p)^n]^2(1 - p)^{n(i-1)}$$

Thus, the amount of geitonogamous selfing, as opposed to crossing, that will occur is a function solely of the rate of pollen exchange that occurs at each flower visit and the number of flowers visited per genet by the pollinator. The latter figure will depend considerably on the number of rewarding flowers presented to the pollinator by a genet at any moment in time. Ramets with many flowers, or genets with many flowering ramets, will receive much more geitonogamous selfing than will few-flowered genets. This will be reinforced by the tendency for the pollinator to fly less far between ramets, and to visit more flowers within a ramet, if the ramet has many rewarding flowers (p. 139). Ironically perhaps, plant genets that perform well are more likely to be self-pollinated. Crawford (1984b) demonstrates this clearly for the musk mallow, *Malva moschata*, in which there is a direct positive relationship between the amount of selfing and the number of flowers per genet. In this case, selfing was estimated using homozygous electrophoretic markers in the parents in experimental populations, and scoring heterozygotes (which were outcrossed) for acid phosphatase in the offspring. Interestingly, Crawford, using acid phosphatase and a flower-colour marker, confirmed this result in a wild population of the same species by showing a clear negative relationship between the amount of outcrossing and the number of flowers per genet.

Most other work uses pollen markers rather than genetic markers to assess pollen carryover. Gerwitz and Faulkner (1972) used source plants marked with $^{32}P$ as a radioactive marker for their work with *Brassica*, and showed that the proportion of radioactive pollen was decreased by about 30% for each subsequent non-radioactive flower visited.

Pollen carryover has been estimated using marked pollen in remarkably few instances. Similar patterns of results have been obtained for *Colias* butterflies visiting *Phlox* (Levin & Berube 1972), *Bombus* species visiting *Delphinium* (Waddington 1981, Waser & Price 1981, 1983), artificial pollination using stuffed hummingbirds (Pyke 1981) and *Bombus* visiting *Erythronium americanum*, *Clintonia borealis* and *Diervilla lonicera* (Thomson & Plowright 1980). Typically, up to 50% of pollen is deposited on the first flower visited after collection, and deposition on subsequent flowers decreases rapidly in a leptokurtic fashion; less than 1% of pollen usually survives on the pollinator after eight visits (Table 5.8). Carryover appears to be greater on butterflies, and less on hummingbirds, than it is on bees, but information is at present very limited. In all such investigations in which single pollen sources are considered, pollen carryover is likely to be underestimated.

Such information on carryover is clearly vital to the understanding of gene travel, and the genetic structure of populations, and it is perhaps surprising that so little information of this kind exists (Handel 1983b). However, work on carryover can only be interpreted with reference to the number of receptive flowers on a plant that acts as a pollen donor, and the number of flowers on subsequent individuals that are visited by the pollinator. This information is given by Waddington (1981) for *Delphinium virescens*. Between two and six flowers are usually visited per inflorescence, the number being highly dependent on the volume of nectar being donated by that inflorescence (p. 139). In this species, 44% of the pollen is deposited on the first flower visited after collection, and 83% is deposited on the first three flowers. Thus, for an inflorescence which is highly rewarding, on which a visitor will visit an average of five flowers, much geitonogamous selfing will occur. For a poorly rewarding inflorescence, on which less than three flowers may be visited on average, much more crossing is likely (Fig. 5.9). Bees also fly further, more than twice as far on average, from poorly rewarding inflorescences. Thus, ironically perhaps, poorly rewarding inflorescences will be the most outcrossed in this species. As the goals of pollination efficiency and maximisation of outcrossing are thus in direct opposition, the plant may adapt to produce a quantity of reward at a threshold point between the optimal points for each goal. However, by varying the quantity of reward with the age of the inflorescence, or even within a daily rhythm, both goals may be met in different time-niches.

We are not told the number of flowers per plant visited on average in the three species enumerated in Table 5.8. It is worth noting, however, that these have very different inflorescence structures. Thus the *Erythronium* will usually only have one to three flowers per plant open at any one time. In contrast, *Diervilla lonicera* is a shrub on which a well-grown individual might have in excess of a thousand receptive flowers open together. *Clintonia borealis* takes an intermediate position, although closer to that of *Erythronium*. This is clearly of vital importance if gene flow is considered. In *Erythronium americanum*, over half the pollen carry is likely to be between plants (it appears to be self-incompatible, at least in cultivation in the UK), whereas in *Diervilla lonicera* nearly all pollen

**Table 5.8** Carryover of pollen on *Bombus* species visiting three North American flowers (after Thomson & Plowright 1980). Figures are numbers of grains.

| Flower | Sequence | Flower number visited after collection | | | | | | | | | | | | | | |
|---|---|---|---|---|---|---|---|---|---|---|---|---|---|---|---|---|
| | | 1 | 2 | 3 | 4 | 5 | 6 | 7 | 8 | 9 | 10 | 11 | 12 | 13 | 14 | 15 |
| *Erythronium americanum* | a | 4 | 6 | 0 | 0 | 0 | 0 | 1 | 1 | 0 | 0 | 1 | 0 | 0 | 0 | 0 |
| | b | 14 | 11 | 2 | 5 | 5 | 0 | 3 | 3 | 2 | 0 | 5 | 0 | 0 | 3 | 0 |
| | c | 8 | 15 | 3 | 4 | 5 | 3 | | | | | | | | | |
| | d | 4 | 0 | 6 | 7 | 3 | 2 | 6 | 0 | 0 | | | | | | |
| | e | 5 | 2 | 1 | 3 | 0 | 0 | 0 | 0 | 0 | 0 | 0 | 0 | 0 | 0 | 0 |
| | f | 4 | 7 | 3 | 3 | 5 | 1 | 7 | 0 | 0 | 0 | 0 | 0 | 0 | 1 | 0 |
| | g | 0 | 3 | 0 | 4 | 1 | 0 | 0 | 0 | | | | | | | |
| *Clintonia borealis* | a | 369 | 148 | 110 | 2 | | | | | | | | | | | |
| | b | 30 | 21 | 153 | 28 | 484 | 12 | 17 | 159 | | | | | | | |
| | c | 24 | 21 | 58 | 47 | 19 | 36 | 2 | 2 | 9 | 1 | 6 | 5 | | | |
| | d | 258 | 178 | 5 | 28 | | | | | | | | | | | |
| | e | 346 | 21 | 34 | 73 | | | | | | | | | | | |
| | f | 131 | 142 | 16 | 68 | 45 | 69 | 51 | 17 | 5 | 22 | | | | | |
| | g | 97 | 20 | 93 | 54 | 49 | 5 | 34 | 6 | | | | | | | |
| *Diervilla lonicera* | a | 80 | 1 | 1 | 5 | 18 | 2 | 0 | 2 | 0 | 1 | 0 | | | | |
| | b | 48 | 10 | 32 | 12 | 21 | 23 | 11 | | | | | | | | |
| | c | 24 | 63 | 8 | 8 | 19 | 0 | 4 | 0 | 10 | 1 | | | | | |
| | d | 32 | 5 | 4 | 8 | 1 | 0 | 3 | 1 | | | | | | | |

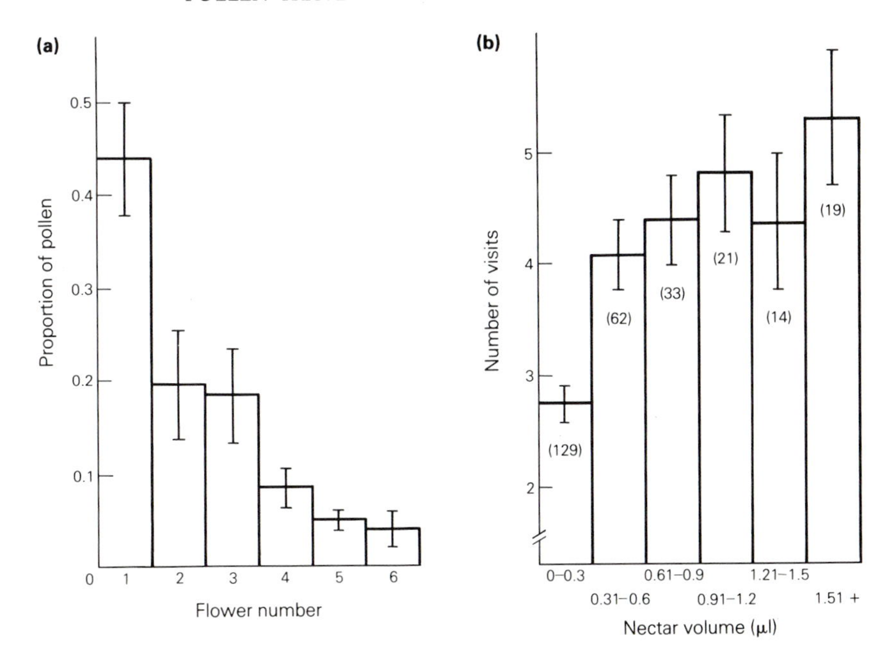


**Figure 5.9** Foraging strategies of *Bombus americanorum* on *Delphinium virescens* in Kansas, USA (after Waddington 1981) (a) The mean proportion of pollen grains deposited on the stigmas of the 1st, 2nd . . ., 6th flower visited after the marked flower; (b) relationship between nectar volume of the bottom-most flowers and the number of flowers visited per inflorescence. Poorly rewarding inflorescences will be more outcrossed than well-rewarding inflorescences.

carry will be within a plant (this species has self-compatible relatives, although I am uncertain as to its breeding system). Clearly, it is more likely to be advantageous for a species with good pollen carry between plants to be self-incompatible than it is for one in which most pollen is selfed. Even when self-incompatible, selfed pollen may be a disadvantage by blocking stigmatic sites.

For a many-flowered plant, herkogamy and dichogamy (Ch. 4) are relatively unimportant, for although within-flower selfing may be prevented, geitonogamy (between-flower selfing) will be rife. In order to understand patterns of gene transport within and between plants in a population, it is not enough to describe pollen carryover. *It is also necessary to know the number of flowers visited per plant, the number of receptive flowers carried by a plant, and the self-incompatibility or self-compatibility of the plant.* It is rare indeed for all this information to be transmitted together.

When pollen carryover occurs between genets, it is very likely that pollinators will carry pollen from more than one male parent. Thus, pollen arriving on a stigma should result in multiple paternity of seeds within a single fruit. However, competition between pollen grains on a stigma may result in a single-sired fruit (Levin 1975). Sexual selection for maternal fitness should favour multiple paternity, for sibling competition will be reduced by increased sibling variability. In contrast, sexual selection for paternal fitness should favour pollen that outcompetes pollen from other sources on a stigma. Although earlier work indicated that multiple paternity of siblings frequently occurs, the quantification of multiple paternity has only recently been achieved by Ellstrand (1984), working with wild radish, *Raphanus sativus*. I am informed that this influential paper will be followed by others from a number of authors working with other plant species.

Ellstrand sampled parental genotypes and seedling genotypes from known parents of seeds within single fruits. An entire population of 59 potential male parents was analysed for six allozyme loci, and 246 seedlings from 59 fruits collected from nine of these parents were similarly examined. Ellstrand identified the minimum number of male parents involved in the parentage of any fruit using logical exclusion techniques derived from human paternity studies; the application of these techniques is clearly explained in an appendix to the paper. All the nine female parents in the population showed evidence of multiple mating, and at least 85% of the fruits contained seeds that had different male parents. The maximum number of parents per fruit identified with certainty was four (8% of fruits), and 20% of the fruits had at least three male parents (fruits contain between two and eight seeds).

Radish is a self-incompatible annual which is pollinated by a variety of pollinators (p. 185). According to Ellstrand's results, little if any pollen competition occurs, and the dictates of maternal selection fitness seem to be followed. It is Ellstrand's view that patterns of multiple paternity are a by-product of the pollination system, rather than the result of sexual selection, and with this I concur.

### *Pollen travel between plants*

In contrast to our very limited information concerning pollen travel within plants, we have a considerable body of information on pollen travel between plants, gleaned from techniques of pollen marking or gene marking (see Table 5.5) or by manipulating plants with various breeding systems (see Table 5.4). Information on pollen travel is vital to our understanding of the genetic structure of populations in terms of levels of heterozygosity, maintenance and breakage of linkage groups and establishment of novel mutants, etc. Gene migration is also mediated by patterns of seed dispersal, and on this we are relatively ill-informed (p. 180). Estimations of gene travel using genetic markers do, of course, incorporate both pollen travel and seed travel, but only in Schaal (1980) is gene travel compared with various components in *Lupinus texensis*.

**Leptokurtic distribution** The outstanding feature of both anemophilous and zoophilous pollen travel is its leptokurtic distribution, a pattern that is common to most biotic movements (Handel 1983b). Thus, the distribution of between-plant flights of bees and butterflies visiting three American *Senecio* species (see Fig. 5.2), and the distribution of between-plant bee flights and pollen flow in *Primula veris* (see Figs. 5.7 and 5.8) are examples quoted earlier in this chapter that clearly show this leptokurtic distribution. Other bee flowers, such as *Aconitum columbianum*, *Delphinium virescens*, *Trifolium repens* and *Lupinus texensis* show very similar patterns of pollinator travel between plants (Fig. 5.10), and patterns of gene travel in *Lupinus texensis* and *Cucumis sativus* (cucumber) are also similar, although further travel is apparent because of carryover (Fig. 5.11).

Hummingbird-pollinated *Heliconia* species (Musaceae) are the only cases known to me where the leptokurtic rule does not always hold for pollen and gene dispersal. Study of the distribution of marked pollen in Costa Rica by Linhart (1973) shows that territorial birds visit *Heliconia* species growing in clumped distributions (*H. latispatha* and *H. imbricata*) and in these the pollen travel is very markedly leptokurtic (Fig. 5.6). However, for randomly spaced *Heliconias* (*H. acuminata*) or those which in this case showed linear distributions (*H. tortuosa*), pollen dispersal showed a linear relationship with distance, or was even bimodal. These species were visited by non-territorial birds which showed a randomised far-ranging feeding pattern quite different from those that held territory.

The leptokurtosis typical of pollen and gene travel in most zoophilous flowers, certainly when visited by bees, butterflies and most birds, is found with respect to both within-plant and between-plant travel, and results from an interaction of several behavioural and physical features:

(a) Most plants have a clumped distribution.
(b) Within a plant population, some patches will be more rewarding than others at a given time, thus exaggerating this clumping.
(c) Pollinators will concentrate on more rewarding patches.
(d) When pollinators visit more rewarding patches, they will travel less far to the next flower or plant.
(e) Pollen carryover itself tends to be leptokurtic in distribution (Table 5.8), thus most pollen is carried to the next flower visited.
(f) For most flower visitors, length of flight between flowers is dependent on their behavioural strategy at that moment. Thus, for a foraging bee, between 90 and 99% of flights are short-distance foraging flights. The remainder are 'escape flights' over much greater

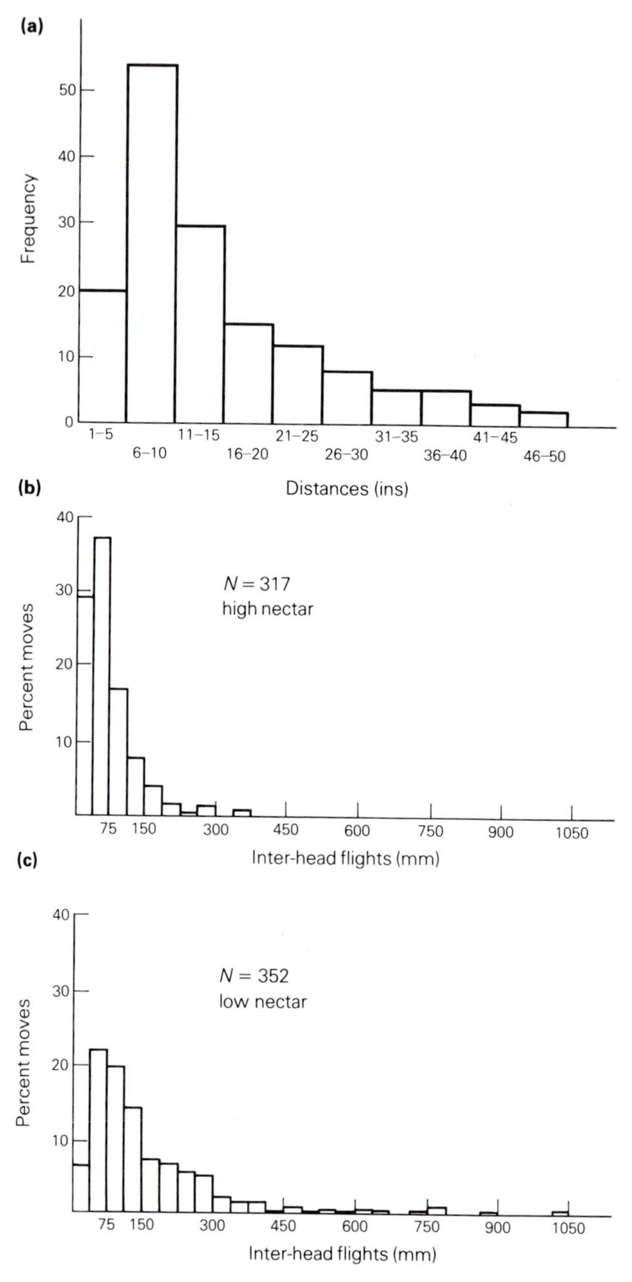


**Figure 5.10** Distribution of distances travelled between inflorescences by foraging bees in (a) *Aconitum columbianum* (after Pyke 1978) and *Trifolium repens* (after Heinrich 1979), showing differences in pollinator travel between well-rewarding (b) and poorly rewarding (c) patches.

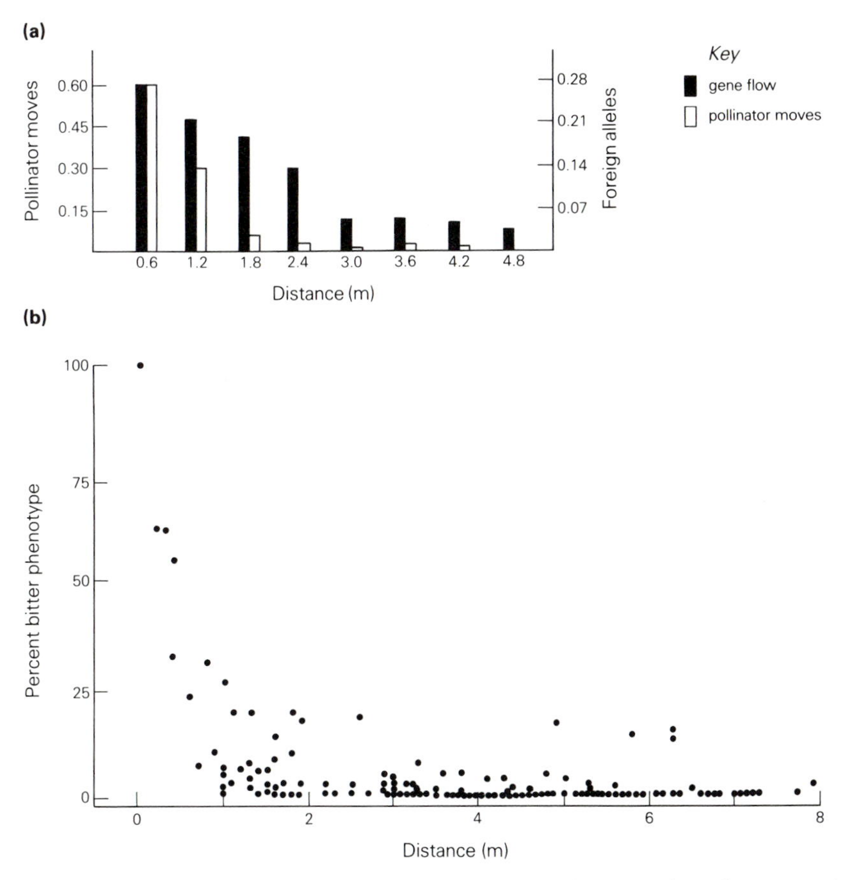


**Figure 5.11** Gene travel as estimated by the distribution of marker genes in seedlings of *Lupinus texensis* (a, after Schaal 1980) and *Cucumis sativus* (b, after Handel 1983).

distances, for the purposes of visiting new patches, escaping predators or returning to the hive or nest. For flower-visiting birds, foraging flights are much shorter than those involved in territoriality, aggression and courtship. Combinations of normally distributed flight patterns with different variances (i.e. foraging and escape flights) will automatically generate leptokurtic flight distributions.

## *Pollen transport by bees*

Although the leptokurtic pattern of dispersal of pollinators, pollen and genes is almost universal, the distances involved do vary a great deal. This variation can be considerable even within a flower species. Thus, for the *Senecio* species studied by Schmitt (1980) mean bee flight distances vary from 0.32 m to 1.06 m, whereas mean butterfly flight distances vary from 2.30 m to 12.39 m. Clearly, the leptokurtosis of pollen dispersal is increased in this case by the different behaviour of the two classes of pollinator. Heinrich (1979a), working with clover, *Trifolium repens*, visited by *Bombus terricola*, shows that for plants protected from nectar foraging for two days, and thus with plenty of nectar, 99% of foraging flights were less than 2 m in distance. For unprotected plants, with less nectar, 99% of flights were less than 4 m (Fig. 5.10).

For bee-pollinated plants, it is usual to find that at least 80% of flights are less than 1 m in distance, and 99% are less than 5 m (Figs. 5.2, 5.7, 5.10). Occasional grains may, however, travel much further, such is the nature of pollinator behaviour and the resulting leptokurtic dispersal distributions. Halijah Ibrahim and I found that occasional grains of *Primula veris* may move at least 13 m, and there is evidence in this species of gene travel of at least 30 m (Ibrahim 1979). Naturally, the mean distance of pollen dispersal is dependent not only on pollinator behaviour, but also on plant density. Travel will be much shorter in areas of high plant density, as originally emphasised by Levin and Kerster (1968, 1969a,b, 1974) and Levin (1979). We found pollen dispersal highly dependent on plant density in *Primula veris*, where in our experimental population, density of plants varied from 0.10 to 22.0 per m$^2$.

In sparsely populated areas, pollen travel will, by definition, be more distant. However, the number of plants which the pollen may reach may be no greater than in dense populations, and indeed may well be less. Also, sparse areas of dense populations will receive far fewer pollinator visits, as these areas will form patches of low reward, and this will minimise gene travel in these areas.

For many species, especially in tropical forest, plant density is always low. It is common to find for a species of tropical tree, and for many tropical epiphytes and lianes, that patches are very small, having a 'steady-state' strategy (p. 151). Thus, a species may only occur once in a thousand individuals of the same growth form, randomly distributed, and nearest-neighbour distances may be hundreds of metres apart. Because dominant species are rare or absent, many pollinators in these warm climates, with energetically rewarding flowers, fly long distances between plants, which, being of a 'steady-state' type, may each have only one or a few flowers open. This plant/pollinator strategy is termed 'trap-lining', and originates from work on Euglossine bees and orchids by Dressler (1968) and Williams and Dodson (1972). Later work by Janzen (1971), Frankie (1976), Frankie *et al.* (1976) and Heinrich (1975) extended these ideas to forest trees and other classes of solitary bees. However, as yet, this work seems to have been confined to neotropical forests, and to bees. A good deal of work remains to be done on the pollination genetics of sparsely distributed rainforest plants in other areas of the world. It is probable that trap-line foraging also occurs in other groups of pollinators such as large butterflies and moths, birds (especially parrots), bats and monkeys. All are highly mobile with efficient thermoregulation, and most have good long-distance vision or smell.

### *Pollen transport by birds*

The only reported work I am aware of concerning pollen transport by birds (Linhart 1973) shows that, although pollen dispersal is leptokurtic for territorial birds visiting clumped *Heliconias*, distances involved can be 50 or 100 times greater than for colonial bees (see Fig. 5.6). In one case (*Amazalia saucerottei* visiting *H. latispatha*) only 1% of the pollen travelled more than approximately 20 m. It is not clear how much these differences are the result of patch sizes, or densities, and how much the result of bird behaviour, but distances are greater than one would expect with a colonial bee, or even a butterfly.

### *Pollen transport by wind*

Quite a lot of information is also available concerning wind-dispersed pollen. This also has a leptokurtic dispersal pattern, but the causes for this are quite different (Gleaves 1973).

(a) The lateral component of pollen travel will be due to wind (Fig. 5.12), which is rarely steady, but blows in gusts; the distribution of speeds of wind gusts is itself leptokurtic, that is, the strongest gusts are the rarest. Thus, most dispersed grains will only be blown a short distance, but a few will meet stronger gusts after dehiscence and will be blown much further.
(b) Tree pollen, in particular, is released from various heights above ground and, as most trees have a somewhat conical canopy shape, the amount of pollen available for release will decrease with height above ground. The potential for lateral dispersal, that is of encountering a gust of wind, will increase with the height of the point of pollen release above ground; this is partly because it is windier higher up, and partly because the pollen is falling for a longer time.
(c) Most wind-dispersed pollen is released in a conspecific canopy (trees) or stand (grasses, sedges) and it will thus tend to lodge on foliage, and stick or fall to the ground a short distance from the point of dispersal. Only a relatively few grains, or grains from relatively few isolated plants, will be able to escape the shelter and obstruction conferred by the canopy and disperse any distance. Handel (1983b) shows that the sedge, *Carex platyphylla*, receives mostly crossed (exogenous) pollen in small clones with less than 10 flowering ramets (culms), but if clones (genets) are larger than this, the amount of

**Figure 5.12** Pollen being distributed by wind from male cones of *Pinus nigra*. Pine pollen has two air sacs (Ch. 3, Fig. 3.3) and is unusually well distributed in the air (×0.33).

selfed (geitnogamous) pollination is considerable, and increases no further with increasing clone size.

As a result of these factors, most wind-dispersed pollen travels relatively short distances only. However, pollen travel is very much further than is usual in animal-dispersed pollen (Table 5.9), although it must be remembered that for trees, grasses or sedges with large clones, individuals may be very large, and densities of genetically different individuals (genets, or clones) very low (Ch. 10).

The data in Table 5.9 must be interpreted with considerable care and caution, for different measuring techniques are used, different units of measure are employed, and dispersal estimates differ from distances from forest or crop edges to distances from isolated trees or marked pollen sources. Although some of these enormous potential biases are overcome by expressing data as a percentage of the total detected, the figures obtained do rely heavily on the distances at which measurements were actually made. Thus, cross-comparisons are dangerous, and the data can

**Table 5.9** Percentage of detected wind-dispersed pollen encountered at various distances from source (adapted from Altman and Dittmer 1964).

| | Distance from source (m) | | | | | | | | | | | | | | | | | |
|---|---|---|---|---|---|---|---|---|---|---|---|---|---|---|---|---|---|---|
| Species | 1 | 2 | 3 | 5 | 10 | 15 | 20 | 50 | 100 | 150 | 200 | 300 | 400 | 500 | 600 | 700 | 800 | 900 |
| **Grasses** | | | | | | | | | | | | | | | | | | |
| *Phleum pratense* | | | | | | | | | 17.8 | | 11.5 | 7.8 | | 3.2 | | | | |
| *Dactylis glomerata* | | | | | | | | | | | 11.4 | | 4.3 | | 3.1 | | 2.2 | |
| *Lolium perenne* | | | | | | | | | | | 17.0 | | | 8.7 | | 5.6 | | |
| *Pennisetum glaucum* | | | | 90.8 | | 8.1 | | | 0.7 | 0.4 | | | | | | | | |
| *Zea mays* | | | 93.7 | | 4.3 | | 1.5 | 0.4 | | | | | | | | | | |
| *Beta vulgaris* | | | | | | | | | | | | 13.1 | | 7.2 | | | 1.9 | |
| **Trees** | | | | | | | | | | | | | | | | | | |
| *Fraxinus excelsior* | | | | | 68.4 | | 27.1 | 3.8 | | 0.8 | | | | | | | | |
| *Populus* sp. | | | | | | | 26.5 | | | 21.2 | | | 18.8 | | | | | 17.1 |
| *P. deltoides* | | | | | 47.3 | | | | 25.5 | 18.9 | | | | 8.2 | | | | 0.1 |
| *Ulmus* sp. | | | | | | | | | | 40.1 | | 53.0 | | | | | | 4.2 |
| **Conifers** | | | | | | | | | | | | | | | | | | |
| *Pinus cembroides* | | | | 93.0 | | | | 5.1 | 0.9 | 0.6 | | | | | | | | |
| *P. echinata* | | | | | | | | | | 17.0 | | 15.0 | | 10.0 | | | | |
| *Cedrus atlantica* | | | | | | 44.2 | | 27.2 | 16.7 | 12.0 | 0.02 | | | | | | | |
| *C. libani* | | | | 52.3 | | | | 25.0 | 14.9 | 8.9 | | | | | | | | |

only safely be used to show the orders of distances that wind-dispersed pollen may travel, and its tail-off with distance. It must be remembered, however, that in most of these cases readings are only of pollen grains that have travelled at least as far as the minimum distance recorded from the source.

Two features perhaps stand out from an examination of this table. Firstly, that appreciable quantities of both grass and tree pollen can in some circumstances travel distances of 1 km or more, as sufferers from pollen allergies ('hay fever') living in plant-less cities will testify. In all the cases quoted here except for one, at least one in 200 grains which travelled did so for at least 100 m. This distance of dispersal is only known for bird pollination amongst zoophilous plants. The exception, *Zea mays* (maize), is a crop plant in which the nature of pollen, and its presentation and release, may have changed markedly since domestication. For instance, it has remarkably large, and somewhat sticky, pollen grains. Secondly, the decline in grain number with distance is remarkably gradual, especially in the meadow grasses and the poplars examined. The leptokurtic curve is very shallow in contrast to that of zoophilous species, or most other wind-pollinated species, although in no case is there a linear relationship with distance. These very gradual 'fall-offs' in pollen travel with distance will render the variances about the means of pollen travel very large, and will lead to very considerable estimates of neighbourhood area and, depending on plant density, neighbourhood size (p. 175).

There is a very strong relationship between the area occupied by an individual clone (genet) and its pollination system. Most trees which are individually large, or which form suckers to make large clones, are wind pollinated. Equally, herbaceous perennials that ramify to make large clones (e.g. grasses, sedges and rushes) are also wind pollinated. Thus wind pollination, with its much greater travel, may be interpreted as being advantageous in allowing more pollen to reach beyond the bounds of the clone (genet), thus enhancing outcrossing and, in the case of self-incompatibility or dicliny, seed-set. In contrast, most zoophilous plants are smaller in size or clonal growth. Where trees are zoophilous, they are usually small (*Sorbus, Crataegus, Malus*) and without vegetative reproduction, or habitually solitary, or non-dominant, as in the very diverse trees of tropical forest, most of which are zoophilous, or in temperate genera such as *Prunus*. It is usual to find that wind-pollinated (anemophilous) species are dominant or codominant in a locality, as this will aid the efficiency of their non-specific pollination.

Thus wind pollination is usual in dominant species in which the individual genets cover large areas, a combination that often goes together. When considering the 'chicken and egg' situation of whether anemophily allows large genets to evolve, or whether it has become advantageous for large genets to become wind pollinated, it is worth noting that early Angiosperms were probably large (trees) but zoophilous, anemophily being secondary. Thus, it may be reasonable to conclude that it has become advantageous for large genets to evolve anemophily, a syndrome from which further evolution of other pollination systems has rarely occurred (but see for instance *Chamaepericlymenum suecicum* as an example of secondary entomophily, Ch. 4).

Very few studies have investigated gene flow, as opposed to pollen travel, in wind-pollinated plants. In rye grass (*Lolium perenne*), over 99% of the genes travelled less than 20 m (Gleaves 1973) and in beet (*Beta vulgaris*), all genes travelled less than 10 m. These values are of an order of magnitude less than the pollen travel observed in these species (see Table 5.9), and this contrasts sharply with entomophilous plants in which gene travel usually exceeds pollen or pollinator travel (p. 169). This difference is probably due to technique and to pollination efficiency. For zoophilous pollen, travel is measured on to a stigma, whereas for anemophilous pollen it is generally trapped in the air. Very little of the air-trapped pollen would have reached a stigma, and this would be most likely to happen over short distances. Also, where pollinator moves are measured for a zoophilous species, pollen carryover is not usually taken into account, so that pollinator moves will always be an underestimate of gene travel. In neither case is seed travel estimated.

## NEIGHBOURHOOD AND GENE FLOW

Neighbourhood is a rather difficult concept introduced by the great theoretical population geneticist Sewall Wright (1938, 1940, 1943a, 1946, 1951). It concerns the area within a population in which panmixis, that is random gene exchange, can be said to occur. A number of factors, including the distribution of pollen travel and of seed travel, and the varying density of interbreeding individuals, may result in panmictic areas, neighbourhoods, which are more limited than the whole population. Wright's aim was to describe a neighbourhood size $N_e$ (number of individuals) that occurred in a neighbourhood area $A$, where $N_e$ described the number of individuals responsible for a decay in genetic variance equal to that in a model population of size $N$. That model population $N$ has constant size, finite limits, non-overlapping generations, total random panmixis within those limits with all individuals interbreeding, equal numbers of both sexes (or hermaphroditism) and an equal chance for each individual to transmit genes to the next generation. If the model population shows decay of genetic variance in polymorphic genes of neutral effect, this will be a consequence of inbreeding effects in populations in which $N$ is smaller than infinitely large (i.e. gene drift, Ch. 2).

Wright showed that $N$ can be estimated as the number of individuals falling within a circle whose radius is equivalent to twice the standard deviation of the gene dispersal distance per generation in that population. As the area of a circle is $\pi r^2$, and $r = 2\sigma$, this circle is described by $4\pi\sigma^2$, where $\sigma$ is the standard deviation of the gene dispersal distance for each sex of a unisexual species. This is approximated by $12.6\sigma^2$ and gives the neighbourhood area $A$. The neighbourhood size (number of individuals) $N$ will be dependent on the density of individuals $d$ (using the same area units) capable of interbreeding at a given time. Thus, $N = 12.6\sigma^2 d$.

To obtain $N_e$ rather than $N$, one must first estimate $N$, and then $N_e/N$, allowing for unequal numbers of males and females, fluctuating population size and unequal parental contributions to the next generation, etc. Wright's equation was originally formulated for unisexual organisms, where $\sigma$ is usually a function of the distances males and females may travel to mate (from birth), and it has been so successfully used, as by Kerster (1964) for the Rusty Lizard, *Sceloporus olivaceus*. However, for most plants, gene dispersal has two components, pollen travel and seed travel, and, furthermore, most plants are not unisexual (dioecious) but hermaphrodite. Levin and Kerster (1969a, 1971) have proposed that for totally outcrossed hermaphrodite plants, neighbourhood size is best described by $12.6(\sigma_p^2/2 + \sigma_s^2/2)d$ where $\sigma_p^2$ and $\sigma_s^2$ describe the variance of dispersal of the pollen and seed, respectively. This was used by Richards and Ibrahim (1978). In the case of a partially selfed species this can be modified to incorporate the amount of outcrossing $r$ as follows:

$$12.6(\sigma_p^2 r/2 + \sigma_s^2/2)d$$

or:

$$6.3(\sigma_p r + \sigma_s^2)d$$

Another modification has been used by Schmitt (1980) where two pollinators A and B with very different flight variances $\sigma_A^2$ and $\sigma_B^2$ share a proportion of the pollinating events $a$ and $1 - a$. In this case:

$$N = 6.3[(a\sigma_A^2 + (1 - a)\sigma_B^2)r + \sigma_s^2]d$$

(this equation is not as it appears in Schmitt's paper).

Kerster (1964) noted that Wright's original equations make the assumption that populations do not move in space. Thus, there must be no net movement of pollen or seeds in any given direction (as could happen with unidirectional winds, for instance on coastal sites; p. 141). He suggests that mean move distances will, by definition, equal zero. Thus if the variance of pollen moves, $p_i$, around a mean $\sigma^2 p = \Sigma(p_i - \bar{p})^2/n_p$, and $\bar{p} = 0$, then the variance can be calculated merely by the mean of the squares of pollen moves, $\Sigma\, p_i^2/n_p$. Naturally, the same holds for seed moves, $\Sigma\, s_i^2/n_s$, where $n_p$ and $n_s$ are sample sizes of measurements of pollen and seed moves respectively.

Levin (1978a) noted that whereas the genetic contribution of pollen moves was haploid, that of seed moves was diploid, and thus twice that of pollen moves. Thus, his modified neighbourhood size equation becomes:

$$\mathrm{N} = 6.3\left(\sum p_i^2 r/2n_p + \sum s_i^2/n_s\right)d$$

(In both Levin (1978a) and Schmitt (1980), $r$, the proportion of outcrossing, is placed outside the equation brackets. However, it seems to me that $r$ relates only to pollen travel and not to seed travel.) What is essentially this equation, but omitting $r$, is used by Schaal (1980); presumably the value of 3.6 in her equation is a misprint for 6.3.

Recently, Crawford (1984a) has suggested that $\sigma^2$, the total variance of plant movement between generations is composed of moves of both the male and female components of reproduction. Male gamete dispersal is via pollen and contributes $\sigma^2 p$, whereas female gametes do not disperse in higher plants. The average gamete dispersal variance is therefore $\frac{1}{2}\sigma^2 p$, to which must be added the seed dispersal variance. Crawford suggests that $\sigma^2 = \sigma_p^2/2 + \sigma_s^2$, and thus $N = 12.6(\sigma_p^2 r/2 + \sigma_s^2)d$. If, as in Levin (1978a), we assume that net moves are zero, this can be translated to:

$$N = 12.6\left(\sum p_i^2 r/2n_p + \sum s_i^2/n_s\right)d$$

or estimates that are twice those of Levin's. This seems to me the best way of interpreting this difficult concept.

The neighbourhood area, $A$, is merely $N/d$. This area is a circle (which assumes that the population is two-dimensional, not necessarily always accurately). A circle of this type should include between 63.2 and 86.5% of the parents of an individual at its centre for non-linear population. An interbreeding population can thus be considered as a very large number of overlapping circles of this type. Neighbourhood is valid in space, but not in time. Thus, although neighbourhood may be a useful way of considering the number of individuals that are able to interbreed panmictically in a given season, and thus give indications of potential inbreeding effects, it is less successful as an indicator of gene migration and recombination. Nearly all plant populations overlap heavily for time as well as space. Generation overlap, and neighbourhood overlap, will allow some genetic recombination to occur throughout individuals in a population capable of some genetic exchange, however non-panmictic they may be with respect to one another in a given generation.

Although I have used the concept of neighbourhood in an earlier paper on *Primula veris* (Richards & Ibrahim 1978), I have become increasingly concerned about its general application and usefulness, for two practical reasons.

First, measurements of pollen or gene travel are always problematical, and in particular tend to vary very considerably from site to site, year to year, day to day, and hour to hour. In *P. veris*, estimates of $\sigma_p^2$ vary by a factor of more than 15 at the same site (Richards & Ibrahim 1978, Ibrahim 1979).

Second, estimations of neighbourhood size are heavily dependent on the density of plants ($d$). In a few species, particularly those of dry places such as dunes and deserts, plants are regularly spaced due to the limitations of water or other resources, and in these density may be a highly repeatable function. However, in the very great majority of plant populations, density varies greatly from one location in a population to another. In our experimental populations of *P. veris*, density of plants and thus estimates of neighbourhood size varied by a factor of over 50, which made any estimate of neighbourhood *size* within a population almost meaningless, although neighbourhood *area* remains valid. Variation in density estimates does of course depend on quadrat size, which in this case was 1 $m^2$ (where clone size averaged about 10 $cm^2$, or one-hundredth of the quadrat size). Larger quadrats would mean less variation in estimates of density, but if quadrat size approaches or exceeds neighbourhood areas, as would be the case for *P. veris* at even 5 $m^2$, estimates of neighbourhood size would become highly misleading. In fact, because of considerations of varying density, every plant in each population will have a different estimate of neighbourhood size appertaining to itself, and overall population estimates of neighbourhood seem to be meaningless in most cases.

Similar results, and conclusions, to ours have recently been published by Cahalan and Gliddon (1985) for the related primrose, *Primula vulgaris*. However, these authors make the point that varying density only affects estimates of neighbourhood size, and estimates of neighbourhood area will be density independent.

Apart from regularly spaced populations, the only instances in which neighbourhood otherwise might be a useful concept is when $A$ equals the limits of the population $n$. This will often be the case in outbreeding plants with good pollen travel in which populations are small spatially because of habitat limitations (biological islands), or because of scarcity (e.g. colonants, relicts, etc.). In these populations, neighbourhood calculations may be realistic, but may not be so informative as straight population counts, as in the work on small populations of *P. veris* reported in Chapter 2. However, population counts take no account of gene dispersal.

With these very considerable reservations in mind, it may nevertheless be instructive to report on a few neighbourhood sizes in plants that have

**Table 5.10** Neighbourhood sizes in trees and herbs.

| | $A$ ($m^2$) | $N_e$ | Reference |
|---|---|---|---|
| **Trees** | | | |
| *Fraxinus pennsylvanica* | 42 m | 16 | Wright (1953) |
| *Fraxinus americana* | 1766 | 4.4 | Wright (1953) |
| *Ulmus americana* | 1681 m | 253 | Wright (1953) |
| *Populus deltoides* | 1528 m | 230 | Wright (1953) |
| *Pseudotsuga taxifolia* | 2101 | 26 | Wright (1953) |
| *Cedrus atlantica* | 33 623 | 208 | Wright (1953) |
| *Pinus cembrioides* | 1766 | 11 | Wright (1953) |
| *Pinus radiata* | 29 550 | 1–3200 | Bannister (1965) |
| | | | |
| *Linanthus parryae* | 2.6–260 | 14–27* | Wright (1943b) |
| *Linanthus parryae* | 30 | 10–100* | Wright (1978) |
| *Phlox pilosa* | 11–21 | 75–282 | Levin and Kerster (1968) |
| *Lithospermum caroliniense* | 4–26 | 2–7 | Kerster and Levin (1968) |
| *Liatris aspersa* | 17–30 | 30–191 | Levin and Kerster (1969a) |
| *Liatris cylindracea* | 33 | 165 | Schaal and Levin (1978) |
| *Viola* spp. | 19–57 | 167–547 | Beattie and Culver (1979) |
| *Primula veris* | 20–30 | 5–200 | Richards and Ibrahim (1978) |
| *Lupinus texensis* | 2.8–6.3 | 42–95 | Schaal (1980) |
| *Senecio* spp. (bees) | 0.7–7.3 | 8–24 | Schmitt (1980) |
| *Senecio* spp. (butterflies) | 57–572 | 993–6154 | Schmitt (1980) |
| *Lupinus amplus* | | 36† | Zimmerman (1982) |
| *Thermopsis montana* | | 132† | Zimmerman (1982) |

* Some allowance made for effective density.
† Spike counts of clonal species.
m, length of neighbourhood on linear model.

been reported in the literature (Table 5.10). These have been extracted from the very useful review by Crawford (1984b), and include examples of both wind-pollinated trees and insect-pollinated herbs. In many cases, estimates of seed travel are scanty, or absent. It will be seen that many estimates are so variable as to be virtually meaningless.

In some cases, neighbourhoods ($N$) may be very small, indeed sometimes less than 10 individuals in extent, especially where populations are sparsely distributed, or limited in extent. Very small plant populations are probably very common, as in *Carex* where a third of all populations may number less than 20 individuals (Ch. 2). In other instances, $N$ may be very much smaller than $n$, as in *Primula veris*, where the study population exceeded 10 000 plants. Similarly, in *Lupinus texensis* and *Linanthus parryae*, populations often exceeded 10 000 and were thus very much larger than neighbourhood size calculations. It is noteworthy that all these species are herbaceous perennial entomophilous outbreeders, and at least some (*Primula veris, Phlox pilosa, Linanthus parryae*) are obligate outbreeders. Yet even in plants of this nature, effective population size, at least within a season, may be very small, and certainly small enough to lead to inbreeding effects, although these have yet to be demonstrated in plants of this nature. There is scope for more work on gene travel and the genetic structure of populations of plants with other means of gene transport, other breeding systems and other growth forms. There can be little doubt that neighbourhoods of trees and grasses with much more efficient pollen transport (see Table 5.9) and good seed transport, may be much larger, despite low genet densities brought about by large clone sizes.

## SEED DISPERSAL

As we have seen, estimations of neighbourhood require information about the variance of seed dispersal, and more recent formulae give seed dispersal twice the weighting of pollen dispersal in neighbourhood calculations. However, relatively little attention has been given to the seed dispersal of plants in population and breeding system genetics, and there is often a tacit assumption (as in Kerster & Levin 1968 or Schmitt 1980) that seed dispersal is much less important than pollen dispersal. Yet neighbourhood estimates are almost doubled in Levin and Kerster (1969a), when seed dispersal variance is included in the calculations, and the estimates of neighbourhood based on gene travel in Schaal (1980) are more than twice those based on pollinator flight distances alone.

The importance of seed travel in determining gene travel and neighbourhood will depend largely on the nature of the seed and the fruit. In *Primula veris*, the variance of seed travel measured by direct means varied from 0.67 to 1.77 $m^2$. This species has rather large seeds with no specialised means of transport, and compares with estimates of the variance of pollen travel from 1.75 to 13.07 $m^2$. Rather similar values are obtained by Levin and Kerster (1968, 1969b) for *Phlox pilosa* and *Liatris aspera*. Yet, one would imagine that the pappus-bearing cypselas of *Senecio* would disperse over long distances, and would add greatly to the neighbourhood estimates of Schmitt (1980). Little information is available for most animal-dispersed fruits. Those that pass through a gut, or are carried on fur and feathers and are dispersed by grooming, may habitually travel long distances, but non-randomly. Indeed, their final dispersal points may be very clumped indeed, as under a bird roost, at the entrance of a mouse burrow (*Primula vulgaris*; M. Wilson personal communication), in an ant nest (*Viola* spp.; Beattie 1978) or at a car park (burrs of *Acaena novae-zelandiae* on Lindisfarne, UK; Culwick 1982). As far as I am aware, algebraic estimates of the effects of far-dispersed clumped seed on neighbourhoods have yet to be derived.

However, Beattie and Culver (1974) show that the variance of the dispersal of ant-dispersed *Viola* seed varies from one-sixth to two-thirds times that of the variance of pollinator flight distances. *Viola* seed is first dispersed ballistically, and then a high proportion of the seed is carried by ants, who are attracted by the gelatinous elaiosomes, to their nests, which provide favourable sites for germination and seedling establishment. Ant carry is, of course, random in direction with respect to the initial ballistic phase of dispersal, and is from one-third to one-sixth the mean dispersal variance of ballistically dispersed seed. The variance of seed dispersal varies from 8.2 to 10.2 $m^2$, much higher than that estimated for *Primula veris*.

The burred fruits of the New Zealand member of the Rosaceae, *Acaena novae-zelandiae*, which has become extensively naturalised on Lindisfarne dunes, in the UK, can be transported in the fur and feather of animals and on the clothes of humans. They mostly disperse a very short distance through the agencies of the weather and small rodents. Culwick (1982) found that the variance of the dispersal of marked single fruits varied from 15 to 30 $cm^2$, or only about 0.02 $m^2$. However, when the distance of 50 isolated burrs from the nearest fruiting plant was measured, the variance of the minimum dispersal distance of these proved to be 20 $m^2$. Some burrs travelled at least 20 m, and the distribution of isolated seedlings suggests that single dispersal events of hundreds of metres may occasionally happen.

The leptokurtosis on the distribution of animal-dispersed fruits may be very extreme indeed. Thus petrels and albatrosses (Procellariidae) have almost certainly carried *Acaena* burrs, found in their downy feathers, thousands of kilometres across southern oceans, and the genus is distributed on all the isolated sub-antarctic islands. In 1981, a Newcastle University botany class discovered a single seedling of *A. novae-zelandiae* beside the path of an upland (300 m) site, 50 km from the nearest site on the coast at Lindisfarne. It immediately occurred to us that during the previous year the equivalent class had visited Lindisfarne and the upland site on successive days, and had been much troubled by the itchy burrs that had adhered to socks etc. from the previous day. This new upland colony has since expanded (1983) to 20 plants distributed over 200 m of track. In recent years, several new coastal populations have also been established, doubtless due to the increased mobility of holidaymakers. Recent dispersal events in the county thus vary from 5 to 50 km, and variance estimates based on these accidental events are exceedingly large.

Quite a lot of evidence is available for the dispersal of seeds and fruits of many kinds (Salisbury 1942, Altman & Dittmer 1964) and there is no doubt that seed dispersal can at times be very distant, even intercontinental, with huge variance estimates. Thus one seedling of an unnamed species of pine can arise every 10 $m^2$, as far as 12 km downwind from the nearest stand of the tree (wind-dispersed seed). In contrast, the berry-forming *Juniperus scopularum* (western red cedar), was recorded as not establishing more than 60 m from the source stand, and nearly all seedlings dispersed less than 30 m. Unfortunately, most of these early measurements are not suitable for the estimation of variance, and are usually made without any corresponding information on pollen travel, so that neighbourhood calculations are impossible.

It is also likely that those accurate measurements made on seed dispersal do not take into account rare, distant dispersal events. Thus, although we found that cowslip (*Primula veris*) seeds had a mean travel of only 11 or 12 cm, and a maximum travel of 80 cm (using sticky tapes 5 m in length), indirect methods (as in the distance of seedlings and hybrids from potential parents) suggested that seed travel might be much further, certainly in excess of 50 m. It is noticeable in the UK how cowslips have colonised motorway verges in recent years, and have often had to travel hundreds of metres, or even a few kilometres, to do so. Presumably such distant travel of relatively large, unspecialised seeds occurs through the chance transportation of mud on feet or hooves, which will, by its nature, be a rare event. Such long-distance events are very important in colonisation, in gene travel, and in breaking down populational or even specific isolation. They have little significance, however, in the consideration of inbreeding effects due to small population size.

## ASSORTIVE MATING

Up to this point, there has been an assumption that pollen is potentially able to fertilise any plant in the population equally, and if panmixis is restricted and populational subdivision occurs, these are functions of spatial separation alone.

However, it has been realised for many years that pollinators may discriminate between different flowers in the same population, and that features that lead to this discrimination are often heritable, as for flower colour (Kay 1978, 1982, Levin & Watkins 1984), shape (Levin & Kerster 1971, 1974) or height above ground (Levin & Kerster 1973, Levin & Watkins 1984, Eisikowitch 1978). It has often been suggested that intra-population variability of this kind, which leads to assortive mating by pollinators, may result in gene barriers that allow sympatric speciation to occur. Many fine books on the processes of plant speciation have been written, and in some of these the rôle played by pollinators is given due consideration, especially when written by Grant (1963, 1981), based on his work on *Aquilegia* (Grant 1952) and particularly *Gilia* (e.g. Grant & Grant 1965).

### *Flower-colour polymorphisms*

Although Darwin (1877) pioneered work on intraspecific discrimination by pollinators, he restricted himself to heterostylous and diclinous species, and to some domesticated flower mutants. The work concerning heterostyly and dicliny is reported elsewhere in this book (Chs. 7 and 8). In contrast, detailed field observations on flower polymorphisms, and especially on flower-colour polymorphisms, are very recent, and have been reviewed by Kay (1978, and especially 1982), who has led the field, and his ex-student Mogford (1978). Although flower-colour polymorphisms were intensively studied as early as 1942 (Epling & Dobzhansky 1942) in *Linanthus parryae*, their pollination seems to have been ignored until Lloyd (1969) worked with *Leavenworthia* and Levin (1969) studied *Phlox*.

Kay (1982) provides detailed evidence of assortive within-population pollination for four species with flower-colour polymorphisms (Mogford 1978 does so for a fifth), and for two species with dioecy, as well as providing a review of the subject. I have no intention of repeating his detailed data here, but will summarise his findings below.

**Occurrence** Flower-colour polymorphisms occur in a wide range of species, especially entomophilous herbs of temperate climates. They are most frequent in flowers that are coloured shades of pink, purple or blue by anthocyanin pigments, and the biosynthetic pathways leading to anthocyanin pigments have many steps at which single mutants may change or stop biosynthesis. Thus, anthocyanin-less morphs, which are white to yellow (depending on additional carotenoid pigmentation, p. 97), may result from one of several different mutations. Such mutations probably occur at low frequencies in all flowers with anthocyanin pigmentation (very well known to the British public in populations of heather, *Calluna vulgaris*). Establishment of flower-colour polymorphisms will depend on disruptive selection pressures, genetic drift, frequency dependent selection or, in a few cases, hybridisations or introgression.

**Distribution** The ecological and geographical distribution of gene frequencies for flower-colour polymorphisms may show contrasting patterns. In the marsh thistle, *Cirsium palustre*, white plants are widespread, but become more common with altitude. This distribution is echoed by some species in the Alps, and other mountainous regions (Mogford 1978).

In the UK, the mountain pansy, *Viola lutea*, in southern and low-level populations is purple, whereas in upland and northern populations it is yellow. Polymorphic populations, including many plants with different coloured petals, occupy broad intermediate zones. In contrast, white flowers (although with purplish veins) of the predominantly yellow-

**Figure 5.13** Purple- and white-flowered plants of the fritillary *Fritillaria meleagris* (Liliaceae) in southern England (×0.5).

flowered wild radish, *Raphanus raphanistrum*, are restricted in the British Isles to the south and east (Kay 1978).

In at least five different species of Himalayan *Primula*, western races are yellow-flowered and eastern races purple-flowered (Richards 1977). This strong correspondence between species suggests a common selection pressure that changes geographically. Populations in intermediate positions on the cline usually have mixed flower colours, perhaps through repeated immigration into this zone, or perhaps in response to disruptive selection.

Polymorphic populations may be very local. Among very many examples, I can instance the white/pink population (60% white) of *Geranium robertianum* in a small wood at Capheaton, Northumberland, the only one I know in this pink-flowered species, or the white/pink population (30% white) of the storksbill, *Erodium cicutarium*, at Seaton Sluice, Northumberland, UK. It is likely in these isolated instances that white-flowered genes have become established in populations by chance events (genetic drift).

In a few cases, all populations of a plant tend to show flower-colour polymorphism. I believe that all British populations of the fritillary, *Fritillaria meleagris*, are polymorphic for purple and white (Fig. 5.13). This bulbous vernal genus displays polymorphism in a high proportion of its 100 species and their populations. Sometimes these polymorphisms are for a wide range of colours between pale green and almost black, as in

*Fritillaria graeca*. It is curious and remarkable that a high proportion of the Mediterranean vernal bulbous and cormous genera, as in *Anemone*, *Ranunculus asiaticus*, *Iris*, *Crocus*, *Scilla* and *Orchis* demonstrate flower-colour polymorphisms in a high proportion of populations. This seems never to have been explained satisfactorily.

**Pollination** Although it is possible that flower-colour polymorphisms may pleiotropically reflect disruptive selection pressures that have nothing to do with pollination, this has never been established. However, for devil's bit (*Succisa pratensis*), marsh thistle (*Cirsium palustre*) (both purple/white), crown daisy (*Chrysanthemum coronarium*) and wild radish (*Raphanus raphanistrum*) (both yellow/white), Kay (1978, 1982) convincingly shows that pollinators discriminate between colour morphs in wild populations. In the last two species in particular, pollinator visits are of crucial importance, for both are annuals, and *Raphanus* at least is self-incompatible.

By marking individual pollinators, Kay has been able to study discrimination not only by a species of pollinator, but by an individual. In so doing, he has revealed, somewhat unexpectedly, that different individuals within a pollinating species at a single site frequently show differential discrimination of flower colours. Thus, individuals of hive bees (*Apis*) visiting *Succisa* may specialise on purple flowers or on white flowers, or may not discriminate between them. Nevertheless, from Kay's results, some generalisations about pollinator preferences can be made. For *Cirsium* and *Succisa*, butterflies of several species nearly always visit purple rather than white flowers. For *Chrysanthemum* and *Raphanus*, hive bees nearly always visit yellow (insect purple) rather than white (insect white) flowers. However, generalisations concerning one plant break down when another is investigated. Butterflies (chiefly *Pieris* spp.) very markedly prefer yellow flowers of *Raphanus*, whereas some individuals also prefer yellow flowers of *Chrysanthemum*, others prefer white flowers, and yet others do not discriminate between the colours. Kay suggests that *Pieris* species may use the yellow flowers of some Cruciferae (such as *Raphanus*) as a recognition signal for choice of plant for egg laying, and thus innately prefer yellow-flowered plants for nectar feeding as well.

Kay (1978) also notes differential discrimination on *Raphanus* between the long-tongued *Bombus pascuorum* (white) and the short-tongued *B. terrestris* (yellow). This he suggests may relate to the more abundant, and short-tubed, Charlock (*Sinapis arvensis*), which is always yellow flowered, but superficially closely resembles *Raphanus*. It is possible that many individuals of *B. terrestris* major on *Sinapis*, and cannot differentiate between this species and yellow *Raphanus*. In contrast, the long-tongued *B. pascuorum* may find the short-tubed *Sinapis* relatively unrewarding, as many flowers will have been visited by other species. It may therefore avoid yellow cruciferous flowers, but recognise the white flowers of long-tubed *Raphanus* as a more rewarding source.

It is clear from Kay's work that pollinator discrimination between flower-colour morphs leading to assortive mating (which has been tested in progeny) is a common and widespread phenomenon in polymorphic populations, although it is probably never complete enough to lead to a total breeding barrier.

**Selection** Thus, it is likely that geographical separation of flower-colour morphs (p. 184) may at times show adaptation to different frequencies of types of pollinators in different regions, although this has never been demonstrated. In other cases, white or yellow morphs may prove attractive to a wider range of pollinators (more generalist), and thus succeed in areas of low pollinator diversity or density, as in northern or upland regions (Mogford 1978). It is well known that the flora of New Zealand has a very high preponderance of white flowers (66% compared to 21% in the UK) (Godley 1979) and I have suggested (Richards 1981) that this may be linked to the low diversity of pollinator species on these remote islands. It is also conceivable that white flowers may prove more visible to pollinators in misty or rainy upland areas (Richards 1970d).

For some species, such as *Chrysanthemum coronarium* which can be polymorphic throughout its range, in some areas (Spain) populations are more usually monomorphic (either yellow or white), whereas in others (Greece) populations tend to be polymorphic. It is possible that polymorphic populations are associated with low densities and/or diversities of pollinators, and are encouraged to become polyphilic by having different flower colours. In other areas richer in pollinators, oligophily, encouraged by colour monomorphy, may be preferred.

### *Other characters leading to population subdivision*

It is possible that intrapopulation variability in scent may also lead to population subdivision, although this seems never to have been shown. In contrast, Levin (1969) gives a clear example of discrimination by pollinating butterflies between cultivars of *Phlox drummondii* with different corolla shapes. He additionally observed discrimination between different flower-colour morphs in this species. Levin (1972) also shows that interspecific differences in corolla outline between two sympatric species of *Phlox*, *P. bifida* and *P. divaricata*, are probably largely responsible for the very considerable discrimination between these species shown by their lepidopteran pollinators. Only 1–2% of outcross pollinations are between species in mixed populations. Very similar values (1.6% interspecific pollinations) were obtained by Levin and Berube

(1972) for artificial mixed populations for another pair of *Phlox* species, *P. pilosa* and *P. glaberrima* pollinated by the butterfly *Colias eurytheme* (a 'sulphur', or 'clouded yellow'). These species have flowers of very similar colour and outline, but very different tube lengths and anther and stigma positions.

Intraspecific discrimination according to the height of the flowers above ground occurs as a result of constancy in foraging height (p. 143). It has been demonstrated by Levin and Kerster (1973) for dwarf and tall forms of purple loosestrife, *Lythrum salicaria*, and by Eisikowitch (1978) for artificially mixed populations of the tall inland love-in-the-mist *Nigella arvensis* ssp. *tuberculata*, and its dwarf maritime ecotype ssp. *divaricata*.

Population subdivision frequently occurs due to differences in flowering time, and such differences are probably among the commonest stimuli to sympatric speciation. Such familiar interfertile species pairs in the British flora as the red and white campions (*Silene dioica* and *Silene latifolia* (=*alba*), the primrose and cowslip (*Primula vulgaris* and *P. veris*) and herb bennet and water avens (*Geum urbanum* and *G. rivale*) are separated by flowering time as well as at least partially by habitat. Hybridisation is commonest in the north of Britain where overlap in flowering time is the greatest (Stace 1975). Mutants with abnormal flowering seasons quite frequently occur in wild populations (the winter flowering form of hawthorn, *Crataegus monogyna*, known as 'Glastonbury Thorn' is a familiar example). Such mutations often respond to 'Wallace Effects' where hybridisation between two ecodemes is disadvantageous, allowing reproductive isolation in time to occur between the ecodemes. This has been found to occur in Welsh mine populations of the grasses *Agrostis tenuis* and *Anthoxanthum odoratum* (McNeilly & Antonovics 1968), where metal-tolerant races flower some six to eight days earlier than non-tolerant plants growing in adjacent non-toxic sites. This difference is sufficient to provide considerable reproductive isolation for the tolerant ecodeme, and serves to preserve the genetic integrity of the tolerant and non-tolerant races. Zinc-tolerant races of the bladder campion, *Silene vulgaris*, similarly flower several weeks earlier than non-tolerant races in Germany (Broker 1963). Unlike the grasses, this species is insect pollinated.

It is difficult to assess the importance of behavioural (ethological) discrimination by pollinators on flower variation within populations in allowing speciation to occur. In *Pedicularis* (Macior 1982) and *Ophrys* (Stebbins & Ferlan 1956), it is probable that pollinator discrimination has played a major rôle in speciation through a great variety of flower shapes and point of pollen release and reception in *Pedicularis*, and by providing pseudocopulatory targets suitable for different species of bee in *Ophrys* (Ch. 4). Flower-colour variation may also have led to speciation in genera such as *Aquilegia* (Grant 1952) and *Mimulus*, where variation between white, blue and red, and yellow and red, respectively, can differentially attract such different classes of pollinators as moths, butterflies and birds, or bees and birds.

A subject that deserves a brief mention in this chapter is the effect that *lack* of pollinator discrimination between two intersterile species with similar flower syndromes can have on the success of one or both. Competition for the same pollinators can result in a minority species receiving most of its outcrossed pollen from the other species. If it is self-incompatible, this may severely reduce its seed-set, and in the case of the self-incompatible annual charlock and white mustard (*Sinapis arvensis* and *S. alba*) this effect may be so severe as to result in the competitive exclusion of the rarer species (M. A. Ford and Q. O. N. Kay unpublished observations).

## CONCLUSION

In most genera, reproductive isolation leading to speciation occurs through chromosomal mechanisms, or temporal or spatial isolation. The evolution of floral traits resulting in differences in pollination strategy more often occurs secondarily. Nevertheless, as this chapter has attempted to show, floral traits play a very important rôle indeed in controlling pollen flow within and between flowers, and within and between plants. Levels of within-population genetic variability, and hence of evolutionary potential, and niche width are highly reliant on floral traits and pollination strategies.

CHAPTER SIX

# *Multi-allelic self-incompatibility*

---

Self-incompatibility (s-i) is best defined as the inability of a fertile hermaphrodite seed plant to produce zygotes after self-pollination (De Nettancourt 1977). It is a mechanism that usually ensures obligate outbreeding, but it may be reproductively inefficient. Self-incompatibility results from the failure of selfed pollen grains to adhere to, or germinate on, the stigma, or the failure of selfed pollen tubes to penetrate the stigma, or grow down the style. With one possible exception (*Borago officinalis*, Crowe 1971), all self-incompatibility operates before fertilisation. This definition thus rules out failure of sexual fertilisation due to sterility, embryonic lethality or breeding barriers.

We must suppose that cultivators have realised for thousands of years that many plants are self-incompatible, that is self-sterile but cross-fertile. However, early botanists were principally concerned with the systematics and medicinal properties of plants, and little attention was paid to their reproduction. Because most flowers were seen to be hermaphrodite, it was assumed that they self-fertilised (Kolreuter 1763, Sprengel 1793). It was Kolreuter, however, who has been credited with the first published observation of self-incompatibility, in *Verbascum phoeniceum*. Casual observations of self-incompatibility were made during the 19th century, but there was a tendency to regard these as the result of inbreeding, and their true significance was missed. Darwin (1876) studied self-incompatibility in five species, but he ignored the existence of cross incompatibility, and the importance of this phenomenon in understanding the inheritance of self-incompatibility. As he was also in ignorance of the principles of genetics, this is perhaps understandable. He did, however, note that self-incompatibility favours outbreeding, and that fertilisation is prevented when the sexual elements are identical. There is a good review of early work on self-incompatibility in Arasu (1968).

The rediscovery of Mendel's paper on inheritance at the turn of the century stimulated further work on incompatibility, and early attempts were made to fit patterns of self-incompatibility into the new genetics (De Vries 1907, Correns 1913). In 1917, Stout coined the world self-incompatibility (self-sterility, which now has a wider meaning, had been used until then). It was not until the 1920s that a group of papers, which were produced independently, first showed how multi-allelic gametophytic self-incompatibility was inherited (Prell 1921, East & Mangelsdorf 1925, Lehmann 1926, Filzer 1926).

It is curious that, even today, little attention is paid to self-incompatibility except by plant breeders and research workers concerned with breeding systems. With the exception of Clapham *et al.* (1962), no regional floras give information on the breeding system in general or self-incompatibility in particular. There have been few concerted attempts to collect information on self-incompatibility in plants in a systematic way, as has been done for chromosome numbers, pollen morphology, anatomy, cuticle structure, secondary chemical products, ecology and chorology. The exceptions are East (1940) and Brewbaker (1957). Textbooks on genetics rarely give self-incompatibility in plants more than a passing mention (Williams 1964 is a distinguished exception). Usually, any treatment in depth refers to the much more uncommon di-allelic, heteromorphic conditions such as that in *Primula* (Ch. 7). Even modern research papers on pollination biology and gene flow usually manage to omit any information about self-incompatibility, a knowledge of which is vital to the understanding of any pollination system. Horrific confusions between self-pollination and self-fertilisation are commonplace.

Yet, at the simplest level, it is extremely easy to establish whether a plant is self-incompatible or not. Isolated genets (single plants or clones) will set little if any seed if they are self-incompatible (s-i). If they are self-compatible (s-c) they will usually set seed. If s-i is suspected, it is readily confirmed by making a cross with another genet (i.e. a plant that has arisen from another seedling). This should usually result in seed-set. However, herkogamy or dichogamy may result in s-c plants setting no seed in isolation if an animal visit is required for pollination to occur. Thus the control test of artificial self-pollination should also be carried out:

| | Seed-set | |
|---|---|---|
| | s-i | s-c |
| isolated genets, open pollination | none | usual |
| artificial crosses between genets | yes | yes |
| artificial selfs within genets (no open pollination) | no | yes |

If no seed is set after both crosses and selfs, disease or predation of seeds (examine when young), or sterility (examine pollen fertility) should be suspected.

Self-compatibility or self-incompatibility are often considered to be synonymous with the breeding system (as in 'breeding system genetics').

However, they form only one function in a suite of characteristics which control the breeding system, all of which are equally important, and all of which must be considered together:

(a) Within-flower position of anthers and stigma (herkogamy); is within-flower pollination (autogamy) possible without an animal visit?
(b) Within-flower timing of anther dehiscence and stigmatic receptivity (dichogamy); is within-flower pollination possible with respect to timing?
(c) Amount of pollen carry to stigmas between flowers of the same plant (geitonogamy), and different plants (xenogamy) (together, allogamy); degree of pollen carryover (Ch. 5).
(d) Number of different plants (genets) normally reached by cross pollination (neighbourhood size, Ch. 5).
(e) Behaviour of the pollen on stigmas of the same genet (s-c or s-i).
(f) Proportion of the ovules behaving sexually (possibility of agamospermy, Ch. 11).

Thus, it is frequently found that an s-i species, which is an obligate outbreeder, undergoes substantial self-pollination, although the selfed pollen is non-functional. Equally, an s-c species may rarely, if ever, be self-pollinated and is thus effectively an outbreeder (this is true of many orchids, Ch. 4). Particular attention must be paid to the little-studied phenomenon of geitonogamy. Many s-c species undergo no within-flower, or even within-ramet, self-pollination, due to herkogamy and/or dichogamy, and may be thought to be outbreeders. Nevertheless, a high proportion of successful pollinations may occur within genets (self-fertilisation), even if made principally between ramets.

Neither is it the case that multi-allelic self-incompatibility is the sole cause of self-sterility. Di-allelic self-incompatibility (Ch. 7) and dicliny (Ch. 8) are other important mechanisms that are discussed in other chapters.

## THE BASIS OF MULTI-ALLELIC SELF-INCOMPATIBILITY

Early workers in the field soon established the principle that recognition of self led to incompatibility, and that recognition factors were heritable (East 1915a,b, 1917a,b, 1918, 1919a,b,c, Correns 1913, 1916a, Baur 1919, Stout 1916, 1917, 1918a,b, East & Park 1917). Work on *Nicotiana* by Anderson (1924) and especially by East and Mangelsdorf (1925) showed in some detail how one form of self-incompatibility (which we now know to have been one-locus gametophytic) was inherited. There is a useful early review by Sirks (1927), and for fruit trees by Crane and Lawrence (1929).

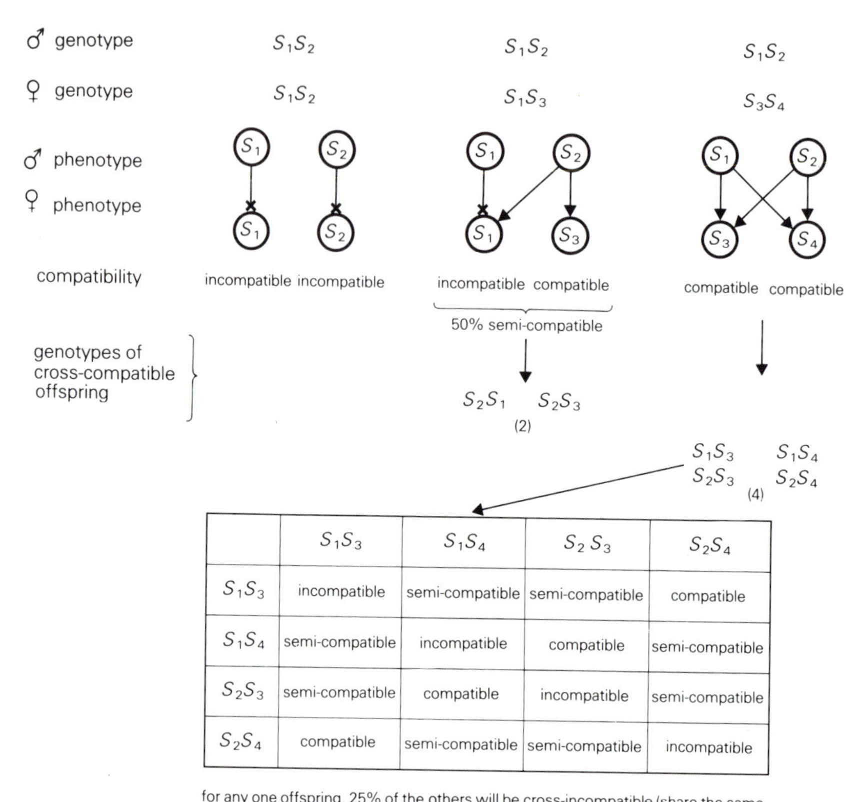


| | $S_1S_3$ | $S_1S_4$ | $S_2S_3$ | $S_2S_4$ |
|---|---|---|---|---|
| $S_1S_3$ | incompatible | semi-compatible | semi-compatible | compatible |
| $S_1S_4$ | semi-compatible | incompatible | compatible | semi-compatible |
| $S_2S_3$ | semi-compatible | compatible | incompatible | semi-compatible |
| $S_2S_4$ | compatible | semi-compatible | semi-compatible | incompatible |

**Figure 6.1** Behaviour and inheritance of *S* alleles in one-locus gametophytic self-incompatibility.

The following principles were established on the basis of this work:

(a) The same recognition factors operated in the pollen and the stigma.
(b) These factors operated independently in the diploid stigma, and in the haploid pollen grains; thus pollen from the same diploid father could show two types of reaction on a single stigma (semi-compatible) (Fig. 6.1).
(c) When the pollen grain and the stigma carried the same factor, incompatibility ensued; thus successful crosses arose between gametes carrying different factors, and plants were by definition heterozygous for incompatibility factors.
(d) Recognition factors were inherited at many alleles at the same locus; this was known as the *S* locus, and alleles were characterised by $S_1$, $S_2$, $S_3$, $S_4$ ... $S_n$.

As we will see, work over the following 60 years has revealed many exceptions to these early principles. We are also much closer to understanding the physiological basis of self-incompatibility, which has proved surprisingly difficult to elucidate.

The experimental basis of early work on gametophytic systems usually took the following pattern:

(a) Cultivation of as many genets as possible of a population or species that showed self-incompatibility.
(b) Creation of as many pair-wise crosses within this material as possible.
(c) Examination of pollen-tube growth and/or seed-set of these pair-wise crosses.
(d) Classification of parents into groups, within each of which plants are cross-incompatible. These are assumed to share the same *S* genotype; if pollen growth is examined, semi-compatible crosses can also be identified (Fig. 6.1).
(e) Determination of the number of alleles residing at the *S* locus by the formula: number of distinct *S* genotypes = $n(n-1)/2$ where $n$ is the number of alleles in the system (Lewis 1955). This equation must be solved quadratically. It is also possible to show that the proportion of all crosses that are incompatible equals $2/n$.
(f) Growing to maturity families of siblings from one or more crosses, and determining the number of cross-compatible classes that occur between siblings of a single cross.
    (i) If the parents have totally different genotypes, four cross-compatible classes should arise amongst the offspring; for each offspring, a quarter of the others will be incompatible, half semi-compatible and a quarter fully compatible (Fig. 6.1):

$$S_1S_2 \text{ (male)} \times S_3S_4 \rightarrow S_1S_3,\ S_1S_4,\ S_2S_3,\ S_2S_4$$

    (ii) If the parents share an *S* allele (semi-compatible), two cross-compatible classes should arise amongst the offspring:

$$S_1S_2 \times S_1S_3 \rightarrow S_2S_3,\ S_2S_1$$

    If more cross-compatible classes arise amongst the offspring family, it can be assumed that more than one incompatibility locus is involved (Fig. 6.2). If fewer cross-compatible classes arise among the offspring family, it is likely that the system is under sporophytic control.

## *Frequency dependence*

One of the features of the *S*-allele system is the very large number of alleles that can coexist, apparently at one locus, within a population. In

*Oenothera organensis*, Emerson (1939) reported the occurrence of 45 alleles in approximately 500 plants. Extrapolations from the proportion of *S* alleles discovered to be novel when new plants were added to the system allowed Lewis (1955) to report that up to 400 alleles may coexist in *Trifolium pratense* and *T. repens*. Lewis (1949a) reviewed the literature concerning the size of polyallelic series at the *S* locus. He concluded that whereas selection pressures will act on most genetic loci to reduce the number of coexisting alleles (most of which will be disadvantageous), the opposite will be true for incompatibility loci. If it is assumed that all alleles are equally viable, frequency-dependent selection will encourage the coexistence of large numbers of alleles, the more the fitter. Maybe, he argued, all loci have the capability for producing many alleles, which is only revealed in the incompatibility system. More recent information on the structure and function of genes, and of the mutation and complementation of *S* alleles, has led to the creation of different models, which are discussed later.

However, it is irrefutable that such an incompatibility system will encourage many alleles to coexist, for the greater the number of alleles, the lower the frequency of cross-incompatible matings, which are disadvantageous. It can also be stated that if all *S* alleles are equally viable, there will be a frequency-dependent selection pressure for each to be equally numerous. Commoner alleles will encounter each other more frequently, and will be disadvantaged. Rarer alleles will be advantaged by meeting each other more rarely. Thus the optimal frequency for each allele should be $1/n$ where $n$ is the number of alleles (but see O'Donnell & Lawrence 1984 for a modern discussion of linkage effects).

## GAMETOPHYTIC AND SPOROPHYTIC SYSTEMS

The conspicuous nature of heterostylous systems (Ch. 7) led to early work. Darwin (1862, 1877) and Hildebrand (1863) showed that the distyly in *Primula* was associated with a self-incompatibility, the phenotypes of which segregated at approximately equal numbers in the offspring of a legitimate cross. Pin selfs gave pins, but thrum selfs gave pins and thrums to the extent that selfing was successful. The significance of these simple results was noted much later, after the rediscovery of Mendel's genetics. Thrum selfs can generate both morphs and must be heterozygous. Thus they must produce pollen grains of two genotypes (*S* and *s*). Yet, all thrum pollen grains behave in the same way, as thrums (S phenotype). Control of pollen grain behaviour is by the heterozygous thrum male parent *Ss*, *S* being dominant. In other words, behaviour is controlled by the diploid sporophyte (sporophytic control). If the haploid pollen grains arising from a heterozygous parent showed independent behaviour, depending on their individual genotype, control would depend on the gametophyte (gametophytic control), as we have already seen for *Nicotiana*.

Work by Correns (1912, 1913) on *Cardamine pratensis* showed that a similar inheritance of a di-allelic sporophytic incompatibility system could occur in the absence of distyly or heteromorphy. However, it was thought for many years that sporophytic systems were di-allelic and gametophytic systems multi-allelic. It was not until 1950 that Gerstel, and Hughes and Babcock demonstrated multi-allelic sporophytic incompatibility in *Parthenium argentatum* and *Crepis foetidus* respectively. Similar systems were later shown to operate in *Cosmos bipinnatus* (Crowe 1954), *Iberis amara*, *Raphanus sativus* and *Brassica campestris* (Bateman 1954) and by later workers in *Brassica oleracea* (cabbage) and *Raphanus raphanistrum* (wild radish). These species all belong to the families Compositae (Asteraceae) and Cruciferae (Brassicaceae) respectively. It seems to be accepted today that multi-allelic sporophytic systems are confined to these two large families of plants, which contain many species of agricultural significance, and that all s-i plants in these families have sporophytic systems. It is remarkable that these two families have no obvious phylogenetic relationship. It has been suggested that sporophytic systems are secondary, having been derived from self-compatible plants, or (which I consider less likely) from gametophytic systems directly. If this is the case, similar systems appear to have arisen independently early in the developmental history of these two plant families.

In contrast, di-allelic sporophytic systems occur in some 13 plant families, usually in conjunction with a floral dimorphy. These are also considered to be secondary in origin, having arisen from self-compatible plants (Ch. 7). It is not possible for a di-allelic system to function under gametophytic control, the minimum number of alleles permissible at a gametophytic locus being three, as shown below:

(a) *Ss* (male) × *ss*: *S* pollen can function giving *Ss* offspring only;
(b) *ss* (male) × *Ss*: no pollen can function.

Gametophytic multi-allelic systems are far more widespread than sporophytic systems. Our knowledge of their distribution is still very incomplete, but Darlington and Mather (1949) estimated that half the species of Angiosperms were s-i, and Brewbaker (1957) recorded s-i in at least 71 families, and in 250/600 genera. In a much smaller sample of 25 species of tropical forest tree, Bawa (1979) showed that 88% were s-i. Certainly, a high proportion of Angiosperms shows gametophytic s-i, and this may be higher than 50% overall. What is certain is that gametophytic s-i is found in most orders of flowering plants, and is found in what are traditionally considered to be the most primitive orders in the dicotyledons and the monocotyledons, the Magnoliales, Winterales,

Hamamelidales and Nympheales. There is thus reasonably strong circumstantial evidence that gametophytic s-i is primitive to the Angiosperms, particularly as comparable systems have been discovered in the Pteridophytes and Gymnosperms (Bateman 1952). Such a view has not been seriously challenged since it was first proposed by Whitehouse (1950), who, however, differed from Bateman about the point of the first evolution of s-i. Whitehouse considered that s-i arose very early in the history of the Angiosperms, but not before, and was associated with the development of the hermaphrodite flower and the closed carpel. He believed that it played a crucial rôle in the early expansion and subsequent success of the Angiosperms, and doubtless in this he was correct.

Generally speaking, it has been considered that all other mechanisms of s-i are secondary, having arisen from one-locus gametophytic systems as follows:

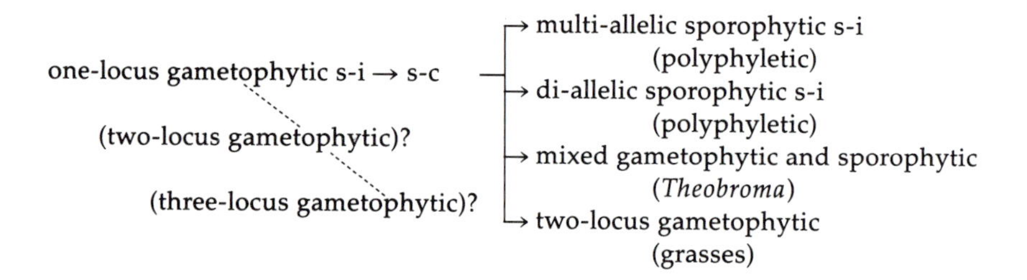


In order to identify the system operating in a self-incompatible plant, attention should be paid to three features:

(a) The number of cross-compatible types arising in a parental population.
(b) The occurrence of semi-compatibility; i.e. does all pollen from a single parent always behave the same way on a single female parent?
(c) The number of cross-compatible classes that arise in the family resulting from a single cross.

Systems can be identified as follows:

(a) Number of cross-compatible groupings among parents:
  (i) one – di-allelic sporophytic;
  (ii) three or more – multi-allelic.
(b) Occurrence of self-incompatibility;
  (i) all crosses fully compatible or fully incompatible – sporophytic;
  (ii) some crosses semi-compatible (examine pollen growth – one-locus gametophytic (Fig. 6.1);
  (iii) some (very few) crosses quarter-compatible – two-locus gametophytic (Fig. 6.2).

(c) Number of cross-compatible groupings in a single family from a fully compatible cross:
  (i) four – one-locus gametophytic;
  (ii) more than four – two-locus gametophytic or three-locus gametophytic (see Fig. 6.2);
  (iii) fewer than four – sporophytic.

The reason why sporophytic systems usually generate fewer cross-compatible classes amongst offspring is discussed later (p. 215). It is due to dominance interactions between *S* alleles.

There are features of pollen, the stigma, and pollen germination and growth in vitro, and in incompatible crosses that separate gametophytic and sporophytic systems (Table 6.1). Although it would involve a circularity to state that these can identify the system, their correspondence with various systems is very high, and these features provide a valuable confirmation as to the system operating. They also provide valuable clues as to how the various systems function (Brewbaker 1959, Heslop-Harrison 1975b).

The close correspondence within gametophytic or sporophytic systems for these various features allows one to suggest the following points:

(a) Gametophytic and sporophytic systems differ fundamentally in their operation.
(b) These features might be functional with respect to the operation of each system and coadapted.

**Table 6.1** Morphological and physical features that differentiate multi-allelic sporophytic and gametophytic systems.

| Features | Gametophytic | Sporophytic* plus Gramineae |
|---|---|---|
| Pollen grain: | binucleate | trinucleate |
| respiration | low | high |
| viability | long | short |
| growth *in vitro* | easy | difficult |
| stigma papillae | 'wet' with gappy cuticle | 'dry' with entire cuticle |
| site of pollen-tube inhibition when incompatible | in style | at stigma surface |
| site of callose deposition in incompatible pollen | intine | exine |

* Di-allelic, heteromorphic sporophytic systems do not fit this pattern; some heteromorphic species, or morphs, can have gappy wet stigma cuticles, easily germinated pollen and stylar inhibition; however, all appear to have trinucleate pollen.

(c) Gametophytic and sporophytic systems have each arisen on only one occasion (monophyletically). Although this may be true for gametophytic systems, the sporadic, scattered occurrence of di-allelic and multi-allelic, heteromorphic and homomorphic sporophytic systems between unrelated taxa strongly suggests that these are in fact polyphyletic. If this is the case, we must assume that these features of a sporophytic system are necessary for it to operate correctly.

(d) It has been suggested that the timing of gene expression for gametophytic incompatibility in the pollen must occur late, binucleate gametophytic pollen only showing expression after the second pollen-grain mitosis in the pollen tube (Brewbaker 1959). In sporophytic systems, it is argued, the second pollen-grain mitosis can occur much earlier, for the phenotype of the pollen grain is expressed by the anther, not the pollen. This view has been challenged by Lewis (1949b) and Pandey (1959, 1970), whose mutagenic work suggests that expression of gametophytic pollen occurs before either pollen-grain mitosis.

(e) The Gramineae (grasses) show all the characteristics of a sporophytic system, with trinucleate-type behaviour. However, they display a two-locus gametophytic control of incompatibility (p. 204). This suggests that self-incompatibility in the grasses (a very derived type of plant family) has arisen secondarily, in the same way as the sporophytic systems, not being part of the evolutionary continuum of gametophytic systems. It is not apparently a precondition of multilocus gametophytic systems to have trinucleate-type behaviour, for members of the Ranunculaceae may have three-locus gametophytic systems with binucleate type behaviour (Lundqvist *et al.* 1973). Perhaps trinucleate-type behaviour is characteristic of all secondary systems, which are otherwise sporophytic, not being part of the ancestral binucleate gametophytic system.

### *Genetic control of gametophytic systems*

The most frequent form of s-i in the Angiosperms, perhaps occurring in half the plant families, is one-locus multi-allelic gametophytic control. This system, first fully elucidated by East and Mangelsdorf (1925), for *Nicotiana sanderae* has been considered primitive to the Angiosperms by Whitehouse (1950), and it certainly occurs in many primitive taxa. Essentially, between 3 and over 400 alleles appear to occur at a single locus. Diploids are heterozygous for two of these alleles, both of which are expressed in the stigma and style. Pollen grains show independent expression, segregating at 1:1, and thus crosses can be incompatible, semi-compatible (with half the grains growing normally) or fully compatible, depending on the genotype of each parent (see Fig. 6.1). Because crosses can only occur between pollen grains and stigmas containing different *S* alleles, offspring must be heterozygous.

There has been a good deal of discussion as to the nature of the *S* gene locus, and its generation of mutants. Two basic models of the structure of the *S* gene have been proposed. In the first, originally put forward by Lewis (1942b, 1949a), and later championed by Lundqvist (1960, 1964), it is assumed that the gene has two equal parts, both of which control specificity, one in the pollen and the other in the style.

After work involving experimental irradiation of pollen at different ages (Lewis & Crowe 1953, 1954), Lewis produced a new model (1960), which anticipated the regulator model of Jacob and Monod (1961). He envisaged a single specificity component of the gene, closely linked to two activator components, one for the pollen and one for the style (see Fig. 6.4). Pandey (1956, 1957), van Gastel (1972), De Nettancourt (1972), and van Gastel and De Nettancourt (1974, 1975) have identified mutants that confer self-compatibility only to the male function (pollen-part mutants) or only to the female function (stylar-part mutants) or to both (specificity mutants) (Table 6.2).

A number of workers have shown that pollen-part s-c mutants tend to be associated with centric chromosome fragments, or with cytologically evident duplications in members of the Solanaceae (summarised in De Nettancourt 1977). It has been suggested (Pandey 1967) that such fragments or duplications contain *S* gene material of a type different from that inherited on the basic genome, which may be dominant to the genomic *S* allele, and thus confers self-compatibility. In other cases (De Nettancourt 1975), duplications of nucleolar organiser chromosome segments may reinstate ribosomal RNA formation in the potentially incompatible pollen tube. Although the association of such phenomena with s-c mutants is common in the Solanaceae, they have not been associated with stylar-part mutants, and they have not been recorded in other plant families.

Although the tripartite model of *S* gene structure and action has considerable logical attractions, and some experimental support, Lewis (1960) was unable to obtain recombinants between the three hypothetical segments of the gene. This may merely suggest that they are very closely linked.

*S* gene mutants are usually detected by noting occasional self-compatible pollen grains or seed-set in an otherwise fully s-i self. Care must be taken that such occasional self-compatibility does not have other causes, such as pseudocompatibility, unreduced pollen grains (which are diploid and may show dominance effects), the effects of timing, high temperatures etc., which are dealt with later in the chapter. Estimates of frequencies of spontaneous *S* gene mutation are given in De Nettancourt (1977) and vary from 0.2 to 4.3 per million grains. Such estimates do not

**Table 6.2** Detection of pollen-part, stylar-part and specificity mutants.

| | Type of mutant | | Specificity | |
|---|---|---|---|---|
| | pollen-part | stylar-part | *S* allele changed | *S* allele lost |
| self | self-compatible | self-compatible | self-compatible | self-compatible |
| cross with another ramet of same genet: | | | | |
| male parent | cross-compatible | cross-incompatible | cross-compatible | cross-compatible |
| female parent | cross-incompatible | cross-compatible | cross-compatible | cross-compatible |
| self on seedling from mutant self | self-compatible | self-compatible | self-incompatible | self-compatible |

distinguish between the three possible types of s-c pollen mutants without investigating the behaviour of selfed seedlings, and comparing rates of s-c in within-ramet and between-ramet selfs (see Table 6.2).

In particular, it is of interest to enquire whether specificity mutants change the specificity of the *S* allele ($S_1$ to $S_3$), or lose it ($S_1$ to $S_0$). Both changes lead to self-compatibility in the first generation, but the progeny of mutant selfs, selfed, will be s-c if specificity is lost ($S_0$). If specificity is changed ($S_1S_2$ to $S_1S_3$, selfed offspring $S_2S_3$ or $S_1S_3$), these offspring will revert to s-i. (This has been called revertible mutation, and as De Nettancourt (1977) shows is only satisfactorily explained by the tripartite model, and changes rather than loses in specificity.)

Although this question has attracted a good deal of experimental investigation, results are confused, and rather unexpected. There is a strong suggestion that we still lack basic understanding of the structure of the *S* locus, and the generation of new alleles at it. The following findings must be explained:

(a) Hundreds of different alleles can apparently coexist at a locus within a population in genera such as *Nicotiana*, *Lycopersicum*, *Trifolium* or *Oenothera*; in such cases, individual alleles have very low frequencies, and it is unusual to encounter the same allele in more than one plant. At such low frequencies, stochastic effects are likely to result in the loss of alleles (Wright 1939), and high levels of *S* allele mutation must be invoked to explain the maintenance of *S* allele diversity (Fisher 1961).

(b) Although mutagens induce s-c mutants (summarised in De Nettancourt 1977), these seem unusually to be mutants that lead to general compatibility, rather than to new *S* allele specificity. They seem often to involve gross chromosomal changes, or to give pollen-part mutants, but curiously may not produce the missense mutations required for the creation of new *S* allele specificity (Nasrallah *et al.* 1969). However, sufficiently detailed genetic analyses (Table 6.2) have rarely been carried out in conjunction with radiation mutagenesis, and *S* allele creation may yet be discovered.

(c) Spontaneous creation of new *S* alleles has indeed been reported in *Trifolium* (Denward 1963, Anderson *et al.* 1974), *Nicotiana* (Pandey 1970), and *Lycopersicum* (De Nettancourt *et al.* 1971, Hogenboom 1972). Most remarkably, however, it seems to be limited to experimental populations that have been subjected for some generations to forced inbreeding (by heat treatment, hormone treatment or bud pollination). For *Lycopersicum*, De Nettancourt *et al.* (1971) showed that a single new *S* allele ($S_3$) arose repeatedly in the styles of a clone homozygous for $S_2$ (because of forced inbreeding), and could occasionally revert to $S_2$ again.

If the tripartite *S* gene structure of Lewis (1960) is to be accepted, as seems necessary, it is simplest to assume that the pollen-part and stylar-part activator components each only have one functional state; mutants will fail to transmit the function of the specificity component, resulting in self-compatibility. Such s-c mutants may at times be selected for, but in a population in which outbreeding is favoured, they will be selected against. Presumably such mutants will not be able to back-mutate to renewed activity. Selection pressures that favour inbreeding might encourage the spread and fixation of pollen-part and stylar-part s-c mutations (p. 330). Thus, isolated populations or species that totally lose activity components of the *S* gene (either pollen-part or stylar-part loss would be sufficient) could only restore self-incompatibility through the evolution of a totally new system (such as a sporophytically controlled one).

New s-i *S* alleles that arise by mutation of the specificity component have two immediate advantages: being able to self in the generation in which they arise they will stand a good chance of immediate establishment (that is minority-type disadvantage effects will be minimised) and being rare, initially they will be able to outcompete the more common established *S* alleles.

It is not clear whether the specificity segment can ever actually generate mutants for loss of specificity ($S_0$); many mutants, including those associated with chromosomal abnormalities, can be assigned to pollen-part mutants, or to combined pollen-part and stylar-part mutations. For specificity mutations that generate novel *S* alleles, which have only been detected with certainty after repeated artificial selfing, several models of origin can be proposed, none of which is entirely satisfactory:

(a) The *S* locus consists of a single cistron. *S* alleles differ by minor changes in the nucleotide sequence within a relatively short section of DNA giving gene products that only differ from each other by a few amino acids. The main problems with this delightfully simple model are that the mutation rate would be expected to increase with exposure to mutagens, and that nonsense and frameshift mutants leading to a loss of specificity would be expected to be frequent; also, it would have to be a highly mutable gene.

(b) The *S* locus consists of many cistrons between which meiotic and somatic (male side) or somatic (female side) recombinations occur. *S* alleles would thus depend on complementation between cistrons, which might be very few. The number of allelic possibilities would be $2^n$ where $n$ is the number of cistrons within the locus. Thus only eight cistrons could generate 256 *S* alleles (Fisher 1961). Such a recombinational model is very attractive, but could only originate in plants heterozygous at the *S* locus. Therefore, it fails to explain why inbreeding generates new *S* alleles, and how *S* homozygotes, specifically, are able to do so.

(c) Mulcahy and Mulcahy (1983) have recently produced an entirely novel explanation of *S* gene control, based on cistronic complementation between loci which they claim need not be linked, and heterotic effects between the pollen and the style. They suggest that gametophytic incompatibility is based entirely on heterosis. The greater the difference between the cistronic components of the *S* gene in the pollen and the style, the more likely the pollen tube is to reach the ovule. This is an ingenious and attractive model which deserves further exploration, for it explains a number of worrying features of self-incompatibility. As stated, however, it seems implausible because of several features:
   (i) gametophytic incompatible pollen usually germinates and grows just as well as compatible pollen in initial phases;
   (ii) the nature of the failure of incompatible pollen, as observed microscopically, strongly suggests an oppositional rather than a complementary mechanism (p. 214);
   (iii) if unlinked, the components of the *S* gene would segregate after meiosis, so that, for instance semi-compatibility, could not occur, and one could not detect the same *S* allele amongst siblings;
   (iv) for loci with many alleles, very many cistrons would be necessary, and, as nearly all would be homozygous between different alleles, heterotic effects would be minimised. Lawrence *et al.* (1985) have recently published a fully-argued rebuttal of the Mulcahys' hypothesis.

(d) In recent years, transposable genetic elements (TGEs) have proved to be very widespread in a variety of organisms (perhaps even all) (Calos & Miller 1980). TGEs are fragments of DNA which are nomadic, having the capability to integrate themselves into, and excise themselves from, the chromosomes. Their activity may be generalised, generating an increased rate of chromosomal structural reorganisation (Woodruff & Thompson 1980), or highly specific, as shown by the pioneering work of McClintock on maize (discussed in Fincham & Sastry 1974). They may have at least three effects:
   (i) inhibition in the expression of a gene with which they become associated;
   (ii) transport of such a gene to another site, giving rise to position effects;
   (iii) high levels of chromosome breakage, chromosomal reorganisation and somatic recombination.

   TGEs are now thought to be responsible for many previously mysterious phenomena, not least the very high mutability and very large number of phenotypic expressions of antibodies in response to various antigenic stimuli in man and other animals.

Thus, it is possible to envisage a model of the specificity component of the *S* locus in which one or more TGEs travel through the locus, giving a different phenotypic response (*S* allele) in each of its different positions. This model is very attractive, and explains many features of the gametophytic incompatibility system. It does not in itself explain why new alleles should only appear after inbreeding. However, mutation rates in general, and TGE-generated changes in the DNA in particular, are themselves under genetic control, and can be influenced, for instance, by hybridity. Extreme heterozygosis can destabilise the genome by releasing TGE activity. It is possible that elements controlling TGE activity at the *S* locus are recessive, only being expressed when homozygous. If they are closely linked to, or part of, the *S* locus themselves, they may only be expressed after a cycle of inbreeding, when *S* alleles become few or homozygous. If this mechanism does exist, and there is no evidence for it, it would be of great selective advantage to outbreeders in which populations become so small that *S* alleles become critically few, as has been discussed by Sewall Wright (1939). It would encourage the rapid creation of new *S* alleles during such crisis points, which probably occur very frequently (Ch. 9). It would also explain the occurrence of very many *S* alleles in very small inbred populations, which appears mathematically very unlikely, as in *Oenothera organensis* (Emerson 1939, Wright 1939).

**Two-locus gametophytic incompatibility systems** Two-locus (sometimes called 'bifactorial') incompatibility systems under gametophytic control were first described in the grasses (Lundqvist 1956, 1961, 1968), and have otherwise only been discovered in a few aberrant members of the Solanaceae (Pandey 1957, 1962). The majority of grasses are self-incompatible (although this is untrue of all cereals except rye), and most that have been investigated have this mechanism (*Lolium* may have a three-locus system, p. 207). A two-locus system is most readily detected by the number of cross-compatible classes arising amongst full siblings. As a fully compatible cross will usually involved two heterozygous individuals containing between them four alleles at any locus (if they are diploid), the number of cross-compatible classes amongst siblings is calculated by $4^n$ where $n$ is the number of loci. Thus in the two-locus grasses, up to 16 sibling classes can occur (Fig. 6.2).

As already noted, two-locus systems in the grasses are most probably secondary, having arisen from self-compatible derivatives of one-locus gametophytic systems. In particular, they share with the secondary sporophytic systems a trinucleate pollen grain, and the various morphological and physiological peculiarities of sporophytic systems with

| ♂ Parent genotypes | Pollen phenotype combinations | | ♀ Parent genotype: $S_1S_2Z_1Z_2$ | | | | Compatibility | | Offspring genotypes |
|---|---|---|---|---|---|---|---|---|---|
| | | | $S_1Z_1$ | $S_1Z_2$ | $S_2Z_1$ | $S_2Z_2$ | | | |
| $S_1S_2Z_1Z_2$ | $S_1Z_1$ | self | incompatible | | | | incompatible | fully self-incompatible self | |
| | $S_1Z_2$ | | | incompatible | | | incompatible | | |
| | $S_2Z_1$ | | | | incompatible | | incompatible | | |
| | $S_2Z_2$ | | | | | incompatible | incompatible | | |
| $S_1S_2Z_1Z_3$ | $S_1Z_1$ | cross | incompatible | | | | incompatible | 50% semi-compatible cross | $S_1S_1Z_1Z_3$ $S_1S_2Z_3Z_2$ |
| | $S_1Z_3$ | | ✓ | ✓ | ✓ | ✓ | compatible | | $S_1S_1Z_2Z_3$ $S_2S_2Z_1Z_3$ |
| | $S_2Z_1$ | | | | incompatible | | incompatible | | $S_1S_2Z_1Z_3$ $S_2S_2Z_2Z_3$ |
| | $S_2Z_3$ | | ✓ | ✓ | ✓ | ✓ | compatible | | (6 cross-compatible classes) |
| $S_1S_3Z_1Z_3$ | $S_1Z_1$ | cross | incompatible | | | | incompatible | 75% semi-compatible cross | $S_1S_1Z_1Z_3$ $S_1S_3Z_1Z_3$ |
| | $S_1Z_3$ | | ✓ | ✓ | ✓ | ✓ | compatible | | $S_1S_1Z_2Z_3$ $S_1S_3Z_2Z_3$ |
| | $S_3Z_1$ | | ✓ | ✓ | ✓ | ✓ | compatible | | $S_1S_2Z_1Z_3$ $S_2S_3Z_1Z_1$ |
| | $S_3Z_3$ | | ✓ | ✓ | ✓ | ✓ | compatible | | $S_1S_2Z_2Z_3$ $S_2S_3Z_1Z_2$ |
| | | | | | | | | | $S_2S_3Z_1Z_3$ $S_2S_3Z_2Z_3$ |
| | | | | | | | | | (10 cross-compatible classes) |
| $S_3S_4Z_3Z_4$ | $S_3Z_3$ | cross | ✓ | ✓ | ✓ | ✓ | compatible | fully compatible cross | $S_1S_3Z_1Z_3$ $S_1S_4Z_1Z_3$ |
| | $S_3Z_4$ | | ✓ | ✓ | ✓ | ✓ | compatible | | $S_1S_3Z_2Z_3$ $S_1S_4Z_2Z_3$ |
| | $S_4Z_3$ | | ✓ | ✓ | ✓ | ✓ | compatible | | $S_2S_3Z_1Z_3$ $S_2S_4Z_1Z_3$ |
| | $S_4Z_4$ | | ✓ | ✓ | ✓ | ✓ | compatible | | $S_2S_3Z_2Z_3$ $S_2S_4Z_2Z_3$ |
| | | | | | | | | | $S_1S_3Z_1Z_4$ $S_1S_4Z_1Z_4$ |
| | | | | | | | | | $S_1S_3Z_2Z_4$ $S_1S_4Z_2Z_4$ |
| | | | | | | | | | $S_2S_3Z_1Z_4$ $S_2S_4Z_1Z_4$ |
| | | | | | | | | | $S_2S_3Z_2Z_4$ $S_2S_4Z_2Z_4$ |
| | | | | | | | | | (16 cross-compatible classes) |

**Figure 6.2** Behaviour and inheritance of two-locus gametophytic self-incompatibility in grasses. Number of haploid genotypes considered (left-hand column) are $4S \times 4Z = 16$.

trinucleate grains (see Table 6.1). It seems possible that the trinucleate system has been necessary for plants that have 'reinstated' incompatibility after self-compatibility (p. 198). The occurrence of trinucleate grains, with trinucleate grain characteristics, in the grasses tells us quite clearly that the trinucleate condition is not a prerequisite for sporophytic behaviour. Grasses have a gametophytic system. This is most simply observed in semi-compatible crosses amongst siblings, in which the pollen grains do not all behave in the same way (a quarter or a half of the pollen grains are incompatible, rather than one-half only for one-locus gametophytic systems; Fig. 6.2). Because sporophytic systems are controlled by the genotype of the anther, all grains from a single male parent will behave in the same way, and semi-compatibility does not occur.

The two loci in grasses are termed *S* and *Z*. They are unlinked, segregating independently from each other after meiosis, and each is polyallelic. There is co-operation between alleles at different loci in the pollen, but independent reactions of *S* and *Z* alleles in the style. Each combination of alleles establishes a specificity in the haploid pollen, and rejection occurs when this specificity is matched by one of the four possible combinations of *S* and *Z* in the diploid style (see Fig. 6.2). Thus, the effects of the two loci are multiplicative, the number of incompatibility genotypes being the product of the number of alleles at each locus. Lundqvist (1964) calculates that for *Festuca pratensis*, with 6 alleles at the *S* locus and 14 alleles at the *Z* locus, 84 haploid genotypes occur. It follows that this system is far more efficient than a one-locus system, where for 20 alleles (6 + 14) only 20 haploid genotypes are possible. It is important for the efficiency of the system that *S* and *Z* are unlinked. If linked, they would behave essentially as a single locus with additive rather than multiplicative effects, although some recombination between the loci might occur. Loosely linked multilocus systems such as this have not been recorded.

The efficiency of the two-locus system means that it is more resistant to decay than a one-locus system. It will suffer less from inbreeding and loss of alleles in very small populations, and the viable number of alleles at each locus is much fewer, probably as low as three at each (nine pollen genotypes). It is also much less likely to acquire self-compatibility mutants. The *S* and *Z* loci act in both a complementary and an independent manner. Thus, if one locus produced an s-c mutant, the other would continue to convey incompatibility, and the s-c mutant would not be advantaged selectively, as would be the case in a one-locus system. However, this intuitive approach is not supported by the models of Mayo and Hayman (1968), who consider that times to extinction of two-locus alleles when inbred are similar to those for one-locus systems.

Lundqvist (1962) has suggested that interaction between the *S* and *Z* loci may occur, giving dominance effects between an *S* allele and a *Z* allele in a pollen grain, as would be the case if the pollen grain were diploid (p. 209). There is as yet no clear proof of this, but if the *S* and *Z* loci originally arose by duplication of a single locus, and thus act in the same way, such interaction might be expected. The apparent similarity of mode of action by the *S* and *Z* loci, and their complementary activity with respect to each other, support the view that they are indeed unlinked duplicates.

De Nettancourt (1977) favours the view that the *S* and *Z* loci have a similar structure to one-locus gametophytic systems, that is that they are probably tripartite in structure. Because of the complementary action of the loci, it is very difficult to undertake mutation studies in two-locus systems, for mutants will usually be unexpressed phenotypically. There is in fact no reason why the structure of the two-locus system should be the same as that of the one-locus system, if they have different origins (p. 204). Lundqvist (1964) followed Lewis (1954) rather than Lundqvist (1960) in assuming a bipartite structure for two-locus systems, but in fact there seems to be little evidence in either direction.

**Three- and four-locus gametophytic incompatibility systems**

Gametophytic systems with more than two multi-allelic loci were first detected by Lundqvist *et al.* (1973) for *Ranunculus acris* and *Beta vulgaris*, and claimed by Spoor (1976) and McCraw and Spoor (1983a,b) for *Lolium* species, and by Murray (1979) for *Briza spicata*. As we have seen already, multilocus systems generate more cross-compatible classes amongst the offspring of a single cross. Using the formula $4^n$ for the number of such classes, where $n$ is the number of loci, it will be seen that three-locus systems can generate 64 cross-compatible offspring classes from a single cross, and four-locus systems can generate 256. A conventional two-locus system can only generate 16 classes. Naturally, it is necessary to grow up large numbers of offspring, and make many crosses before such effects are detected. Thus, for *Ranunculus acris*, families with 20, 19 and 18 cross-compatible classes were detected (only slightly more than 16 and many fewer than 64).

Other distinctive effects of multilocus systems are percentages of successful pollen-tube growth in semi-compatible crosses which depart from 75% or 50% (in three-locus systems they may include crosses with 62.5%, 75% or 87.5%), and one-way incompatibility (non-reciprocity). One-way incompatibility, in which a reciprocal cross between two individuals may be incompatible in one direction but not the other, may be caused by incompatibilities arising in the absence of a total match of alleles between the pollen and the style (McCraw & Spoor 1983a). This would probably occur through dominance effects between alleles at different loci within the pollen, resulting in a limitation of male, as against female phenotypes. It only apparently occurs in multilocus systems. However, one-way incompatibility in *Lolium*, and indeed the occurrence of three-locus systems in *Lolium* has been severely criticised by Lawrence *et al.* (1983).

It must be emphasised that the existence of multilocus (more than two loci) systems is based on rather slender evidence, and other interpretations could explain the existence of more than 16 cross-compatible offspring categories arising from a single cross involving a two-locus system. The simplest would be if dominance interactions occurred between alleles within a locus in the style (as below).

Possible effects in a two-locus system of dominance interactions in the style:

Conventional self: $S_1S_3Z_1Z_3 \times S_1S_3Z_1Z_3$

| Pollen genotypes | Female genotype | |
|---|---|---|
| $S_1Z_1$<br>$S_1Z_3$<br>$S_3Z_1$<br>$S_3Z_3$ | $S_1S_3Z_1Z_3$ | incompatible |

Self with stylar dominance in $S$ and $Z$ loci, $S_1$ and $Z_1$ dominant: $S_1S_3Z_1Z_3 \times S_1S_3Z_1Z_3$

| Pollen genotypes | Female phenotype | |
|---|---|---|
| $S_1Z_1$ incompatible<br>$S_1Z_3$ compatible<br>$S_3Z_1$ compatible<br>$S_3Z_3$ compatible | $S_1Z_1$ | 75% semi-compatible |

(i.e. more apparent cross-compatibility than expected would occur between siblings)

For this model to be acceptable as it stands, it would be necessary for at least some selfs of the sibling family to be semi-compatible. McCraw and Spoor (1983a,b) do indeed find that half the selfs in the sibling families of *Lolium* are semi-compatible, using their *in vitro* technique. The semi-compatibility largely disappears when conventional selfs are made, and it is more than possible that the large number of cross-compatible classes obtained, which indicate a three-locus system, are merely an artefact of their *in vitro* system.

Other phenomena in a two-locus system that might give more than 16 cross-compatible classes in offspring families of a single cross are dominance interactions on the male side, or on both the male and female sides, or interactions of different $S$ and $Z$ alleles with the same cytoplasm. Any of these could give the one-way incompatibilities observed by McCraw and Spoor, and Lundqvist *et al.* (1973). Yet another possibility is the rather mysterious phenomenon of certation, in which certain combinations of different *S* alleles in the pollen and stigma may apparently give rise to full self-compatibility.

Whether or not three- or four-locus systems operate in these genera, any explanation of the results obtained must at least require a two-locus system. Pollen in *Lolium* is trinucleate, and the system is otherwise similar to those in other grasses, from which it can be presumed to have developed. In *Ranunculus*, the pollen grains are binucleate and their mode of action is apparently of the binucleate type (Table 6.1). The rarity of such apparently multilocus systems renders it highly unlikely that such systems were aboriginal to the Angiosperms, as Lundqvist *et al.* (1973) have suggested. It is more likely that multilocus systems have arisen from one-locus systems by duplication of the locus, or have arisen *de novo*, than that multilocus systems have lost loci, to give rise to, polyphyletically, many one-locus gametophytic systems (p. 196).

**Polyploidy in gametophytic systems** Up to this point, only diploids, or at least plants disomic for the incompatibility locus, have been considered. For these, the independent expression of incompatibility alleles in the diploid style is firmly established for one-locus, and most grass two-locus, systems. As discussed above, *Lolium* may provide an exception. However, the pollen grain is haploid, and thus interactions between alleles of one-locus gametophytic systems are not possible. Each pollen grain has its own distinctive single-factor control.

For polyploids in which the pollen grain is disomic, or even polysomic for the *S* locus, a different condition operates. The different *S* alleles within the pollen grain interact showing either dominance with respect to each other or independent expression. Thus, the pollen grain $S_1S_2$ has the possibility of three different phenotypes, $S_1$, $S_2$ and $S_1S_2$. It follows that partial homozygotes can arise, as from the cross $S_1S_2S_3S_4$ (male) × $S_1S_3S_4S_5$ where $S_2$ is dominant to $S_1$ in the pollen. Thus, a pollen grain $S_1S_2$ would have the phenotype $S_2$, and would be compatible with $S_1S_3S_4S_5$. It could therefore cross with a female gamete $S_1S_3$, giving the offspring $S_1S_1S_2S_3$, partially homozygous for $S_1$. In such a system, recessive alleles, which are less likely to recognise self in the style, are favoured and will tend to be both commoner and more often partially homozygous, with a concomitant loss of dominant alleles. If many alleles are lost, fertility of the population will be impaired and s-c will be favoured.

In practice, new polyploids, such as those artificially induced by the use of colchicine, of species with one-locus gametophytic systems seem usually to lose incompatibility on the male side (Crane & Lawrence 1929, Crane & Lewis 1942, Stout & Chandler 1942). It is typical of these to be s-c as male parents, but s-i as female parents, as exemplified by differences in reciprocal crosses between tetraploids and their diploid parents. This raises an interesting point with respect to occasional self-compatibility in diploids, which may be caused by non-reductional (diploid) pollen grains. The offspring of these selfs will of course be sterile autotriploids.

Autotetraploids will be partially homozygous (e.g. $S_1S_2 \rightarrow S_1S_1S_2S_2$) and thus may give rise to either homozygous ($S_1S_1$ or $S_2S_2$) or heterozygous ($S_1S_2$) pollen. Lewis (1947, 1949a) showed that only heterozygous grains showed such self-compatibility, there being semi-compatibility (50%) in the newly formed autotetraploid when selfed. There seems to be no rational explanation for this finding at present, which suggests that we are still some way from a complete understanding of the mechanism of gametophytic incompatibility. Although Lewis has suggested that the different alleles (for instance $S_1$ and $S_2$) may compete for substrate in the diploid pollen grain and that this competition allows neither to become expressed, this does not explain how the homozygous grain (e.g. $S_1S_1$) can obtain enough substrate to express its incompatibility. A convincing model would also have to explain why autotriploids fail to show self-compatibility (van Gastel 1974), and why autopolyploids in the monocotyledons do not apparently show self-compatibility (Annerstedt & Lundqvist 1967).

Autopolyploidy may lead to partial self-compatibility immediately, and increasing homozygosity for *S* alleles and loss of *S* alleles may eventually encourage a total loss of self-incompatibility in autoploids. However, autopolyploidy is a rare condition in natural populations. Most polyploids, perhaps half of all Angiosperm species, are alloploids that have obtained genomes, for each of which they are usually diploid, from two to several different species. Whether the polyploids are disomic or polysomic for the incompatibility locus/loci will, of course, depend on how many of the parents carry functional self-incompatibility loci, and whether these loci are genetically and positionally homologous between parents (Fig. 6.3). Only those loci that are functionally disomic in a complex polyploid will be likely to survive. However, gametophytic multi-allelic incompatibility is frequently found in allopolyploids, and we must assume that in these the *S* locus is disomic.

Nevertheless, self-incompatibility is much less common in polyploids than in diploids (e.g. Stebbins 1950). Allopolyploids will usually arise from a single founder, caused by chromosomal doubling in a sterile diploid hybrid, or by the fusion of non-reductional gametes. Backcrosses of allopolyploids to their diploid parents will give rise to sterile triploids, so polyploids may suffer from minority-type disadvantage and may thus be unable to reproduce sexually and establish, unless self-fertile. As we have seen, polysomy at the incompatibility locus is likely to generate such male self-fertility. However, if the *S* locus is disomic in the newly

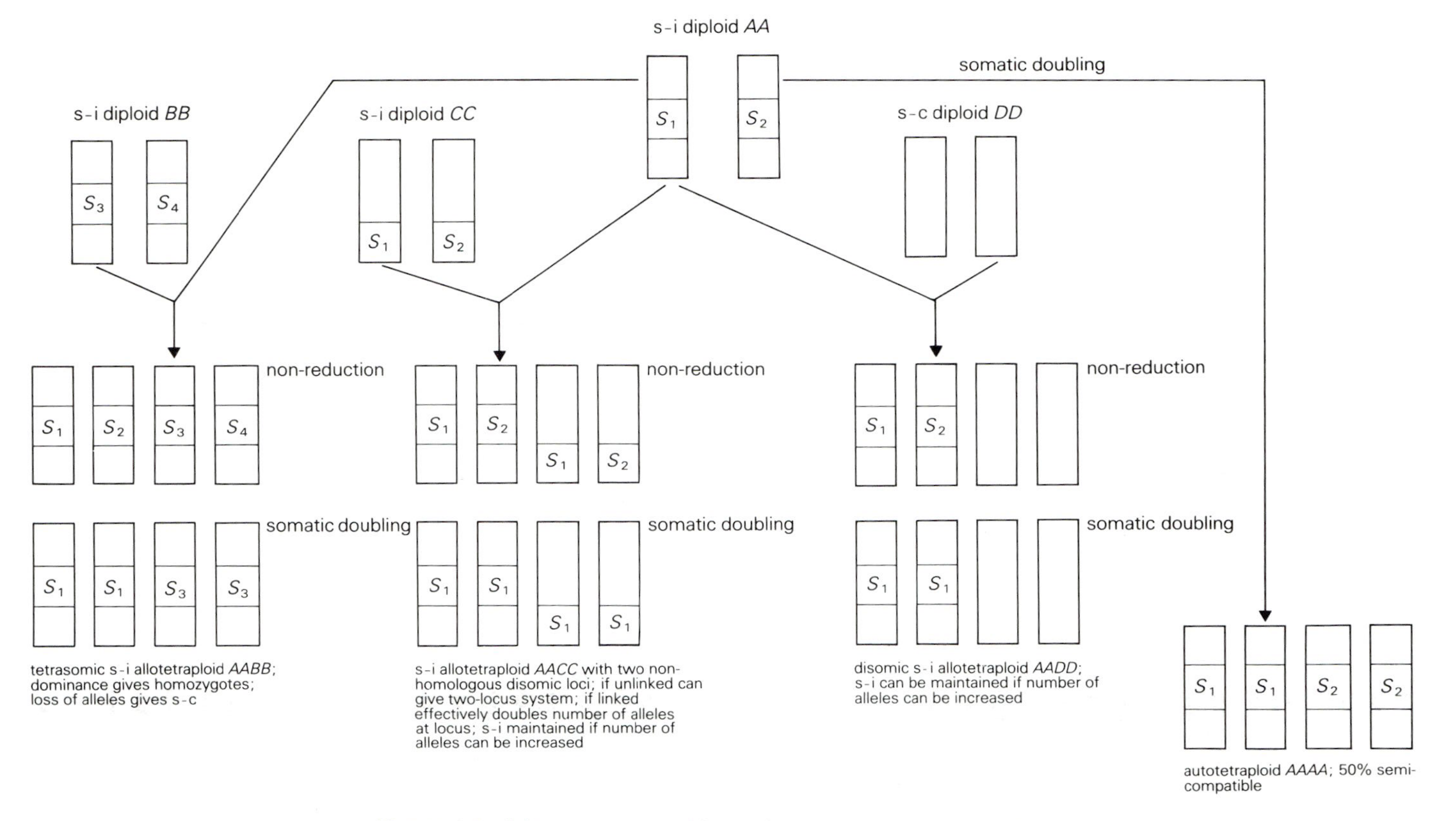


**Figure 6.3** Maintenance and loss of gametophytic s-i in polyploids.

formed allopolyploid, the plant will remain self-sterile, and unable to produce fertile offspring in the absence of other tetraploids.

A further problem facing the new s-i allopolyploid is that only two alleles will have entered the new allopolyploid gene pool (see Fig. 6.3). These are too few for an s-i system to function correctly (p. 195). This problem will only be overcome if more than one different alloploid, each with different *S* alleles, occur together, or if a new alloploid crosses with unreduced gametes from parental diploids exhibiting other *S* alleles in the surrounding population. Even then, the new population will be extremely short of *S* alleles, but their number may increase by mutation, or by occasional introgression with unreduced gametes from diploids. That such processes do happen is substantiated by the frequent occurrence of self-incompatible polyploids, including such well-known species as *Trifolium repens* and *Lotus corniculatus*.

Nevertheless, self-compatibility has proved more successful in polyploids than diploids, perhaps due to the greater capacity of polyploids for storing variability, and releasing variability more slowly, when selfed (Ch. 9). Thus, polyploids may be capable of tolerating more selfing than diploids.

**Breakdown of gametophytic incompatibility** Up to this point, it has been assumed that gametophytic self-incompatibility is normally total in its effect, only being lost through self-compatible mutation (mostly pollen-part mutation) or diploid pollen (from tetraploids, or from non-reduction in diploid male meiosis).

However, there are other circumstances through which an otherwise functional gametophytic s-i system can fail but continue to be inherited in the offspring. These have been called pseudocompatibility (Pandey 1959) and are well reviewed in de Nettancourt (1977), from which this account is taken, and Pandey (1979). They include the following conditions.

(a) Bud pollination. In *Nicotiana* (Pandey 1959) and *Petunia* (Shivanna & Rangaswami 1969) substantial quantities of selfed seed can be obtained by pollinations with mature pollen on to immature stigmas. It is concluded that the *S* gene is not yet expressed in the young style, so allowing the growth of incompatible pollen tubes.
(b) Delayed pollination. Ascher and Peloquin (1966) working with *Lilium* have shown that fresh selfed pollen on overmature stigmas can also achieve the breakdown of self-incompatibility, although similar effects cannot be obtained in some other genera (they do, however, occur in the sporophytic genus *Brassica*). Once again, decay of *S*-gene

products in the ageing style may permit some incompatible pollen-tube growth.

(c) End-of-season effects. It has been noted for several members of the Solanaceae that abnormally late flowers, often on a moribund plant, can be self-compatible.

(d) Stylar irradiation. Linskens *et al.* (1960) for *Petunia*, and Hopper and Peloquin (1968), for *Lilium* have shown that X-radiation of styles immediately after self-pollination prevents self-incompatibility. The optimum dose for *Petunia* was 2000 rads, but above 6000 rads for *Lilium*. Radiation before pollination, or 24 hours after pollination did not give this effect. It is suggested that temporary gene inactivation is involved, although there is no indication as to exactly how this can operate.

(e) Mentor effects. These describe the induction of self-compatibility in incompatible pollen when it is mixed with foreign pollen. The foreign pollen does not achieve fertilisation as it has been previously killed (usually by doses of radiation, Knox *et al.* 1972) or is distantly related and thus interspecifically incompatible, although sharing the same self-incompatibility system (as in *Lotus*, e.g. Miri & Bubar 1966). Although it is not unexpected that such effects may occur in sporophytic systems, in which the incompatibility factor is known to be carried on the outside of the pollen grain (p. 228), it is very surprising that it should operate in gametophytic systems such as *Lotus*. It must be presumed that a diffusable substance from the non-functional compatible pollen overcomes the incompatibility inherent in the selfed pollen (see below).

(f) High temperatures. Many workers have discovered that high temperatures, usually in excess of 30°C and as high as 60°C, remove self-incompatibility. This appears to be a widespread and general phenomenon when high temperatures are applied to the style immediately after self-pollination, and occurs in both gametophytic and sporophytic systems. There is some suggestion that the temperature effect is specific to certain isozymes in the style (Pandey 1973) and that variation occurs between plants with respect to heat sensitivity.

(g) Biological inhibitors. Inhibitors of RNA synthesis such as actinomycin D and 6-methylpurine, and enzyme inhibitors such as puromycin and *p*-chloromercuribenzoate, have all been shown to limit self-incompatibility to various extents.

(h) At least in *Petunia*, fertilisation *in vitro* can break self-incompatibility when pollen grains are allowed to germinate in juxtaposition with cultured ovules. Other mutilative and manipulative experiments have shown equally clearly that self-incompatibility is mediated by various fractions of the stigma and style.

All of these methods of overcoming self-incompatibility show that self-incompatibility is an active phenomenon, which must be bypassed or inhibited in some way if it is to be broken. This argues in favour of an oppositional system of control, and against the complementary system of Bateman (1952) and Mulcahy and Mulcahy (1983). The latter requires a system that promotes compatibility. In its absence, pollen-tube growth is inadequate and incompatibility results passively. Although many experimental results of different kinds encourage us to espouse an active, oppositional system of incompatibility, a final example of experimental inhibition of self-incompatibility suggests otherwise.

If inhibitors of floral abscission such as auxin (3-indole-acetic acid) or its analogue, $\alpha$-naphthalene acetic acid are applied to the calyx or the base of the flower in solution at regular intervals after an incompatible pollination, the flower remains in good condition for a longer time. Incompatible pollinations can then often result in seed-set. This is apparently due to the ability of the slow-growing incompatible pollen tubes eventually to reach the ovules (in gametophytic systems). Applications of auxin to the style or stigma do not promote self-fertilisation, and indeed inhibit compatible fertilisation, so there is no direct effect of auxin on incompatibility. The effect is indirect, apparently by allowing the incompatible tubes sufficient time to reach the ovule. These results, which have been obtained for many plants with gametophytic systems (de Nettancourt 1977), are very surprising in view of the observed behaviour of incompatible tubes in the style (p. 223). It is possible that effects similar to those for delayed pollination (d) operate, in that *S*-gene products decay in the ageing style, permitting growth of late-germinating incompatible pollen. Thus, although the effects of auxin appear to indicate the operation of a passive system of incompatibility, it may merely indicate the breakdown of an active, oppositional system with time.

## *Genetic control of multi-allelic sporophytic systems*

As already noted, sporophytic systems differ basically from gametophytic systems in that the control of the behaviour of the pollen grain comes from the sporophytic anther that gave rise to it, not from the grain itself. This will have the following genetic consequences:

(a) All the pollen grains from a single male parent will show the same behaviour on a given female parent (i.e. semi-compatibility does not occur).

(b) Because the control of the behaviour of the pollen grain comes from the diploid anther, dominance will usually be expressed, i.e. the compatibility phenotype of the pollen grain will express only one of the alleles in the anther. Thus, the individual haploid pollen grain may carry an allele different from its incompatibility phenotype. For instance, if $S_1$ is dominant to $S_2$ in the male parent $S_1S_2$, a grain carrying $S_2$ will have the phenotype $S_1$. [If there is incomplete dominance, both paternal alleles may be expressed (i.e. independent interaction).]

(c) One of the consequences of this is that homozygotes will occur for the $S$ locus. Thus if $S_1 > S_2 > S_3$, the cross $S_1S_2$ (male) $\times$ $S_2S_3$ will be compatible as the pollen will show the phenotype $S_1$. But $S_2$ carrying grains can fertilise $S_2$ carrying egg cells to give $S_2S_2$ homozygous offspring.

(d) Another consequence is that the number of cross-compatible classes amongst the offspring of a single cross will be less than $4^n$, where $n$ is the number of loci (i.e. four classes for one-locus diploid gametophytic systems). Thus, if the following crosses occur, the number of cross-compatible classes amongst the offspring will be as shown ($S_1 > S_2 > S_3 > S_4$):

| Male | | Female | Offspring genotypes | Offspring male phenotypes | Number of cross-compatible classes in offspring |
|---|---|---|---|---|---|
| $S_1S_2$ | $\times$ | $S_3S_4$ | $S_1S_3$ | $S_1$ | |
| | | | $S_1S_4$ | $S_1$ | 2 |
| | | | $S_2S_3$ | $S_2$ | |
| | | | $S_2S_4$ | $S_2$ | |
| $S_1S_2$ | $\times$ | $S_2S_3$ | $S_1S_2$ | $S_1$ | |
| | | | $S_1S_3$ | $S_1$ | 2 |
| | | | $S_2S_2$ | $S_2$ | |
| | | | $S_2S_3$ | $S_2$ | |
| $S_1S_2$ | $\times$ | $S_2S_2$ | $S_1S_2$ | $S_1$ | |
| | | | $S_1S_2$ | $S_1$ | 1 |
| | | | $S_2S_2$ | $S_2$ | |
| | | | $S_2S_2$ | $S_2$ | |
| $S_1S_1$ | $\times$ | $S_2S_2$ | $S_1S_2$ | $S_1$ | |
| | | | $S_1S_2$ | $S_1$ | 0 |
| | | | $S_1S_2$ | $S_1$ | |
| | | | $S_1S_2$ | $S_1$ | |

Up to this point, a model to describe the operation of multi-allelic sporophytic systems has been described which assumes a simple dominance hierarchy between alleles in the anther, but independent expression of alleles in the stigma. Such a simplistic system is probably never true for any sporophytic system, although it is approached most closely in *Brassica* (Richards & Thurling 1973, Ockendon 1974). More frequently, a complex series of interactions between *S* alleles occurs in both the anther and the stigma, the details of which commonly differ between the anther and stigma even in the same population. Four types of interactions between alleles are recognised, which are listed in order of importance; all can occur within the same species:

(a) dominance $S_1 > S_2 > S_3 > S_4$; genotype $S_1S_2$; phenotype $S_1$;
(b) independent $S_1 = S_2$; genotype $S_1S_2$; phenotype $S_1S_2$;
(c) interaction $S_1.....S_2$; genotype $S_1S_2$; phenotype $S_3$;
(b) mutual weakening $S_i$------$S_2$; genotype $S_1S_2$; phenotype $S_{(1)}S_{(2)}$ (more or less self-compatible).

It is most common to find dominance as the most important system in the anther, and independent action as the commonest mechanism in the stigma (as in the examples given above), but considerable variation occurs. To take two simple examples which are frequently quoted, for instance by de Nettancourt (1977), from *Iberis amara* (Cruciferae) (Bateman 1954), and *Cosmos bipinatus* (Compositae) (Crowe 1954) respectively (*S* alleles in circles):

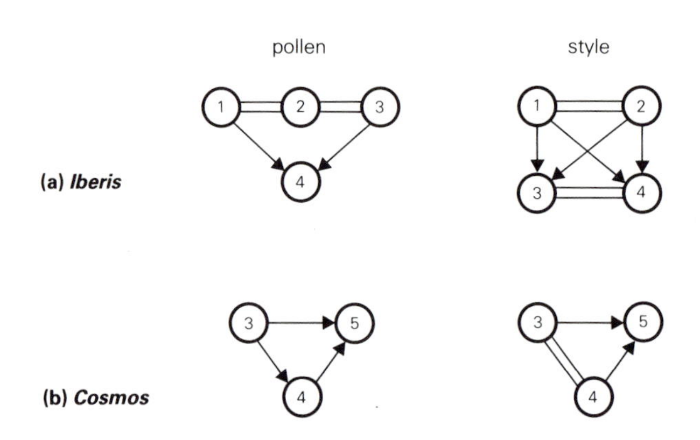


To illustrate how these systems work, let us take by way of illustration one reciprocal cross, and one reciprocal self from each example:

| | Male | | Female | | |
|---|---|---|---|---|---|
| | Genotype | Phenotype | Genotype | Phenotype | Compatibility |
| *Iberis* | | | | | |
| cross | $S_2S_3$ | $S_2S_3$ | $S_3S_4$ | $S_3S_4$ | incompatible |
| | $S_3S_4$ | $S_3$ | $S_2S_3$ | $S_2$ | compatible |
| self | $S_3S_4$ | $S_3$ | $S_3S_4$ | $S_3S_4$ | incompatible |
| *Cosmos* | | | | | |
| cross | $S_3S_4$ | $S_3$ | $S_4S_5$ | $S_4$ | compatible |
| | $S_4S_5$ | $S_4$ | $S_3S_4$ | $S_3S_4$ | incompatible |
| self | $S_3S_4$ | $S_3$ | $S_3S_4$ | $S_3S_4$ | incompatible |

From this we can make the further following deductions:

(a) Self-pollinations will always be self-incompatible unless the dominance order between alleles differs in direction between the pollen and the style (a rare occurrence, if indeed it ever does occur).

(b) Reciprocal differences in compatibility will commonly occur between the same two parents (such reciprocal differences never occur in diploid gametophytic systems, but can occur in polysomic gametophytic systems).

(c) Cross incompatibility will frequently occur between two individuals with different genotypes (cross incompatibility will only occur in diploid gametophytic systems when individuals share the same *S* genotype).

(d) Although semi-compatibility cannot occur, reciprocal differences and cross incompatibility between different genotypes will only occur when the two parents share one *S* allele.

(e) Dominant alleles will be at a selective disadvantage in comparison to alleles of independent action, which will be at a disadvantage with respect to recessive and mutual weakening alleles. This is a complicated relationship, because of chains of dominance relationships between many alleles, which may differ between the pollen grain and the style. Alleles that show dominance to the greatest number of alleles on the male and female side will have the greatest disadvantage. This is because they are most frequently expressed, phenotypically, and are thus most likely to show a cross-incompatible reaction through self-recognition.

Ockendon (1974, 1977, 1980) in an elegant analysis of the distributions of *S* alleles amongst commercial strains of Brussels sprouts and broccoli (*Brassica oleracea*) showed that *S* alleles are few in number within varieties, very uneven in distribution between varieties, and that the many scarce alleles were dominant, and the few common alleles were recessive to them. These factors all militated against high seed-set, and hampered the successful breeding of certain new strains. From a breeding standpoint, a great premium was placed on the introduction of one or more rare dominants into a strain. This uneven distribution, so typical of a sporophytic system, contrasts vividly with the even frequency of *S* alleles in a diploid gametophytic system, in which every allele should have the same frequency.

It is clear that the genetic inheritance and breeding behaviour of diploid gametophytic, and of sporophytic, systems differ markedly in the presence of semi-compatibility, the presence of homozygotes for the *S* locus, number of cross-compatible classes amongst offspring of a single cross, reciprocal differences in compatibility, cross incompatibility between different genotypes, and frequency distribution of *S* alleles. In all of these, a diploid gametophytic system tends to be more efficient; that is, although self-incompatibility is general in both systems, there will be much less cross incompatibility in a gametophytic system – more individuals in the populations will be available for crossing. However, by restricting sibling mating, a sporophytic system will reduce mating between close relatives and thus may lead to more outbreeding than a gametophytic system.

It also follows that the identification of a system should be relatively straightforward if it can be stated with confidence that the subject is diploid (p. 209). However, there are many features in common between polyploid gametophytic systems and sporophytic systems, in particular co-occurrence of dominance and independent expression between alleles in the pollen (p. 216). Thus, most of the features listed above which differentiate sporophytic systems from diploid gametophytic systems (and disadvantage the former) are also true of polyploid gametophytic systems. Perhaps the major distinction between the latter and sporophytic systems, from a genetical standpoint, is that self-fertilisation is much more frequent in polyploid gametophytic systems (p. 210).

As has been noted above (Table 6.1), gametophytic systems differ markedly from sporophytic systems in the number of pollen-grain nuclei, pollen-grain physiology, stigma-papilla ultrastructure, and the site of incompatibility. Thus, in a plant showing genetic characteristics of a sporophytic system, but in which the level of 'ploidy of the *S* locus is uncertain, reference can be made to these markers. It should, however, be noted that the grasses resemble sporophytic systems in these characteristics, and furthermore do not show self-compatibility when polyploid. It would seem that the only way of knowing that a possibly polyploid grass has a gametophytic system when it shows sporophytic-type genetic behaviour is by reference to the genetic behaviour of its diploid relatives, in other words by knowing that it is a grass!

As far as is known, all sporophytic systems are controlled by one locus only, or at least by one linkage group only, for the di-allelic 'locus' of

*Primula* is recombinable (Ch. 7). The only way by which multilocus sporophytic systems could be detected is by finding more than two cross-compatible classes in the offspring of a single cross between parents that in other genetic, cytological and physiological features were clearly sporophytic. This has not occurred so far.

It is difficult to estimate the number of *S* alleles occurring within a population, or a species, in a sporophytic system. Although a sporophytic system resembles a gametophytic system in that there is nearly always total self-incompatibility, one cannot assume, as in the latter, that incompatible crosses are between the same genotype. They will often be between genotypes that share one allele. At the same time, crosses between genotypes sharing an allele will often be compatible in one direction, and if the common allele is recessive to both the others, such a cross will be compatible in both directions. Other factors that must be allowed for are interaction between alleles (giving more compatibility) and homozygotes (which will also tend to give more cross compatibility, through reducing the frequency of allele sharing).

The only genotypes that can be relied upon to be always cross compatible in a sporophytic system are those that share no *S* alleles. These are groups of plants that are always fully cross compatible with all others in the system for both reciprocal crosses. These will equal $n/2$, where $n$ is the number of alleles. Thus, for a system containing eight *S* alleles, the following genotypes will always be fully cross compatible: $S_1S_2$, $S_3S_4$, $S_5S_6$, $S_7S_8$, (4 out of 28 possible different genotypes), the number of possible genotypes being calculated by $(n^2 - n)/2$ (that is where $n = 8$, $64 - 8/2 = 56/2 = 28$). The remaining genotypes should show cross incompatibility with some of these four, and with each other due to dominance or independent action, namely, $S_1S_3$, $S_1S_4$, $S_1S_5$, $S_1S_6$, $S_1S_7$, $S_1S_8$, $S_2S_3$, $S_2S_4$, $S_2S_5$, $S_2S_6$, $S_2S_7$, $S_2S_8$, $S_3S_5$, $S_3S_6$, $S_3S_7$, $S_3S_8$, $S_4S_5$, $S_4S_6$, $S_4S_7$, $S_4S_8$, $S_5S_7$, $S_5S_8$, $S_6S_7$, $S_6S_8$ (24 in all).

A very thorough series of reciprocal crosses between all genotypes present is necessary if the number of fully cross-compatible genotypes present is to approximate to $n/2$. In fact, due to interactive effects, and homozygotes, this equation will always overestimate the value of $n$ (the number of *S* alleles), as will dominance effects unless these are very fully understood. In practice, more sophisticated techniques for the estimation of the number of alleles are used, which vary according to the system, and the level of understanding of interactive effects between alleles within each system. Nevertheless, the numbers of *S* alleles identified in the few cases where this has been attempted are high (Table 6.3).

The $n/2$ relationship breaks down when less than four (two or three) alleles are involved, notably in two-allele systems, where simple dominance renders the number of cross-compatible classes equal to the number of alleles (Riley 1936, and Ch. 7).

**Table 6.3** The number of *S* alleles identified in various plants with sporophytic incompatibility.

| Species | Number of plants analysed | Number of *S* alleles estimated | Reference |
|---|---|---|---|
| *Iberis amara* | 47 | 22 | Bateman (1954) |
| *Brassica oleracea* (Brussel sprouts) | 488 (16 cultivars) | 19 | Ockendon (1974) |
| *Raphanus raphanistrum* | 45 (5 populations) | 9 | Sampson (1967) |

The elegant, readily analysed nature of the gametophytic system allows the investigator to estimate the number of alleles involved more readily (p. 193), and to assign accurately self-compatible breakdown to polyploidy (p. 210), or to mutation (p. 199). In particular, it has been possible to confirm the probability of a tripartite gametophytic *S*-gene structure through mutation studies. In contrast, the much more complex nature of the control of sporophytic systems, has meant that little progress has been made towards the understanding of the structure, mutation, or polysomic behaviour of sporophytic *S* genes, when multi-allelic. Much more is known about di-allelic sporophytic *S* genes, and these are discussed in Chapter 7.

However, there has appeared a considerable body of work about the physiological breakdown of sporophytic systems, particularly in *Brassica*. Early work showed that self-incompatibility can be overcome in this genus by both bud pollination (Attia 1950) and delayed pollination (Kakizaki 1930), as is also found in some gametophytic systems (p. 212). Such self-fertilisations are routinely carried out in plant breeding stations today to obtain inbred lines from which $F_1$ hybrids can be made. Self-incompatibility can also be broken by high temperatures (Richards & Thurling 1973), perhaps because enzymes that mediate the action of the *S* gene in *Brassica* have lower optimal temperatures than those that control pollen-tube growth and fertilisation. In *Primula*, with a di-allelic sporophytic system, Lewis (1942b) has described the breakdown of self-incompatibility at high temperature, and I have found (Richards & Ibrahim 1982, and A. J. Richards unpublished observations) that thrum *Primula* × *polyanthus* shows high levels of self-fertilisation at 35°C for 24 hours after pollination (Ch. 7). Once again, deactivation of one or more enzymes at high temperatures can be plausibly implicated. De Nettancourt (1977, p. 109) makes the telling point that processes involved in the incompatibility system seem as a rule to be more sensitive to abnormal external stimuli than those that control pollen germination and fertilisation in general, so that it is possible for the incompatibility to be successfully broken.

Other external, but artificial, stimuli implicated in the breaking of *Brassica* self-incompatibility include high concentrations of gaseous $CO_2$ (at 3–5%, Nakanishi and Hinata 1973), electrical stimuli (100 V, Roggen *et al.* 1972) and stigmatic abrasion using a wire brush (Roggen & van Dijk 1972). In the last cases, damage to, or chemical transformation of, the cuticle of the stigmatic papillae may promote the penetration of selfed pollen tubes.

## THE OPERATION OF SELF-INCOMPATIBILITY

There has been a great deal of work on the physiology and ultrastructure of self-incompatibility since about 1965. Despite this, a final comprehension of the mechanisms by which it functions has proved peculiarly difficult to achieve, and we are still some way from this goal. There seem to be two reasons for this. Firstly, the mechanisms appear to be complex, with a number of different functions interacting. Secondly, it is clear that different systems (one- and two-locus gametophytic, multi-allelic sporophytic and the various di-allelic sporophytic) have quite different mechanisms, which reinforces the supposition that these systems arose independently of one another. Thus, it is necessary to discuss the operation of these different systems separately.

It is very important that we achieve a thorough understanding of incompatibility mechanisms. From a theoretical standpoint, we need to know how plant cells recognise each other. Animal cells, which are naked, surrounded only by a membrane, tend to accept cells of the same genotype but to reject or attack foreign cells of the same or different species. Thus, for a successful mating, the innate antagonism of the female environment against the male sperm must be suppressed. Mechanisms of antagonism, which are mediated through surface-carried immunoglobulins, are quite well understood, and the science of immunology has made great advances in manipulating these mechanisms, leading to success in combating disease, transplanting organs and overcoming cross sterility.

In contrast, plant cells are enclosed in cellulose boxes. Membranes in different cells are most likely to transmit and receive chemical messages via water-soluble substances, although stigmatic and stylar cell contents may connect by plasmodesmata, most notably in the so-called 'key junctions' in the style of *Petunia*. Plant cells can recognise foreign cells, for example when they are attacked by a fungal pathogen. They can also distinguish between different types of foreign cells, so that an orchid root will accept invasion by an endomycorrhizal fungus but will reject other soil-borne fungi. A legume root nodule will accept invasion by the nitrogen-fixing *Rhizobium*, but may reject other bacteria.

However, the self-incompatibility mechanism is without parallel, for it *rejects* cells of the *same* genotype (at a specific locus), but *accepts* cells of different genotypes. The various mechanisms of self-incompatibility all achieve this rejection of self by different means, so demonstrating a very widespread and evolutionarily repeated requirement for obligate outcrossing.

What is very remarkable about self-incompatibility is that plant cells are continuously in contact with cells of the same genotype *without* rejecting them. It is only the pollen grain or the pollen tube that is rejected. Thus, to understand self-incompatibility, we must search for chemical and physical features that are unique to the pollen grain and pollen tube. It is then necessary to discover how these may react with the stigmatic papillae (Ch. 3).

The understanding of the mechanisms of self-incompatibility should produce great practical benefits in plant breeding. Successful plant breeding depends above all on an ability to manipulate the breeding system. Selfing allows a genotype to be fixed; a 'pure breeding line' to be developed. Crossing allows attributes to segregate so that they may be selected for, allows different genotypes to be combined, and often leads to improved vigour and yield. Within a breeding programme, there are times when selfing is needed, and times when crossing is preferred. Although it is possible to cross selfers laboriously through emasculation, or efficiently using sterile males, it is much more difficult to self-fertilise self-incompatible strains. This can usually only be achieved with difficulty using manipulative techniques (p. 212), or not at all. It would be very convenient if a chemical treatment that leads to instant, reversible self-compatibility could be developed. This, perhaps, is the final practical goal of studies on the mechanisms of self-incompatibility.

This book is not about plant physiology or cell recognition; neither is it a manual to plant breeding. It would be possible to write another book of this size about work on the mechanisms of self-incompatibility without reaching a final conclusion. Thus, present knowledge on these mechanisms are presented here very briefly, and very incompletely.

### *Gametophytic mechanisms*

**One-locus gametophytic systems** The basic features of the one-locus gametophytic incompatibility system can be briefly listed as follows:

(a) Control of the behaviour of the individual pollen grain depends on the genotype of that haploid grain.
(b) Pollen grains become hydrated externally from the 'wet' stigmatic exudate.
(c) Pollen tubes germinate, grow and penetrate the stigma equally well

in compatible and incompatible pollinations; tubes grow over the surface of the papilla and penetrate the stigma at the base of the papillae by dissolving the middle lamella of the cell wall.

(d) In incompatible pollinations, pollen tubes grow between cells into the style (except in *Oenothera* in which pollen tube inhibition takes place in the stigma). In the style, at least with the Solanaceae, rough endoplasmic reticulum (ER) appear in concentric circles in the tip of the tube; the walls of the tube become thinner, and the callosic inner wall disappears; at the same time, numerous particles, about 0.2 $\mu$m in diameter appear in the cytoplasm at the tube tip, the tip swells markedly, and then bursts, releasing the particles; the tube then ceases to grow (de Nettancourt *et al.* 1974, de Nettancourt 1977). This behaviour closely resembles that which occurs in a compatible tube at the time of fertilisation within the embryo-sac (Ch. 3, p. 56), when the apex of the tube bursts to release the sperm cells. Thus, de Nettancourt (1977) very plausibly suggests that the gametophytic self-incompatibility reaction 'may perhaps be equated to an anticipation of a release-phenomenon scheduled to take place, upon a signal from a synergid, at the time the pollen tube has reached the ovule'. Unfortunately it is not yet clear how widespread s-i through tube apex rupture is outside the Solanaceae. It does not apparently occur in *Lilium longiflorum*.

It is worth noting that the incompatibility reaction does not occur until after the second pollen mitosis has taken place in the tube (Ch. 3). All the features exhibited lead to the suggestion, at the simplest level, that an incompatibility factor(s) is synthesised by the pollen tube as it grows down the style, and that this recognises another factor present in the stylar tissue. Most evidence suggests that the stylar factor is already potentially present in the unpollinated style, but that transcription and translation via RNA to implement the incompatible reaction takes place during pollen tube growth (Pandey 1967, Linder & Linskens 1972, van der Donk 1974, de Nettancourt 1977). Nevertheless, it is possible to simulate the effects of incompatibility *in vitro*, using crude stylar extracts, at least in *Petunia* (Sharma & Shivanna 1982). In *Lilium*, similar results can be obtained using secretion from the stylar canal (Dickinson *et al.* 1982).

As discussed above, it seems clear that gametophytic self-incompatibility is oppositional in character, rather than complementary. Although complementary effects operate in the implementation of normal fertilisation (by hydrating the pollen grain, and thus allowing it to germinate, and by providing stylar resource for the pollen tube to grow), these also obtain for incompatible pollinations, which are later impeded, oppositionally. Failure of the pollen grain to germinate, or to grow, which is found in some di-allelic sporophytic system (Ch. 7, p. 272) can, however, be interpreted as breakdown in complementarity. [It is interesting, however, that in a survey of the Polemoniaceae by Plitmann and Levin (1983), widespread correlations between many different species were found for pollen-grain diameter, pollen-tube diameter, stigma–papilla width and stylar length, suggesting that pollen-grain size is associated with features of the gynoecium down which the pollen grain must grow (i.e. complementary).]

Thus, a search for incompatibility factors in a gametophytic system should not concern those enzymes involved in mediating pollen germination, pollen-tube penetration, or pollen-tube growth (such as cutinase-type esterases or pectinases), but some factors that directly inhibit growth in the style. It must also look for a system that accounts for the known genetic features of gametophytic incompatibility: a tripartite locus with a specificity segment, and very closely linked activity segments for the pollen and style; the ability of the specificity segment to code for one of very many allelic forms; absence of interactions between compatible and incompatible pollen tubes in the same style (i.e. 'non-diffusibility', compare p. 229); loss of incompatibility in a heterozygous diploid pollen grain.

A number of models have been proposed to account for the observed morphological and genetic facts of gametophytic self-incompatibility. The search for the actual compounds involved still goes on. It seems to be generally agreed that, despite the interesting serological results of Lewis (1952) and Lewis *et al.* (1967), these are not antigens or seroproteins in the animal sense. It seems very possible that they are in fact enzymes which may be glycoproteins, but unfortunately the exciting initial results of Pandey (1967) concerning the apparent specificity of electrophoretic bands of peroxidase to *S* genotypes in *Nicotiana alata* could not be repeated by Bredemeyer (1973, 1975). Peroxidase has long been suspected as a likely agent, as it participates in the destruction of the growth hormone IAA (auxin), and Bredemeyer and Blaas (1975) do indeed show that peroxidase number 10 inhibits the growth of *Nicotiana* pollen tubes *in vitro*.

It does seem, however, that more than the cessation of pollen-tube growth is implicated in at least some gametophytic incompatible reactions in which the apex of the tube swells and thickens and may burst, as during fertilisation. As yet, the factor in the embryo-sac which is transmitted from the synergid(s) to the pollen tube and promotes the tip of the tube to burst and release the sperm cells during fertilisation has yet to be identified. (It is not calcium, which forms a gradient 'direction-finder' for the tube, nor yet gibberillic acid, which promotes synergid disintegration immediately beforehand, thus probably releasing the 'synergid message'.) Once the nature of the 'synergid message' is discovered, the incompatibility factor might prove to be the same. *S* allele-related differ-

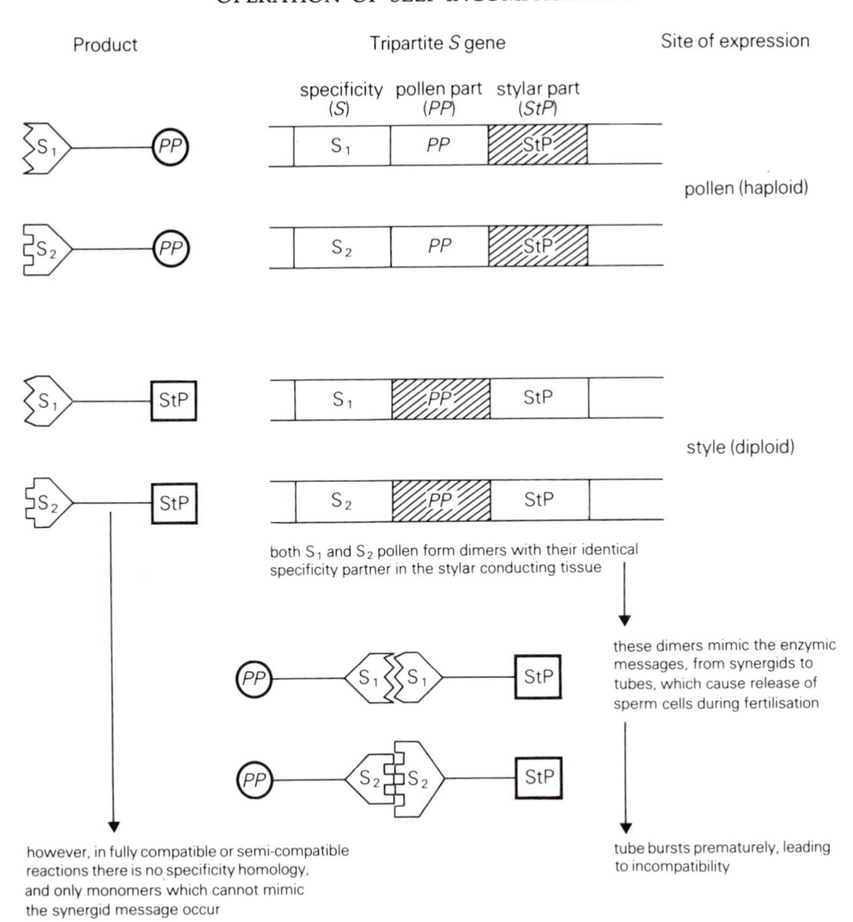


**Figure 6.4** A model to account for gametophytic self-incompatibility gene action, in the self of the genotype $S_1S_2$.

ences in the styles of *Nicotiana* and *Prunus* have proved to be glycoproteins. Glycoproteins restricted to stylar conducting tissue have also been discovered in *Trifolium* and *Lilium* (reviewed in Heslop-Harrison 1983).

The most popular current model to account for *S* gene action is based on Lewis (1964), modified by Ascher (1966), Linskens (1967, 1975) and Heslop-Harrison (1978). I have further modified it (Fig. 6.4) to include the concept of synergid message mimicry. The basic features of this model are:

(a) Monomers, which may be glycoproteins, are formed in the stylar conducting tissue prior to tube growth, and in the tube tip subsequent to tube growth. These each have the same specific segment (from the specificity part of the locus), if selfed, and a non-specific segment. The non-specific coded segments are different on the male and female sides, being coded by the pollen part and the stylar part of the locus, respectively; they have an 'activator' function which only operates in tube or stylar tissue respectively; the other non-specific gene segment is 'switched off' during transcription.

(b) These monomers are non-functional; however, if 'male' and 'female' monomers with identical *S* allele specificity segments meet, they form dimers, due to the correspondence in their morphology at specific sites, and these dimers attach to the apex of the pollen tube.

(c) These dimers mimic the synergid message, causing the tube apex to swell, thicken, and sometimes to rupture prematurely (at least in the Solanaceae).

**Two-locus gametophytic systems in grasses** It has already been noted that grasses (Gramineae) differ from other gametophytic s-i systems, not only in having two-locus control, but also in showing cytological and physiological features of the pollen grain which are much more similar to those in sporophytic systems (i.e. they are trinucleate, with trinucleate-type behaviour; Table 6.1). I suggest therefore that gametophytic s-i may have arisen independently in the grasses in the same way as in sporophytic systems, that is secondarily from s-c plants. Consequently, it would be hardly surprising to discover that the operation of the s-i system in grasses differs fundamentally from that in other gametophytic s-i plants.

This does in fact appear to be the case, and the morphology and physiology of grass s-i has been elegantly reviewed by Heslop-Harrison and Heslop-Harrison (1982b) and Heslop-Harrison (1982). There are two major morphological differences between the function of s-i in grasses and in plants with one-locus gametophytic s-i:

(a) Although s-i pollen germinates well, and the pollen tubes start to grow normally, tube growth usually ceases as soon as the tube touches the stigma papillae; occasionally, penetration of the stigma cuticle occurs, but then tube growth ceases soon afterwards, in the intercellular spaces of the stigma.

(b) The s-i tube does not swell and burst at the apex; instead contact of the tube with the papilla is rapidly followed by the appearance of nodules, probably formed from microfibrillar pectins in the wall at the extreme apex of the tube. From this initial response, there follows a rapid accumulation of pectin at the apex of the tube until the whole apex of the tube is occluded. As in one-locus gametophytic systems, the subsequent cessation of growth is accompanied by a secondary concentration of inner-sheath callose around the tube tip, which forms a useful marker when fluorochrome dyes and ultraviolet microscopy are employed. As the arrested tube can be shown to continue to respire normally, it appears that cessation of growth is implemented solely by the pectic occlusion of the tube tip.

These facts have led the Heslop-Harrisons to propose the following model of gene action in s-i grasses:

(a) The incompatibility factors on the female side are proteins (probably glycoproteins) with binding properties, which are located on the papilla surface.
(b) These glycoproteins bind specifically to sugar moieties present in the long-chain carbohydrates in the wall of the apex of the pollen tube; such binding is highly specific to incompatible pollen tubes.
(c) Such tight binding of the tube tip mechanically prevents the growth and stretching of the tube tip, thus interfering with the extension of polysaccharide microfibrils. As a result, there is a build-up of wall precursor particles, including microfibrillar pectins. These cannot be dispersed into the wall, which has stopped growing. The apex, the wall of which is already trapped and stuck fast by the papilla, is thus internally 'gummed up' with pectins. Thus the apex, being disorganised, stops growing. As Heslop-Harrison (1983) states 'the key event is the disturbance of pectin insertion in the apical cap of the tube, and this leads to a new view of the nature of the response'.

The basis of this model is the lectin-like properties of the glycoproteins which specifically recognise sugar moieties associated with the pectin of the pollen-tube wall. It is interesting that Heslop-Harrison should consider that sufficient diversity can occur in such sugar arrays to account for the action of multiple *S* alleles. Neither does this model account for the necessity that the same *S* allele can code for both a glycoprotein and a non-protein related sugar, although the sugar moiety could, I suppose, be the same in both. However, the position of the reaction, on the stigma surface, does allow for very precise ultrastructural analysis of the phenomenon.

It is surprising that Heslop-Harrison (1983) goes on to suggest that a similar binding reaction between a glycoprotein and a sugar might be common to all gametophytic reactions, the glycoprotein otherwise occurring in the stylar conducting tissues. Although style-specific glycoproteins do indeed occur in stylar conducting tissue in *Trifolium* (Heslop-Harrison & Heslop-Harrison 1982a) and *Lilium* (Dickinson *et al.* 1982), it is not clear that they have sugar-binding properties or functions. They could equally well be enzymic and react with other similar glycoproteins present in the pollen-tube apex. It is not necessary to invoke the same mechanism for grasses and one-locus gametophytic reactions, for they almost certainly have separate origins.

### *Sporophytic mechanisms*

Most work on the function of multi-allelic sporophytic incompatibility refers to the Cruciferae, and in particular to *Brassica*. Because the sporophytic mechanisms are almost certainly secondary and independent of one another in origin (p. 196), there is no particular reason to suppose that similar mechanisms operate in the Compositae or in the various di-allelic sporophytic genera. The latter, and in particular *Primula*, are dealt with in detail in Chapter 7.

In *Brassica*, inhibition of incompatible pollen occurs at, or very near to, the stigma surface (de Nettancourt 1977, Heslop-Harrison 1983). Thus, specific agents of incompatibility are probably borne superficially on the stigma, as in grasses. *S*-specific glycoproteins, which have lectin-like properties, have been detected in the stigma and may correspond to the substances with antigenic properties identified by Lewis (1952) and Linskens (1960). However, these did not vary with the *S* genotype according to Heslop-Harrison *et al.* (1974), although *S*-specific antigenic activity has been noted by Nasrallah and Wallace (1967) and Sedgely (1974).

The outstanding feature of sporophytic systems is that the phenotype of the pollen is controlled from the diploid anther. Thus, all pollen grains from the same plant exhibit the same phenotype, and gene interaction occurs in the diploid phenotype (although actually haploid) pollen. It therefore seems quite clear that the phenotype of the grain has nothing to do with its genotype, but is imposed upon it from the anther. There is a great likelihood that such a mechanism would operate by the anther physically endowing the pollen grain with a substance externally, after pollen-grain formation.

This is indeed what happens. Cytochemical, ultrastructural and electrophoretic observations all show that complex substances manufactured by the tapetum of the anther are later discovered associated with the exine of the pollen-grain wall. The pollen-grain wall is essentially constructed of the exine (composed of the polymer sporopollenin) externally and the intine (composed of pecto-cellulosic compounds) internally (Ch. 3). Many proteins are associated with both layers, but those of the intine only show activity after pollen-grain hydration, and are presumed to be of gametophytic origin. Thus, they are not implicated directly with the sporophytic incompatibility mechanism, although they very probably play a rôle in governing hydration and tube growth of the pollen (Knox *et al.* 1976, Heslop-Harrison *et al.* 1975, de Nettancourt 1977, Heslop-Harrison 1978).

Proteins derived from the anther tapetum are diverse. Seven proteins have been identified in the exine of *Cosmos bipinnatus* (Howlett *et al.* 1975) and in *Brassica oleracea* (Knox *et al.* 1976). In the Compositae, the proteins are inserted into micropores in the tectum (between the extine and intine), whereas in the Cruciferae they are deposited into cavities in the exine. They are sealed into place by sticky coatings of lipoprotein, also derived from the tapetum, termed 'tryphine' or, by German authors, especially when coloured by caroteins, 'pollenkitt'. This tryphine, which also surrounds the germinating pollen tube, plays a vital rôle in the attachment of the pollen grain to the 'dry' stigma. Attachment of incompatible pollen may not be as effective as that of compatible pollen, so some binding recognition by the tryphine may well occur. The tryphine almost certainly also plays a rôle in the hydration of the pollen grain.

However, in *Brassica* and most other sporophytically controlled s-i species, some incompatible grains do adhere, and these usually germinate. As shown by Ferrari *et al.* (1981), who closely analysed the stages of *Brassica* pollen germination and penetration, *Brassica* pollen will germinate almost anywhere at high humidity, including petals and leaves, as well as in water.

Incompatibility results from the failure of the tube successfully to penetrate the stigma papilla. This is almost certainly mediated by proteins independent of the tryphine, despite the suggestion of Dickinson and Lewis (1975) that tryphine may be implicated in the incompatibility of *Raphanus*. Several exine-held proteins in *Brassica*, which diffuse rapidly from the exine on hydration, are glycoproteins. It has proved possible to mimic the incompatibility response in *Iberis*, using exine-held proteins, which results in the deposition of callose in the stigma, the characteristic secondary response of incompatibility (Heslop-Harrison *et al.* 1974). Cytochemical tests using the fluorescent protein 'probe' 1-anilinonaphthylsulphonic acid clearly show that these proteins originate from the tapetum.

The diffusibility of these proteins contrasts sharply with those of most gametophytic systems (p. 224) in that 'mentor effects' can occur. When dead compatible pollen is mixed on a stigma with live incompatible pollen, incompatibility is frequently overcome (reviewed in de Nettancourt 1977). Such a mechanism is expected in a sporophytic system in which all the pollen from one parent has the same effect. In gametophytic systems, pollen from a single parent shows different phenotypes (i.e. semi-compatibility), which could not be achieved if active factors were diffusible.

The stigmas of *Brassica* and *Raphanus*, and probably of many Angiosperms with dry stigmas, consists of an outer proteinaceous pellicle, which ensheathes an inner cuticle without being in close contact with it. Unlike gametophytic systems, in which tube penetration takes place between cell walls, the tube of the sporophytic pollen penetrates the pellicle and cuticle before growing between the cell-wall lamellae (Ch. 3). Thus, whereas pectinase activity is required by the successful gametophytic tube, cutinase activity is also required by the sporophytic tube. It is very probable that failure of cutinase activity is the one cause of sporophytic incompatibility. However, as yet it is not clear whether this failure results from the absence of an heterodimeric cutinase enzyme formed from the exine and the stigma pellicle. It is possible that a cutinase enzyme can only act if the tube is held in juxtaposition to the papilla by a binding (lectic) association between glycoproteins of these tissues.

A curious feature, common to gametophytic systems, has been the failure to identify identical proteins in the pollen exine and the stigma pellicle, as for instance in *Iberis* (Heslop-Harrison *et al.* 1974). As the incompatibility recognition, genetically, takes place between the same *S* alleles, one might expect the same gene product to be identified on the male and female sides of the same individual. However, the tripartite gene model suggested for gametophytic systems may code for glycoproteins, the specificity segment-coded parts of which are identical on the male and female sides, but which have pollen-part and stylar-part coded moieties of different weights. This would cause male and female products of the same *S* allele to behave differently during electrophoretic or serological analyses. The failure to identify identical glycoproteins on the male and female side in sporophytically controlled plants thus suggests that genetic control may also be by complex loci which show different gene expression of the same allele on the male and female side.

As has been pointed out by Kanno and Hinata (1969), and discussed by de Nettancourt (1977), a typical sporophytic incompatibility reaction, as seen in *Raphanus*, cannot be simply explained by failure of cutinase (esterase) activity. In fact, there are three points in the fertilisation process at which self-incompatibility is expressed (Table 6.4):

(a) Pollen attachment to the stigma. The pollen grain can only lodge on the dry stigma pellicle if it is firmly attached, a process that probably occurs through the binding properties of the tryphine coat of the pollen grain to the stigmatic pellicle proteins. The poorer rate of attachment of incompatible pollinations suggests that self-recognition leads to less efficient binding, although significant levels of self-attachment still occur. Successful binding of the grain to the stigma allows hydration and germination of the grain, although this will also occur in moist environments away from the stigma.
(b) Penetration of the stigma by cutinase activity. In incompatible pollinations, some close binding of grains to the stigma still occurs, thus enzymic activity might only occur if heterodimers formed from different *S* alleles expressed in the exine and the stigma pellicle can form.
(c) Successful pollen-tube growth within the stigma. In incompatible pollinations the few tubes that successfully penetrate the stigma

**Table 6.4** Possible sites and functions for self-recognition and rejection in multi-allelic s-i systems.

| | One-locus gametophytic | Two-locus gametophytic (grasses) | Sporophytic (at least Cruciferae) |
|---|---|---|---|
| pollen-grain attachment, hydration and germination | | | lectic poor pollen attachment (?tryphine) |
| penetration of papilla by pollen tube | | lectic failure of pollen tube wall growth | ?enzymic homodimers fail to elicit cutinase reaction? |
| pollen-tube growth in style | ?enzymic homodimers mimic part or whole of gamete release prematurely? | | (callose from enzymic recognition reaction blocks tube path) |

papilla are rapidly blocked by the formation of large quantities of free callose, usually while still in the papilla. This callose is almost certainly formed as a by-product of the incompatible reaction at the stigma surface. A callose sheath of the pollen tube in the style is an omnipresent feature of all successful pollinations in the gametophytic and sporophytic systems, and the production of excess callose is the usual by-product of all incompatible reactions at the point where incompatibility acts. However, in the Cruciferae at least, the callose appears to take an active rôle in blocking the pollen tube, which is not the case in other systems.

## CONCLUSION

The following generalisation may be made about the function of incompatibility systems:

(a) Recognition and incompatibility takes place between identical *S* alleles; however, it appears that the gene products of these are not identical on the male and female sides.
(b) The action of incompatibility is oppositional rather than complementary. It seems to occur either through a failure (sporophytic) or malfunction (grasses) of lectic (binding) functions, or through a

failure (sporophytic) or malfunction (gametophytic) of possible dimeric enzyme systems. However, in compatible reactions, both lectic and enzymic systems are required for normal function (Table 6.4). Both lectic and enzymic systems may involve glycoproteins (except for sporophytic pollen attachment).

(c) One-locus gametophytic systems, two-locus gametophytic systems (grasses) and the various sporophytic systems almost certainly have independent origins. It is therefore likely that they have different modes of function, which is borne out by their genetic control.

CHAPTER SEVEN

# *Di-allelic self-incompatibility and heteromorphy*

---

Most self-incompatible plants have multi-allelic gametophytic incompatibility. This system is probably primitive to the Angiosperms. As discussed in Chapter 6, many *S* alleles must coexist at a locus for it to succeed. In contrast, sporophytic systems are probably secondary in origin, having evolved polyphyletically from self-compatible, partially inbred populations, encouraged by renewed selection pressures for outbreeding. A small majority of sporophytic systems is also multi-allelic, and these are also discussed in Chapter 6.

However, unlike gametophytic systems, sporophytic systems can function with only two alleles (di-allelic), which are termed *S* (dominant) and *s*. This is possible because diploid heterozygous male parents produce pollen of only one incompatibility phenotype *S*, although the pollen grains have the genotypes *S* and *s*. Likewise, the heterozygous stigma *Ss* has the phenotype S. Thus there are only two phenotypes in the population, S (genotype *Ss*) and s (genotype *ss*). As incompatibility works on a self-rejection basis, S phenotypes can only mate with s phenotypes; illegitimate matings S × S and s × s (cross or self) are incompatible. The compatible crosses are genetically *Ss* × *ss* (legitimate), and should yield seedlings with S and s phenotypes in equal ratios. It follows that only half the individuals in a population are available for outcrossing.

A di-allelic system is less efficient than a multi-allelic system in which most genets are cross compatible.

In common with gametophytic systems, multi-allelic sporophytic systems have cryptic phenotypes; that is, it is not possible to determine the incompatibility phenotype of a plant by its morphology, but only by experimental crosses. In remarkable contrast, in almost all di-allelic systems the two incompatibility phenotypes S and s have different floral morphologies, i.e. they are heteromorphic. In most cases, this heteromorphy takes the form of distyly or heterostyly, in which one morph has a long style and short stamens (and very often smaller pollen) and is termed 'pin', and the other has a short style and long stamens (often with larger pollen) and is called 'thrum'. In *Primula*, the most familiar case (Fig. 7.1), the form of the stigma papillae also differs, being longer in the pin (Figs. 7.2 and 7.3), as does the cell shape in the stylar conducting tissue (Fig. 7.5).

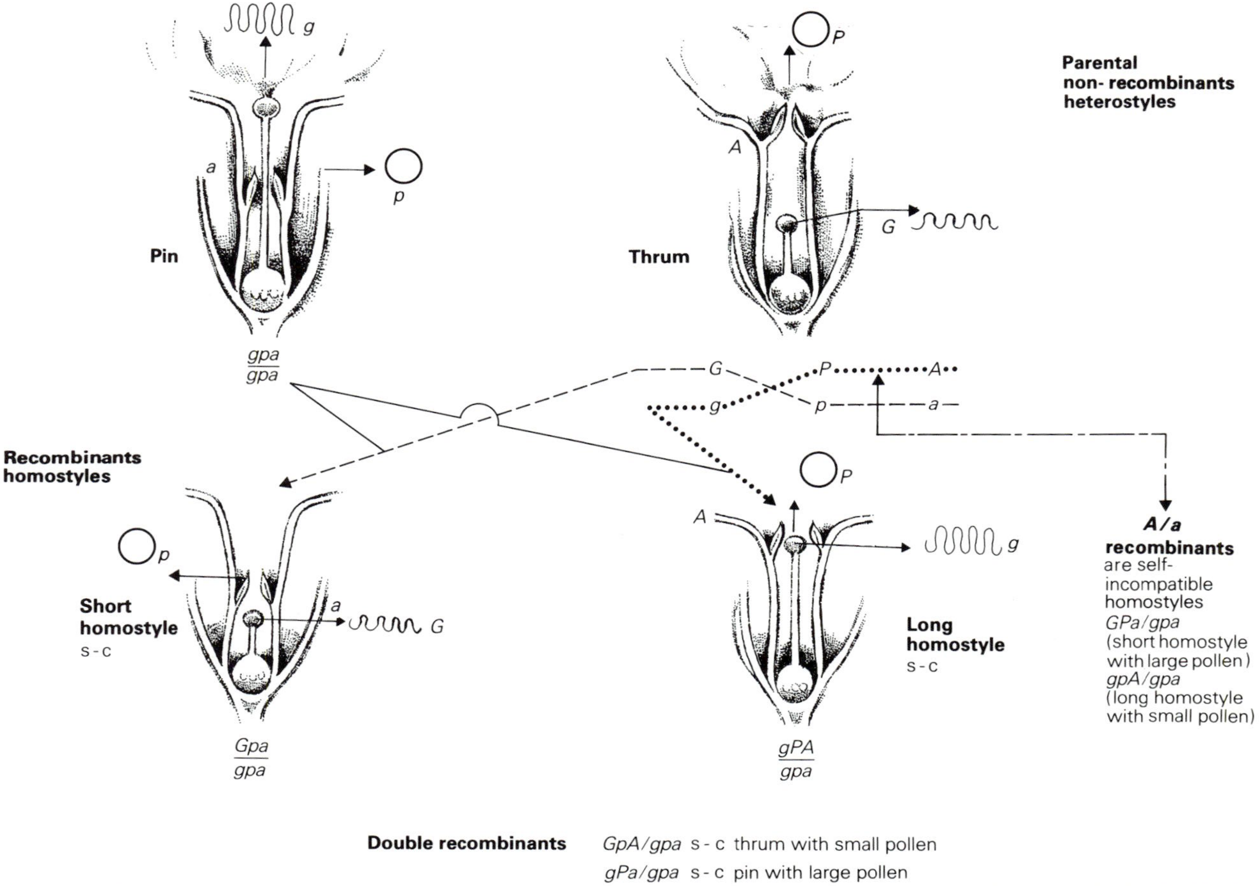

Pin
g
a
p
gpa/gpa
Thrum
P
A
G
Parental
non- recombinants
heterostyles
G
P
A
g
p
a
Recombinants
homostyles
Short
homostyle
s - c
p
a
G
Gpa/gpa
Long
homostyle
s - c
P
A
g
gPA/gpa
A/a
recombinants
are self-
incompatible
homostyles
GPa/gpa
(short homostyle
with large pollen)
gpA/gpa
(long homostyle
with small pollen)
Double recombinants
GpA/gpa s - c thrum with small pollen
gPa/gpa s - c pin with large pollen

**Figure 7.1** The *Primula* heteromorphy linkage group. The common non-recombinant morphs are pin (top left) and thrum (top right). These are between-morph compatible, but within-morph incompatible. Heteromorphy and incompatibility are controlled by two linkage groups, *GPA* and *gpa* in which the thrum is heterozygous *GPA/gpa* and the pin homozygous *gpa/gpa*. *G* donates thrum female characters of short style, short stigmatic papillae and thrum female incompatibility; *g* donates pin female characters of long style, long stigmatic papillae and pin female incompatibility. *P* donates thrum male characters of large pollen and thrum male incompatibility; and *p* donates pin male characters of small pollen and pin male incompatibility. *A* donates high anthers (as in a thrum) and *a* donates low anthers (as in a pin).

Recombinants can only arise from the heterozygous thrum. Recombinants between *G/g* and *P/p* are short homostyles *Gpa/gpa* (bottom left) and long homostyles *gPA/gpa* (bottom right); these are self-fertile, automatically self-pollinate, and are easily detected morphologically.

Recombinants between *P/p* and *A/a* are also short homostyles and long homostyles, but have different pollen sizes from *G/g–P/p* recombinant homostyles, having large pollen and small pollen respectively. Unlike *G/g–P/p* recombinants, they are self-sterile.

Double recombinants are very rare. They will be morphologically thrum and pin, but self-fertile, and with small and large pollen respectively. Thus, *G/g–P/p* and double recombinants are cryptic, and can only be detected by microscopic examination of the pollen, or breeding tests.

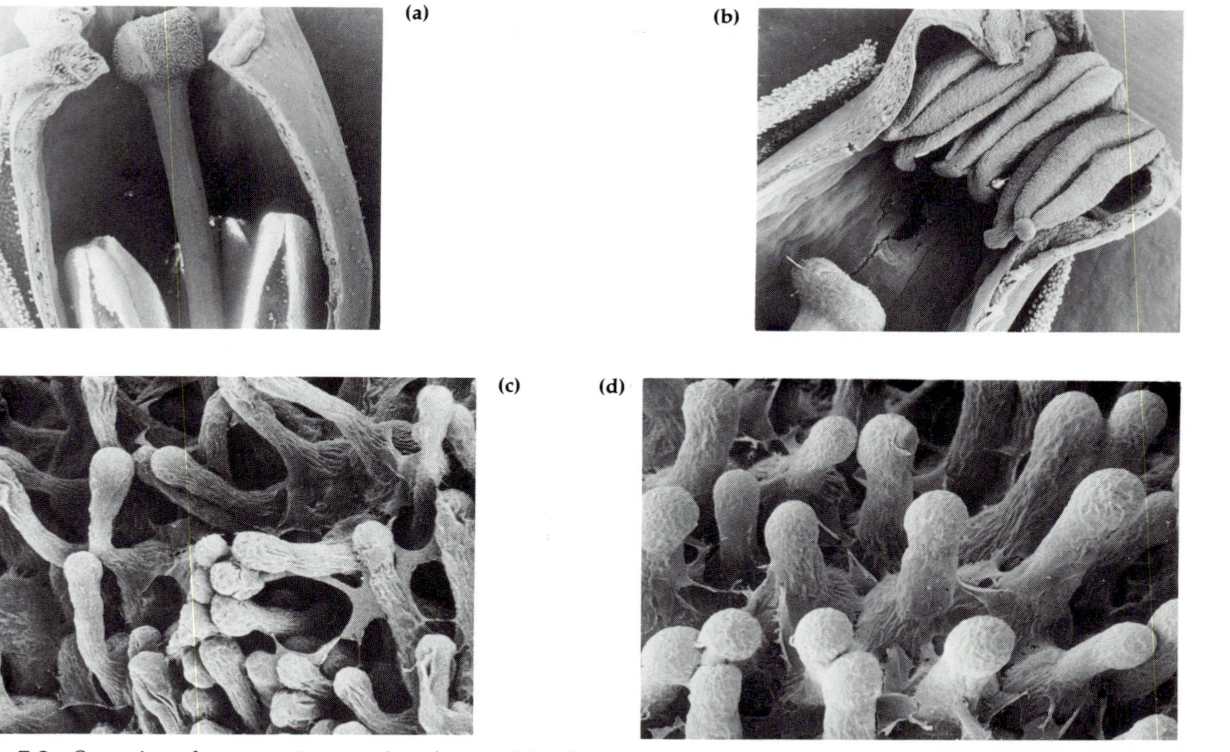


**Figure 7.2** Scanning electron micrographs of part of the flower of (a) pin (stigma at top, anthers at base; ×12) and (b) thrum (anthers top right, stigma bottom left; ×12) *Primula modesta* and stigmatic papillae in (c) pin (×250) and (d) thrum (×400) *Primula modesta*.

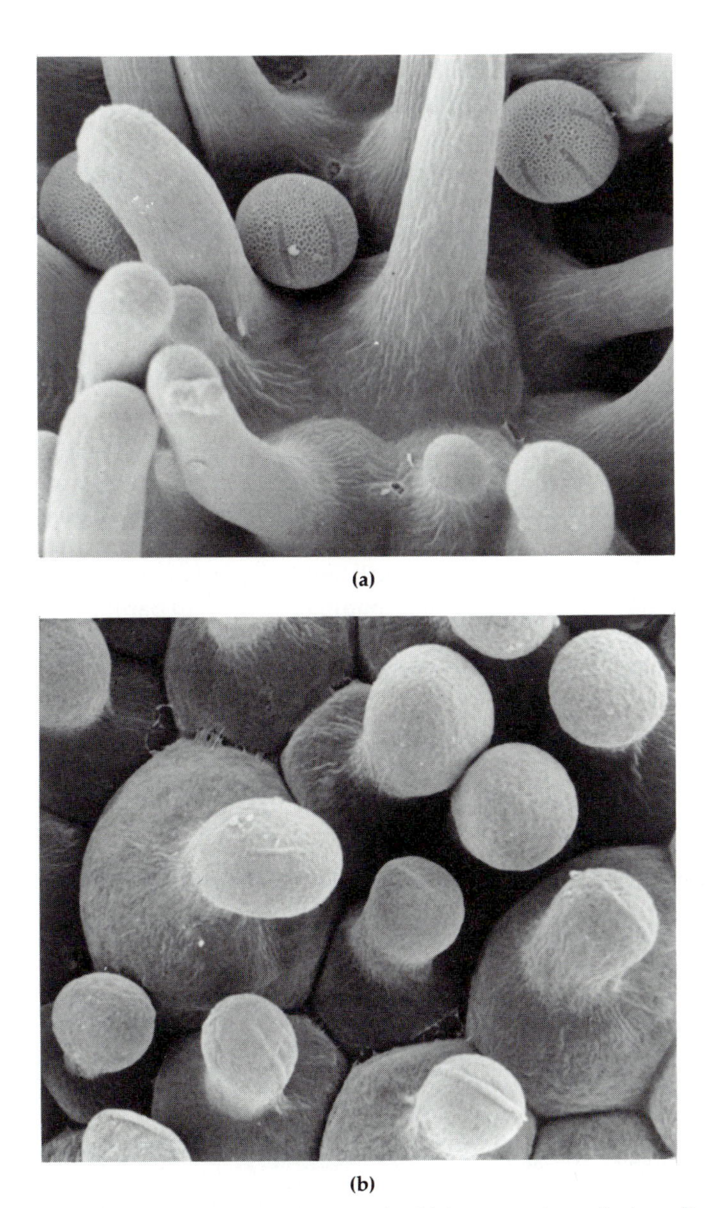

(a)

(b)

**Figure 7.3** Scanning electron micrograph of (a) ungerminated pin pollen grains on the pin stigmatic papillae of *Primula veris* (×800) and (b) stigmatic papillae in thrum *Primula veris* (×800).

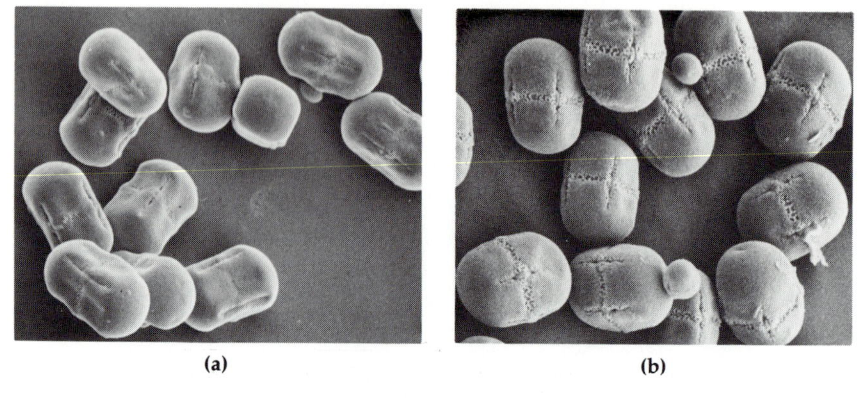

**Figure 7.4** Scanning electron micrographs (×500) of pollen of (a) pin and (b) thrum (two grains have started to germinate) *Primula modesta*.

The heteromorphy of the S and s phenotypes takes various forms. Between two and seven different morphological characters vary between phenotypes, depending on the genus. Thus, in *Armeria maritima* (thrift), there is no heterostyly or difference in pollen-grain size, but the stigmatic papillae differ between morphs with 'cob' (named after the resemblance of the stigma to a maize cob) and 'papillate' (Fig. 7.6) papillae. These two morphs also have markedly different sculpturing to the pollen exine, that of papillate being uniformly reticulate, but those of cob being coarser and tricolporate (Fig. 7.7).

In most cases, there is a very close correspondence between the incompatibility phenotype and the heteromorph. Thus one morph is always S phenotype, and the other is always s phenotype. This might

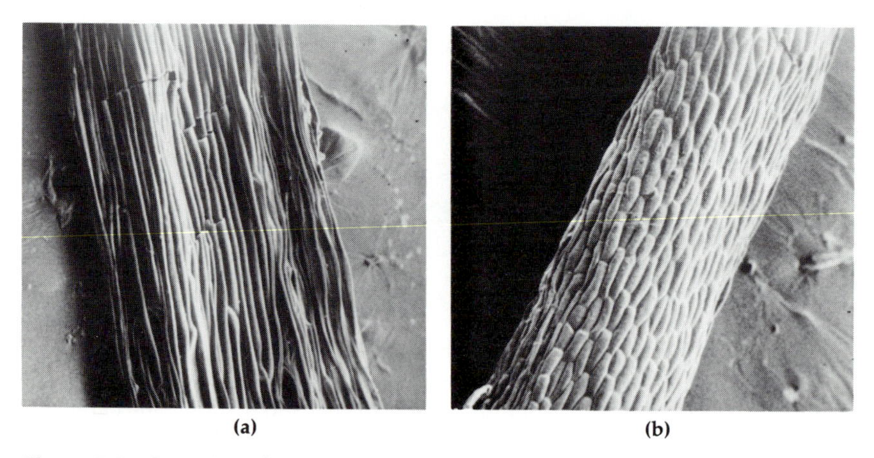

**Figure 7.5** Scanning electron micrographs (×100) of stylar tissue of (a) pin and (b) thrum *Primula veris*.

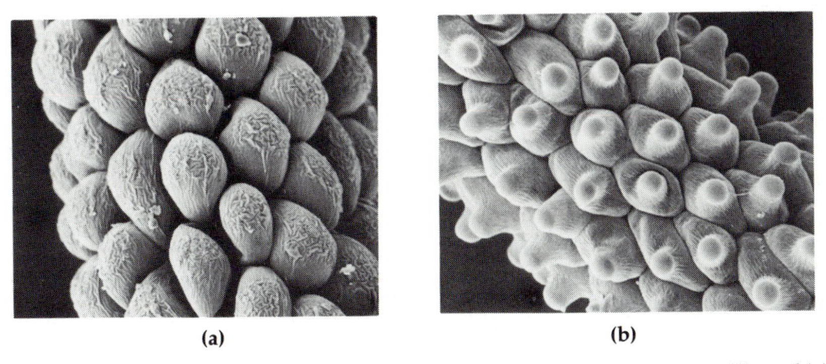

(a) (b)

**Figure 7.6** Scanning electron micrographs (×500) of the stigmatic papillae of (a) cob and (b) papillate *Armeria maritima*.

suggest that the incompatibility phenotype and the heteromorph are pleiotropic expressions of the same gene. But in several genera, recombinants between the incompatibility phenotype and the heteromorph, and between character states within the heteromorphy, are known. These recombinants are best studied in *Primula*, some of which are 'homostyles', and these are described below. It is probable therefore that most cases of heteromorphy are controlled by two linkage groups, in which the genes that control the heteromorphic characters are closely linked to the *S/s* incompatibility locus as a supergene. Curiously, in most cases, amongst unrelated plants, the heterozygote (S phenotype) is thrum with large pollen, whereas the *s* supergene is pin with small pollen. Only in *Hypericum aegypticum* (Ornduff 1979a) is this relationship known to be reversed (pin heterozygote) (Ganders 1979, p. 616).

The rarity of recombinants within the linkage group, between the

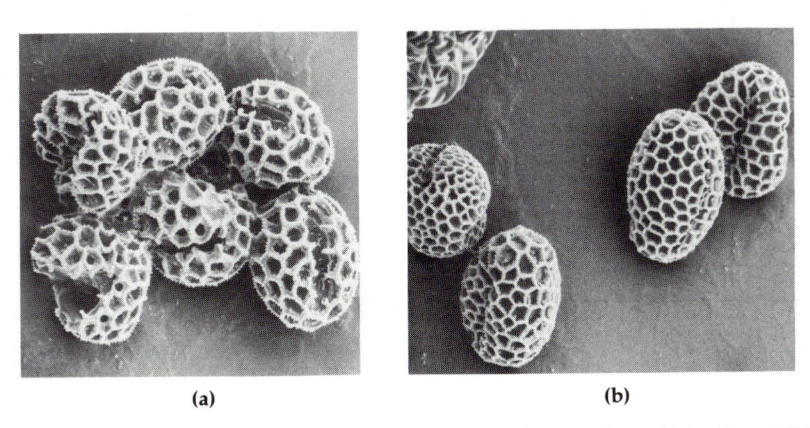

(a) (b)

**Figure 7.7** Scanning electron micrographs (×450) of the pollen of (a) cob and (b) papillate *Armeria maritima*.

incompatibility phenotype and the heteromorph or between the various characters involved in the heteromorphy, strongly suggests that the various features controlled by the loci of the supergene are strongly coadapted. As early as 1877, Darwin proposed that the correspondence of position in the flower between the anthers and stigma of cross-compatible morphs was an adaptation that maximised pollen transfer between morphs. Thus, an insect visiting a pin flower would receive pin pollen apically, and on visiting a thrum flower, and adopting the same position in the flower, this pollen would be positioned on the pollinator in such a way as to pollinate the thrum stigma. The same argument applied to thrum pollen, most of which would be deposited relatively distally on the pollinator. Several workers, in particular Ornduff, have examined legitimate (between-morph) and illegitimate (within-morph) stigmatic pollen loads in *Primula* and other heterostylous genera to verify this assumption. There is evidence that heterostyly does promote legitimate pollen transfer, as will be examined in more detail later, but it is not always convincing.

It also seems intuitively likely that heteromorphic features of the pollen (size, sculpturing), and of the gynoecium (papilla shape, stylar conducting tissue) play a functional rôle in the incompatibility. Although Lewis (1942b) suggested that the faster growth rate of large thrum pollen down the long pin style of *Primula obconica* might be a function of grain size, it now seems to be established that the pollen tube is self-supporting within the style (Ch. 2). Thus pollen size is probably irrelevant to the functioning of the incompatibility, which as we will see has different but manifold bases. In fact, there is only one case (*Armeria*, p. 262) in which heteromorphic features clearly function in the incompatibility. Dulberger (1975a) argues that such heteromorphic features as pollen size, exine sculpturing and papilla length are intimately involved in *S* gene action (di-allelic incompatibility) in heteromorphic species, but I find her arguments unconvincing.

The strength of the within-morph incompatibility is variable. Within *Primula* it can vary from total to very weak between different species. However, self-incompatibility seems never to be greater than within-morph incompatibility. Crosses between pins, or between thrums, are as incompatible as selfs. There is a strong suggestion that in some species relatively high levels of within-morph compatibility can be selected for. However, in most species the incompatibility is strong, and the plants are obligate outcrossers. Thus, homostyles that are self-compatible and automatically self-pollinate have aroused much interest when they occur in predominantly heterostylous genera, species or even populations. Several different homostylous conditions can be identified, which have different origins and explanations, and it is important to distinguish between them.

## DISTRIBUTION OF HETEROMORPHY

In a thorough and useful review, Ganders (1979) shows that heteromorphy (which he calls heterostyly although he includes non-heterostylous heteromorphs such as *Armeria*) occurs in 24 families of flowering plants, spread among 18 orders. He lists 155 genera in which heteromorphy has been reported, of which no less than 91 belong to the Rubiaceae. Heteromorphy has a scattered distribution in the Angiosperms. Some major families have only single heteromorphic representatives (*Jepsonia* in the Saxifragaceae, *Sebaea* in the Gentianaceae). Other families apart from the Rubiaceae, which contribute many heteromorphic genera and species, are the Boraginaceae, Plumbaginaceae and Primulaceae. *Primula*, with over 600 species, at least 90% of which are heterostylous, is the genus with the largest number of heteromorphic species, and is certainly the best known.

Ganders expressly omits species that are heteromorphic, but in which it is known that a di-allelic incompatibility is lacking. He mentions only *Narcissus tazzeta* (multi-allelic incompatibility; Dulberger 1964), which has two morphs with different style lengths, and the variably heterostylous *Amsinckia*. In the latter genus, there may be some slight incompatibility through competition between selfed and crossed pollen; the control of heterostyly is complex and probably polygenic, with pin-type and thrum-type plants occurring in different ratios and with different levels of heterostyly (Ganders 1975). *Narcissus triandrus* can be tristylous but is self-compatible, at least in gardens, whereas *Anchus hybrida* (Dulberger 1970) and *A. officinalis* (Schou & Philipp 1983) are distylous but have an unlinked multi-allelic self-incompatibility system. Such exceptions to the association between heteromorphy and di-allelic incompatibility which are known are few. It should, however, be pointed out that the breeding system of many heterostylous genera is not known, particularly in the tropical woody Rubiaceae, some of which are not in cultivation (an account of the operation of di-allelic incompatibility in some heteromorphic Rubiaceae has recently appeared; Bawa & Beach 1983).

Equally, it is possible that there are many genera or species with a di-allelic sporophytic incompatibility system which is cryptic, i.e. without heteromorphy. As yet, such a condition is only known in two familiar members of the Cruciferae, *Capsella* and *Cardamine* (Ch. 6). It is also possible that many more genera have a semi-cryptic heteromorphy, as is observed in the thrift (p. 262, Figs. 7.6, 7.7); in this and some other members of the Plumbaginaceae it is only possible to distinguish between morphs microscopically (or with a good lens). Such differences might well remain undetected on herbarium material of little known genera.

The distribution of heteromorphism is scattered amongst 18 orders of flowering plants, and is found in all continents and in all habitats from desert to aquatic and from the tropics to the arctic. Nevertheless, this condition is associated with certain major taxonomic groups. It is absent from those orders of flowering plants considered to be the most primitive, and it is very rare in the monocotyledons (some Pontederiaceae are tristylous and the irid *Nivenia* from southern Africa is distylous).

Heteromorphy is also associated with certain growth forms and floral morphologies. Nearly all heteromorphic species are perennial (the annual buckwheat *Fagopyrum esculenteum* is a familiar exception , and is the only heteromorphic economic crop of any significance). It is noticeable that in some partially heteromorphic genera, annual or short-lived species tend to be homomorphic and self-compatible (*Lithospermum*, *Primula*).

Heteromorphic flowers may have evolved in order to promote pollen transfer between the two di-allelic incompatibility phenotypes S and s (but see p. 280). They are therefore only likely to be favoured in species with relatively specialised, animal-mediated cross-pollination with accurate pollen presentation and reception (Ch. 4). Thus, they nearly all have fused petals which form a bell, trumpet or tube. For pollination to be effective, especially on thrum stigmas, pollinators with narrow mouthparts which can penetrate a tube to feed on basally presented nectar will be encouraged. Thus, most pollinators of heteromorphic flowers are social bees, moths, butterflies or birds; these are also likely to be oligolectic, promoting oligophily (Ch. 4, p. 74) in the plants (but see p. 281). Such mechanisms will encourage the efficiency of cross-pollination which is inherent in the success of a heteromorphy. Tube flowers are particularly frequent in heteromorphic systems; the 4 cm tubes of *Pentas* are quoted by Ganders (1979), but those of *Primula sherriffae* can reach 6 cm. Typically, the diameter of the limb of the flower is relatively small; zygomorphy is rare in heteromorphic flowers, as are large numbers of stamens. It is notable that in heteromorphic plants without heterostyly, such as *Armeria*, the corolla is bowl-shaped, and the petals are almost unfused, not forming a tube.

## FORMS OF HETEROMORPHY

As has been stated already, by far the most common conditions in heteromorphy are morphs with long styles, short stamens, small pollen and long stigmatic papillae ('pin') and those with short styles, long stamens, large pollen and short stigmatic papillae ('thrum'). The former are homozygous (recessive) and the latter are heterozygous (dominant). However, many characters have been shown to vary between heteromorphs in one or more genera (Table 7.1).

**Table 7.1** Characters that can vary between heteromorphs (after Ganders 1979).

| Character | Comment | Examples |
|---|---|---|
| style length | not always exactly reciprocal to stamen length, difference longer in pin *Pentas*, in thrum *Cordia* | *Primula* and most others including *Linum* |
| stylar conducting tissue | different cell shape and area | *Primula* |
| style pubescence | in pins | *Oxalis* |
| style colour | | *Eichhornia* |
| stigma size | bigger in thrum | *Jepsonia* |
| stigma shape | flatter in thrums | *Primula* |
| papilla shape | almost always longer in pins | *Primula* and most others (39/53, Dulberger 1975a) |
| stamen length (anther position) | see style length | *Primula* and most others, but not *Linum* |
| anther size | thrum bigger | *Lithospermum*, *Pulmonaria*, *Hottonia* |
| pollen grain size and number | pin smaller, more numerous | *Primula* and most others (50/55 Dulberger 1975a) |
| pollen sculpturing | | Plumbaginaceae, *Linum* |
| pollen shape | thrum ovoid, pin dumb-bell shaped | *Lithospermum* |
| pollen colour | thrum green, pin yellow | *Lythrum californicum* |
| starch in pollen | thrum with starch | *Lythrum*, *Jepsonia* |
| corolla diameter | thrum larger | *Fagopyrum* |
| corolla pubescence inside tube | pin pubescent | *Lithospermum obovatum* |

Of these 16 characters, between 2 and 7 can vary within a species (most *Primulas* vary for 6). Thus, there is a basic pattern of characters varying between morphs which is common to most genera (and furthermore which shows common relationships to the breeding system), yet there are many additional modifications within and between genera. This pattern of distribution strongly suggests that heteromorphy has evolved in connection with a di-allelic incompatibility system independently on many occasions. Only rarely do taxonomic relationships between heteromorphic genera, or the constancy of heteromorphy and types of

**Figure 7.8** Tristyly in *Lythrum salicaria*, mid-styled morph. Photo by M. C. F. Proctor (×5).

heteromorphy within families, allow us to postulate single origins of heteromorphy between different genera (*Dionysia*, and perhaps *Hottonia*, derived from *Primula*; *Armeria*, *Limonium* and *Acantholimon* in the Plumbaginaceae; and various related genera in the Boraginaceae are examples).

Yet the correspondence in basic features between so many polyphyletic systems suggests that very strong pressures favour the coexistence of certain heteromorphic features if the heteromorphy is to be successful.

This argument is reinforced by an examination of tristyly, which I have not as yet mentioned. Tristyly is found in only three families, all unrelated, two of which are dicotyledons (Oxalidaceae, Lythraceae) and one of which is monocotyledonous (Pontederiaceae). It has thus certainly arisen on three occasions (Ganders 1979). Yet, in all examples, there is a strong correspondence in basic features (Fig. 7.9). The three style lengths are each accompanied by the same patterns of stamen position in all three families. There is a correspondence in pollen size (long stamens always have the largest pollen), in papilla lengths (longest styles have stigmas with the longest papillae) and in cross-compatibility relationships (only pollen from stamens of the same position as the stigma is compatible on it).

Two loci (supergenes) control the tristylous systems (Fig. 7.9). These are termed *S* and *M*. Short-styled plants are *Ss*. Plants that are homozygous *ss* are epistatic to *M* with *ssM-* being mid-styled forms and *ssmm* being long-styled morphs. In *Lythrum*, and some *Oxalis*, *S* and *M* are unlinked. The frequency of the long- and mid-styled morphs will thus depend on the product of the frequencies of *ss* homozygotes and the allele *M*. In *Oxalis rosea*, *S* and *M* are rather closely linked. In this case, there is in effect an extension of the supergene, there being two extended linkage groups *sM* and *sm* which are relevant to the phenotypes in the system (frequencies of *SM* and *Sm* are irrelevant). The mating system should ensure that *sM* and *sm* chromosomes are equally frequent, and that *S* has a frequency of 0.33.

The tristylous system is also remarkable in that heteromorphic characters such as pollen size, papilla size, style length and stamen length are each apparently controlled by two linkage groups *S* and *M*, and thus presumably loci which control each of these features are repeated in both linkage groups. This leads to the suggestion that the *M* linkage group is a duplicate of the *S* group, and that tristylous plants have thus evolved from heterostylous (distylous) systems (see Charlesworth 1979 for another opinion). This is very probably the case in both *Oxalis* and *Lythrum*, both of which also have distylous species. However, the Pontederiaceae have no distylous counterparts, and tristyly in this family appears on the face of it to have arisen *de novo*.

The correspondence between different tristylous species and between tristylous and distylous species seems strong (the long-styled morph is usually recessive, and has small pollen and long stigma papillae in both tristylous and distylous systems). Also, compatible crosses are only made between anthers and stigmas of the same level in both systems. However, it is claimed that one species of tristylous *Oxalis* (*O. rosea*) has a recessive short-styled morph (Weller 1976). As this is also the species in which the

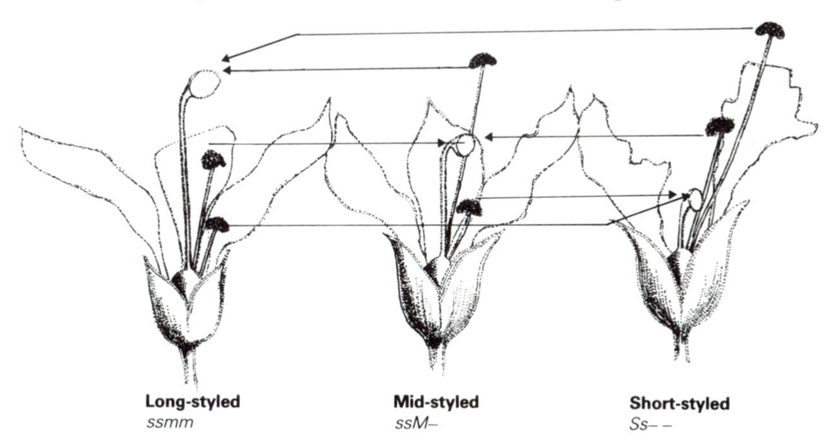


**Figure 7.9** Tristyly in *Lythrum salicaria*. Stigmas are pale and anthers dark. Arrows indicate compatible pollinations between morphs.

*S* and *M* loci are linked, it is very possible that this species has a different tristylous origin from the other tristylous *Oxalis*. Endoduplication of the *S* locus within one chromosome may have occurred in this species.

### *Recombination in the di-allely/heteromorphy supergene*

Recombination of different heteromorphic features has only been studied in *Primula* (Ernst 1933, 1955, Mather & De Winton 1941, Mather 1950, Dowrick 1956) and *Limonium* (Baker 1966), and even then not intensively. More modern work, which accumulates recombinant types, makes test-crosses and maps the supergene, is badly needed in this and other heteromorphic genera which present splendid opportunities for the analysis of coadapted linkage groups rarely possible in an advanced eukaryote.

In *Primula*, three subunits of the supergene have been identified:

(a) *G/g* style length, stylar conducting tissue, stigmatic papilla type, female incompatibility.
(b) *P/p* pollen size, male incompatibility.
(c) *A/a* anther height.

As the thrum is heterozygous *Ss*, thrum characteristics are dominant, and the thrum has the genotype *GPA/gpa* (the order could also be *GAP*). Thus, the pin will be *gpa/gpa*. Recombination within the supergene can therefore only occur in the heterozygous thrums, and will appear in legitimate crosses to the recessive homozygote pin, which thus act as test backcrosses. By far the commonest recombinants to be discovered are *Gpa* and *gPA*, the so-called homostyles. These occur at quite high frequencies in the offspring of cultivated strains of *Primula* × *variabilis* (derived from hybrids between the primrose, *P. vulgaris* and the cowslip *P. veris*). *Gpa* recombinants have thrum female characteristics, and pin male characteristics (see Fig. 7.1). Both the stigma and the anthers occur at the same level down in the floral tube, and thus such plants are known as short homostyles. Because they have thrum S phenotype female incompatibility, and pin s phenotype male incompatibility, they are self-fertile, automatically self-pollinating, and set high levels of selfed seed.

In contrast, *gPA* recombinants have pin female characteristics, and thrum male characteristics. The stigma and the anthers are produced at the mouth of the floral tube, and such plants are known as long homostyles (Fig. 7.10). They have pin s phenotype female incompatibility, and thrum S phenotype male incompatibility. Thus once again they are self-fertile, and show high levels of self-pollination and selfed seed.

Homostyles are much rarer in the field, although other workers and I have found long homostyles in wild populations of *Primula vulgaris*, *P.*

**Figure 7.10** Pin (left), thrum (right) and long-homostyle (centre) flowers of the primrose, *Primula vulgaris*. In the homostyle, both anthers and stigma are at the mouth of the floral tube. Photo by M. Wilson (×1).

*veris*, *P. farinosa* and *P. latifolia* (=*P. viscosa*), and they probably occur in all heterostylous wild *Primula* populations at low frequencies (usually less than 1%). Unlike cultivated populations of *Primula* × *hortensis* (Ernst 1936) and *P. obconica* (Dowrick 1956), short homostyles are rarely if ever found in the wild. Three reasons can be put forward to explain this absence of *Gpa* chromosomes (which presumably arise at the same frequency as *gPA* chromosomes as a result of recombination).

First, it is assumed that both long and short homostyles produce entirely selfed seed (which is open to question as we shall see below); however, as male parents, long homostyles which have thrum male incompatibility can mate with pin females. The long homostyle chromosome *gPA* is *dominant* to the pin chromosome *gpa*. If the long homostyle is homozygous *gPA/gPA*, all its offspring when crossed to a pin female will be self-fertile long homostyles *gPA/gpa*. If the long homostyle male parent is heterozygous *gPA/gpa*, half its offspring when crossed to a pin female will be long homostyles.

Short homostyles that have pin male incompatibility can mate with thrum females. The short homostyle chromosome *Gpa* is *recessive* to the thrum phenotype GPA. If the short homostyle is homozygous *Gpa/Gpa*, half its offspring, when crossed to a thrum female *GPA/gpa*, will be short homostyle *Gpa/gpa* and half thrum *GPA/Gpa*. If the short homostyle is heterozygous *Gpa/gpa*, when crossed to a thrum female *GPA/gpa*, only one-quarter of its offspring will be short homostyle *Gpa/gpa*, one-quarter will be pin *gpa/gpa*, and one-half will be thrum *GPA/Gpa* or *GPA/gpa*. Thus as outcrossing male parents, long homostyles will generate more self-fertilising homostyle offspring than will short homostyles. Thus, if the self-fertility of homostyles is of selective advantage, long-homostyle recombinant chromosomes will be selected in preference to short-homostyle chromosomes (Charlesworth & Charlesworth 1979, Ganders 1979).

Second, although Crosby (1949) assumed that long-homostyle primroses were entirely self-fertilised as parents in the wild, this was queried by Bodmer (1958). Bodmer found an excess of long homostyles in offspring of heterozygous long-homostyle primroses *gPA/gpa* in an experimental plot in which homozygous long homostyles *gPA/gPA* also occurred. If the heterozygous long homostyles *gPA/gpa* entirely selfed as female parents, they would be expected to give 25% pin *gpa/gpa* seedlings. Bodmer (1958) attributed the excess of homostyle seedlings to crossing with homozygous *gPA/gPA* homostyles, and estimated outcrossing rates onto long homostyles of up to 80%. More convincing genetic ratios from controlled crosses, and from open-pollinated wild-type parents led Crosby (1959) to suggest that more realistic figures for outcrossing onto long homostyles were 5–10%. Similar levels of outcrossing have been discovered recently by the use of more sophisticated techniques (Piper *et al.* 1984, Richards 1984).

As yet, nobody has estimated outcrossing rates onto short homostyles. However, it is reasonable to assume that they are very low indeed, for both the anthers and the stigma are hidden deep in the corolla tube. Figures of crossed pollen from pin anthers, and of pollination onto thrum stigmas in heterostyles, show that the organs hidden in the corolla tube suffer a severe disadvantage with respect to pollen donation and pollen reception (p. 258). Thus, although short homostyles will automatically self-fertilise with high efficiency, they may outcross only rarely. Perhaps the short homostyles experience more inbreeding depression than do long homostyles.

Third, Eisikowitch and Woodell (1975) show that the erect flowers of the primrose, *P. vulgaris*, frequently fill with water after rain. If the anthers are dehiscing, about 30% of the pollen germinates in the water, if it contains the anthers. In the absence of the anthers, germination in the water is 90%. Such water-germinated grains are presumably unavailable for fertilisation. It is assumed that the anthers contain a water-soluble inhibitor to pollen germination. (*Primula elatior*, the oxlip, with pendant flowers, contains no such inhibitor, and flowers that are artificially held erect show 90% water-germinated pollen after rain in the presence of anthers.) It is likely that the germination inhibitor in the anthers is associated with the mechanism of incompatibility, for water extracts of anthers and pollen can be shown to elicit incompatible-type responses on the same-morph stigmas (Richards & Ibrahim 1982, Shivanna *et al.* 1981). Thus, it can be assumed that such inhibition is absent from self-fertile homostyles. If this is the case, most of the pollen will germinate and become unavailable for fertilisation after rain in the short homostyle, in which the anthers are near the bottom of the corolla tube. However, in the long homostyle, the anthers will be immersed much more rarely, for they occur at the mouth of the corolla. This may be an important selection factor against short homostyles.

Although recombination between the hypothetical loci *G*/*g* and *P*/*p*, *A*/*a*, which gives self-fertile long and short homostyles has been most frequently reported, this may be because they are the recombinants most readily detected by examination of the flower. Ernst (1933, 1936, 1955; reported and analysed by Dowrick 1956), working with *P. latifolia* and *P.* × *hortensis*, also noted other types amongst the large experimental progenies that he grew. These were self-incompatible long and short homostyles and self-compatible pins. He did not find self-fertile thrums. In all cases, the abnormal incompatibility behaviour of the pollen was associated with pollen size. Thus, self-fertile pins and self-incompatible short homostyles not only had thrum-type pollen incompatibility behaviour, but also large thrum-type pollen grains. Self-incompatible long homostyles had small, pin-type pollen. These types have led Dowrick to propose a locus *P*/*p* which controls pollen size and behaviour. It is probable that such plants are therefore recombinants, or *gPa*/*gpa* (self-fertile heterostyles), or *GPa*/*gpa* and *GpA*/*gpa* (self-incompatible homostyles). There is a suggestion from Ernst's data that self-fertile heterostyles are the rarest. If this is the case, and they are double recombinants, *gpa* is the correct order, and the self-fertile heterostyles are *GpA*/*gap* and *gPa*/*gap*.

Thus, the following eight types of recombinant chromosome can be expected, of which Ernst found seven (each will occur with the pin chromosome *gpa*):

| | Thrum | Pin |
|---|---|---|
| self-incompatible | *GPA* | *gpa* |
| self-compatible | *GpA* | *gPa* |
| self-incompatible | long homostyle *gpA* | short homostyle *GPa* |
| self-compatible | long homostyle *gPA* | short homostyle *Gpa* |

Ernst had in fact assumed that each of the aberrant types he discovered were mutants to a complex single locus *S*, but as Dowrick points out, these are far better explained as recombinants between different but tightly linked loci in a supergene.

Although I have also noted *P*/*p* recombinants in progenies of *P.* × *variabilis* it seems likely that they are rare in the field. In a survey of a large population of *P. veris*, Ibrahim (1979) found no heterostyles with strong self-compatibility or aberrant pollen size. I am not aware of any report of self-incompatibility amongst wild homostyles.

As yet, no other recombinants have been reported in *Primula*. Dowrick (1956) noted that seven different components of the *Primula* supergene can be identified, suggesting, in a single sentence, that 'factors for style and pollen incompatibility reaction . . . for length of stigmatic papillae and for area of conducting tissue also exist'. Without any recombinational evidence she produced a hypothetical supergene with seven, rather than three, loci. She did not mention two additional heteromorphic features, stigma shape (see Table 7.1) and area of style section (which is smaller in pins). There are seven heteromorphic features, and two incompatibility functions, making nine components in all.

The way that this modestly stated hypothesis, in the absence of *any* experimental evidence, has been absorbed into the basic genetic literature as the classic example of a coadapted linkage group 'supergene' should be a lesson to us all. Thus in recent years, Darlington (1971) and Berry (1977) have been among those texts that have quoted this supergene in detail. It seems that Ford (1964, p. 177) was first responsible for its adoption: 'these have enabled Dowrick (1956) to analyse and predict its component parts more fully. Thus, when expanded, $S = CGLI^{s}I^{p}PA$. There seems fairly good evidence for this order with respect to $P$, $G$ and $A$' ($C$ is thrum area of conducting tissue; $L$ is papilla length of thrum and $I^{s}$ and $I^{p}$ are female and male incompatibilities respectively, according to Ford).

It cannot be stated too strongly that there is in fact *no* evidence for $C/c$, $L/l$, $I^{s}/i^{s}$ or $I^{p}/i^{p}$ as separate loci, and that even the order of $G$, $P$ and $A$ is very much open to question.

## TYPES OF HOMOSTYLY

As we have seen, the association of incompatibility factors with heteromorphic loci, apparently closely linked on the $S$ and $s$ chromosomes, has allowed self-pollinating, self-fertilising homostylous recombinants to occur, probably on repeated occasions and in different species. Many superficial accounts assume that only one kind of homostyle can occur, but as Ganders (1979) emphasises, several essentially different phenomena have been termed homostyly. Of course, most species of flowering plant are homostylous in one sense, that is, they are not heterostylous. Thus, homostyly must be reserved for non-heterostylous individuals in species that are mostly heterostylous, or non-heterostylous species in genera that are mostly heterostylous. Homostyles should also be self-compatible, thus excluding genera such as *Capsella* and *Cardamine* with di-allelic incompatibility but no heteromorphy.

Several types of homostyly can be identified and these will be detailed below.

(a) Those due to recombination in the heteromorphy supergene, as in the celebrated Somerset populations of the primrose, *P. vulgaris* (Crosby 1949). Similar cases are known in *Limonium* (Baker 1953a,b, 1966), but seem not to be widespread.

(b) Those in which homostyly is associated with a loss of incompatibility, presumably of mutational origin. In *Linum lewisii* (Baker 1975), crosses with the apparently ancestral heterostyles *L. austriacum* and *L. perenne* showed that the pollen of *L. lewisii* is thrum in reaction, but that the stigma can receive pin, thrum and homostyle pollen equally readily. Thus, the stigma has lost incompatibility, rather than recombining it. It is not clear whether the homomorphy itself is of mutational or recombinational origin. If Lewis's (1943) model of incompatibility in *Linum grandiflorum* is correct (and it seems from Ghosh and Shivanna 1980 that this may not be so), it may be that the same mutation that conferred homostyly also promoted self-compatibility.

(c) Those in which only one heteromorph survives in an agamic clone. Such cases are well recognised in *Limonium* (Baker 1966). *Limonium* agamospecies are generally characterised by being either cob or papillate, which strongly argues for a monophyletic origin (and obligate agamospermy) in each case (Ch. 11).

(d) Those that have apparently been encouraged by polyploidy, or in which polyploidy has been encouraged by homostyly. Stable allotetraploid heteromorphs are well-known in *Limonium* (Baker 1966) and *Linum* (Ockendon 1968, Heitz 1973). In these, the *S* locus may be amphidiploid, being present in only one parental genome, or suppressed in the other. However, Dowrick (1956) shows that stable heteromorphy can be maintained in autotetraploid *Primula obconica* with a simplex control (*Ssss*/*ssss*). Duplex and triplex genotypes will only occur through thrum selfing, or irregular chromosomal disjunction.

Nevertheless, there is a striking correspondence between 'ploidy and homostyly in *Primula* section Aleuritia, as was pointed out by Bruun (1930, 1932) (Table 7.2; Figs. 7.11 and 7.12). This is also roughly correlated with latitude in the Northern Hemisphere.

There are other correlations between breeding system and 'ploidy in this large genus, for instance in the candelabra *Primulas* (section Proliferae), where 18/26 species have $2n = 22$ and are heterostylous, but the isolated Japanese *P. japonica* has $2n = 44$, and is homostylous, as are, however, *P. chungensis* ($2n = 22$) and *P. cockburniana* ($2n = 22$). However, it must be made clear that such relationships do not hold for the genus as a whole. In the 20 species of section Auriculastrum which are confined to the European mountains, chromosome counts have been made of $2n =$ (62)–66–(73), 122–128, and c.198. Presumably $2n = 66$ must be considered diploid to this group now, but in view of the widespread aneuploidy it displays, and the regular occurrence of diploids with $2n = 18$, 20 or 22 in all other sections, these species must be regarded as being ancestrally $6\times$,

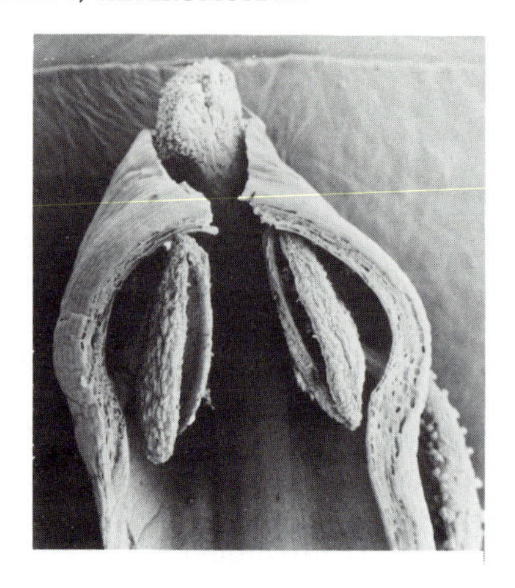

**Figure 7.11** Scanning electron micrograph of a section of part of the flower of the homostylous species *Primula stricta* (×15). Anthers and stigma are both at the top of the floral tube. The petal limb has been excised.

12× and 18×. Yet all are usually heterostylous today, and almost certainly originally evolved from a diploid heterostylous ancestor.

There is certainly little indication that polyploidy *per se* leads to homostyly. Rather, it is likely that polyploidy, with its attendant advantages (Ch. 9) was more able to become established in occasional self-fertilising diploid homostyles, thus overcoming the minority-type disadvantage of the newly arising polyploid. Once established, the increased opportunities for variability in the polyploid, even when selfed, would put the polyploid inbreeder at less of a disadvantage in contrast to its outbred heterostylous diploid parents. All homostylous polyploid

**Table 7.2** Diploid chromosome count, breeding system and latitude of species in *Primula* section Aleuritia.

| Species | Latitude range (°N) | 2*n* | Breeding system |
|---|---|---|---|
| *P. farinosa* | 40–60 | 18 | heterostylous |
| *P. frondosa* | 42 | 18 | heterostylous |
| *P. longiscapa* | 48–53 | 18 | heterostylous |
| *P. exigua* | 43–44 | 18 | heterostylous |
| *P. modesta* | 35–45 (Japan) | 18 | heterostylous |
| *P. daraliaca* | 42–44 | 18 | heterostylous |
| *P. halleri* | 46–48 | 36 | homostylous |
| *P. farinosa* | 57 (Gotland) | 36 | homostylous |
| *P. egaliksensis* | 55–66 | 36 | homostylous |
| *P. scotica* | 58–59 | 54 | homostylous |
| *P. scandinavica* | 64–67 | 72 | homostylous |
| *P. incana* | 40–55 (North America) | 72 | homostylous |
| *P. laurentiana* | 45–55 (North America) | 72 | homostylous |
| *P. magellanica* | 45–55°S. (South America) | 72 | homostylous |
| *P. stricta* | 67–71 (Scandinavia)<br>55–70 (North America) | 126 | homostylous |

*Primulas* are long homostyles, which probably arose originally as a result of *gPA* recombination.

(e) In *Amsinckia spectabilis* (Ganders 1975, 1979), *Linum austriacum* agg. (Heitz 1973), and *Primula verticillata*, variable levels of structural homostyly and heterostyly occur within and between populations which are self-fertile. Also, in *Anchusa officinalis*, variable heterostyly, including some homostylous morphs, is associated with multi-allelic incompatibility (Phillip & Schou 1981). Such conditions have usually been interpreted as resulting from the breakdown of di-allelic incompatibility, through mutation or recombination in the heterostyly supergene in response to conditions that favour increased amounts of selfing (Dulberger 1975a, Ganders 1979). However, this is clearly not the case in *Anchusa* (p. 241).

The situation in *Primula floribunda* and *P. verticillata* is interesting and invites further discussion. These species, frequently grown as decorative plants for the cool greenhouse, are placed in section Sphondylia, which is considered by Wendelbo (1961) and Schwarz (1968) as the most primitive section in *Primula*. There are six taxa known in the section; all have $2n = 18$, and all may be self-compatible.

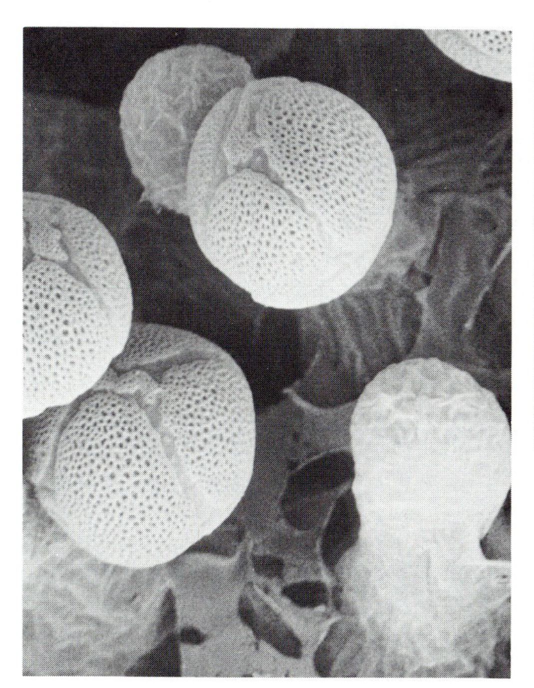

**Figure 7.12** Large thrum-type pollen selfed on the pin-type papillae of the homostyle species *Primula stricta*, which is self-fertile (×1400).

The following results are taken from Wendelbo (1961) and have been checked against extensive, including type, material in the Kew herbarium. They differ importantly from those of Ernst (1955):

| *Primula* spp. | Distribution | Homostyle/heterostyle |
|---|---|---|
| *davisii* | south-east Turkey | homostyle |
| *gaubeana* | west Iran (Luristan) | homostyle |
| *verticillata* | Yemen, Aden | homostyle to somewhat heterostyle |
| *boveana* | Sinai | somewhat heterostyle |
| *simensis* | north-east Ethiopia | homostyle to heterostyle |
| *floribunda* | west Himalaya | homostyle to heterostyle |

Ernst (1955) noted homostyle, pin with long homostyle and heterostylous populations in cultivation, but claimed that most wild populations were heterostylous and self-incompatible.

I have made a survey of 40 dried herbarium specimens of *P. floribunda* which originate from 15 Himalayan localities with altitudinal notation. These range from homostyles with stigma and anthers at the same level in the flower (near the mouth and thus 'long homostyles') (scored zero) to pins and thrums in which the anthers and stigma are displaced reciprocally by 7 mm (scored '4'). Every intermediate level of thrum and of pin heterostyly also occurs (scored 1–3). Although there is some variation in the level of heterostyly within a location, there is a good relationship with altitude (Fig. 7.13). Below 1400 m, all plants were fully heterostylous. At 2000 m, all plants were fully homostylous. Between these altitudes, variation occurred, with a distinct tendency for the level of heterostyly to decrease with altitude. All plants had 'pin'-type stigma papillae, which nevertheless varied in length a good deal; pollen diameter varied from 16 to 21.5 $\mu$m in mean diameter (within the pin-type pollen range of most species) without any clear relationship as to morph. However, homostyles had distinctly small pollen, and certainly did not have the thrum-type pollen characteristic of recombinant long homostyles.

Heterostyly, being so closely correlated with altitude, appears to be adaptive. Self-pollination appears to be at an advantage at higher altitudes, perhaps due to a shortage of pollinators during the winter flowering season. Similar results have been obtained by Ganders (1979) for self-compatible *Amsinckia* species which are homostyle in more marginal locations. If this is the case, heterostyly may have evolved in *Primula before* the di-allelic incompatibility. This directly opposes the view held by Charlesworth and Charlesworth (1979), who produced models which suggest that heterostyly and the other morphological features of heteromorphy, become linked to the di-allelic incompatibility locus to maximise legitimate pollination. It does, however, agree with the views of Charnov (1984) concerning sexual selection for heterostyly

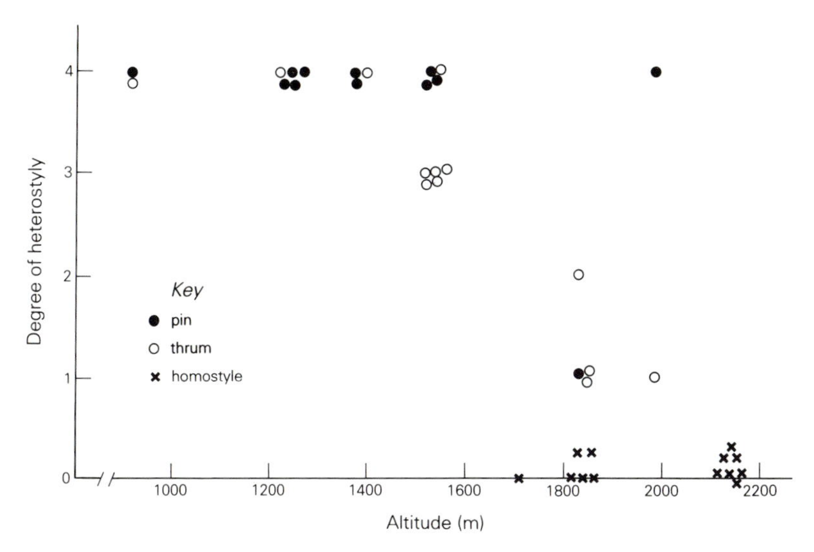


**Figure 7.13** Relationship between altitude ($x$ axis) and the degree of heterostyly ($y$ axis, scale 0 to 4) for individual herbarium specimens of *Primula floribunda* collected in the wild in the western Himalayas. Plants scoring 0 are homostyles, those scoring 4 have anthers and stigma displaced by 7 mm or more; those scoring 1 to 3 are intermediate. Closed circles are pins, open circles thrums, and X denotes homostyles.

(p. 282), which explain the occurrence of heterostyly in genera such as *Amsinckia*, *Anchusa* and *Narcissus* that are not di-allelic.

(f) Finally, it should be pointed out that in at least one species of *Primula* with di-allelic incompatibility, *P. sinensis*, Mather and de Winton (1941) and Mather (1950) have shown that homostyly is controlled polygenically. Thus, it is possible to breed experimentally for increasing or decreasing amounts of homostyly by selfing (this species shows relatively high levels of self-compatibility) and artificial selection. The several loci involved in the control of style length and anther height are unexpectedly not linked to the incompatibility locus. Maybe *P. sinensis* represents an intermediate condition between *P. floribunda*, which may have polygenically controlled heterostyly which responds selectively to the environment, and the majority of *Primula* species, including *P. vulgaris*, in which the di-allelic incompatibility and the control of heterostyly have become closely linked.

## *Survival and success of homostyle* Primula vulgaris

Since Crosby (1949) identified and examined recombinant-type long-homostyle populations of the primrose in Somerset and the Chilterns, southern England, these populations have been the centre of scrutiny and speculation. They present an unusually clear opportunity to study evolutionary pressures working on related plants with strikingly different breeding systems in the field. Additionally, they should be able to show why the heteromorphy supergene has evolved in a similar way in so many plants, and why it has been successful.

Crosby showed that long homostyles occurred in Somerset and Chiltern woods at frequencies between 10 and 80%. At low frequency they are accompanied by pins, and smaller numbers of thrums; at high frequency they are usually only accompanied by pins. Crosby points out that the scarcity of thrums is predictable because long homostyles have thrum-type pollen $S^h$ (dominant linkage group *gPA*) which competes on pin stigmas with thrum pollen *S* (*GPA*). The $S^hs$ offspring (homostyle) are fitter than *Ss* offspring (thrum), because they self-pollinate and are self-fertile. When selfed, the homostyle heterozygotes $S^hs$ only give rise to 25% pins *ss*, whereas the offspring of all thrums which must be outcrossed will give rise to 50% pins *ss*. Thus, Crosby predicted that homostyles should increase in frequency, and spread geographically. During the subsequent decade it became evident that no such 'take-over' by homostyles was occurring. If anything, the homostyles were retreating. Crosby's calculations had assumed that all homostyle seed was selfed. As we have seen, Bodmer (1958) challenged this assumption with data purporting to show up to 80% outcrossing onto homostyles. This he explained by his observation of a two-day protogyny in the homostyles, during which time the stigmas were available for outcrossing but could not be selfed. Crosby's more carefully gathered estimates from three sources (Crosby 1959) showed that crossing onto long homostyles was in the order of 5–10%, low enough to promote homostyly by selfing advantage, and by male outcrossing in competition with thrums. He explained the lack of success of the homostyles in terms of lower homostyle viability, at about 0.65 of heterostyles. He suggested that the homostyle homozygotes $S^hS^h$ might be particularly unviable. However, these suggestions were generated numerically and through an early, pioneering computer study (Crosby 1960), and lacked experimental support.

A quarter of a century elapsed before John Piper, Brian Charlesworth's student, produced field data which go a long way to settling the argument as to how much outcrossing homostyles receive in the field (Piper *et al.* 1984, Richards 1984). This information comes in three forms:

(a) Selfed pollen loads on homostyle stigmas are massively greater than thrum loads on pin stigmas; in fact 35 times greater on average. It should be noted, however, that the homostyle pollen load is on average 20 times greater than is needed for all the ovules to be fertilised. If one extrapolates from thrum loads on pin stigmas, 60%

of the thrum pollen needed to outcross all the homostyle ovules might arrive on homostyle stigmas.

(b) Homostyles in the field produce 65% more seed per plant, and 62% more seed per flower on average than do heterostyles (among the heterostyles, pins generally produce more seed than thrums, having more ovules, and being more efficiently pollinated, see p. 258). They do not, however, set more seed than artificially crossed heterostyles. Thus the increase in seed-set can be attributed entirely to increased pollination efficiency. It will be noted that the increased seed-set per homostyle plant can be entirely attributed to increased seed-set per flower. This agrees with Mark Wilson's unpublished observations (Ch. 9, Table 9.7) that homostyles do not set more fruit than heterostyles in the field, and that fruit-set is in fact rather poor, varying between 20% and 80% in different populations and years (John Piper *in litt.*). The self-pollinating self-fertilising homostyles might be expected to set 100% of fruits. Mark Wilson (*in litt.*) attributes the poor fruit-set of homostyles to increased predation, mostly by small mammals. The entrances to the burrows of field mice (*Apodemus*) and voles (*Clethrionymus*) in these woods are littered with the remains of primrose capsules. Thus it seems as if the poor fruit-set of heterostyles, attributable to low levels of insect visits to flowers, is balanced by the increased attractiveness of the more bountiful homostyles to predators. The homostyles only gain by the better seed-set of surviving capsules. However, John Piper (*in litt.*) found no difference in levels of seed predation between homostyles and heterostyles in mixed population.

(c) Estimates of outcrossing were based on frequencies of heterozygotes in the offspring of wild-set seed from homostyles and heterostyles, using two isozyme markers. The $S$ estimate (where $S$ = proportion of selfing) for homostyle seedlings was 0.92 for both systems, whereas that of pooled heterostyles was statistically inseparable from zero. These estimates fit in quite well with Crosby's estimates of outcrossing onto homostyles ($S = 0.90$ to 0.95). There is an inherent danger in estimating the breeding system (i.e. degree of outcrossing onto homostyles) from the genetic structure of the population, which may be influenced by other factors apart from that parameter in which one is particularly interested. Thus not only limited, non-panmictic gene flow, but also selectively non-neutral isozyme markers can bias estimates of outbreeding (Chs. 9 and 11).

Nevertheless, Piper *et al.* (1984) clearly substantiate Crosby's claim that the homostyles are largely selfed. We are still left wondering why the homostyles do not spread, as Crosby had predicted. It may be that they do indeed have a lower fitness, especially as homozygotes $S^hS^h$, perhaps due to inbreeding depression (Ch. 9). This does not explain why the recombinant long-homostyle chromosome *gPA* has partially established in two limited areas of the country, but nowhere else. There seems a strong presumption that these woods are in some way ecologically different from other primrose habitats. Yet environmentally or geographically they are not exceptional in any obvious way.

One clue may reside in a suggestion made by Crosby on a BBC television programme some years ago. Primrose flowers are subject to predation by molluscs, slugs and snails. Observations by Mark Wilson show that mollusc grazing in the Somerset woods (Sparkford) is very heavy. Slugs and snails crawl over the surface of the petals, which they may eat, but they particularly relish the anthers and stigma at the mouth of the tube and thus thrums become female and pins become male. However, the long homostyles that are eaten are neutered and so these flowers cannot act as parents. Such an ingenious explanation certainly requires further detailed investigation. At present there remains no convincing reason why the homostyle chromosome has become established in only these limited areas, or why it fails to spread.

## FUNCTION OF HETEROMORPHIC CHARACTERISTICS

In considering the function of heteromorphic characters, two striking features of heteromorphy should be kept in mind:

(a) That in most cases there is very close, and apparently coadapted, linkage between the di-allelic incompatibility, and other heteromorphic features, so that it is very unusual to find plants (such as homostyles or self-compatible heterostyles) in which the linkage disequilibrium has been lost.
(b) That in most cases, unrelated genera with di-allelic incompatibility have heteromorphic features that show an uncanny correspondence from genus to genus (e.g. reciprocal stigma and anther positions, and homozygous pins with small pollen and long stigma papillae).

Such correlates between unrelated genera in which di-allelic incompatibility and heteromorphy must have arisen independently, very strongly suggest that the heteromorphic features of pollen size, papilla length, style length and anther position, at least, must be very important in the functioning of the di-allelic incompatibility. The question is, how are they important? There are two contrasting schools of thought concerning the function of heteromorphic characters. In the first, Darwin (later supported by Ganders 1974, 1976, 1979, Ornduff 1970, 1975a,b, 1979b and others) suggested that the reciprocal placing of the stigma and anthers in pin and thrum flowers maximised the flow of legitimate pollen between flowers of two incompatibility types.

Pin stigmas, being prominent, receive far more legitimate and illegitimate pollination (Ch. 5, Table 5.7) than thrum stigmas in all heterostylous plants that have been investigated (Ganders 1979). In nearly all cases, they have been found to receive adequate amounts of legitimate thrum pollen. However, the thrum stigma is hidden in the flower (as are pin anthers). In order to maximise thrum female function, there will be selection for the production of more pin pollen. This could be achieved by bigger pin anthers. This would, however, increase the male load on pins, a situation that Charnov (1984) has suggested selects for female function. Thus, it is suggested, the size of pin pollen grains have decreased with the effect that more pin pollen than thrum pollen can be produced for the same input in resource. This does not explain why thrum pollen is larger, however.

As yet, there is no reasonable explanation for the correspondence between pin styles and longer pin papillae. It is not apparently a developmental correlate of the longer pin style (which may also have longer cells), for non-heterostylous Plumbaginaceae show papilla length heteromorphy. In contrast, Lewis (1942b, 1943, 1949a, 1954), Mather and de Winton (1941), Dulberger (1974, 1975a,b), Dickinson and Lewis (1974), Ghosh and Shivanna (1980) and others have suggested that heteromorphic features are intimately concerned with the functioning of the di-allelic incompatibility, as will be discussed below.

Ganders (1974) and Yeo (1975) have taken the middle ground, by suggesting that some heteromorphic features (e.g. reciprocal heterostyly) function primarily in promoting legitimate pollination, whereas others (e.g. papilla length, exine sculpturing) are concerned with the function of the incompatibility. In the face of conflicting evidence, this may be the most sensible viewpoint. Accordingly, I shall present evidence concerning the different components of heteromorphy in turn.

### *Heterostyly (reciprocal anther and stigma positions)*

Heterostyly (two or three different style-length morphs) is almost always associated with reciprocal anther positions, although this is not the case in *Linum grandiflorum*. It has never been suggested that different anther positions could be directly involved in the functioning of the incompatibility, and it is difficult to see how this could be so. The reciprocity of anther position and stigma position therefore strongly suggests that style length is also not involved in the functioning of the incompatibility.

Indeed, the only serious suggestion that this might be so was made by Lewis (1942b) with respect to *Primula obconica*. He claimed that thrum pollen tubes grew down the long pin style twice as quickly as pin grains grew down the short thrum style. Thus he suggests that pin incompatibility is a product of slow tube growth and a long stigma. Clearly, thrum incompatibility cannot be of this type, and Stevens and Murray (1982) confirm that although thrum selfed grains germinate, they do not penetrate the stigma. It is obviously attractive to suggest, as Lewis did, that the larger thrum pollen promotes faster tube growth in the longer pin style. Indeed, Stevens and Murray do show that thrum grains *in vitro* do produce longer tubes than pins. However, the behaviour of selfed pin tubes suggest that they are positively inhibited a short distance down the style. The great variation in incompatible grain and tube behaviour found in *Primula* (p. 268) suggests that different style lengths have quite another function in most *Primula* species.

Thus, we must look to Darwin's original suggestion of the enhancement of legitimate pollination if we are to explain the function of heterostyly. Ganders (1979, pp. 620–624), in a very useful review of the rôle played by heterostyly in promoting legitimate pollination, makes an important point. If pin and thrum flowers coexist in a population in equal frequencies, and if they produce equal amounts of pollen (which as we have seen they generally do not), random between–flower pollination should result in equal frequencies of pin and thrum pollen on stigmas. Only if significant excesses of legitimate (other morph) pollen is found on stigmas can the heterostyly be said to be functional. However, this does not take account of within–flower pollination. For this, with only one morph involved, the heterostyly cannot be expected to mediate legitimate pollen transfer, and within-flower pollination, often brought about by the flower visitor, may swamp the effects of the between-flower pollination that is being examined. Thus, legitimate/illegitimate pollen loads should be examined on the stigmas of experimentally emasculated flowers. So far this has been successfully attempted only for the tristylous *Pontederia cordata* (Price and Barrett 1984) and the distylous *Jepsonia heterandra* (Ganders 1974). The latter provides the most elegant examples of the pollination efficiency of heterostyly (Table 7.3).

Thus, in this species there is noticeable disassortive pollination onto thrum stigmas and slight, but not significant, disassortive thrum pollination onto pin stigmas, thus bearing out Darwin's contention. The

**Table 7.3** Pollination efficiency of heterostyly in *Jepsonia heterandra* (after Ganders 1974).

| | Stigmatic pollen load | | | | |
|---|---|---|---|---|---|
| | Legitimate | | Illegitimate between-flower (emasculated) | | Illegitimate within-flower (non-emasculated minus emasculated) |
| Stigma | observed | expected | observed | expected | |
| pin | 19.9 | 15.2 | 30.7 | 35.4 | 49.4 |
| thrum | 75.0 | 63.1 | 15.3 | 27.2 | 9.7 |

**Table 7.4** The effect of heterostyly on pollination efficiency in the presence of within-flower selfing (from Ganders 1979) (for explanation of asterisk, see text).

| Pollen on pin stigmas | Pollen on thrum stigmas | | |
|---|---|---|---|
| | excess pin* | random | excess thrum |
| excess thrum* | | | |
| random | *Amsinckia douglasiana* | | *Hypericum aegyptium*<br>*Lythrum lineare* |
| excess pin | *Jepsonia heterandra*<br>*Amsinckia grandiflora*<br>*A. vernicosa*<br>*Primula elatior* | *Lithospermum californicum*<br>*Jepsonia parryi*<br>*Menyanthes trifoliata*<br>*Primula veris* | *Lithospermum carolinense*<br>*Primula vulgaris*<br>*Lythrum californicum*<br>*Pulmonaria obscura* |

amount of within-flower selfing is much greater in pins, as it also seems to be in *Primula*. This is the only example of possible excess thrum pollination onto pin stigmas, although excess pin pollination onto thrum stigmas occurs frequently, perhaps because of the smaller and more plentiful pin pollen. In the absence of experimental emasculation (i.e. including within-flower selfing), from the relationships found (Table 7.4), we can conclude that only the five species in the left-hand column (marked *) show that heterostyly has increased the efficiency of legitimate pollination onto thrum stigmas beyond random, and that in no case is the efficiency of legitimate pollination onto pin stigmas better than random. However, as efficient disassortive (legitimate) pollination has been shown in the only case in which the effects of within-flower selfing have been removed by emasculation, these results should be treated with caution. This is particularly true of pins, which act as very efficient mothers (pollen receptors), doubtless due to the exposed position of the pin stigma.

Perhaps the most convincing example of the disassortive effect of pollen travel in heterostyles also derives from the work of Ganders (1979). Pollen loads were compared on the pin stigmas of forms of *Lithospermum californicum* which differed in the degree of heterostyly:

| Style length (mm) | Stigma/anther distance (mm) | % pin pollen | % thrum pollen |
|---|---|---|---|
| 9.2 | 1–3 | 99.5 | 4.5 |
| 10.3 | 4–5 | 86.4 | 13.6 |

These results, which are reminiscent of those of Schoen (1982a), on the effects of herkogamy on the amount of outcrossing in *Gilia* (and heterostyly is nothing but reciprocal herkogamy), certainly suggest that disassortive pollen travel between morphs is augmented by increasing levels of heterostyly. This is shown despite most pollinator moves being between flowers of the same plant, there being very high levels of geitonogamy.

Although Darwin noted that pin and thrum pollen tended to be deposited on different positions on the bodies of bees and wasps visiting *Primula veris*, there have been very few later attempts to verify this presumed basis to disassortive pollen travel onto heterostyles. The most convincing are given in Rosov and Screbtsova (1958), for buckwheat, *Fagopyrum esculentum* (Table 7.5).

### *Pollen size*

Two contrasting ideas about the function of the pollen-size dimorphism present in most heterostylous species have already been discussed, namely (a) thrum pollen is larger in order to provide the reserves and/or the faster growth rate necessary to grow down the long pin style and (b) pin pollen is smaller in order that more pollen grains are produced in order to pollinate efficiently the less exposed thrum stigma.

In addition to the arguments already presented, it should be noted that Ganders (1979) found no relationship between pollen-size ratios and style-length ratios in 24 heteromorphic species. This strongly indicates that it is not necessary for thrum pollen grains to be larger in order to grow successfully down long pin styles. It is, however, possible that the assymetry of pollen flow noticeable in most heterostylous plants (less pin pollen going to thrum stigmas) has selected for the production of more and thus smaller pin grains, leading to a dimorphy in grain size and hence in pollen reserves. This difference in carbohydrate reserve between pin and thrum pollen might secondarily augment the heterostyly at least in some cases. For instance, in *Lythrum* and *Jepsonia*, thrum pollen has detectable starch reserves, which are apparently absent from pin pollen (Ganders 1979).

**Table 7.5** Position of pin and thrum pollen grains on honeybees (*Apis*).

| Position of pollen on *Apis* | Mean numbers of pollen grains pin | thrum |
|---|---|---|
| thorax | 402 500 | 252 500 |
| abdomen | 180 000 | 762 500 |

### *Pollen sculpturing*

In some cases, notably in the Plumbaginaceae, a marked difference in exine sculpturing between morphs occurs, which may or may not be accompanied by pollen-size dimorphism (see Figs. 7.7a,b). A similar exine dimorphism also occurs in the unrelated heterostylous *Linum* species (Dickinson & Lewis 1974, Dulberger 1974). In both *Armeria* and *Linum*, it has been shown that the apertures in the exine tectum contain lipid and protein.

Mattsson (1983) showed that the lipid is borne relatively externally in the finely reticulated pollen of papillate stigma plants in the thrift *Armeria maritima*, and this allows the pollen to stick to smooth surfaces such as glass. In the pollen of cob-stigma plants, with much more coarse reticulation and deeper apertures, the lipid is sunken in the apertures. Pollen hydration, and subsequent germination, depends on contact being made between the pollen lipid and the stigma papilla cuticle. Whereas this takes place externally in legitimate crosses of papillate pollen onto the relatively smooth cob stigma, the success of the reciprocal cross depends on the prominent narrow papillate stigma papillae penetrating an aperture of the coarsely reticulate cob pollen grain. Only then does the papillate stigma cuticle come into contact with the cob pollen lipid. Of course, this does not explain why the finely reticulate papillate grain, with external lipid, fails to adhere to the papillate stigma papillae in an illegitimate pollination in papillate morphs (Fig. 7.14). That some lipid/cuticle recognition is involved is strongly suggested by experiments in which lipid extracts from legitimate pollen are associated with illegitimate pollinations. When cob lipid extract is associated with papillate selfs, the percentage of illegitimate grains that produce pollen tubes is elevated from 0.8 (control) to 10.3.

In *Armeria*, there is a 'dry' cuticle reaction type (Ch. 6). In a recent paper, Schou (1984) showed that *Primula obconica* had a wet pin stigma, and a dry thrum stigma. This has not been previously reported in this genus, which usually have 'dry' stigmas, and does not seem to occur in our material of *P. obconica*. Dickinson and Lewis (1974) and Ghosh and Shivanna (1980) have shown that the *Linum* pin stigma is of the dry type, with a continuous cuticle, but the thrum stigma, which has proteinaceous exudates and higher levels of esterase and acid phosphatase reactions, is of the wet type, with an interrupted cuticle. Pin pollen does not adhere to the pin style. Both pin and thrum pollen adhere to the wet thrum style and germinate, but the illegitimate thrum pollen is unable to penetrate the thrum papilla cuticle. Thrum pollen is apparently able to adhere to the dry pin stigma by virtue of pollen surface-carried lipids, which are absent from the pin pollen. Thus, according to Heitz (1973), who provided scanning electron micrographs of the surface of the pin and thrum pollen

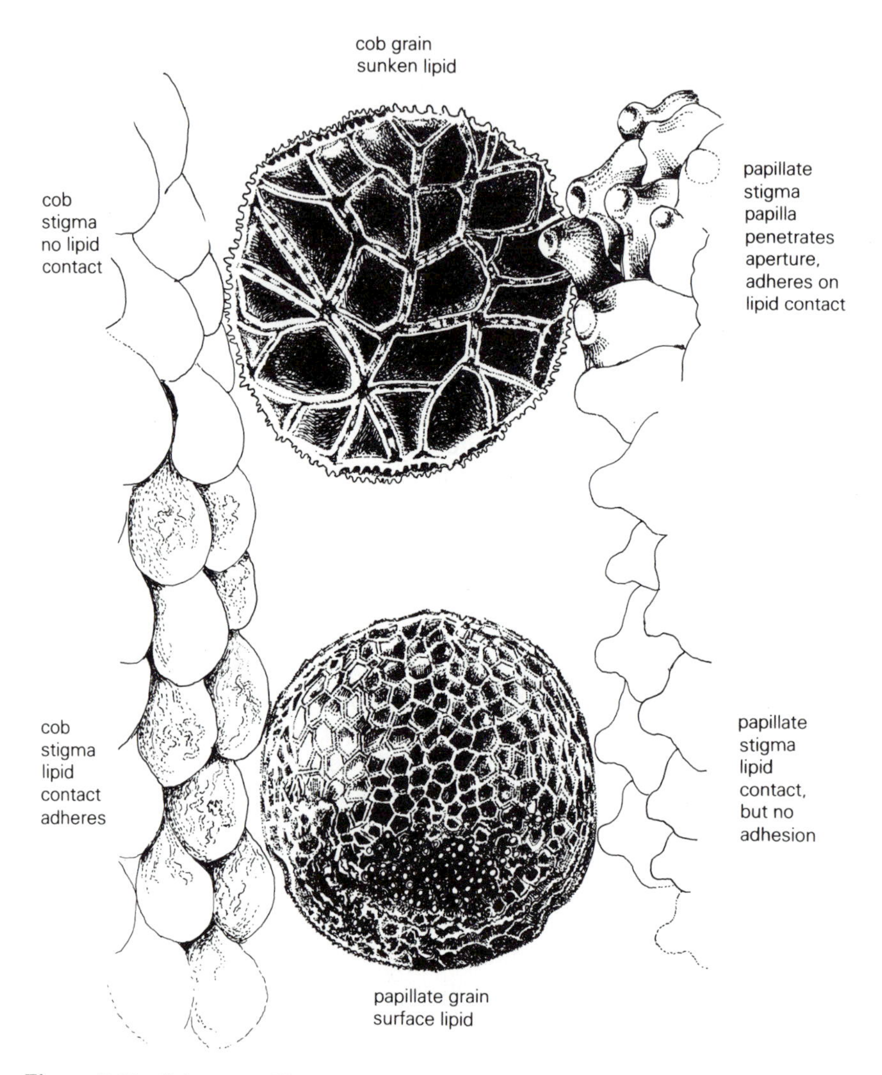


**Figure 7.14** Scheme to illustrate the operation of within-morph incompatibility in the thrift, *Armeria maritima*. Lipid is borne external to the pollen grain in the finely reticulated papillate grain (below), and is able to adhere to the smooth cob stigma, but less well to the papillate stigma. In the coarsely reticulated cob grain, the lipid is sunken in the tectum apertures, where it can only make contact with the papillate stigma.

types, it is always the thrum morph that provides the adhesion necessary for successful pollen lodgement and germination. Thrum pollen, which is larger than pin, also differs in having larger, and uniform, reticulations to the exine (those of pin pollen are mostly smaller and of two sizes). It is not yet clear whether these sculpturing differences play a rôle in the purveyance of pollen surface lipids in thrum pollen.

Thus, we can concur with Dulberger (1975a) in concluding that, unlike pollen-size dimorphisms, pollen-sculpturing dimorphisms are probably directly involved in the functioning of the di-allelic incompatibility.

### *Papilla length*

In most heterostylous genera, pin styles have longer papillae. There have been two suggestions as to the cause or function of this correlation, which extends to tristylous genera.

First, it is a developmental correlation; longer pin styles are achieved by longer pin-style cells and, as a result, terminal style cells (stigmatic papillae) are also longer (Ganders 1979). This apparently sensible suggestion is brought into question by the cob/papillate Plumbaginaceae such as *Limonium* and *Armeria*, which are not heterostylous (both morphs have styles of the same length), but in which the two stigma morphs are very comparable with those found in pins (papillate) and thrums (cob) of other genera.

Second, differences in the ultrastructure and chemistry of papilla cuticles are brought about by the differential elongation of the cell wall of pin and thrum papillae (Dulberger 1975a). This interesting suggestion has some support from ultrastructural studies of the papilla cuticle in the Plumbaginaceae (Dulberger 1975b, Ghosh & Shivanna 1980), but is refuted by *Linum*, in which, as already recounted, it is the longer pin papillae that have a continuous cuticle.

Pollen grains have to lodge and adhere to the stigma papillae before they can germinate (Ch. 6). Whereas adhesion is a chemical reaction, the initial lodgement is a mechanical function. Presumably, the larger thrum grains lodge with greater difficulty than the smaller pin grains, and the longer pin papillae may help thrum grains to lodge on the stigma (as in *Armeria*, Figs. 7.6b and 7.14). There is some evidence that this may be the case in the Plumbaginaceae (Dulberger 1975b, Mattsson 1983), and in *Primula* (F. Wedderburn personal communication). In *Pulmonaria* (Figs. 7.15a,b and 7.16a,b) and *Anchusa* (Philipp & Schou 1981), the stigma papillae are crowned by a ring of grapnel-like processes. Pollen grains which lodge and germinate (and both legitimate and illegitimate grains lodge and germinate on both stigma types in *Pulmonaria*) do so beneath the ring of grapnel hooks. Pollen tubes enter the papilla in a basal position. These hooks undoubtedly play a very important function in

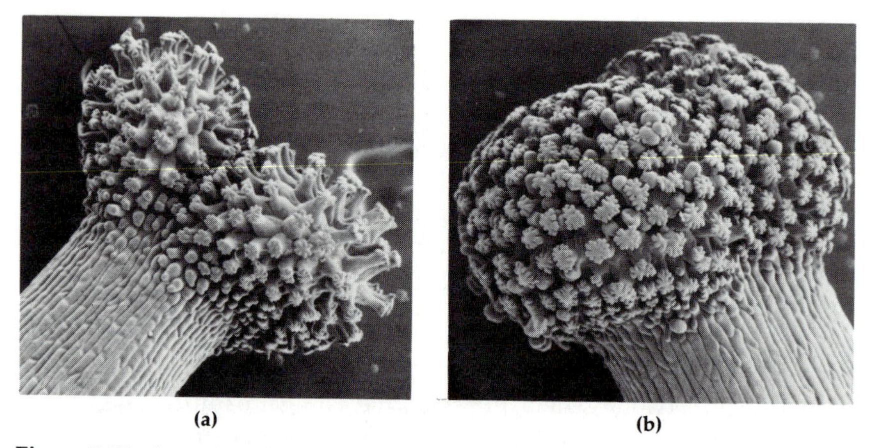


**Figure 7.15** Scanning electron micrographs (×65) of the stigma of (a) thrum and (b) pin *Pulmonaria affinis* (Boraginaceae).

trapping the pollen on the stigma, and the papilla is longer, and the hooks longer and sharper (although fewer), in the pin stigma which has to trap the larger thrum pollen.

The occurrence of longer stigma papillae in pins in many unrelated genera suggests that such longer papillae play an important rôle in the functioning or the efficiency of the di-allelic heteromorphy, and that this rôle is the same in all cases. As yet, there is no clear indication as to which, if any, of the above functions is the most important. It is possible that in *Primula vulgaris* yet another function operates, for the longer pin papillae may provide a moister microclimate at the stigma surface which encourages the thrum grain to rehydrate and germinate (p. 275).

### *Stylar transmission tissue*

As yet, little attention has been paid to heteromorphic differences in the histology of the style, although differences in the styles of pin and thrum *Primulas* was noted by Darwin (1877). In *P. veris*, pin styles have longer and narrower cells than thrum styles (see Fig. 7.5a,b). This might be considered to be a developmental correlate of the longer extension of pin styles (see above). However, Modilbowska (1942) and Dowrick (1956) show that pin *P. obconica* has a much smaller area of stylar conducting tissue in cross-section (127 $\mu m^2$ compared with 285 $\mu m^2$ in thrums). This, they claim, limits the growth of both legitimate and illegitimate pollen on pin, but not thrum, styles, in such a manner that insufficient tubes can reach the ovary for all pin ovules to be fertilised.

Other heteromorphic differences (see Table 7.1) are restricted to one or a few related genera, and are thus presumably adaptive to conditions

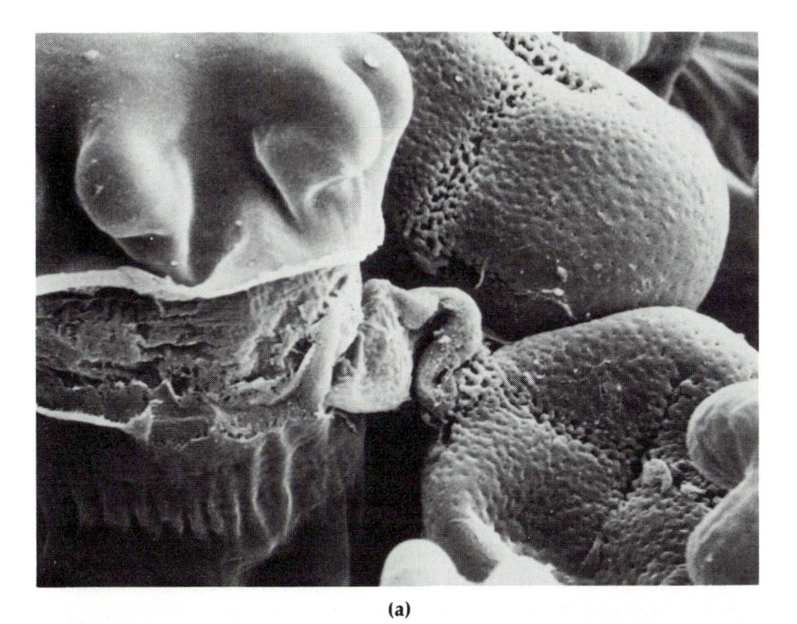

(a)

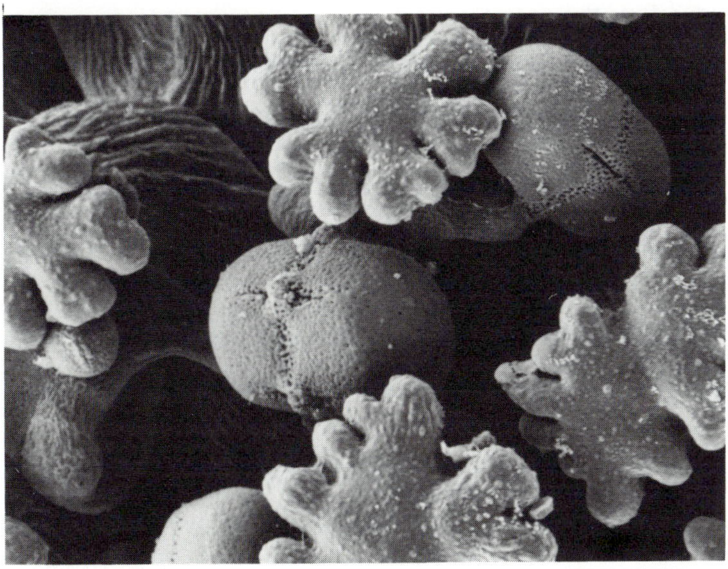

(b)

**Figure 7.16** Scanning electron micrographs of pollen tubes from (a) pin (×1500) and (b) thrum (×1000) *Pulmonaria affinis*. In (a) the pin pollen tube is penetrating a stigmatic papillae of a thrum flower beneath the apical 'corona' of hooks. In (b) the thrum pollen tube is penetrating a stigmatic papillae of a pin flower. It is considered that the apical corona of hooks on the papillae, which are more marked in the pin, help the lodgement of the pollen grains, which are larger in the thrum.

special to those cases. Thus, heteromorphic differences in corolla size, corolla pubescence, stylar pubescence, style colour or pollen colour may well all be concerned with disassortive or asymmetrical pollen flow in those genera.

## OPERATION OF DI-ALLELIC INCOMPATIBILITY

### *Timing and strength of incompatibility*

Di-allelic incompatibility is under sporophytic control, and thus it might be expected that its physiological basis is similar to that of species with multi-allelic sporophytic incompatibility (Ch. 6). However, di-allelic systems probably have an independent origin from multi-allelic systems, and from each other, so one cannot assume that all sporophytic systems, or even all di-allelic systems, function the same way. For instance, in multi-allelic sporophytic systems incompatible pollen tubes rarely penetrate the stigma, but in some di-allelic plants, incompatible tubes may grow some distance into the style.

Most work on the operation of di-allelic systems has been undertaken in *Primula* (Heslop-Harrison *et al.* 1981, Shivanna *et al.* 1981, 1983, Pandey & Troughton 1974, Stevens & Murray 1981, 1982), and in *Linum* (Dickinson & Lewis 1974, Ghosh & Shivanna 1980). Much additional work on *Primula* incompatibility has been done by F. Wedderburn (Wedderburn *pers. comm.*). This clearly shows that even in the one genus *Primula*, the operation of the incompatibility takes a number of different forms, which presumably bear an evolutionary relationship to each other. Variation in the incompatibility system between species in such a large genus provides an ideal opportunity to study the evolutionary development of incompatibility. In order that the operation of incompatibility in *Primula* can be discussed, it is first necessary to show how it varies within the genus.

In *Primula*, within-morph incompatibility can occur at a number of stages in the fertilisation process, which can be simplified as:

(a) Lodgement, adhesion and germination of pollen on the stigma.
(b) Penetration of the stigma papilla by the pollen tube.
(c) Growth of the pollen tube in the stigma.
(d) Growth of the pollen tube in the style.

At any of these stages, incompatibility is rarely total, so that consideration of each stage is quantitative, rather than qualitative. As pointed out by Shivanna *et al.* (1981), the operation of the incompatibility at each of these stages thus has a cumulative effect, which usually, but by no means always, results in total within-morph incompatibility (Richards & Ibrahim 1982). Because of the technical and statistical difficulties, we have found it most useful to record the operation of incompatibility at each of these stages by scores on the scale 0–4, in which 0 is no pollen tubes, 1 is 1–3 tubes, 2 is 4–10 tubes, 3 is 11–40 tubes and 4 is more than 40 tubes. In all cases, it can be assumed that many legitimate pollen grains (well in excess of 100) can be found on stigmas (Table 7.6). Although pollen-grain germination is given as a percentage, it should be emphasised that these values concern the pollen grains that remain on the stigma after fixation and staining, so that the actual percentage germination of all grains which reach the stigma may be much lower. All crosses are made onto emasculated flowers, and grains of a legitimate type are never found on stigmas after illegitimate crosses. Percentage germination of pollen is assessed by conventional light microscopy, with fuchsin staining. Pollen tubes are examined under ultraviolet radiation after staining with aniline blue.

It is found that behaviour of related species varies relatively little, and thus results are given under the sections of the genus. For every reading given, at least 20 crosses involving at least five individuals were examined (often many more). Little variation occurs within samples. In most cases, legitimate crosses were examined as well. In all legitimate crosses, pollen germination was over 80% and large numbers of pollen tubes could be found in the stigma and at the top of the style (score 4). However, in some cases, few if any tubes could be found at the bottom of the style or in the ovary. As is shown in Table 7.6 (bottom), a similar situation occurs in seven homostyle species examined. In some (*P. halleri*, *P. laurentiana* and *P. scotica* in section Aleuritia), many tubes could be detected in the ovary. In almost all cases, legitimate crosses in heterostyle and selfs in homostyles set abundant seed, including those in which tubes could not be found in the ovary. One must conclude that technical problems render the detection of tubes in the ovary difficult in some, but not all, species. The results of heterostyle species in Table 7.6 must be viewed in this light.

Examination of Table 7.6 shows that at least four different types of incompatibility can be identified in the genus *Primula*, which can be briefly listed as:

(a) Good within-pin and within-thrum germination, with quite good within-pin, but poorer within-thrum, stigmatic penetration; stylar inhibition of remaining tubes; this is similar to the mechanism reported for *P. obconica* (which is in this group) by Lewis (1942b) (Figs. 7.17 and 7.18) (this may be the most primitive mechanism).

(b) Good within-pin germination, but poor within-pin stigmatic penetration, followed by stylar inhibition; in strict contrast to (a) there is reduced, or in most cases zero, within-thrum germination.

**Table 7.6** Germination and tube growth of pollen in illegitimate and homostyle crosses of *Primula* (for details, see text).

| Section | Number of species | Mean pollen germination (%) | | Pollen tubes (range 0–4) | | | | | |
|---|---|---|---|---|---|---|---|---|---|
| | | | | stigma | | style | | ovary | |
| | | P* × P | T × T | P × P | T × T | P × P | T × T | P | T |
| (a) | | | | | | | | | |
| Auriculastrum | 3 | 67 | 71 | 3 | 1–2 | 0 | 0–1 | 0 | 0 |
| Monocarpicae (*malacoides*) | 1 | 40 | 44 | 3 | 1 | 0 | 1 | 0 | 0 |
| Obconicolisteri (*obconica*) | 1 | 68 | c70 | 2–4 | 1–2 | 3 | 1–2 | 1–2 | 1–2 |
| Auganthus (*sinensis*) | 1 | 38 | – | 2 | – | 0 | – | 0 | – |
| (b) | pin germination better than thrum | | | | | | | | |
| Cortusoides | 2 | 90 | 39 | 2 | 2–3 | 0 | 0 | 0 | 0 |
| Aleuritia (Farinosae) | 7 | 74 | 0 | 1 | 0 | 0 | 0 | 0 | 0 |
| *P. luteola* | 1 | 90 | 0 | 1 | 0 | 0 | 0 | 0 | 0 |
| Sibiricae (*nutans*) | 1 | 42 | 0 | 0 | – | 0 | – | 0 | – |

| | | | | | | | | | |
|---|---|---|---|---|---|---|---|---|---|
| (c) | thrum germination better than pin, pin stigmatic inhibition | | | | | | | | |
| Petiolares | 8 | 45 | 61 | 0–3 | 0–3 | 0–1 | 0–2 | 0 | 0 |
| Denticulatae (*erythrocarpa*) | 1 | 35 | 78 | 1 | 3 | 0 | 1 | 0 | 1 |
| Proliferae (Candelabria *poissonii*) | 1 | 25 | 92 | 2 | 4 | 0 | 2 | 0 | 1 |
| Capitatae (*capitata*) | 1 | 7 | 93 | 1 | 3 | 0 | 2 | 0 | 1 |
| Oreophlomis (*rosea*) | 1 | 5 | 90 | 1 | 2 | 0 | 2 | 0 | 2 |
| (d) | thrum stigmatic inhibition | | | | | | | | |
| Primula (Vernales) (*veris*) | 1 | 30 | 67 | 4 | 1 | 1 | 0 | 0 | 0 |
| *P. megaseaifolia* | 1 | – | 52 | – | 1 | – | 0 | – | 0 |
| | homostyle species | | | | | | | | |
| Aleuritia | 3 | 86 | | 4 | | 4 | | 4 | |
| Sphondylia | 3 | 80 | | 4 | | 2 | | 0 | |
| *P. japonica* | 1 | 94 | | 4 | | 0 | | 0 | |

* Abbreviation: P, pin; T, thrum.

**Figure 7.17** Penetration of a stigmatic papillae of pin *Primula veris* by a germinating pollen-tube (centre).

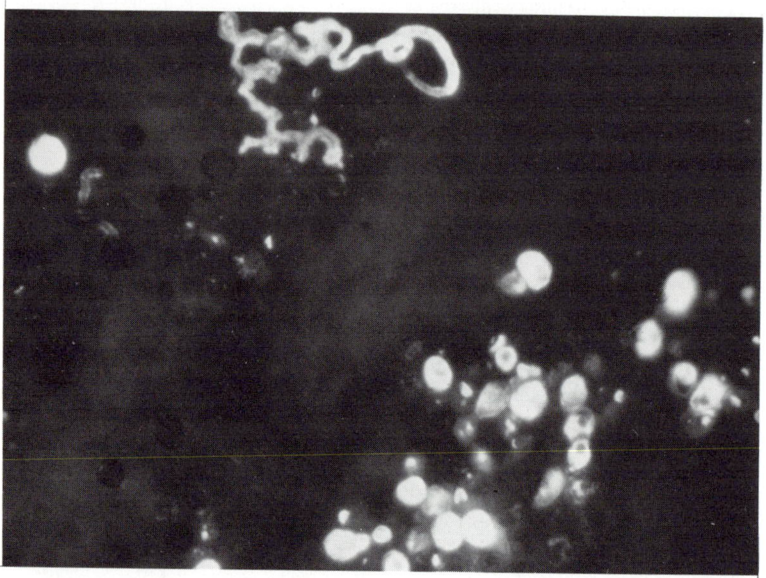

**Figure 7.18** A selfed pollen grain on the stigma of thrum *Primula veris*. The pollen grain has germinated, but the thick and irregularly swollen pollen tube wanders aimlessly over the surface of the stigma, and does not penetrate a stigma papilla. Microphotograph under ultraviolet light (×300). Note stigmatic papillae fluorescing in neighbourhood of the illegitimate grain, apparently due to the deposition of callose.

(c) Good within-thrum germination and stigmatic penetration, but with progressive stylar inhibition, although some tubes are usually found at the ovary and there is some within-thrum seed-set. In strict contrast to (a) and (b) there is reduced, and often very poor, within-pin germination. Pin tubes penetrate the stigma but are inhibited in the style.

(d) As for (c), but adequate within-pin germination is followed by good stigmatic penetration and stylar inhibition, whereas the good within-thrum pollen germination mostly fails to penetrate the stigma (this may represent the most advanced mechanism). This report differs from that of Richards and Ibrahim (1982) in which pollen germination (as against stigmatic penetration) was inadequately reported.

We can thus state that:

(a) The mechanism of compatibility varies between species.
(b) Within a species, the mechanism of incompatibility *usually* varies between morphs.
(c) Failure of fertilisation can be caused by failure of the pollen to germinate, failure of the pollen tube to penetrate the stigma papilla, failure of the pollen tube to grow down the style or, most often, by a combination of all three. It seems likely that stylar inhibition (which is most likely to fail) represents primitive conditions, while stigmatic inhibition represents a more advanced state.
(d) In the majority of *Primula* species, within-morph incompatibility is total; that is, no seed is usually set as a result of pin × pin or thrum × thrum pollinations. This confirms reports of Lewis (1942b) for *Primula obconica*, and Richards and Ibrahim (1982) for *P. veris*. Richards and Ibrahim (1982) have reviewed earlier reports concerning the strength of incompatibility in *Primula*, for instance by Darwin (1877) who claimed that in five species, within-morph compatibility varied between 20 and 50% of legitimate seed-set. However, it is probable that the only reliable reports of high levels of within-morph compatibility in heterostylous species refer to *Primula sinensis* (Mather & de Winton 1941) in which within-pin pollinations result in 63% seed-set of legitimate crosses. This high level of within-morph compatibility of cultivated forms of this species is probably the result of intense artificial selection for self-fertility over many years.

Nevertheless, it is noteworthy that some within-thrum compatibility does occur in our sample in some species in group (c), which show high levels of within-thrum pollen germination, and good within-thrum stigmatic penetration. It appears that stylar inhibition *alone*, in the absence of germination or penetration inhibition, is not strong enough to give complete incompatibility.

## PHYSIOLOGY OF INCOMPATIBILITY

### *Pollen germination*

Unlike many multi-allelic sporophytic and two-locus gametophytic systems of incompatibility, the lodgement of pollen on the stigma in *Primula* does not apparently depend on lectic functions of lipid or glycoprotein. As is shown by Shivanna *et al.* (1981) and Stevens and Murray (1982), the germination of both legitimate and illegitimate grains occurs without the pollen grain being firmly glued to a papilla. The stigma is of a dry type, without a liquid exudate, and with a continuous cuticle. The cavities in the exine of the pollen are filled with proteinaceous material, including several different glycoproteins and esterases, as well as lipid. In view of the sporophytic nature of the incompatibility, it is certain that at least some of this material is involved in the control of pollen germination.

Shivanna *et al.* (1981, 1983) attach great importance to the ability of the pollen to rehydrate and thus germinate. In *P. vulgaris* (primrose) they show that the germination and penetration of illegitimate pollen tubes is highly dependent on external relative humidity (Table 7.7). Thus, at high relative humidities (which will often be experienced by primroses in April woodlands) high levels of illegitimate pollen germination, and appreciable levels of illegitimate stigma penetration will occur. The higher levels of within-thrum germination and the higher levels of within-pin penetration, match our own results for the related cowslip, *P. veris* (see Table 7.6).

There is a nice correspondence between the greater response of within-thrum germinations to high humidity, and the *in vitro* germination of previously desiccated pin and thrum grains. Desiccated pin pollen germinates well, but desiccated thrum pollen does not germinate, although it can do so after carefully controlled rehydration, and in fact it rehydrates faster than pin pollen. Thus, we can conclude that the germination of thrum pollen on a pin stigma is a function of the relative humidity in the immediate environment of the grain at the stigma. If the

**Table 7.7** The effect of external relative humidity on germination and penetration of illegitimate pollen tubes in *Primula vulgaris* (after Shivanna *et al.* 1983).

| Relative humidity (%) | Germination of pollen (%) | | Penetration of stigma by pollen (%) | |
|---|---|---|---|---|
| | P* × P | T × T | P × P | T × T |
| 5–10 | 0 | 0 | 0 | 0 |
| 56–65 | 5 | 25 | 3 | 0 |
| 95–100 | 47 | 75 | 25 | 13 |

* Abbreviations: P, pin; T, thrum.

relative humidity at the illegitimate thrum stigma is also high, the pollen will also germinate. Perhaps the longer pin papillae function in providing a microenvironment with higher relative humidity for the successful germination of the legitimate thrum pollen? However, the germination of pin pollen is not a function of relative humidity, at least to the same extent.

How then does this model correspond to the sporophytic control of thrum × thrum pollen germination? Shivanna *et al.* (1983) have shown that if the tapetal proteins carried on the thrum pollen are removed using cyclohexanes, the dehydrated thrum pollen loses its ability to rehydrate, and thus cannot germinate on a thrum stigma. If, however, dehydrated thrum pollen is treated with cyclohexane, and is then placed on a neutral stigma (of red campion, *Silene dioica*) it can rehydrate and germinate. One can conclude from these experiments that rehydration, and germination of thrum pollen is a function of externally carried tapetal proteins, but that this function can be replaced by at least some (although not thrum primrose) stigmas.

Shivanna *et al.* (1981) have also conducted experiments on primroses into the effect of protein extracts from styles and stigmas on pollen germination *in vitro*. They show that total extracts affect rates of pollen germination, so that germination of both pin and thrum pollen is depressed by both pin and thrum extracts, but the greatest depression in germination is experienced by thrum pollen receiving thrum extracts. This also suggests that there is a stigmatic component in the stimuli which encourages thrum pollen to rehydrate and germinate. Dialysates containing protein fractions only over 10 000 daltons in size do not have this effect, suggesting that the stigmatic component is a protein of relatively small size, which can pass through membranes.

It seems possible that for those species of *Primula* that show poor or zero illegitimate thrum germination (Table 7.6, group b) a stigmatic component has an overriding effect in blocking the rehydration of thrum pollen. In other species (groups a, c and d), the stigmatic component is relatively ineffectual, and thrum germination is a function of relative humidity and may relate to the weather and/or the microtopography of the stigma surface. As yet, there is no indication as to what factor depresses pin illegitimate pollen in those species classed in Table 7.6 groups (c) and (d).

### *Penetration of the stigma papilla and stylar growth of pollen tubes*

Heslop-Harrison *et al.* (1981) have shown that legitimate penetration of stigma papillae differs in the two morphs of the primrose. Thrum grains enter the long pin papillae subapically (Fig. 7.17), where the cuticle is relatively thick, but pin tubes enter the shorter thrum papillae basally, and intercellularly, where the papilla cuticle is thin.

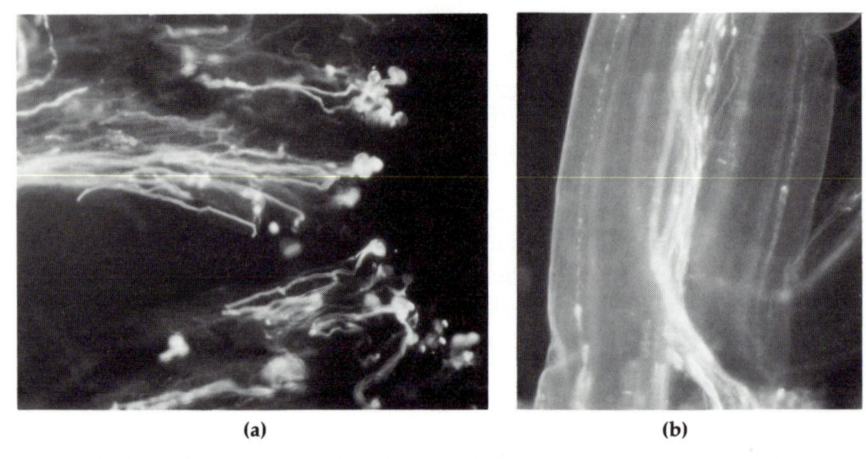


**Figure 7.19** Microphotographs taken under ultraviolet irradiation of (a) good pollen germination and pollen-tube growth into the stigma in a self in the homostylous *Primula japonica* (×100), and (b) good pollen-tube growth in the central style of the homostylous *Primula scotica*, after a self (×40).

They also demonstrate a major difference in behaviour of illegitimate pollinations, which is essentially similar to that found for the cowslip by Richards and Ibrahim (1982), and for *Primula obconica* by Stevens and Murray (1982). That is, although the majority of germinated illegitimate pin grains penetrate the pin stigma, very few thrum tubes penetrate the thrum stigma. Instead, the thrum tubes, which are thick, distorted and swollen apically, wander around the thrum stigma surface (Fig. 7.18). In the cowslip, Richards and Ibrahim (1982) showed that callose is deposited in thrum papillae in the neighbourhood of thrum grains (Fig. 7.18) and that this callose response is also elicited by pollen-free extracts of thrum pollen water. Callose responses have not been reported in other species, nor in pin stigmas of cowslip. It is probable that callose responses are secondary, not being involved in the mechanics of the incompatibility. Both Richards and Ibrahim (1982) and Shivanna *et al.* (1983) have suggested that the abnormal morphology of illegitimate thrum tubes may prevent illegitimate thrum stigma penetration. Illegitimate thrum tubes that do penetrate have an abnormal morphology, that is they are much broader than pin tubes at the base, but narrow rapidly to about 7 μm in width, similar to that of pin tubes, as if in response to an external stimulus.

Shivanna *et al.* (1981) showed that style and stigma protein extracts which have been dialysed so that they contain only units over 10 000 daltons in size, have a marked effect on pollen-tube growth *in vitro*, although they do not affect germination. Thrum stigma-style extracts

**Table 7.8** Characteristics of water-soluble pin and thrum sigma-style extracts in *Primula obconica* (from Golynskaya *et al.* 1976).

| | Pin extracts | Thrum extracts |
|---|---|---|
| Position of extract with maximum agglutination function | style | stigma |
| Agglutinating function of extract with greatest inhibition of illegitimate pollen-tube growth | zero | high |
| Molecular size of total protein fraction | smaller | larger (but with 90% overlap) |
| Molecular size of agglutinating protein fraction | larger | Smaller (no overlap in Sephadex fractions) |

inhibit the growth of thrum pollen tubes, and pin extracts inhibit the growth of pin pollen tubes. In both cases, inhibited tubes have swollen vesiculated ends similar to those observed in thrum illegitimate pollinations. Golynskaya *et al.* (1976), working with *P. obconica,* also demonstrated that water-soluble pin and thrum stigma-style extracts affect the growth of pin and thrum grains respectively *in vitro*. They assembled further interesting information concerning these extracts (Table 7.8). From these results we may conclude that in, *P. obconica,* agglutinating proteins, which Golynskaya *et al.* (1976) tentatively identified as phytohaemagglutinins, probably play an important rôle in the inhibition of the thrum pollen tubes on the thrum stigma. These are probably the same as the larger part of the dialysate fraction of Shivanna *et al.* (1981), and probably elicit the swollen thrum-tube response. In contrast, the protein fraction involved in inhibiting illegitimate pin grains in *P. obconica* is concentrated in the styles, and does not have an agglutinising function (although agglutinins do also occur in pin styles). This is, of course, exactly as one would predict, for in *P. obconica* (and indeed *P. vulgaris* and *P. veris,* but not some other species) pin inhibition takes place principally within the style, whereas thrum-tube inhibition is mostly confined to the surface of the stigma.

Further work on *P. obconica* by Stevens and Murray (1982) concerned the growth of legitimate and illegitimate pollen grains in decapitated styles, in which the stigma had been removed. Performance was assessed as the percentage of pollen tubes at the top of the style which grow to the bottom of the style:

| | Pin | Thrum |
|---|---|---|
| legitimate | 33% | 66% |
| illegitimate | 21% | 72% |

Thus, it is the nature of the style, rather than the nature of the pollination that governs pollen-tube growth in decapitated styles. These results strengthen the suggestion that in *P. obconica* (and probably in *P. vulgaris*) most if not all the within-thrum incompatibility takes place at the stigma. If the stigma is removed, pollen-tube growth of illegitimate and legitimate tubes alike is quite good. With respect to pin stigmas, it is clear that much inhibition takes place in the style. However, the poor growth of legitimate thrum pollen on the excised pin style suggests that the pin stigma plays a positive rôle in encouraging thrum pollen growth. This is the only suggestion so far of a complementary, rather than oppositional, function in the operation of di-allelic incompatibility (compare Ch. 6). This is supported by the results of bud pollinations, in which pollen grains germinate but fail to penetrate the stigma of small buds in any pollinations, including both legitimate and illegitimate pollinations onto pins. Clearly, the mature pin stigma contains something that is lacking in the immature stigma and which encourages pollen tube penetration by both legitimate and illegitimate pollen.

It should be emphasised that the fairly consistent picture of the operation of incompatibility which has been obtained for *P. obconica* and *P. vulgaris* so far, does not apparently hold for other species of *Primula*, for instance those in Table 7.6, group (c). In these, not only is there good within-thrum germination of pollen, suggesting that there is no stigmatic or environmental block to the rehydration of thrum pollen in these species, but there is also good within-thrum penetration of the stigma and style. Thus, for these species, there is apparently stylar control of thrum incompatibility, which may or may not be of the same type as that found in the stylar control of pin incompatibility in other species. Also, in this group, there is poor within-pin stigmatic penetration. Whether this is of the within-thrum type found in other species (in which thrum stigmatic extracts cause thrum pollen-tube distortion) has yet to be discovered.

Although there is clearly much more to learn about the operation of di-allelic incompatibility in *Primula*, it is possible to make some suggestions about its operation in *P. vulgaris* and *P. obconica*, which appear to be similar (Table 7.9). As yet, little if any effort has been made to interpret these findings with respect to our knowledge of the genetics of *Primula* incompatibility. It should be re-emphasised that:

(a) Genetic control is sporophytic, and operators should thus reside on the outside of the pollen grain, and in the stigma or style.

**Table 7.9** The operation of di-allelic incompatibility in *P. vulgaris* and *P. obconica*.

| Site of operation | Within-pin | Within-thrum |
|---|---|---|
| illegitimate pollen germination | relatively poor, operation unknown, responds to increased humidity; desiccated grains hydrate well | relatively good only at high humidity possibly provided by long papillae of pin stigma; rehydration of desiccated grains dependent on external (tapetal) proteins, and/or pin (but not thrum) stigma proteins of less than 10 000 daltons |
| illegitimate penetration of illegitimate pollen tube | most germinated tubes penetrate mature (but not immature) pin stigma apically on papilla | most germinated tubes thick, sinuous, swollen apically, fail to penetrate thrum papilla, apparently caused by stigma proteins of over 10 000 daltons; pollen tapetal proteins also affect papillae, causing callose response |
| growth of illegitimate pollen tube in stigma and style | growth very poor eventually failing due to non-agglutinating proteins located in style | growth very poor, rapidly failing due to larger, agglutinating proteins mostly located in stigma (probably the same as those that affect illegitimate tube form outside the stigma) |

(b) One morph is homozygous (*s* genes only), and the other heterozygous (*S* and *s* genes); thus features restricted to the thrum are presumably linked to the *S* complex, but thrums may also be able to call on *s* factors on the male or female side. Features restricted to pins must be *s* complex factors that are switched off in the thrum heterozygote.

(c) Male incompatibility factors (*P*/*p*) and female incompatibility factors (*G*/*g*) are recombinable, and thus presumably have different physical locations on the chromosome. Thus, unlike gametophytic incompatibility mechanisms (Ch. 6, Fig. 6.4), it may well be that the incompatibility recognition genes for a morph differ male and female function, although the same gene might of course be duplicated at *P*/*p* and *G*/*g*.

There is apparently one embarrassing feature of the function of *Primula* incompatibility which is very difficult to explain. Multi-allelic sporophytic systems operate at the stigma surface by impeding pollen germination, or papilla penetration, which is predictable if the male factor is carried externally on the pollen grain (Ch. 6, Table 6.4). However, it appears that an important component of the sporophytic control in *Primula* operates internally, in the stigma (within-thrum) or in the style (within-pin). It is difficult to understand how externally carried pollen factors can operate in this way, and it is necessary to propose that tapetal lipoprotein is carried into the stigma by the penetrating pollen tube in some way. Although this may indeed occur, there is no evidence for it so far, but such evidence may prove very difficult to obtain.

If such a model of operation is correct, there may be as many as six gene products involved in the incompatibility reactions of *P. vulgaris* and *P. obconica*, the thrum cistrons *G* and *P* (part of *S*) each providing two, and the pin cistrons *g* and *p* (part of *s*) each providing one, as follows:

| *G* (thrum female) | *P* (thrum male) | *g* (pin female) | *p* (pin male) |
|---|---|---|---|
| (1) papillar cuticular protein less than 10 000 daltons, non-agglutinating, discourages T pollen hydration (P pollen hydrates easily) | (1) tapetal protein only allows T pollen germination at high humidity and controls rehydration | stylar protein more than 10 000 daltons, non-agglutinating, inhibits P tubes (switched off by *G*) | presumptive factor in pin tapetum carried by tube and reacting with pin stylar protein (switched off by *P*) |
| (2) stigmatic protein, more than 10 000 daltons, agglutinating, inhibits T tubes | (2) presumptive factor in T tapetum carried by tube and reacting with T stigmatic protein | | |

Thus, some progress has been made in the understanding of the physiological and genetical control of di-allelic incompatibility in *Primula* but it is clear that much interesting work remains to be done. Although a good deal of variation in function occurs within the one genus, this variation may finally lead to a greater understanding to how di-allelic systems in particular, and incompatibility systems in general, evolve. There are features of the *Primula* incompatibility system which render it particularly charismatic and attractive to evolutionary workers. In particular, it presents an outstanding example of coadapted linkage groups in action which deserves even more thorough attention than it has already received, especially with respect to the measurement of selective forces working on the units of the linkage group in the field. As I have said on a previous occasion (Richards 1984), *Primula* presents 'an irresist-

**Table 7.10** Ratio of turgor pressures (given in parentheses) in pollen and styles of *Linum grandiflorum* (from Lewis 1943).

| | Pin pollen (4) | Thrum pollen (7) |
|---|---|---|
| pin style (1.75) | 2.86 | 4 |
| thrum style (1) | 4 | 7 |

ible invitation to investigate the complexities of organic evolution (which) seems almost God-given. Like *Drosophila, Cepaea, Papilio* or phage, the pin/thrum/homostyle system has attracted a string of evolutionary geneticists; their names read like a roll of honour for the science: Darwin, Bateson, Gregory, Haldane, Fisher, Mather, Ford, Dan Lewis, Darlington.'

Although *Primula* has attracted much of the experimentation on the function of heteromorphic systems, there is little doubt that other heteromorphic genera would also repay detailed study, although the nature of the function might prove to be quite different. For instance, a repetition of Lewis's (1943) results concerning the turgor pressures in pollen and stigmas of pin and thrum *Linum grandiflorum* is badly needed. Lewis claimed that legitimate pollinations both gave pollen : style turgor pressure ratios of 4, whereas illegitimate pollinations gave ratios that were less, leading to plasmolysis, or more, leading to bursting of the pollen tube (Table 7.10). Although this result is very interesting, it needs to be viewed in the light of more modern knowledge of the function of incompatibility systems, and in particular the findings of Ghosh and Shivanna (1980) that pin and thrum stigmas differ in this species, being of dry and wet types respectively.

## HETEROMORPHY AND SEXUAL SELECTION

During the past few years, our conception of the processes that influence the evolution of breeding systems has been enriched by ideas of functional gender and sexual selection (Chs. 1 and 2) which have been promoted particularly by Charnov (1982, 1984). As early as 1877, Darwin, who as usual was first in such matters, noted the asymmetry of female function in pin and thrum *Primulas*. He hypothesised that selection had rendered pins relatively female and thrums relatively male, which he considered to be part of the function of heterostyly in this genus. The idea that selection encourages plants to commit various resource to male or female sexual function in an adaptive way has been applied to heteromorphic species more recently by Lloyd (1979), Beach and Bawa (1980),

Casper and Charnov (1982), Charnov (1984, pp. 374–7) and Taylor (1984). This has resulted in some rather inconclusive algebraic modelling, which has been partially hampered by a paucity of useful field data to choose or uphold various assumptions. However, it is agreed that evidence for functional gender selection can result from various sources as outlined below.

**Unequal seed-set in pins and thrums** This is frequently found (Ganders 1979) in a variety of heteromorphic genera and, in all cases, pins set more seed. In theory, such differences could be caused by the greater production of ovules by pins, by the more efficient legitimate pollination of pins, by higher levels of selfing in pins, or by the greater acquisition or commitment of resource to seeds by pins (this could also result in larger pin seeds, which have not been found). In practice, pin stigmas usually receive more legitimate grains than do thrum stigmas, apparently due to their more exposed position, and this probably accounts for the better pin seed-set. As a result, pins are often considered to be better female parents, and thus thrums to be better male parents, than vice versa.

**Unequal pollen production** As already discussed, pins frequently produce more, and smaller, pollen grains than thrums. It seems that the greater production of pin pollen occurs in response to the inaccessibility of the thrum stigma, and the poorer accessibility of pin pollen. Ornduff (*in litt.*) has shown in *P. veris* and *P. vulgaris* that approximately 50% of thrum pollen is removed from anthers but for pins the figure is only 20%. The small size of pin pollen can be viewed as a stratagem to reduce male expenditure in pins. However, the total male resource output of pins and thrums has never been accurately measured in any species. It can be approximated by anther size, which does not significantly differ between pins and thrums in many species, but which is larger in thrum *Lithospermum*, *Amsinckia*, *Hottonia*, *Nymphoides* and *Pulmonaria* and smaller in thrum *Linum flavens* and *Forsythia suspensa* (Ganders 1979).

**Unequal pollen flow** Asymmetrical pollen flow, with more illegitimate and legitimate pollen travelling onto pin stigmas, is commonplace, as has been discussed above, and has the effect of increasing pin female-gender function and thrum male-gender function. Beach and Bawa (1980) have suggested that effective legitimate pollen carried onto stigmas may be mediated by different pollinators for pins and thrums of some species. They point out that legitimate pollen transport onto thrum stigmas in many species requires a specialist pollinator with long narrow mouthparts such as a bird, long-tongued bee or lepidopteran, and that these may be scarce, or inefficient pollinators. However, thrum anthers and pin stigmas may be visited by non-specialists including pollen-eaters such as small bees, syrphids and beetles, which achieve legitimate pollination onto pin stigmas, but not onto thrums, and these may be much more frequent flower visitors. Such differences in effective pollinators may amplify the asymmetry of pollen flow, emphasising the male function of thrums and the female function of pins. They quote evidence from Ornduff (1975a) on *Jepsonia* which shows that halictid bees are better pollinators of thrum stigmas than are syrphid flies, the other main flower visitors. Beach and Bawa (1980) also suggest that a change in pollinator frequency, such that specialist pollinators become rare, may encourage male function in thrums and female function in pins to the extent that thrum female sterility and male pin sterility are favoured, leading to dioecy.

This appears to have occurred in several rubiaceous genera such as *Coussarea* and *Mussaenda* in which dioecious species show less specialised (shallower) flowers than heterostylous relatives, and have rudimentary thrum-positioned anthers in the long-styled females and rudimentary pin-type gynoecia in the long-stamened males. A similar situation is described by Lloyd (1979) for *Cordia* (Boraginaceae), for which he computes the male gender function and the female gender function for five species algebraically, using data on pin : thrum ratios and fruit-set for each morph. Increasing levels of differential gender function in four species apparently show evolutionary trends towards total gender differentiation (full dioecy) in the fifth.

**Unequal pin : thrum ratios** Although heteromorphic species without di-allelic control, or with weak control resulting in much selfing may often show pin : thrum genet ratios differing from unity, this seems to be rare in the majority of heteromorphic species in which within-morph incompatibility is strong (Ganders 1979). There has been some controversy as to whether cowslips *Primula veris* and primroses *P. vulgaris* show morph frequencies which differ from 0.50 (summarised in Richards & Ibrahim 1982), but it can be safely asserted that no consistent departures from this frequency are found in either species, nor do such departures consistently favour pins or thrums. For five species of *Cordia*, however, Opler *et al.* (1975) record significant levels of anisoplethy (departures from expected 1 : 1 ratios) in all cases, with an excess of thrums in four species, and an excess of pins in the other. These data have been incorporated by Lloyd (1979) in his calculations of gender function. Individuals of the rarer morph make a relatively larger contribution to gender function per individual when seed can only arise from legitimate (between morph) crosses. If, for example, pins represent 0.25 of the population (and thrums 0.75 of the population), and thrum seed must be fertilised by pin pollen, the relative male contribution to the setting of a seed per individual male parent is three times greater for pins than for thrums.

If one morph is habitually scarce, there will be selection for gender fitness in the gender(s) by which it primarily functions. Thus, high resource allocation to female function (numbers and sizes of ovules and seeds) in pins should be especially favoured when pins are in the minority. It remains to be seen whether such selection actually occurs.

## CONCLUSIONS

As yet, we lack thorough energy-budget studies which can demonstrate conclusively whether most heteromorphic species adjust gender expenditures in a way that matches the relative gender function in each morph, and whether such adjustments enhance reproductive efficiency (seed output per plant and seed quality).

Above all, it is not yet clear to what extent gender-function theory can explain the evolution of heteromorphy in the first place. At present, we seem to be faced with a chicken and egg situation. Did heteromorphy, the linkage of style-length, stamen-length, papilla-length and pollen-size polymorphisms to di-allelic incompatibility, result from selection pressures partially to separate expenditure on male and female function onto different individuals? In other words, has heteromorphy resulted from the same selection pressures for outbreeding and gender maximisation which seems to have led to dioecy? Or are the morphological features of heteromorphy primarily involved in the efficiency and function of the between-morph pollen carry necessary in a di-allelic incompatibility system? If this is the case, differential gender function has been imposed upon pin/thrum systems secondarily by the asymmetry of pollen flow between morphs resulting from pollinator behaviour. Perhaps the separation of gender function in at least some distylous species must for the present be viewed as an addition to the complex suite of selective pressures which act upon the heteromorphy supergene. As the study of this supergene enters its second century, it promises to be just as fascinating and enlightening as it was for Darwin, more than a hundred years ago.

# CHAPTER EIGHT
# *Dicliny*

## DEFINITIONS

A flower is considered to be male when it bears an androecium (stamens, the microsporangia) with viable pollen (microspores) capable of forming fertile pollen tubes (male gametophytes). A flower is considered to be female when it bears a gynoecium (= pistils), the megasporophylls, including the stigma(s), style(s) and an ovary containing ovules (the megasporangia with integuments). The ovules when fertile contain viable embryo-sacs (female gametophytes), each with a fertile egg cell (Ch. 3).

Dicliny is said to occur when all members of a population are not regularly hermaphrodite (bearing male and female flowers). There are a number of conditions with respect to the distribution of male and female organs on individual plants (genets), and these have rather confusing names, which are explained below. They include both hermaphrodite and diclinous situations.

| | Hermaphrodite |
|---|---|
| all flowers hermaphrodite | [⚥] |
| all flowers monoecious | [♂ ♀] |
| gynomonoecious | [⚥ ♀] |
| andromonecious | [♂ ⚥] |
| | **Diclinous** |
| subgynoecious | [♂] [⚥ ♀] |
| * subandroecious | [♂ ⚥] [♀] |
| * gynodioecious | [⚥] [♀] |
| androdioecious | [♂] [⚥] |
| polygamous (including trioecious) | [♂] [♂ ⚥] [⚥] [⚥ ♀] [♀] (not all may occur) |
| * dioecious | [♂] [♀] |

For each of these categories, the symbols within brackets represent the sexual states that can be represented within an individual genet. In all cases except full hermaphrodity, some flowers have only one sex. For andromonoecy and androdioecy, these are male, and all flowers with female organs are also male. These conditions share with hermaphrodity the capability for all female organs to be pollinated by the same flower (i.e. autogamously). In all the other conditions, there are some flowers which are only female, and which therefore have to receive pollen from another flower in order to set seed. Thus, they may receive pollen from another plant, and be outcrossed, but in some cases they may receive pollen from another flower on the same plant, and if self-fertile they can be geitonogamously selfed (Ch. 4). Only in the three diclinous conditions marked with an asterisk (*) are there at least some individual genets which are entirely female, and which must therefore be cross pollinated (outcrossed) for seed to be set. Two of these three conditions, gynodioecy and dioecy, are the commonest forms of dicliny, and are those which most commonly lead to outbreeding. The other conditions are mostly of interest in so far as they represent intermediate evolutionary steps towards dioecy, or result from environmentally triggered variations in the expression of genetically controlled dioecy or gynodioecy.

It should be noted that dioecy in seed plants is not homologous with so-called dioecy in bryophytes, nor does it have the same genetic consequences. In bryophytes, it is the gametophyte (haploid) generation which may be unisexual ('dioecious'); it is always unisexual in seed plants. Variations in sexual expression in seed plants (above) occur in the sporophyte generation, which is dominant. In bryophytes, the sporophyte generation is always uniform in morphology, producing only one morphological type of sporangium and spore. However, the sporophyte of a dioecious bryophyte will produce genetically different spores, segregating out to form male and female gametophytes.

Thus, bryophytes lack a mechanism for preventing mating between meiotic products of the same sporophyte (selfing), which is achieved by dioecy in seed plants. Pteridophytes are also unable to prevent selfing by the distribution of sex organs. Those few pteridophytes which are heterosporous (*Isoetes*, *Selaginella*, *Azolla*) produce both spore types from the same sporophyte. Although microspores give male gametophytes, and megaspores give female gametophytes, mating can still occur between products of the same sporophyte. The term dioecy should be reserved for those systems in which selfing in the products of sporophytes is impossible by virtue of the unisexuality of the sporophyte. This is restricted to seed plants, and bryologists should seek another term.

## DIOECY

### *Distribution of dioecy*

The following discussions refer only to dioecy. Gynodioecy is dealt with separately later in the chapter, as are other conditions, which are relatively unimportant, and are mentioned only briefly.

**Gymnosperms** Of the five main orders of Gymnosperms, the Ginkgopsida, Cycadopside and Gnetopsida are almost always dioecious. These orders contain relatively few species which are mostly tropical in distribution, and they are considered to be primitive, and relict from earlier eras when they played a more important rôle in the flora. A fourth order, the Taxales, also contains some dioecious species, including the familiar yew (*Taxus baccata*). The remaining, dominant group, the Coniferopsida are predominantly hermaphrodite but monoecious, and the Pinaceae are entirely so. There are, however, dioecious members of the Cupressaceae (some junipers including the common juniper, *Juniperus communis*), Podocarpaceae (the Australasian *Podocarpus*) and Araucariaceae (*Agathis*, such as the Kauri *A. australis* from New Zealand). The small family Taxodiaceae (redwoods, swamp cypresses) is, however, entirely monoecious. Viewed as a whole, it seems that the more tropical groups of Gymnosperms are more likely to be dioecious than those from temperate and boreal regions.

Little work has been done on the genetic control, energetics, pollination biology or niche differentiation of dioecious Gymnosperms. Intermediate conditions such as gynodioecy have rarely been reported (although some gynodioecious *Juniperus* occur). As far as is known the sex expression is stable. At least some populations of British *Juniperus communis* have sex ratios far from equality, with females predominating. The cause of this is unknown, but it can have severe consequences in relict populations. The five remaining bushes recently rediscovered in Castle Eden Dene National Nature Reserve in Durham, UK, are all female, and thus are unable to fruit. Such unequal sex ratios are not necessarily typical of this species, for I recently found the following numbers of males and females in a thriving population at Valessard, Auvergne, France: males 92, females 81; male frequency 0.53, $\chi^2 < 1$ (not significant).

**Angiosperms** In the flowering plants, dioecy is a rather unusual condition which is found in about 4% of all species. It is scattered through many plant families at a low frequency, and few families are entirely dioecious. The willows and poplars (Salicaceae) form a familiar example of an entirely dioecious family (Fig. 8.1). This distribution pattern strongly suggests that dioecy has arisen secondarily from other breeding systems, on many occasions (polyphyletically), and that it is reversible, or frequently becomes extinct. Dioecy rarely seems to last long enough in evolutionary time, or to be successful enough, to establish a dynasty; i.e. to dominate a higher taxonomic category such as a family, tribe or even a genus.

It must be noted here that a minority view persists that a dioecious, wind-pollinated condition is primitive to the Angiosperms (A. D. J.

**Figure 8.1** Male (left) and female (right) inflorescences of the dioecious goat willow, *Salix caprea* (×0.25).

Meeuse 1973, 1978). I prefer to follow the majority view that the original Angiosperm condition was a hermaphrodite, beetle-pollinated flower with gametophytic self-incompatibility (Chs. 4 and 6). Dioecy is a mechanism that ensures total outbreeding, but it is also inefficient in that only about half of the genets in the population bear seeds (Chs. 1 and 11). The distribution of dioecy within the Angiosperms suggests that dioecy has arisen secondarily from self-compatible inbreeders (Fig. 1.3) in response to renewed selection pressures for outbreeding, or separation of male and female function.

It is difficult to discover taxonomic patterns with respect to the distribution of dioecy in the flowering plants. In contrast, dioecy is associated with certain life forms and habitats. Bawa (1980) has gathered information as to the proportion of dioecious species in the floras of 10 areas (Table 8.1). Yampolsky and Yampolsky (1922) discovered that 3–4% of species in many parts of the world are dioecious. Bawa's figures agree with this assessment, but with two notable exceptions. These are tropical forests and oceanic islands.

**Tropical forests** The figure of 9% dioecious species in the wet tropical forest of Barro Colorado Island, Panama, may be typical of many tropical forests. A high proportion of species in this habitat are trees, and trees tend to be dioecious, especially in tropical forest (Table 8.2). Bawa's own work in a Costa Rican forest (Bawa 1979) gives similar figures, about 20% of tree species being dioecious.

**Table 8.1** Incidence of dioecy in ten different floras (after Bawa 1980).

| Area | Type of flora | % dioecious species |
|---|---|---|
| Barro Colorado Island, Panama | tropical forest | 9.0 |
| Ecuador | tropical, various | 3.0 |
| India | tropical, various | 6.7 |
| Hawaii | subtropical, oceanic | 27.7 |
| New Zealand | various, oceanic | 12–13 |
| south Australia | warm temperate to savannah to desert | 3.9 |
| North Carolina | warm temperate | 3.5 |
| south-west Australia | Mediterranean | 4.4 |
| south California | Mediterranean | 2.5 |
| British Isles | cool temperate | 3.1 |

Bawa (1980) has argued that the association between dioecy and tropical trees is a product of problems faced by the seedling tropical tree. This must germinate and grow at very low light levels on the forest floor. If it is to succeed, and become a sexually reproducing forest giant, it must survive until a gap appears in the canopy caused by the death of an older tree. Therefore, there is a premium on large seeds with large maternal investments of energy; trees producing such seeds are most likely to pass their genes on to the next generation as these offspring are most likely to persist and survive. Bawa argues that a high maternal investment in the fruit will also be selectively advantageous. Nutritious fruits which are attractive to birds, bats and monkeys are more likely to be dispersed away from the competitive environment of the parent tree, perhaps to conditions that have been made favourably open by the animals' roost or nest. He shows a striking association between dioecy and animal-dispersed fruits in two areas of Costa Rica (Table 8.3) although, as animal-dispersed fruits preponderate, it would be more accurate to state that the fruits of dioecious species are rarely wind dispersed.

**Table 8.2** Frequency of dioecy in different growth forms in three floras (after Bawa 1980).

| | % dioecious species | | |
|---|---|---|---|
| Life form | North Carolina | Barro Colorado Island | California |
| trees | 12 | 21 | 20–33 |
| shrubs | 14 | 11 | 0–23 |
| lianes | 16 | 11 | – |
| herbs | 1 | 2 | 4–9 |

**Table 8.3** Numbers of animal and wind-dispersed species that are dioecious or hermaphrodite in two Costa Rican floras (after Bawa 1980).

| Type of forest | Breeding system | Fruit dispersal | |
|---|---|---|---|
| | | Animal | Wind |
| dry deciduous | dioecious | 30 | 3 |
| | hermaphrodite | 60 | 26 |
| wet evergreen | dioecious | 66 | 0 |
| | hermaphrodite | 222 | 29 |

Large seeds and nutritious fruits will cause a considerable drain on maternal resources. Bawa considers that such plants may be more successful when these heavy maternal loads are separated from male loads, in the dioecious condition. Givnish (1980) has provided a similar argument to explain the high levels of dioecy amongst tropical forest Gymnosperms, which also tend to have very large, animal-dispersed seeds (e.g. cycads).

There is another line of argument concerning the occurrence of dioecy amongst tropical trees. Within a species, individuals usually occur at a very low density, and are distant from one another, with many other species in between. As a result, hermaphrodites will be very largely self-pollinated. If they are self-compatibile, most seed will be selfed, with an accompanying loss of vigour and variability (Ch. 9). If they are self-incompatible, stigmatic sites will tend to be blocked by infertile selfed pollen, so that the rarer pollen grains that are incoming from another plant will be at a disadvantage. In a dioecious species, stigmatic sites will be left free for incoming pollen from distant males, brought by far-flying 'trap-lining' pollinators (Ch. 5), and self-fertilisation cannot occur. It will pay a plant that invests heavily in female reproduction (large seeds, nutritious fruits) to produce outbred offspring of high genetic quality; the law of the jungle should favour the production of few high-quality seeds, rather than many low-quality seeds (i.e. '*K*' rather than '*r*' strategy, Ch. 2). Some of the former may survive, whereas none of the latter are likely to. Dioecy is reproductively inefficient, in that only half the individuals are available for seed production (Ch. 11), and pollen flow between males and females may be limited. It is thus likely to be favoured when the quality rather than quantity of offspring is selectively advantageous.

Dioecious species are inconvenienced in another way, when insect pollinated. Although males and females both produce nectar, only males produce pollen. Kevan and Lack (1985) show that the Indonesian *Decaspermium parriflorum* produces only pollen as a reward; females have sterile pollen, which doubtless rewards visiting bees. If cross pollination

**Table 8.4** Most important pollinators of dioecious and hermaphrodite tree species in a Costa Rican dry deciduous forest (after Bawa 1980).

| Main pollinators of species | % of tree species | |
|---|---|---|
| | Hermaphrodite ($n = 94$) | Dioecious ($n = 28$) |
| medium or large bee | 25 | 1 |
| small bee | 26 | 80 |
| beetle | 14 | 3 |
| moth | 19 | 9 |
| other | 16 | 7 |
| total | 100 | 100 |

is to be effective, the plant will be most successful when it attracts visitors which are primarily nectar feeders. It is to be expected that pollen feeders will be poorly represented amongst visitors to tropical dioecious trees, and indeed beetles and large social bees are much rarer visitors than they are to comparable hermaphrodite flowers (Table 8.4). The dominant visitors are small, non-social bees which tend to be entirely nectar feeders, and to fly long distances between nectar sources ('trap-line', Ch. 5). Wind pollination is common amongst dioecious woody species in temperate latitudes (Ch. 5), but is rare in tropical forests, where the density of the vegetation is likely to impede wind-borne pollen flow between plants. It is perhaps surprising that more tropical dioecious trees are not visited by vertebrate nectar feeders such as birds and bats; maybe the heavy energetic load of producing large, nectar-rich flowers suitable for vertebrate flower feeders is unsuccessful in combination with expensive fruit production.

**Oceanic islands** The highest levels of dioecy amongst species of flowering plants are to be found on oceanic islands of volcanic origin, such as New Zealand and Hawaii (see Table 8.1) (Godley 1975, 1979, Carlquist 1974). It is assumed that such islands have been colonised from neighbouring continents, and that intense speciation has subsequently occurred into different niches, and onto different islands, as on the Galapagos, made famous by Darwin. Baker (1955) suggests that most successful colonisers will be hermaphrodite and self-fertile. Most founders will arrive as single individuals, which could not set seed if self-sterile or unisexual (dioecious). It is thus remarkable that such areas should today be notable for the high levels of dioecious and gynodioecious species.

It is reasonable to suppose that oceanic islands have low biological diversity and incomplete vegetational cover (Carlquist 1974). Ample opportunities exist for genetic diversification of founders into different ecological niches to occur and for speciation to result. Such diversification will depend on high levels of genetic variability in populations and offspring families, which will result from outbreeding (Ch. 2). If founders are self-fertile inbreeders, secondarily arising outbred mutants are likely to be favoured. Self-fertile inbreeders seem unable to generate new gametophytic self-incompatibility systems (Ch. 6), and secondarily arising outbreeders have sporophytically controlled incompatibility systems, or are diclinous. Thus male-sterile mutants should be favoured after a self-fertile hermaphrodite founder has arrived on an oceanic island. Although reproductively inefficient, such diclinous mutants should be more outcrossed than their progenitors. Their variable offspring should be able to diversify into a wider variety of niches than the less variable offspring of the inbred progenitors, and thus their outbred parents will be fitter than selfing parents.

Although this theory has many attractions, it is by its nature difficult to prove. It also seems that many diclinous species on oceanic islands have close continental relatives (from which they presumably evolved) which are also diclinous. As discussed in Chapter 2, some evolutionists oppose the 'niche-width' concept of outbreeder fitness on the grounds of group selection. However, the high levels of dicliny in oceanic islands appear to be a very real phenomenon. Perhaps a new generation of population modelling, in which the complexities of niche diversity are more realistically described, will provide further insights into this problem, which lies at the heart of outbreeding theory.

### *The control of dioecy*

In all dioecious species, the expression of sex is basically controlled by special chromosomes, the sex chromosomes. One sex is usually heterogametic XY with two types of sex chromosome, and the other is always homogametic XX with one type of sex chromosome. As will be seen later, it is important to differentiate between the sex genotype, based on the sex chromosomes, and the sex phenotype, in which sex expression may be modified by environmental influences on levels of growth substances which transmit genetic messages. In most plants, like most animals, the male is heterogametic XY and the female homogametic XX, but there are some exceptions.

Because males are bound to cross with females, and the heterogametic sex should generate X gametes and Y gametes in equal numbers, sexes should arise from seed at a ratio of 1 : 1 (a frequency of 0.50 for each). This is a parallel to heteromorphy (Ch. 7) in which a heterozygote for a breeding system linkage group usually crosses with a homozygote so that frequencies of each morph from seed are 0.50. Exceptions to equal frequencies of sexes in the field and from seed do, however, commonly occur, and these will be discussed below.

Dioecy is secondary in origin, and in many species vestigial sex organs in unisexual individuals plainly indicate a hermaphrodite origin (Fig. 8.2). Unisexuality is thus caused by the suppression of one sex. A female is male sterile, and a male is female sterile. As will be discussed below, many diclinous populations represent intermediate conditions in the evolution of dioecy. Some populations have male-sterile, but not female-sterile, forms and these are gynodioecious, with hermaphrodites and females. Others (less frequently) have female steriles, but not male steriles, and these are androdioecious, with hermaphrodites and males.

In a dioecious species, we expect about half the individuals to be male sterile (female), and the other half to be female sterile (male). These primary sex characters are controlled by sexual sterility genes which are located on the sex chromosomes. One sex (most commonly female) is homogametic XX. In this, male sterility is homozygous. The other sex (most commonly male) is heterogametic XY. Female sterility must be carried on the Y chromosome. However, the heterogametic sex also has an X chromosome, which carries male sterility. To function correctly, the Y-carried female sterility must be dominant to the X-carried male sterility. Thus, it is reasonable to suppose that in most dioecious plants with heterogametic males, that femaleness (male sterility) arose first. Secondarily arising maleness (female sterility) could only be expressed if it became dominant to the femaleness already present. Gynodioecy, with females and hermaphrodites (no males) is indeed much more common than androdioecy today, and probably formed the main evolutionary pathway to full dioecy.

**Figure 8.2** Male flowers of the dioecious tetraploid form of the shrubby cinquefoil, *Potentilla fruticosa*. The female bears anthers which are sterile (×0.6).

In some plants the sex chromosomes are visually distinct from the other chromosomes (autosomes). Typically, they are smaller, with less active DNA (more heterochromatic), and although they may associate with each other at meiosis, they fail to form chiasmata, or chiasmata are limited to specific regions of the chromosomes. Often, the X and Y chromosomes can also be distinguished visually. The Y chromosome is frequently even smaller and more heterochromatic than the X chromosome.

However, in many dioecious plants (Table 8.5) it is difficult or impossible to identify the sex chromosomes, especially when all the chromosomes are very small, and segments of the sex chromosomes regularly form chiasmata at meiosis. If dioecy is to be maintained, crossing over cannot occur at meiosis in the heterogametic sex between those segments of the X and Y chromosomes that carry the primary and secondary sexual character loci. It is probable that these sex character linkage groups are protected from crossing over by one being inverted with respect to the other (e.g. Darlington 1939). Often, much of the length of the chromosomes differs by a large pericentric inversion. Crossovers that do occur within the inversion loop at meiosis will lead to non-viable products.

**Heterogametic females** In three genera of flowering plants, females of dioecious species, rather than males, have been shown to be heterogametic. These are *Fragaria* (Staudt 1952), *Potentilla fruticosa* (see Fig. 8.1; Grewal & Ellis 1972) and *Cotula* (Lloyd 1975a). Cytological differentiation of the sex chromosomes is absent from each of these examples. However, when females are crossed with hermaphrodite relatives, the offspring vary in sex expression. The reciprocal crosses, using males, yield offspring that are uniform in sex expression. This reciprocal difference is explained by the donation of either X or Y chromosomes to the offspring from the heterogametic female.

**Heterogametic males** There are three plausible explanations why most flowering plants have heterogametic males. Haldane (1922) has suggested that where one sex is weaker or less viable than the other, especially when hybrid, it is always the heterogametic sex that is so. Whereas the Y chromosome is always exposed to selection in the hemizygous (XY) state, recessive X-linked genes are protected from selection in the homogametic sex XX. However, these are exposed to selection in the heterogametic sex XY, which may suffer in consequence. There is presumably a greater premium on vigour and longevity in the seed-bearing female, and such vigour and longevity may be more associated with the homogametic sex.

Lloyd (1974b) suggests that when females are lost from gynodioecious populations, hermaphrodites (which may later evolve to maleness,

**Table 8.5** A list of some dioecious species indicating differentiation of sex chromosomes, and sex of the heterogamete (after Westergaard 1958, Williams 1964).

| Species | Differentiation of sex chromosomes | | Heterogametic sex |
|---|---|---|---|
| *Cannabis sativa* | XY | XX | male |
| *Humulus lupulus* | XY | XX | male |
| *H. japonicus* | XYY | XX | male |
| *H. lupulus* var. *cordifolium* | XXYY | XXXXX | male |
| *Rumex acetosella* aggregate | XY | XX | male |
| | 3XY | 4X | |
| | 5XY | 6X | |
| | 7XY | 8X | |
| *R. acetosa* | XYY | XX | male |
| *R. hastatulus* | XYY | XX | male |
| *R. paucifolius* | 3XY | 4X | male |
| *Silene latifolia* (*alba*) | XY | XX | male |
| *S. dioica* | XY | XX | male |
| *Asparagus officinalis* | XY | XX | male |
| *Elodea canadensis* | XY | XX | male |
| *Salix* spp. | XY | XX | male |
| *Populus* spp. | XY | XX | male |
| *Urtica dioica* | XY | XX | male |
| *Spinacia oleracea* | XY | XX | male |
| *Coccinea indica* | XY | XX | male |
| *Dioscorea* spp. | XO<br>XY | XX | male |
| *Fragaria elatior* | XX | XX | female |
| *Silene otites* | XX | XX | male and female? |
| *Valeriana dioica* | XX | XX | male and female? |
| *Sedum rosea* | | | ? |
| *Thalictrum fendleri* | | | male |
| *Mercurialis annua* | | | male |
| *M. perennis* | | | ? |
| *Vitis vinifera* | | | male |
| *Carica papaya* | | | male |
| *Bryonia dioica* | | | male |
| *Empetrum nigrum* | | | ? |
| *Ecballium elaterium* | | | male |
| *Potentilla fruticosa* | | | female |
| *Cotula* spp. | | | female |

Note where XY and XX are both given, the X and Y chromosomes can be distinguished visually (are heteromorphic). Where XX XX is given, the sex chromosomes can be distinguished from autosomes, but not told apart.

p. 304) can only regenerate new females if the hermaphrodites are heterogametic.

In the third case, if as it seems, dioecy usually evolves from gynodioecy, male sterility evolves first. Thus, female sterility must be dominant to male sterility, and the heterogametic sex must be female sterile, i.e. male. If this is the case, we might expect those plants with female heterogamety to have evolved from androdioecy, although we have no way of proving it.

If dioecy arose from gynodioecy in most cases (p. 302), X factors (male sterility) preceded Y factors. In any case, X factors will have preceded Y factors, for the heterogametic sex will always evolve after the homogametic sex. The Y chromosome has two functions: to carry female-sterile (male) genes, and to suppress male-sterile (female) genes. However, primary male characteristics (female sterility) and secondary male factors can also function if they are encoded on the other chromosomes (autosomes), as long as male-sterile (female) genes are suppressed by the Y chromosome. Even the female suppression function of the Y chromosome can be replaced by gene dosage effects between the X chromosome and the autosomes.

In some dioecious genera a clear evolutionary progression can be observed in the reduction in size and function of the Y chromosome. As the sex chromosomes evolved from autosomes, both are initially large and rich in active DNA. The X-linked factors, which were first to evolve, remain linked to the X chromosome, which remains relatively large and rich in DNA. However, the Y-linked factors have often been translocated to the autosomes, which thus behave as one large Y unit, balanced against the X chromosome. In time the Y chromosome may disappear completely (XO), or lose all function. In the absence of a functional and dominant Y chromosome, changes in balance between the X chromosomes and the autosomes may allow hermaphrodity to re-establish, particularly in polyploids.

This evolutionary progression is elegantly observed in the sorrels (*Rumex*). In the *R. acetosella* polyploid complex, diploids ($2n = 14$), tetraploids ($2n = 28$), hexaploids ($2n = 42$) and octoploids ($2n = 56$) occur. In females, the X chromosome occurs 2, 4, 6 and 8 times respectively; in males the X chromosome occurs 1, 3, 5 and 7 times, respectively, but the Y chromosome occurs only once (Löve 1944). This mechanism gives rise to some puzzling and unexplained questions about the formation of these polyploids. However, not until the decaploid level ($2n = 70$) is reached do plants with a single Y chromosome exhibit any female tendency. Here is an overdominant Y chromosome with very strong X suppressant features, which may be regarded as rather primitive.

In the field sorrel, *R. acetosa*, wild males have $2n = 15$, with two Y and one X chromosomes, whereas females have $2n = 14$ with two X chromosomes. In male meiosis, a trivalent forms with the X chromosome in a central position. It orientates on the spindle so that the X chromosome passes to one pole at anaphase I, and the two Y chromosomes pass to the other pole.* In artificially induced higher polyploids, in hybrids and in chromosome mutants, the number of Y chromosomes may vary in the male. This variation has no effect on the sex of the plant, which remains male, and is entirely controlled by the ratio of X chromosomes to autosomes (Ono 1935, Yamamota 1938, Zuk 1963).

| | | X | XX | XXX | XXXX |
|---|---|---|---|---|---|
| autosomes | 2*n* | male | female | | |
| | 3*n* | male | hermaphrodite | female | |
| | 4*n* | | male | hermaphrodite | female |

In polyploids, the number of Y chromosomes varies, but it has no effect on sex expression. The Y chromosome is inert, and the autosomes act collectively as the male factor.

The genus *Rumex* exhibits the full range of Y chromosome control in plants. An intermediate condition has been found in the American *R. hastulatus* by Smith (1963). Here the diploid sex control is by X chromosomes and autosomes, as in *R. acetosa*, but in artificial polyploids the Y chromosome modifies intersexes towards maleness. This condition has been interpreted as an intermediate stage in the progression towards total Y inertness. It is very similar to the control described for *Silene dioica* (Fig. 8.3) by Winge (1931), Westergaard (1940, 1946, 1948) and Warmke (1946). It contrasts with the report of Löve and Sarkar (1956) for the octoploid ($2n = 56$) *R. paucifolius* in which males are XY or XO, and females XX. Y chromosomes are inert, but sex chromosomes are only present disomically (twice) rather than polysomically as is the case in other polyploid *Rumex*.

## *Dioecy and polyploidy*

Dioecy is most common in diploids. Where a primitive dominant Y occurs, as in *Rumex acetosella* (above), dioecy also persists in polyploids. Where male control has passed to the autosomes, it is rare for dioecy to remain stable unless the sex chromosomes are disomic (only present twice) as in *Rumex paucifolius*. Gametes from the tetrasomic male XXYY or XXOO will tend to contain one X chromosome if there is regular

* A similar condition occurs in the Japanese hop, *Humulus japonicus*, although not in the common hop *H. lupulus*. In *H. lupulus* var. *cordifolius* the diploid male has four sex chromosomes $X_1Y_1X_2Y_2$ which form a chain quadrivalent at meiosis with regular $X_1X_2$ and $Y_1Y_2$ disjunction.

**Figure 8.3** Male (left), female (right) flowers of the dioecious red campion, *Silene dioica*. The central flower, genetically female, has been infected by the smut fungus *Ustilago violacea* which has caused it to produce stamens, the anthers of which are filled with smut spores. Photo by M. Wilson (×2).

disjunction, and thus trisomic intersexes XXXY or XXXO will commonly occur in the next generation. Hermaphrodites commonly occur amongst the offspring of artificially polyploid sorrel *Rumex acetosa*, or white campion *Silene latifolia* (*S. alba*).

It is often observed in diploid/tetraploid 'species pairs' that the diploid is dioecious, but the tetraploid is hermaphrodite, as in the crowberries *E. nigrum* (diploid) and *E. hermaphroditum* (tetraploid). In other genera, the relationship between polyploidy and sex expression can be more complex. In the annual mercuries (*Mercurialis annua*), the following is found (Durand 1963):

| 2× | 4× | 6× | 8×, 10×, 12×, 14× |
|---|---|---|---|
| dioecious | dioecious<br>androdioecious | dioecious<br>androdioecious<br>hermaphrodite | hermaphrodite |

In *Salix* (willows) all the species in a highly polyploid complex are dioecious (see Fig. 8.1), and we must presume that all still possess a highly dominant Y chromosome, of the *Rumex acetosella* type.

In two of the genera with heterogametic females, the usual situation is reversed, that is diploids are hermaphrodite and polyploids are dioecious. In *Potentilla fruticosa* (see Fig. 8.2), diploids ($2n = 14$) from the Alps, Pyrenees and North America are hermaphrodite, whereas tetraploids and hexaploids are dioecious, including the tetraploid British race (Elkington 1969). In the strawberries (*Fragaria*), all the diploid species ($2n = 14$) are hermaphrodite, and all the polyploids ($2n = 28, 42, 56$) are dioecious. The sex chromosomes cannot be identified in these genera. However, it seems likely that the sex chromosomes are disomic, with only a single pair occurring, even in high polyploids.

### *The stability of sex expression*

In many dioecious plants, such as *Salix* or *Carex dioica* and its relatives, dioecy is absolute, and intersexes or reversal of sex are never found. Other plants are fickle in their sex expression, which is subject to environmental influence. The sexual genotype, as evidenced by the chromosomes and inheritance of sex in the offspring, may not be the same as the sex expressed. In *Cannabis sativa*, spring-sown well-grown material expresses the genotypic sex, with males and females in stable 1:1 ratios. However, material grown at low light intensities or in conditions of water stress is sexually very unstable, and hermaphrodite, gynomonoecious and andromonoecious individuals all occur. Indeed, the sex genotype may even be reversed in expression, with genotypic females becoming male and vice versa (Schaffner 1921, 1923). Heslop-Harrison (1957) showed that sex expression in this species is governed by auxin concentrations in shoots. Auxin (indole-acetic acid) is an important plant growth substance, which responds to a number of environmental influences, such as the quantity and quality of light.

An opposite trend is found in many Cucurbitaceae (melons, squashes, cucumbers, etc.). These are typically monoecious, with both male and female flowers on an individual, and they are genotypically hermaphrodite. However, environmental conditions can result in only male flowers or only female flowers being produced, and thus they become effectively unisexual. Nitsch *et al.* (1952) showed that high temperatures and long days encourage the production of male flowers, whereas low temperatures and short days promote female flowers. Other plant growth substances such as maleic hydrazide, which encourages male sterility, are known to influence the expression of sex in addition to auxin. No doubt day length and temperature influence sex expression by causing variation in levels of several growth substances. In those plants that have evolved mechanisms whereby the expression of genotypic sex is mediated by growth substances, environmental influences will tend to override accurate genotypic expression. These include *Cannabis sativa*, hop (*Humulus lupulus*) and the sorrel *Rumex hastulatus* (Westergaard 1958). Dioecious species with no variation in sex expression presumably use other mechanisms to transmit the message of genotypic sex, which are less subject to environmental influence.

Freeman *et al.* (1981) have suggested that those monoecious (hermaphrodite) species which may become phenotypically (but not genotypically) dioecious, such as the Cucurbitaceae (see above) are more likely to become male under stress (e.g. xeric) conditions. Under mesic conditions, they tend to become female. This may tend to form a general rule for monoecious 'quasi-dioecious' species.

Sex expression can be also altered by disease. It has been known for many years that infection of females of the red and white campions (*Silene dioica* and *S. latifolia* (=*S. alba*) by the smut *Ustilago violacea* induces the formation of anthers. These may produce some pollen, but they are mostly filled with smut spores, thus promoting dispersal of the smut (Fig. 8.3) (Westergaard 1953, 1958). The smut, which is very common in the British Isles, causes species that are usually fully dioecious to become androdioecious.

**The occurrence of homogametic males (YY)** In cases (listed above) in which the phenotypic expression of genotypic sex is unstable, it is possible for crosses to occur between heterogametic (XY) hermaphrodites. This will result in a proportion of the offspring being YY homogametes. In various species of meadow-rue, *Thalictrum* (Kuhn 1939), and in cultivated *Asparagus officinalis* (Rick & Hanna 1943) such plants are male, and do not differ in appearance from XY males. In such cases, genes that are vital to the well-being of males do not occur on the X chromosome, which must be primarily concerned with female determination (male sterility).

In other dioecious species, such as *Mercurialis annua* (Kuhn 1939, Gabe 1939, Durand 1963), YY plants are weak and rather sterile; in these the X chromosome must contain genes which are important to the metabolic efficiency of the male, and indeed to its sexual function.

## GYNODIOECY

So far, this chapter has been concerned with full dioecy, in which genets are usually totally male, or totally female. When the genotypic sex is phenotypically unstable, hermaphrodites that are capable of selfing occur. When only some of the flowers on a female genet become hermaphrodite, they are termed subgynoecious [♂] [⚥ ♀], or subandroecious [♂ ⚥] [♀] if only the male sex is affected. These conditions usually arise as the result of unstable sex expression in a genetically dioecious plant. When all the flowers on a genet become hermaphrodite, they are termed androdioecious [♂] [⚥] or gynodioecious [⚥] [♀]. As we have seen, these conditions can also result from unstable sex expression in a genetically dioecious plant. However, gynodioecy, in particular, is a common diclinous condition in its own right. It is more frequently the result of the regular (polymorphic) occurrence of male sterile genes in otherwise hermaphrodite populations. It can often also form a successful intermediate stage in the evolution of full dioecy. Because female steriles (males) are absent, full dioecy (genetically) is also absent.

The remaining conditions of sexual expression, which are listed at the start of this chapter, are gynomonoecy [⚥ ♀] and andromonoecy [♂ ⚥]. These are not diclinous conditions, because they do not involve genets which exhibit only a single sex. Thus, in no case are individuals obligatorily outcrossed. Rather, they represent intermediate conditions between within-flower hermaphrodity and monoecy. They may be important to the plant with respect to how much within-flower selfing, rather than outcrossing and geitonogamy occurs (Primack & Lloyd 1980) (Chs. 1 and 9), but they will not be considered further here.

It should, however, be pointed out that in subgynoecious [♂] [⚥ ♀] and androdioecious [♂] [⚥] populations, the dicliny only operates with respect to *males*. All females are also male, and can thus potentially be self-fertilised. Because the next generation is transmitted via the female, it is *female dicliny* which is genetically important. With female dicliny, the offspring of unisexual females must be outcrossed. Conditions with diclinous females are dioecy [♂] [♀], gynodioecy [⚥] [♀] and subandroecy [♂ ⚥] [♀]. As the last condition usually occurs in genetically dioecious species, I will restrict the following discussion to gynodioecy.

Gynodioecy has been regarded as an important breeding system since Darwin (1877). It has a rather different distribution from dioecy. In temperate floras, such as those in Europe it is roughly twice as frequent as dioecy. Delannay (1978) shows that 7.5% of the species native to Belgium and Luxemburg are gynodioecious, and lists 223 European gynodioecious species, belonging to 89 genera and 25 families. Similar surveys do not seem to have been made for other temperate floras, such as that of North America. Gynodioecy is a less significant outbreeding mechanism than dioecy in the well-studied oceanic island floras of New Zealand and Hawaii, but it is found in at least seven genera in Hawaii and eleven in New Zealand. In New Zealand, several important genera have many gynodioecious species, and *Hebe*, *Pimelea*, *Cortaderia* and *Fuchsia* amongst others have attracted attention (Ross 1978, Godley 1979). It may be that the occurrence of gynodioecy on oceanic islands can be usefully explained by similar arguments to those adduced for dioecy; i.e. pressures favouring outbreeding in self-fertile colonants.

However, in strict contrast to dioecy, gynodioecy appears to be very rare in tropical forests. No examples were discovered by Bawa (1979) in a sample of 309 species of forest tree in Costa Rica. This is consistent with the arguments of Bawa (1980), and other arguments concerning the high frequency of dioecy amongst forest trees. If hermaphrodites have their female function disadvantaged by male load, and by self-pollination, hermaphrodites should disappear from gynodioecious populations, leading to full dioecy.

### *Stable and unstable gynodioecy*

Ross (1978) and Bawa (1980) distinguish between stable and unstable gynodioecy. In stable gynodioecy, male sterility is usually controlled by at least two unlinked genetic factors, one of which is often cytoplasmic; it is never controlled by a single nuclear locus (Lewis 1941). Gynodioecy is widespread within a species, and often constant for that species; within a population, frequencies of hermaphrodites and females show little variation with time. As Lewis points out, stable gynodioecy is a feature of certain families in which dioecy is rare. Stable gynodioecy can be considered as a well-established breeding system and not an intermediate step in the evolution of dioecy in such families as the Lamiaceae (mints and relatives), Asteraceae (daisies and relatives) and Dipsacaceae (scabiouses and relatives).

In unstable gynodioecy, male sterility is controlled by a single factor, which may be under nuclear or cytoplasmic control. It is thought to represent an intermediate condition in the evolution of dioecy when under nuclear control. If unsuccessful, male-sterile genes will not become permanently established at stable frequencies in hermaphrodite populations. If successful, pressures may arise to encourage hermaphrodites to become male, and thus for full dioecy to evolve, as will be more fully argued below (Fig. 8.4). Cytoplasmic male-sterile genes cannot become linked to a sex chromosome, and thus such genes cannot lead to full dioecy. However, they may lead to stable gynodioecy if associated with nuclear male-sterility genes (Fig. 8.4).

Male sterility, whether cytoplasmic or nuclear, is a common phenomenon amongst flowering plants, and thus there is no shortage of variation on which evolution can act to give rise to gynodioecious populations. Male sterility is not always under genetic control; flowers that produced abnormally early or late in a hermaphrodite flowering season are frequently male sterile, but this condition is not inherited. Most male-sterile mutants do not persist in populations, and as the male-sterility gene does not become polymorphic, we do not consider such mutants to confer gynodioecy. However, such male steriles may be selected for plant breeding purposes, for they offer ideal opportunities to obtain outcrossed (hybrid) seed. Most maize seed, and much seed of a number of other crops including rice, barley and tomatoes is nowadays $F_1$ hybrid seed obtained from true-bred male-sterile maternal lines grown together with true-bred hermaphrodites which act as fathers (Simmonds 1976). Many male-sterile genes can be accumulated, from which a suitable strain can

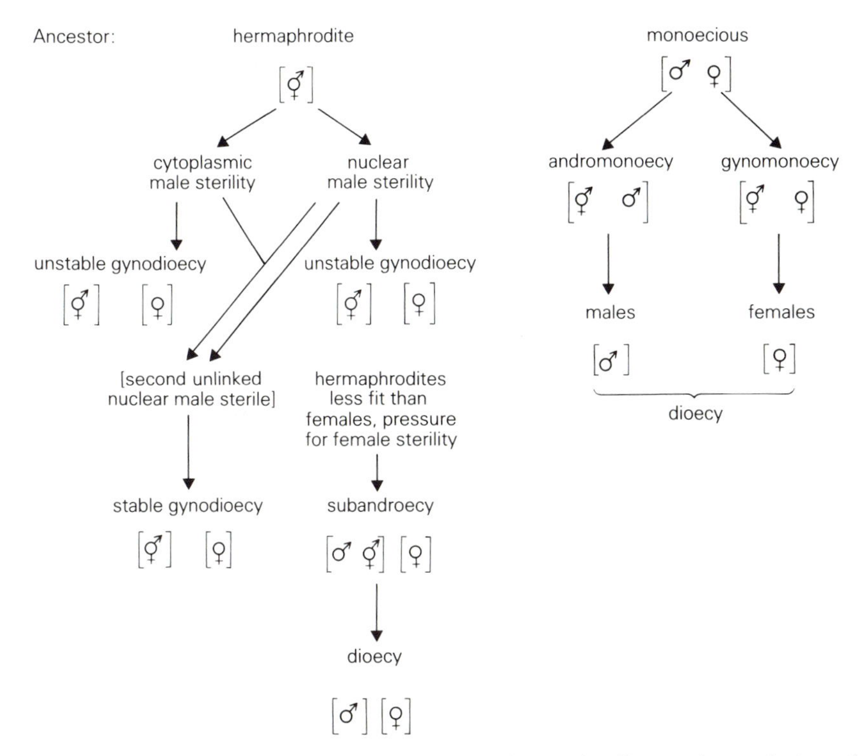


**Figure 8.4** The most important suggested pathways leading to the evolution of gynodioecy and dioecy (based on Ross 1978, 1980 and Bawa 1980).

be chosen. For instance, at least 37 separate male-sterile nuclear genes are recognised in tomatoes (Clayberg *et al.* 1966).

Male-sterile genes are inevitably passed to the next generation by the mother. If the inheritance is cytoplasmic, all the offspring of the male-sterile mother will be male sterile (female). If the inheritance is nuclear, and male sterility is dominant, the mother is bound to be heterozygous at this locus; half the offspring will be hermaphrodite, and half will be female. If the inheritance is nuclear and recessive, the offspring will either be half hermaphrodite and half female, or all hermaphrodite, depending on whether the father is homozygous or heterozygous (Fig. 8.5). Unlike cytoplasmically inherited male sterility, the offspring of a male-sterile female are never all male sterile.

It is frequently found that interspecific hybrids differ between reciprocal crosses with respect to male sterility, as in *Epilobium* (Michaelis 1954). Such non-repricocity is explained by interactions between hybrid nuclei and non-hybrid (maternal) cytoplasm. This forms a general indication that male fertility is reliant on highly specific cytoplasmic factors. Presumably, these are abnormal in cytoplasmic male-sterile mutants.

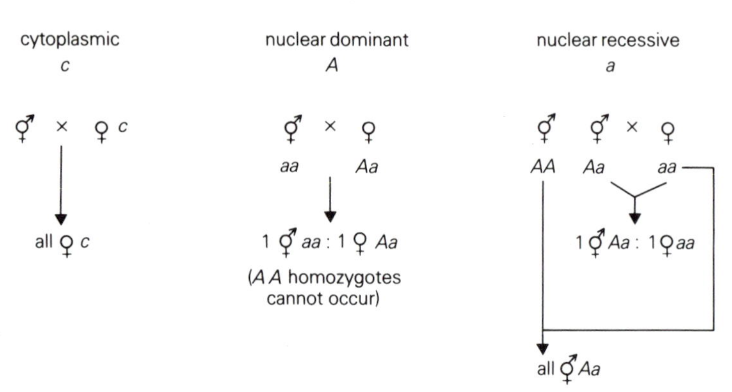


**Figure 8.5** Inheritance of single-factor male sterility.

## *The genetic control of stable gynodioecy*

Early examples of stable gynodioecy which were studied by Darwin (1877) and Correns (1928) seemed to be controlled entirely by cytoplasmic factors, in such a way that female mothers always gave female offspring, and hermaphrodite mothers always gave hermaphrodite offspring. However, Ross (1978) suggests that a reappraisal of the original data in the light of modern knowledge indicates an interaction between nuclear and cytoplasmic factors. Ross argues that gynodioecy will be favoured when females make fitter mothers than hermaphrodites, for reasons that will be examined below. As a result, he considers that hermaphrodites will be relatively unfit mothers, and that female-sterile mutants, which render hermaphrodites male, will be favoured, leading to the evolution of full dioecy (Fig. 8.4). However, dioecy can only evolve when female (male-sterile) factor(s) segregate together when linked on a single sex (X) chromosome. If female factors segregate independently, they will generate hermaphrodites (Table 8.6), and dioecy cannot be fixed. Thus, in stable gynodioecy, Ross predicts that unlinked nuclear genes, or the interaction between one or more nuclear genes and a cytoplasmic gene (which are by their nature unlinked), will always occur. Otherwise, evolution should proceed towards full dioecy as female-sterile genes arise. However, full dioecy will never be achieved by cytoplasmic factors alone, as they cannot form a sex chromosome.

One-locus nuclear male-sterile systems which interact with a cytoplasmic gene for male sterility have indeed been reported, as in *Chenopodium quinoa* (Simmonds 1971). Two-locus nuclear male-sterile systems with cytoplasmic interactions are probably more common. A number of examples are quoted by Ross (1978), as for instance in marjoram, *Origanum vulgare*. A recent case has been analysed in my laboratory by David Stevens (Stevens & Richards 1985) for the meadow saxifrage,

*Saxifraga granulata.* This species is usually hermaphrodite, but Stevens has discovered three gynodioecious populations in upland English hill pastures. This complex system of control of male sterility probably involves a cytoplasmic gene, and two unlinked nuclear loci showing complementation:

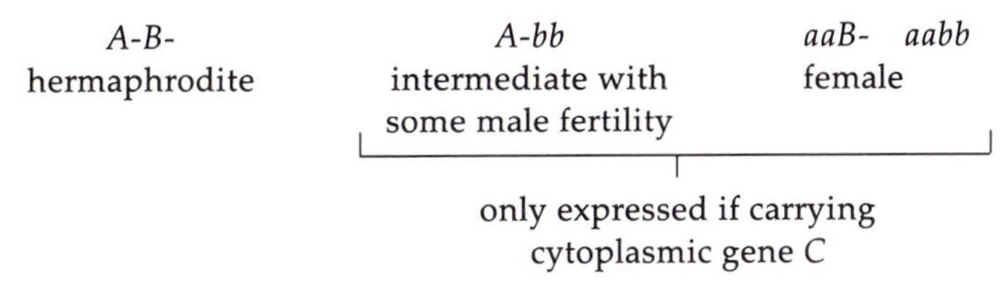


Thus, *A* and *B* interact to override the cytoplasmic gene, but if only *A* or only *B* or neither dominant is present, there is some, or total, male sterility in the presence of *C*, but not in its absence. Offspring sex ratios of controlled crosses of females and hermaphrodites, and of controlled selfs of hermaphrodites fit this model very well.

The commonest system seems to be a two unlinked locus nuclear control of male sterility, in which some complementation also characteristically occurs. Typically, selfed hermaphrodites yield 13:3 hermaphrodite:female ratios. This mode of inheritance is explained in Table 8.6.

In the ribwort plantain, *Plantago lanceolata,* dominant complementation occurs, so that 15:1 ratios are obtained from crosses between hermaphrodites (this species is self-sterile). Both alleles controlling male sterility are recessive. This mechanism may be uncommon, as it will tend

**Table 8.6** Explanation of 13:3 ratio of hermaphrodites: females in families from hermaphrodite selfs in stable gynodioecious populations.

| | Dihybrid *SsMm* selfed | | | |
|---|---|---|---|---|
| gametes | SM | Sm | sM | sm |
| SM | SSMM | SSMm | SsMM | SsMm |
| Sm | SSMm | SSmm | SsMm | Ssmm |
| sM | SsMM | SsMm | *ssMM* | *ssMm* |
| sm | SsMm | Ssmm | *ssMm* | ssmm |

To explain 13:3 ratios, it must be assumed that two unlinked loci control male sterility in which a dominant allele *M* and a recessive allele *s* interact to give sterility. The other phenotypes MS, mS and ms give male fertility, and will be hermaphrodite. To arrive at this result, two true breeding lines are obtained (a) which only yields hermaphrodites by removing all females from the line until only alleles *S* and *m* are left, and (b) which yields only females by crossing females with hermaphrodite siblings until only alleles *s* and *M* are left. These lines are then crossed to yield the dihybrid of known genotype *SsMm* which is then selfed. Stable gynodioecious systems which nearly always give this result are unable to progress to subandroecy and full dioecy as the female determinants are not linked on a single sex chromosome.

**Figure 8.6** Gynodioecy in *Plantago lanceolata* (ribwort plantain). Inflorescences in this species are protogynous, the flowers opening from the bottom of the spike. The central inflorescence is in the female stage. The other two inflorescences are in the male stage (although with some flowers still in the female stage at the top). That on the left has yellow, narrow, collapsed anthers, which have no pollen, and is a female. The inflorescence on the right has plump creamy-white anthers full of pollen, and is a hermaphrodite (×1.5).

to produce few females. Krohne *et al.* (1980) report female frequencies of only 0–6% from inland Californian populations, although in wetter coastal areas females are more common at 2 to 31% of populations. A population at Newcastle, UK had 30% and 21% females in 1983 and 1984, respectively. Females in this species have well-formed anthers, which are, however, yellow rather than cream in colour because they are empty (Fig. 8.6). Experience resulting from microscopic examination allows females to be assessed with a high degree of certainty from anther colour alone.

### *The evolution of gynodioecy*

Gynodioecy is favoured when females are fitter mothers than hermaphrodites. They may produce more seed; or they may produce better quality seed, and thus more viable offspring than do hermaphrodites. Selection for females will be frequency dependent; if hermaphrodites become too infrequent, female fitness will suffer from a lack of pollen. Most gynodioecious plants are self-compatible, but are more than 50% outcrossed when insect-pollinated herbs (Ross 1978). A notable exception is *Plantago lanceolata* (Ross 1973) which is self-incompatible, and largely wind pollinated (Stelleman 1978). Self-incompatible species will not be more outcrossed when female than when hermaphrodite. However, females may benefit by not having stigmatic sites taken by nonfunctional selfed pollen.

Although usually self-compatible, and thus liable to selfing, inbreeding depression and low genetic variability, gynodioecious species are usually substantially outcrossed as hermaphrodites. If they are highly selfed, females are unlikely to receive adequate cross-pollination. Estimates of outcrossing onto gynodioecious hermaphrodites vary from 83% in *Cirsium palustre* (Correns 1916b) to only 25% in *Cortaderia richardii* (Connor 1973). Valdeyron *et al.* (1977) have shown from 51 to 90% outcrossing in different populations of the culinary thyme, *Thymus vulgaris* in France.

Ross (1978) gives two reasons why females may be fitter mothers than hermaphrodites:

(a) Selfed offspring of hermaphrodites will be less variable than outcrossed offspring of females and may suffer from inbreeding depression.
(b) Females avoid the energy drain of producing stamens and pollen.

Although it is usually found that females *are* better mothers than hermaphrodites, producing more seeds of better quality, it can be difficult to separate the effects of post-zygotic vigour (heterosis) from prezygotic maternal resource allocation (removal of male load). To give an example, better seed-set in females will usually be interpreted as the result of greater maternal resource allocation in females rather than hermaphrodites. However, if there is cytoplasmic inheritance of male sterility, females will always give females, whereas hermaphrodite mothers will always give hermaphrodites. Thus, good seed-set in females may result from heterotic vigour expected in an outbred line of females.

Superior maternal function in females may be expressed in a number of ways, which are summarised by Godley (1979), and in Table 8.7. It will be seen that an exception occurs for seed-set in female *Saxifraga granulata*, a case that will be discussed in more detail below. Females may show superior performance in other attributes, for instance ovule size in *Cortaderia richardii*, fruit-set in *Pimelea prostrata* or seed germination in *Fuchsia* (Arroyo & Raven 1975).

Lewis (1941), Ross (1978) and others have pointed out that females normally suffer a disadvantage against hermaphrodites, in that they cannot act as male parents to hermaphrodites, but hermaphrodites can act as male parents to females. However, this disadvantage does not apply to male sterility under cytoplasmic control (see above). Also, Lloyd (1975a) has pointed out that the disadvantage is frequency dependent. If females are rare, hermaphrodites gain little advantage by acting as male parents to them. If females are common, hermaphrodites can spread more quickly.

Charlesworth and Charlesworth (1978) predict equilibrium proportions ($p$) of females by:

$$p = \frac{(f + 2sd - 2)}{2(f + sd - 1)}$$

**Table 8.7** Reproductive performance of females in gynodioecious species, expressed as a ratio to hermaphrodites (based on Godley 1979 and Stevens 1985).

| Species | | Flowers/ plant | Ovules/ flower | Seed/ ovule | Seed/ fruit | Seed/ plant | Weight/ seed |
|---|---|---|---|---|---|---|---|
| *Thymus vulgaris* (Assouad *et al.* 1978) | | 1 | 1 | 2.24–3.25‡ | 2.24–3.25*† | 3.14 | – |
| *Plantago lanceolata* (Van Damme 1984a) | MS1 | 1.69* | 1 | 1‡ | 1 | 1.69‡ | 1.15* |
| | MS2 | 1.69* | 1 | 0.70‡ | 0.70* | 1.18‡ | 0.90* |
| *Stellaria longipes* (Phillip 1980) | | 1.90† | 1 | 0.92‡ | 0.84 | (1.60) | – |
| *Geranium sylvaticum* (Vaarama & Jskelinen 1967) | | 0.69 | – | – | 1 | (0.67) | > |
| *Hirschfeldia incana* (Horovitz & Beiles 1980) | | – | 1.08 | 1.03‡ | 1.11 | – | 1.17 |
| *Leucopogon melaleucoides* (McCusker 1962) | | – | 1 | 1 | 1 | – | – |
| *Origanum vulgare* (Lewis & Crowe 1956) | | – | 1 | 1.19‡ | 1.19* | – | – |
| *Iris douglasiana* (Uno 1982) | | – | 1 | 1‡ | 1 | – | 0.95 |
| *Saxifraga granulata* (Stevens 1985) | | 1 | 1 | 0.57* | 0.58* | 0.54‡ | 1.17* |

* Stated to be significant at $p < 0.05$ (results stated to be insignificant recorded as 1, otherwise reported as given by the authors).
† From cultivated plants.
‡ Calculated directly from other values in this table.
– No data.

where $f$ is seed-set: per genet in females compared with hermaphrodites, $s$ is the proportion of seeds that result from self-fertilisation (on a scale from 0 to 1) and $d$ is the inbreeding depression, expressed as:

$$1 - \frac{\text{(the fitness of the offspring of hermaphrodites)}}{\text{(the fitness of the offspring of females)}}$$

For a male-sterile gene to spread, they predict that $f > 2 - 2sd$ is necessary for nuclear-inherited male sterility, but that $f > 1 - sd$ is required for cytoplasmically inherited male sterility.

Thus, for nuclear male sterility where selfing does not occur in hermaphrodites, $f$ should exceed 2 (females will have twice the seed-set of hermaphrodites). In practice, this rarely seems to be the case (see Table 8.6, where the highly outcrossed *Thymus vulgaris* is the only example), and in fact $f$ often only equals 1, or even less. Where $f$ equals 1, for nuclear control, $sd$ must be 0.5, and thus levels of selfing and inbreeding depression will both be high. For instance, if $s$ (the proportion of selfing) equals 0.7 (which as we have seen is rarely the case), $d$ will also equal 0.7. That is to say, the selfed offspring of hermaphrodites will only have 30% the fitness of the offspring of females. The mean fitness of the offspring of hermaphrodites will be 50% that of females.

However, for single-factor cytoplasmic control of male sterility (which is likely to give unstable gynodioecy), female seed-set need not be superior to hermaphrodite seed-set ($f = 1$ or less) for females to survive. Indeed where $sd$ equals 0.5 (the mean fitness of the offspring of hermaphrodites is 50% that of the offspring of females), females only need to set half the seed of hermaphrodites to survive ($f = 0.5$).

Thus, when hermaphrodites are highly selfed, or pollen flow to females is otherwise restricted (for instance by low levels of visiting by pollen-eating insects to females), gynodioecy is most likely to persist when it is under cytoplasmic control. Male sterility may first become established in many populations through cytoplasmic mutants; other modifying genes which are under nuclear control may then be favoured when they arise, if they allow the outcrossing/selfing versatile gynodioecious system to stabilise. Cytoplasmically controlled systems may be particularly favoured during the establishment of the rare gynodioecious, self-incompatible species (p. 306). For these, $sd$ must be 0, and females will establish only when they have twice the seed-set of hermaphrodites for nuclear systems, but when they have the same seed-set as hermaphrodites for cytoplasmic systems.

Unfortunately, we do not have an algebraic estimate of equilibrium levels for $f$ and $sd$ for cytoplasmic–nuclear systems. However, we can assume for these that when $sd$ equals 0 (no inbreeding depression in the offspring of hermaphrodites) $f$ should lie between 1 and 2.

Such a cytoplasmic-nuclear system is found in the meadow saxifrage, *Saxifraga granulata*, and various reproductive attributes of hermaphrodites and females of gynodioecious populations of this species have been estimated by Stevens (1985). Estimates of fecundity were made in the field, and other attributes were measured after cultivation in constant conditions (Table 8.8). We can conclude that, at least for one population, females have small anthers, shorter filaments and smaller petals than hermaphrodites, and produce no pollen. These savings in resource in

**Table 8.8** Reproductive attributes of a gynodioecious population of *Saxifraga granulata* (only characters that showed statistically significant differences between hermaphrodite and female mothers are listed) (from Stevens 1985).

| Character means | Hermaphrodite (parent) | Female (parent) | Female : hermaphrodite ratio |
|---|---|---|---|
| Parental characters in the field | | | |
| anther (mm) | 1.25 | 0.97 | 0.776 |
| anther width (mm) | 1.05 | 0.70 | 0.666 |
| filament length (mm) | 12.91 | 11.83 | 0.916 |
| petal length (mm) | 11.09 | 10.61 | 0.957 |
| petal width (mm) | 5.91 | 5.50 | 0.931 |
| seed/ovule ratio | 0.381 | 0.217 | 0.570 |
| seed/capsule ratio | 203.68 | 118.85 | 0.583 (= $f$) |
| seed volume ($mm^3 \times 10^{-3}$) | 4.24 | 4.96 | 1.170 |
| style length (mm) | 4.27 | 4.83 | 1.131 |
| Offspring characters in cultivation | | | |
| final seed germination (%) | 44.0 | 57.6 | 1.309* |
| survival of seedlings to flowering (%) | 59.3 ($n$ = 123) | 76.3 ($n$ = 156) | 1.287 |
| proportion of seedlings flowering after 1 year (%) | 75.3 ($n$ = 73) | 84.0 ($n$ = 119) | 1.116 |
| weight of oversummering bulbils of seedlings from different mothers (g) | 1.49 | 1.85 | 1.242 |

* This difference was not statistically significant at 5%.

comparison with hermaphrodites probably allow females to invest more energy in seed size than do hermaphrodites. However, the larger seeds of females might also be a function of their poorer seed-set, which could allow more energy to be invested per seed, or a combination of the two. It is likely that the superior fitness of the offspring of females, as shown in the better germination of seed from female mothers, the better survival of offspring from female mothers, and the more rapid flowering of offspring from female mothers, is at least partly due to larger seeds being produced by female mothers in comparison with hermaphrodite mothers. All these superior fitness attributes could also be due to heterosis (inbreeding depression in the selfed offspring of hermaphrodites), and perhaps both functions contribute to the superior fitness of the offspring of females. Certainly, we must assume that the larger oversummering bulbils produced by the offspring of females (regardless of the sex of the offspring itself) is a function of heterosis, and not of seed size, in the mothers. It is likely that bulbil size is an important fitness attribute, for the vegetative offspring of large bulbils are more likely to flower next year than are the vegetative offspring of small bulbils.

The best measure of $f$ (relative seed-set) in this population is 0.583 (see Table 8.8). The poor seed-set of females in comparison with hermaphrodites is the lowest yet recorded in a gynodioecious species (see Table 8.7), and must present a formidable obstacle to female survival. It may explain why gynodioecy is an unusual condition in this species. It suggests that hermaphrodites may be highly selfed, although Stevens has no direct measure of the amount of selfing that takes place. It is unlikely that the slightly smaller flowers cause much disassortive flower visiting between hermaphrodites and females. Most flower visitors are small unspecialised pollen eaters, and it is possible that pollen flow to females is restricted by the absence of pollen in females. However, females have quite conspicuous stamens, in common with nearly all gynodioecious, and many fully dioecious, species. It probably pays females to retain 'dummy' stamens of this kind, if visited by pollen eaters.

In the three British populations of *S. granulata* with gynodioecy, females occur at rather low frequencies (0.25, 0.20 and 0.04 of flowering stems). Intermediates (p. 305) with a mixture of male-fertile and male-sterile flowers are scarcer, and have not been found at a frequency above 0.04. We do not know whether male sterility is at equilibrium in these populations. If we assume that it is, we can attempt to ascertain whether conditions that would favour the spread of male sterility operate in Stevens' population, following the Charlesworths' models (1978). Unfortunately, neither of these models is appropriate to a cytoplasmic–nuclear mode of inheritance, as occurs in this species, so only rough approximations can be made.

As we have seen, we can make an estimate of $f$ (relative fecundity) in wild parents of 0.583.

We can obtain an estimate of $sd$ (inbreeding depression) from the relative performance of seedlings in cultivation from the following attributes:

| | | |
|---|---|---|
| differential seed germination | 0.764 | product of these is 0.478 |
| differential seedling survival | 0.777 | |
| differential bulbil weight of seedlings | 0.805 | |

Then $sd = 1 - 0.478 = 0.522$.

For the Charlesworths' model for nuclear inheritance, $f > 2 - 2sd$, $sd$ must be greater than 0.708, if $f = 0.583$. This is a very high estimate, much higher than the value of 0.522 obtained.

For the model for cytoplasmic inheritance, $f > 1 - sd$, $sd$ must be greater than 0.417, as is the figure obtained, 0.522.

If we assume that for a cytoplasmic–nuclear system of control with $f$ equal to 0.583 an estimate for $sd$ will lie between 0.417 and 0.708, which it does, it is arguable that conditions operate which allow male sterility to persist at equilibrium in this population of *Saxifraga granulata*. Despite the unusually high disadvantage in seed-set that females carry, females may be able to persist due to the high level of inbreeding depression in the offspring of hermaphrodites, and the partially cytoplasmic system of control of male sterility which will allow female heterosis to be partially heritable. We can predict high levels of selfing in hermaphrodites, both from the high estimate of inbreeding depression $sd$, and the poor seed-set on females (which must be outcrossed to set seed). We can also predict that female frequencies will remain low, and the male-sterile gene will not spread ($p$ will be low). As already pointed out, female frequency will be restricted by availability of pollen, and by the frequency-dependent advantage to hermaphrodites. From the equation on pp. 307–8, female frequency $p$ is likely to be greatest when the seed-set of females relative to hermaphrodites, $f$, and hermaphrodite reduced fitness $sd$ attributes are greatest. The low values for $f$ in *S. granulata* ensure that $p$ will never be very high, whatever the genetic control of male sterility, and however great the reduced fitness $sd$ of hermaphrodites. Indeed, it is likely that male sterility could only have become established in this species through a cytoplasmic mechanism evolving first.

These analyses have reduced the impact of the earlier suggestions of Ho and Ross (1973) who show that overdominance (heterozygote advantage) can maintain gynodioecy in a stable condition even when under single-locus nuclear control. They hypothesise a dominant allele *M* for male sterility, in which the heterozygote shows better reproductive performance than either *MM* (female) or *mm* (hermaphrodite) homozygotes. There is some evidence for such a mechanism in marjoram, *Origanum vulgare* (Lewis & Crowe 1956), although this has a complex cytoplasmic/nuclear control.

Ross (1978) shows that, in the absence of such heterozygote advantage, one-locus gynodioecious systems will be unstable. If male sterility is advantageous, females are bound to be fitter, by virtue of better seed-set and/or fitter offspring than hermaphrodites (p. 306). Consequently, hermaphrodites will be favoured which direct more resource into male function, and thus maximise their fitness as parents through pollen rather than through seeds. This selective trend leads inexorably to subandroecy and to full dioecy.

## ANDRODIOECY

In contrast to gynodioecy, androdioecy, with males and hermaphrodites, seems to be a very rare phenomenon, although reported long ago by Correns (1928) in *Pulsatilla* and *Geum*. Androdioecy is not strictly an outbreeding mechanism, for all female function occurs in hermaphrodites, and pollen is not more likely to be outcrossed if it comes from a male. It is just as readily outcrossed from another hermaphrodite. However, it is possible that androdioecy may be favoured in an obligately outcrossed species in which seed-set and fitness is restricted by the poor flow of pollen between genets (Lloyd 1975a). The disadvantage of reduced frequency of female function might be outweighed by the advantage of increased seed-set of hermaphrodites due to the increased frequency of male function. This seems not to have been shown in practice. It may be that other apparently androdioecious species have sterile pollen in 'hermaphrodites' and are thus functionally dioecious, as Kevan and Lack (1985) have shown for *Decaspermum parviflorum* in Indonesia.

## SUBANDROECY, SUBGYNOECY AND POLYGAMY

As stated above, subandroecy seems to represent an intermediate position between unstable monofactoral gynodioecy, and full dioecy. Females occur together with hermaphrodites with reduced female function, so that some flowers on hermaphrodites are entirely male. Another cause of subandroecy is unstable expression of genetically (X/Y) controlled full dioecy due to environmental modification (p. 299). Subandroecy will occur in a dioecious species, particularly when flowering during unusual day-lengths, when male plants produce some gynoecia. Similarly, abnormal environmental stimuli may cause genetically female plants to produce some stamens (subgynoecy), or both males and females may show unstable sex expression (polygamy).

We have seen that male sterility almost always becomes established before female sterility, because females are outcrossed, but males do not result in more outcrossing. Unstable gynodioecy will favour subandroecy, with reduced female function in hermaphrodites. It follows that subgynoecy is unlikely to form an intermediate stage between hermaphrodity and dioecy, for it would follow from androdioecy, which rarely if ever occurs (see Fig. 8.4). Subgynoecy and polygamy usually result from unstable sex expression in a genetically dioecious species.

Subandroecy may be obvious, as in the New Zealand umbellifer *Gingidia montana*, in which primarily male plants often produce a few fruits (Webb 1979a, Lloyd 1980a) or it may be cryptic. Arroyo and Raven (1975) showed that two American *Fuchsias*, *F. thymifolia* and *F. microphylla*, are not gynodioecious as they superficially appear to be. In both species, about 90% of the apparently hermaphrodite flowers are female sterile (although bearing full-formed gynoecia), and are thus male. Cryptic subandroecy may also occur in apparently dioecious species. In the familiar pest creeping thistle, *Cirsium arvense*, Q. O. N. Kay (personal communication) finds that occasional male-fertile florets bear achenes, although most are male only. Fully female florets never show sporadic male fertility, but in South Wales, Kay has encountered occasional clones which appear to be fully hermaphrodite. This condition is probably best described as polygamous, although variably male females are absent. In the similar condition termed trioecy (which is very rare), males, females and true hermaphrodites coexist.* Trioecy probably operates when both male sterile and female sterile mutants are polymorphic in a population, but unlinked, so that when both are missing, hermaphrodites occur. It may represent a stage in the evolution of full dioecy (see Fig. 8.4), but is unlikely to persist once male-sterile and female-sterile factors become linked. In contrast, subandroecy and polygamy typically show unstable sex expression, with both male and hermaphrodite flowers on the same plant.

A quite different condition has been described by Lloyd (1972a,b, 1975b) for New Zealand species of *Cotula*. In this member of the Compositae, florets are usually unisexual; in hermaphrodites the outer florets in a head are generally male and the inner female (monoecious). When bisexual (hermaphrodite) florets occur, they do so peripherally; the inner florets are always female (gynomonoecious).

All gradations between gynomonoecy and full dioecy are reported in *Cotula*. In some plants the number of peripheral male florets becomes very reduced and inconstant; in others the number of central female florets becomes reduced and inconstant. Plants with few or no female florets tend to have longer florets, perhaps as an aid to pollen transfer. Although a great deal of variation in sex expression occurs between populations, even within a species, increasingly female plants are usually accompanied by correspondingly increasingly male plants. Thus, gynodioecy and androdioecy do not usually occur, nor do subgynoecy or subandroecy. Intermediate states between monoecy and dioecy are best called subdioecious, in which there is some inconstancy in the expression of both males and females. Inconstancy of sex expression is not under environmental influence, and often as much inconstancy can be found at the same time within one individual as in the whole population. Some apparently dioecious populations in one year may prove inconstant in sex control the next year. This has allowed Lloyd to suggest (1980a) that

* Trioecy and polygamy are described by Darwin (1877) for the common spindle tree, *Euonymus europeaus*, although Webb (1979b) concludes that populations introduced into New Zealand are more properly termed gynodioecious. Males are reasonably, but variably, female fertile.

dioecious *Cotula* have reverted to monoecy on at least three occasions, perhaps in response to the rarity or extinction of one sex. Populations of alpine scree *Cotula* are frequently very small and subject to catastrophe, and individual clones can spread over a large area and live to a considerable age. Enforced unisexuality in populations would encourage a sexually inconstant system.

In *Cotula* there is a conventional XY genetic control of sex, so that crosses between males and females give male and female offspring at ratios approaching unity. However, females rather than males are heterogametic. Lloyd (1975b) suggests that unisexuality in a head is achieved by control of timing of the sequential production of male and then female florets, rather than by sterility. Doubtless this provides an explanation that satisfactorily accounts for the ready reversion to monoecy. It also explains why femaleness does not precede maleness in the evolution of dioecy; pressures favouring outbreeding do not lead to gynodioecy, with total females and hermaphrodites, but to gynomonoecy in which females still have some male function. Sexual selection should encourage maleness in andromonoecious lines in direct correspondence with the development of femaleness in gynomonoecious lines. Sexual selection for gender function should result in the eventual evolution of full dioecy, as long as dioecy remains reproductively efficient.

Bawa (1980) and Ross (1978, 1982) have recognised the symmetrical development of dioecy from monoecy, as exemplified by *Cotula*, as one of five mechanisms by which dioecy is potentially able to evolve. By far the most important numerically is the asymmetrical mechanism involving male and then female sterile mutants, via gynodioecy and subandroecy (see Fig. 8.4). A mechanism in which sexual selection for gender function leads from heteromorphy to dioecy has been discussed elsewhere (Ch. 7).

The control of floral timing, which allows monoecious plants to change towards dioecy, is probably under hormonal control. Thus, it is not surprising that Freeman *et al.* (1981) have noted that the ratio of flower genders in three North American trees varies with the environment. They discovered that male flowers (andromonoecy) predominate on xeric sites and female flowers (gynomonoecy) predominate on mesic sites, for the same species. This mechanism might be adaptive, for if female reproductive loads are greater than male loads, femaleness may best be maximised on mesic sites.

In subdioecious states, such as subandroecy, many of the genetic mechanisms of full dioecy seem already to have been evolved (Westergaard 1958). In several species of meadow-rue, *Thalictrum*, and in the economically important *Asparagus officinalis*, the selfing of mostly male hermaphrodites gives females and 'males' at the ratio 1:3. This strongly suggests that 'males' are XY which give 1 XX (females) : 2 XY : 1 YY (males) when selfed. The assumption that YY 'males' are viable and fertile is confirmed as some 'males' when crossed to females (XX) yield only 'males' (XY). The identification of YY 'males' (actually andromonoecious hermaphrodites [♂ ⚥]) in *Asparagus* has allowed commercial breeding in which only the higher-yielding 'males' are produced (p. 320).

YY males also occur in the annual mercury, *Mercurialis annua*, but they are largely male sterile. In the paw-paw, *Carica papaya*, 1 female : 2 'male' ratios are found on selfing andromonoecious males, and it is assumed that YY males are inviable. When the selfing of andromonoecious males gives only males, as seems to be the case in the spanish catchfly, *Silene otites*, it is assumed that the females are heterogametic (Sansome 1938).

In the grape, *Vitis vinifera*, complex patterns of inheritance occur in what was probably a fully dioecious species before cultivation. Hermaphrodite mutants have been favoured by man, for males also bear grapes, and do not take up valuable space non-productively. The following types of offspring result from the selfing of hermaphrodites in different grapes:

| | |
|---|---|
| all hermaphrodite | ? hermaphrodites XX due to breakdown of male sterility in females |
| 3 hermaphrodites : 1 female | XY hermaphrodites with viable hermaphrodite YYs |
| 9 hermaphrodites : 3 females : 4 males | two-locus unlinked dominant control with complementation |

In grapes, hermaphrodity seems to have arisen from dioecy on at least three occasions, and by three different mechanisms:

(a) Loss of male sterility in females (XX hermaphrodites).
(b) Loss of female sterility in males (XY and YY hermaphrodites).
(c) Loss of linkage of male and female sterility (? translocation of primary sex control onto an autosome).

Selection for hermaphrodity by man has also occurred in another important crop that was originally dioecious, *Cannabis sativa*. Genetic sexual constitution can be be judged by a number of secondary sexual characters, or by chromosome karyotype, for Menzel (1964) has distinguished XY from XX karyologically. At least some monoecious hermaphrodites are genetically female XX, and thus there may have been a reversion of male sterility in some strains. The picture is, however, confused by the influence of the environment on the expression of genotypic sex. Selfing of monoecious or andromonoecious males can result in many types of sex expression in the offspring, as is the case for grapes; translocation of the primary male and female factors onto unlinked autosomes has clearly occurred, and some secondary sex characters also show segregation, suggesting that they have also been translocated onto autosomes in some hermaphrodite (monoecious) hemps.

Yet another crop plant, spinach (*Spinacia oleracea*) is normally dioecious with equal proportions of male and female plants. Many gradations of monoecism and hermaphroditism are known, enabling the selection of 'highly male' and 'highly female' lines. These lines can subsequently be used to produce 'hybrid' varieties (Simmonds 1976, see also Janick & Stevenson 1955).

Clearly, man has tended to find dioecy an inconvenient trait in crop plants, and has selected for hermaphrodites of several different types of origin, even within one species.

## SECONDARY SEX CHARACTERS

We are accustomed to the idea that animals often have sexes with very different external appearances. Such secondary sex characters, which have nothing to do with primary sexual function, have been selected for by one or more of a variety of types of selection pressure.

In the case of the bright plumage assumed by male birds such as ducks, pheasants, finches, chats, birds of paradise, etc., mating is achieved after females select displaying males. In this way, sexual selection may directionally favour fantastic features that are selectively disadvantageous in other ways (bright plumage may render the male, or even the whole brood, more liable to predation; large antler size selected for by threat displays in some harem-collecting deer may be otherwise harmful in several fashions). Interactions are complex and may infer group selection. If brightly plumaged males are not involved in reproductive activity after mating, they may draw the attention of predators away from breeding females. Plumage differentiation may provide breeding barriers between potentially interfertile populations. Size differences in total, or in one organ alone (for instance the beak of the Curlew, *Numenius aquaticus*, which is longer in females), may reduce competition for resources between sexes, as well as functioning in sexual recognition and display. Sexual dimorphism for size is common in many mammals, birds, fish and insects. It reaches its limit in mantis, and some spiders, in which the much smaller male aids the female reproductive load after mating by being eaten by the female.

The relative 'disposability' of males is commonly observed. Males do not bear young. As a result they tend to have lighter reproductive loads, and to be evolutionarily irrelevant after mating. The male is usually the heterogametic sex (XY) with X-linked genes borne hemizygously and exposed to selection. Haldane (1922) has observed that this may make the heterogametic sex inherently weaker (but *female* birds are heterogametic). If it benefits the species to be dimorphic in size (feeding niches) or colour (display), then males usually have a size and plumage which is less individually successful. In some cases of Batesian mimicry, as in many butterflies, only females mimic harmful models. Having non-mimic 'disposable' males reinforces the frequency-dependence of mimicry (mimics must be rarer than models if they are to succeed). (These popular arguments are robustly and unashamedly group-selectionist!)

Of course, not all animals display striking sexual dimorphisms, and it is difficult to sex many birds such as the European robin (*Erithacus rubecula*) without dissection, or by observation of territorial behaviour and song. Secondary sex characters exist in the majority of animals, although they may be only behavioural.

Plants do not behave, and thus conspicuous secondary characters relating to territoriality or mate choice are lacking. Secondary sexual characters in plants might be related to:

(a) Reduction in competition between sexes, allowing sexes with different reproductive loads to coexist harmoniously, and at reproductively advantageous ratios.
(b) Differentiation in reproductive load distribution, so that the sex with the higher reproductive load receives resource usage compensation with respect to the pattern, productivity or timing of vegetative growth.

Although secondary sexual characters in dioecious and gynodioecious plants have been recognised since von Mohl (1863) and Darwin (1877), it has been a much neglected subject, although there is a very useful later review (Lloyd & Webb 1977). A continuing problem is the distinction between secondary sex characters that are causal, resulting from differential reproductive loads, and those that are adaptive. The latter should be under independent genetic control. However, as secondary sex characters under genetic control should be determined by loci that are linked on a sex chromosome (usually the X chromosome), or which are triggered by a sex chromosome, they do not usually show independent segregation from primary sex features. Thus, they can be difficult to distinguish from causal effects of the primary sex characters, which are not under independent control.

In contrast with many animals, it is rarely possible to identify the sex of a dioecious plant without examining the flowers for stamens and gynoecia. Secondary sex characters in plants are more subtle, and are usually expressed in terms of growth, resource allocation, timing or longevity. There are exceptions, for instance in *Cannabis*, in which male and female plants have a different overall shape, and other cases are reviewed by Ornduff (1969). According to Lloyd and Webb, secondary sex characters have been found in all dioecious species which have been carefully investigated. Most of these can be interpreted as adaptations to or results of the different reproductive loads of the sexes. Only in a few cases do sexual differences become obviously apparent before flowering commences.

Darwin (1877) showed that in many gynodioecious species, females tend to have smaller flowers, and Baker (1948) provides a list of no less than 73 species of the north-west European flora for which female flowers are smaller than hermaphrodite flowers (some are gynomonoecious). In contrast, he only provides seven examples of fully dioecious species with smaller female flowers. Bawa and Opler (1975) found that 14 out of 20 dioecious Costa Rican species studied have larger female flowers.

Bawa (1980) comments that males frequently also produce more flowers than females. This can be a feature of gynodioecious as well as dioecious species (see Tables 8.7 and 8.8). However, in some gynodioecious species, for instance *Stellaria longipes* (Philipp 1980), females produce more flowers than hermaphrodites. It is possible to interpret superior male function in size and number of flowers as a consequence of cheaper resource expenditure on pollen rather than seeds and fruits. However, Bawa (1980) suggests that superior male function has resulted from intrasexual selection. He argues that female fitness is maximised by expenditure on resource allocation to seeds and fruits, and has components that respond to the size of fruits and seeds, as well as to their number. However, male fitness is maximised only by the production and dissemination of pollen grains. Larger flowers may bear more pollen, and are certainly more likely to receive animal visits. There will be a direct relationship between flower number and pollen production.

Sexual selection on males and females is disharmonious, and indeed may act in directly opposing ways. Put oversimply, males should select for quantity, and female for quality of reproductive effort (see also p. 290). In some conditions, such as a tropical forest tree in which this disharmony becomes great, dioecy may become advantageous, however inefficient a mechanism it may be. However, if male and female effort is compartmentalised into different genets, as in dioecy, males and females will compete with each other. Even when their respective sexual efforts are maximised, the total energy loads for reproduction per sex will almost certainly differ. Most male loads, despite larger male flowers and a greater number of male flowers, will usually be lower than female loads. This difference in resource expenditure may result in three side effects:

(a) More male ramets and/or larger male ramets than female ramets; either or both may result in the production of more male flowers.
(b) Females may flower later, for a shorter period, and at a greater age for first flowering than males.
(c) Hormonal control of primary sex may have, adaptively or incidentally, an influence on other phenotypic features. Placke (1958) relates the larger size of male flowers to high levels of gibberellin in the developing stamen.

Disharmony resulting from intrasexual selection may affect flower visiting by insects. Most flower visitors accept pollen as well as nectar as a reward. Female flowers have no pollen. Deceit is commonly employed by female flowers towards small generalist pollen eaters, through the provision of dummy stamens (see Fig. 8.2). A lack of pollen, and smaller size and numbers of female flowers may be compensated for by greater production of nectar in females, as found by Bawa and Opler (1975) for five out of six Costa Rican species and by Kay *et al.* (1984) in *Silene dioica* (red campion). Earlier and longer male flowering periods should aid unidirectional pollen flow towards the scarcer, less attractive female flowers.

A representative example of secondary sex characters is found in the cultivated *Asparagus* (Table 8.9). Males consistently show a heavier yield, largely due to the greater number of male spears (ramets) produced per plant (genet). In fact, female spears are consistently slightly heavier than males. The greater production by males makes them agronomically more desirable.

Dioecious species of strawberry (*Fragaria*) show a similar relationship between the sexes, with males producing far more runners, as originally recorded by Darwin (1877). Superior male vegetative productivity is not always constant within a species. It can vary with:

(a) Density; male dog's mercury, *Mercurialis perennis* is more vigorous in competition with females at intermediate density, but is less so at low and high densities (Wade 1981b).
(b) Frequency; in the sheep's sorrel, *Rumex acetosella*, the rarer sex shows the better productivity at a range of densities (Putwain & Harper 1972).
(c) Overall vigour; in bog myrtle, *Myrica gale*, there is a positive relationship between the mean height and shoot number of plants and the ratio of male height to female height (Lloyd & Webb 1977).

**Table 8.9** Performance of male and female *Asparagus officinalis* (after Lloyd & Webb 1977).

| | Year | Male | Female | Male : Female ratio |
|---|---|---|---|---|
| mean numbers of spears per plant | 1925 | 2.98 | 1.95 | 1.53 |
| | 1926 | 15.68 | 8.61 | 1.82 |
| mean weight of spears (g) | 1925 | 18.74 | 21.90 | 0.86 |
| | 1926 | 23.77 | 27.73 | 0.86 |
| mean weight of spears per plant | 1925 | 55.90 | 42.70 | 1.31 |
| | 1926 | 372.80 | 238.80 | 1.56 |

In other life forms, superior male vigour may find other modes of expression:

(a) In trees, such as yew, (*Taxus baccata*), *Ginkgo biloba* and poplars (*Populus*), males tend to be taller.
(b) In the New Zealand spaniard, *Aciphylla scott-thompsonii*, males produce more rosettes per genet than females (each rosette can make a terminal inflorescence).
(c) Rosettes may be larger on males than females, as in the white and red campions, *Silene latifolia* and *S. dioica* (see Fig. 8.3).

The consequences of such sexual differences in vigour are both reproductive and ecological. Males very frequently produce more flowers than females, per genet, and often do so over a longer period. In 16 species of diclinous New Zealand umbellifer investigated by Lloyd and Webb (1977), all produce more flowers per male inflorescence than per female inflorescence, and in all species except one more inflorescences are produced by male genets than by female genets. Similar results have been obtained by Kay *et al.* (1984) in *Silene dioica*. As stated on p. 318, it is less clear whether such differences are merely a causal by-product of different reproductive loads, or that they have been selected for in response to pressures for gender function maximisation, or reproductive efficiency.

## *Niche differentiation between sexes*

Intraspecific competition can be assumed to occur in all species, and indeed is an essential component of the process of natural selection. In hermaphrodites, competition between individuals will not normally result in sexual selection, or interfere with reproductive efficiency, although where variation in gender resource expenditure occurs, sexual selection might encourage phenomena such as heteromorphy (Ch. 7).

In diclinous conditions such as dioecy and gynodioecy, however, sexes differ from each other by a major linkage group, and in reproductive load. As a result, one sex may very often be at a disadvantage with respect to the other in direct competition. If one sex shows poorer survival and lower vigour than the other, this may seriously impede sexual fertility. In long-lived perennials with efficient vegetative reproduction and dispersal, low sexual fertility may not be disadvantageous. The canadian pondweed, *Elodea canadensis*; and various species of butterbur, *Petasites*, are among many examples of dioecious species which may be very successful when only one sex is present (Ch. 10). Dioecy is unusual in annual and other monocarpic life forms, perhaps as a result of the effects of intersexual competition on reproductive efficiency in plants in which fitness is maximised by optimal seed-set.

Although intersexual competition may encourage the evolution of niche differentiation between sexes, such evolution may be constrained by the physical scale of niche heterogeneity in comparison to distances of pollen travel. Put simply, one would not anticipate great reproductive efficiency if one sex succeeded in woods, and the other in fields, when woods were several kilometres across. Seed-set would be limited to the edges of the wood. Niche differentiation can be expected to respond to small-scale and constant environmental heterogeneity, such as glades in forests, and might be most successful in wind-pollinated plants with distant pollen travel (Ch. 5). Just such an example is dog's mercury, *Mercurialis perennis*, in which Mukerji (1936) claimed that males predominate in better illuminated areas of woodland, and females in more shaded areas. Some confirmation is provided by Wade *et al.* (1981). Wade (1981a,b) shows that males grow better at high density, under high illumination, and in competition with the creeping soft-grass, *Holcus mollis* (its usual companion in British woods). Competition experiments between the sexes of *Mercurialis* show that competitive effects are lower between sexes than within a sex at a range of frequencies, densities and illumination. In this species, some niche differentiation has occurred between the sexes, although the relationship between sexes and niche is not a simple one.

In another forest-floor herb, the North American liliaceous *Chamaelirium luteum*, Meagher (1980, 1981) showed that male : female ratios were heterogeneous for comparisons between quadrats of areas between 1 $m^2$ and 5 $m^2$. This spatial heterogeneity was presumed to exist with respect to different niches on the forest floor which were not identified.

It may be typical for males to predominate in habitats in which both sexes perform well. This seems to occur in the culinary thyme, *Thymus vulgaris* (Dommée 1976), which is gynodioecious. Male-fertile individuals predominate in the driest and most open habitats, and Dommée finds a positive correlation between the frequency of male fertiles and total performance of the species per area of ground. In bog myrtle, *Myrica gale*, plants perform with the greatest vigour in wetter habitats, and in these males also predominate (Davey & Gibson 1917, Lloyd & Webb 1977). Males, with better reproductive growth, and a lower reproductive load should be able to colonise a habitat in which the species is successful more rapidly than females. This advantage will be less apparent in more marginal habitats.

Such differential advantages may be minimised by frequency-dependent effects. In two species for which detailed within- and between-sex competition experiments have been conducted, the rarer sex in an experiment tends to perform better in vegetative and reproductive productivity than it does at more equal sexual frequencies (Putwain &

Harper 1972 for *Rumex acetosella* and *R. acetosa* and Wade 1981b for *Mercurialis perennis*). In *Rumex acetosa*, field sorrel, the sexes differ in productivity and timing. Males grow earlier, flower earlier, and produce more, smaller shoots than females. These rather characteristic differences between the sexes may lower competition between the sexes in species in which vegetative growth patterns lead to the intermingling of the ramets of the two sexes. If differences in timing and productivity do minimise competitive effects between sexes (and there seems to be no direct proof of this), a rare sex will compete with its own sex rarely, and should do well. When frequencies of the sexes are equal, intrasexual contacts will be disadvantageously common for both sexes. Such frequency-dependent effects will act to minimise spatial effects of niche differentiation.

Niche differentiation, even if expressed in terms of differential vigour between the sexes in better habitats, may have apparently adaptive effects with respect to reproductive efficiency. Once again, it can be difficult to perceive whether such effects are adaptive, or merely causal. On a hill in Shropshire, England, I have observed that female stems represent only 40% of the population of *Rumex acetosella* in mixed woodland at the base of the hill at 100 m. On the exposed summit at 300 m, females formed 80% of stems. The frequency of male stems decreased steadily with altitude and exposure. It is arguable that in this wind-pollinated species, fewer males are required for successful cross pollination in exposed areas. This might be a useful side-product of the general phenomenon that males predominate in areas of high density and good growth (i.e. at the base of the hill).

An exception to the general picture of male supremacy in optimal niches may occur in the shrubby cinquefoil, *Potentilla fruticosa*, in one of its few British populations, in Upper Teesdale where it is dioecious (see Fig. 8.2). It is interesting to note that this is a species in which the female is heterogametic. This shrub can form large clones by woody rhizomes. In stable habitats, away from the River Tees, males and females did not differ in mean age, and females tended to be more vigorous, resulting in more *female* reproductive shoots per unit area. In unstable habitats by the edge of the river, flooding apparently causes a more rapid turnover of genets, for here the mean age of plants is much lower than in stable habitats. In these unstable habitats, male genets predominate, although they are no more vigorous than females (Richards 1975a). A family of 80 seedlings was grown to flowering in the Newcastle Botanic Garden. When two years old, 64 of these had flowered, and a significant excess (66% of the total) were female. At four years, all had flowered, and the remaining 16 all proved to be male, giving 42 females to 38 males, which does not differ significantly from 1 : 1. This seedling population is now 12 years old, and there are 59 survivors of the original 80. Of these 30 are females, so there has been no differential mortality in this population in garden conditions. We can conclude that females tend to flower earlier than males. This may counteract the better male survival among young plants in the field. Earlier female flowering and greater female vigour (but not establishment) are both contrary to the findings of Lloyd and Webb (1977) for 16 species, most of which are known to have male heterogamety. The possibility that superior vigour is associated with heterogamety, rather than reproductive load, should be examined in other female heterogametic genera such as *Cotula* and *Fragaria*.

## SEX RATIOS

Theoretical sex ratios have been examined by Lloyd (1974a,b, 1975a, 1976) and discussed by Webb (1981a) and Webb and Lloyd (1980) for gynodioecious species. Lloyd concludes that the frequency of females $p$ will be a function of the relative fecundity (in terms of seed-set per genet) of hermaphrodites to females ($f$) in the absence of differential selection within or between sexes. Thus:

$$f = \frac{1 - 2p}{2(1 - p)}$$

The structure of this model is such that $p$ cannot reach 0.5, and will only approach 0.5 when female fecundity in hermaphrodites is very low. [Lloyd (1976) and Webb (1981a) replace $f$, relative seed fecundity, by $C$, relative ovule contribution to the next generation by males and females. $C$ includes estimates of seed size and seedling fitness (thus incorporating inbreeding effects *sd*). Here, the original assumption of no differential selection is adhered to, and '$f$' is retained.]

This model can also be applied to fully dioecious species, where if $p$ is 0.5 (equal ratio of males and females) and $f$ is 0 (i.e. there is no male seed-set). Webb has fitted this model with considerable success to a number of New Zealand Umbelliferae in which $p$ and $f$ are known. In practice, in most gynodioecious species, the frequency of females is less than 0.35, and is often less than 0.10. Female frequency will depend on the relative seed-set of females and males (hermaphrodites) ($f$), but also on the relative fitness of females and males as parents ($p$), estimated by inbreeding depression (*sd*) in the offspring of males.

For dioecious species, the heterogametic sex (XY) must mate with the homogametic sex (XX). At the simplest level, one should expect that males and females will be produced at equal frequencies (1:1). However, there are a number of points in the sexual cycle at which differential selection on X and Y chromosomes can occur:

(a) Competition between meiotic products in the heterogametic sex (i.e. usually in pollen formation).
(b) Competition between gametophytes in the heterogametic sex (i.e. usually between pollen grains or pollen tubes at the stigma, or in the style and ovary).
(c) Competition between zygotes (i.e. between heterogametic XY and homogametic XX embryo seeds in the developing fruit, or between XY and XX seedlings or adults).

It is very important to distinguish between the sex ratios of genets, and the sex ratio of ramets. In order to do so, it is necessary to be able to identify the limits of genets, a subject that will be discussed in Chapter 10. As we have seen when examining secondary sex characters, sexes commonly differ from each other in the production of ramets, males usually producing more ramets. Such differences in the production of ramets may be very important with respect to maximisation of gender function and reproductive efficiency. However, the sex ratio of genets may not differ from equality.

Early studies (Correns 1928, Lewis 1942a) considered that female-predominant ratios were common in dioecious species. In four species (*Silene latifolia*, *Rumex acetosa*, *Humulus lupulus* and *Cannabis sativa*) it was found that sex ratios from seed differed from equality when there was a high density of pollen on stigmas. At low pollen densities, males and females were produced from seed at a ratio of 1 : 1. At high densities, an excess of females was produced. This was considered to result from competition between X-carrying pollen grains and Y-carrying pollen grains at the stigma or in the style. Microscopic examination of pollen tubes indicated that one-half germinated and grew more quickly than the other half. These, it was considered, were from the X-carrying grains. When the number of germinating pollen grains per stigma was lower than the number of ovules to be fertilised per stigma, competitive effects were negated, and equal numbers of male and female seedlings resulted.

This model is very attractive, for it carries an elegant negative feedback element. If superior male vegetative reproduction creates an excess of male ramets, there will be surplus pollen available for cross pollination, and stigmas will receive pollen at high density. As a result, an excess of females will occur in the next seed generation, to a point when males are sufficiently scarce for pollen grains received by stigmas to drop below the level where all ovules are fertilised. Sex ratios from seed will return to equality, allowing a corresponding build up of male-ramet frequency. Frequency-dependent selection should result in an equilibrium frequency of ramet sex, which will depend on the relationship between ramet sex frequency and pollination saturation.

This system, if it operates, should result in a negative correlation between male: female ramet ratios in the field, and male: female genet ratios from seed. Mulcahy (1967, 1968) was unable to discover such a relationship in *Silene latifolia* (=*S. alba*). However, sex-ratio equilibrium points with respect to pollination efficiency might differ between populations, or even between seasons for the same population. More modern work on the effect of gametophyte competition on sex ratio is undoubtedly needed.

Lloyd (1974b) challenges the assumption that fitness in a dioecious population will necessarily be directly related to reproductive efficiency and seed-set. He points out that as seed-set in females increases, so does the disparity in reproductive load between males and females. High levels of seed-set may result in inefficiently high male: female ramet ratios. He considers (1974a,b) that the sex ratio of genets will depend on:

(a) Competition between X and Y-carrying male gametophytes at the stigma and the resulting sex ratios in seedlings (he views the poor performance of Y-carrying gametophytes as the unavoidable by-product of dioecious systems in which half the male gametophytes lack X-linked genes).
(b) Differential survival and growth of male and female genets.
(c) Efficiency of pollen-flow from male to female flowers.

High male-genet ratios will result from low differential gametophyte performance, low differential genet survival and growth, and low efficiency in pollen transfer (which will increase individual male fitness). Low male-genet ratios (female predominance) will result from high differential gametophyte performance (resulting in an excess of females from seed), high differential genet survival and growth (superior male genet performance), and high efficiency in pollen travel. Both the latter trends will *reduce* individual male fitness.

However, all of the three attributes considered above are dependent on male : female ramet (*not* genet) frequencies. Genet frequencies (sex ratios) will occur at an equilibrium point dependent on all three attributes. Because all three attributes will also differ from species to species (and to a certain extent from population to population), it is not surprising to find that genet sex ratios are female predominant in some species and male predominant in others.

One uncanny correspondence in sex ratios between two species has been found for the arctic willows, *Salix polaris* and *S. herbacea* (Crawford & Balfour 1983). In both species, the former in Svalbard and the latter in Iceland, the frequency of female genets is 0.59. The authors attribute this disparate ratio to the greater ecological tolerance of females, as exemplified by the greater range of resistances to water loss found in the leaves of females compared with males. They consider that pollination efficiency is high in these wind-pollinated species which are found on open windy sites. They demonstrate that female frequency can be increased to 0.62 of the population with no diminution of effective population number $N_e$ (Ch. 5), but above this level of female frequency, the effective population number falls with increasing rapidity, possibly resulting in disadvantageous genetic consequences (Ch. 9). Thus, they consider that the female ratio of 0.59 is a product of superior female ecological tolerance (and hence survival) and high efficiency of pollen flow, essentially panmictically up to a female ratio of 0.62 (points b and c above). However, Q. O. N. Kay (*pers. comm.*) has discovered constant 3:1 ratios of females to males in several populations of sallow (*Salix cinerea*).

Lloyd (1974a) points out that male-predominant ratios are rather more common than female-predominant ratios, in contrast to Correns (1928) and Lewis (1942a). In 16 New Zealand dioecious species, Godley (1964) found male-predominant ratios in 10, and a female-predominant ratio in only one, whereas Webb and Lloyd (1980) surveyed 53 populations of New Zealand Umbelliferae, which are dioecious or gynodioecious, and found male-predominant ratios in 43. As already discussed, this is expected of gynodioecious species, but in 18 populations of 11 fully dioecious species, all except one have male-predominant ratios, and these significantly differ from equality in 14 (78%). The authors assign these male-predominant ratios to superior survival of male genets due to their lower reproductive loads. It is noteworthy that in most of these species vegetative reproduction is absent, and so males do not reduce their individual fitness by producing a large excess of male ramets (in half of the populations, a significant excess of male flowers per genet is produced, but this never exceeds 1.7 times the number of female flowers per genet).

If Lloyd (1974a) is correct, sex ratios cannot themselves be selected for, but are the product of gametophytic (pollen-tube competition and pollen flow) and sporophytic (genet survival and growth) individual fitnesses. Other workers such as Lewis (1942a), Mulcahy (1967) and Crawford and Balfour (1983) have implied group selection for optimal sex ratios, as pointed out by Prentice (1984). Pollen (gametophytic) competition between X and Y chromosomes can only occur in heterogametic males, and one might expect equal proportions of males and females to arise from seed for female heterogametic species. This does seem to be the case in *Potentilla fruticosa* (as discussed above) and *Fragaria* species (Darwin 1877). *Potentilla fruticosa* presents a warning with regard to deducing sex ratios in seeds or in wild populations from the evidence of limited wild genet sampling (Richards 1975). Various subpopulations of this species present both male-predominant and female-predominant ratios in Upper Teesdale, England, the Burren, Ireland and Öland, Sweden (Elkington & Woodell 1963). It seems that the sex ratio of a subpopulation is a function of the longevity of that subpopulation. Short-lived populations tend to be male predominant, and long-lived populations tend to have predominantly female genets (and ramets).

Opler and Bawa (1978) reached similar conclusions about the sex ratio of 23 species of tropical forest tree from Costa Rica. Of these, eight had male-predominant sex ratios and two had female-predominant ratios. Males tend to flower earlier, and to live longer. Sex ratios were thought to be the product of differential times to maturity and life span of the sexes, which themselves probably reflect the different reproductive loads of the sexes. Certainly, a wealth of circumstantial evidence tends to confirm the position of Lloyd (1974a), that sex ratios are not in themselves adaptive and the result of group selection, but are rather the result of differential selection on individual X and Y chromosome-carrying gametophytes and sporophytes.

## CONCLUSION

Theoretical studies on diclinous systems of reproduction in plants, and their experimental ratification, have been of interest to evolutionary biologists for more than a century, since the publication of Darwin's book *The different forms of flowers on plants of the same species* (1877). The variety in the patterns of distribution of male and female organs shown among closely related species of flowering plant is exceptional, and has no parallel in other groups of living things. It offers exciting opportunities for the investigation of selection pressures working in favour of, and against, the partitioning of different sexual functions in different individuals, and outbreeding, and allows us to calculate the reproductive loads and reproductive advantages encountered by each sex.

Work culminating in the major review of Westergaard (1958) concentrated on the genetic effects of the outbreeding which diclinous systems promote. Since then, a considerable shift in emphasis has been led by the New Zealanders, Godley, Lloyd, Ross and Webb. They have taken full advantage of their rich native heritage of diclinous plants to marry sophisticated algebraic modelling with extensive field observations in an admirable way which is outstanding amongst modern studies of breeding system genetics.

We can now view the evolution of diclinous systems such as gynodioecy and dioecy in the light of reproductive resource allocation to males and females, the genetic input into the generation from individuals of different sexes, and reproductive efficiency. We now realise that the evolution of gynodioecy and dioecy in flowering plants may have had less to do with the promotion of outbreeding (although this may play a part) than previously considered, and may have had more to do with an increase in the function of both sexes, which have disparate requirements, that separation of the sexes onto different genets can achieve.

We can see that the overriding importance to female reproduction is for the number, and especially the size and resource of the seeds and fruits, to be optimised with respect to seedling dispersal and establishment. As male-sterile genes that favour such trends for females become established in gynodioecious populations, so hermaphrodites will experience selection in favour of male function, and will reduce female function towards full dioecy. If full dioecy is to function efficiently, male features must become linked to one chromosome and female features to another, in such a way that one sex is dominant over the other, and this forms the heterogametic (XY) sex. Because maleness usually follows femaleness, it must acquire dominance to function, and thus males are usually heterogametic.

Dioecy provides interesting parallel examples to the disadvantageous results of directional selection for reproductive attributes exhibited by some animals (for instance the huge antler size of the extinct Irish elk). In maximising individual sexual function through dioecy, plants may suffer the effects of uncertain reproductive efficiency; competition between sexes leading to highly disparate and inefficient ramet ratios; low competitive success of the Y chromosome, especially gametophytically; and variable sex expression due to interaction between the environment and the 'hormonal' control of sex. Whereas gynodioecy or dioecy may be favoured in one environment, a relatively slight change in environment or population size may render dicliny disadvantageous. Here diclinous plants may be offered evolutionary opportunities denied to the elk. Various pathways, genetic, chromosomal and physiological, can allow an escape from the trap of dioecy back towards the broader reproductive options of hermaphrodity. Many dioecious plants may have taken this escape route. Others may have succumbed to the trap, and have become extinct. Gynodioecy, and especially dioecy, remain unusual reproductive systems in flowering plants, which have rarely succeeded over long periods of evolutionary time.

## CHAPTER NINE

# *Self-fertilisation*

It is generally considered that the earliest flowering plants had hermaphrodite self-incompatible flowers, and were therefore outbreeders (xenogamous). Today we find that perhaps 95% of all flowering plant species are hermaphrodite, and it is considered that over half of these may be self-incompatible. Thus, the original breeding system has been judged a success by evolutionary experience. Many flowering plants are xenogamous, and may be presumed to benefit from the maintenance of genetic variability typical of xenogamy (Ch. 2). However, xenogamous plants may suffer in three ways:

(a) Copious self-pollination with incompatible pollen may block stigmatic sites, leaving few gaps for later arriving and less frequent crossed pollen.
(b) Inefficiency of pollen transfer between individual plants may severely limit seed-set; isolated plants may fail to set seed.
(c) In monocarpic species, in which fitness is closely associated with the maximisation of female function (quality and quantity of fruit and seed-set), heavy expenditure on male function, required for successful cross-pollination, may be disadvantageous.

These constraints have favoured mechanisms in many plants which have led to partial, or at times even to almost total, self-fertilisation, and it is these plants that will be considered in this chapter.

## FEATURES OF SELF-FERTILISATION

### *Self-compatibility*

One essential feature of self-fertilisation is self-compatibility. It seems likely that self-compatibility can readily arise from gametophytic self-incompatibility, for nearly all mutations observed in an *S* allele system are towards self-compatibility, as is discussed in Chapter 6. Thus self-compatibility 'mutants' may be common features of gametophytic self-incompatibility systems. Sporophytic self-incompatibility systems are usually to some extent 'leaky', and the degree of self-compatibility in such systems will be susceptible to selection. Heteromorphic di-allelic incompatibility systems may give rise to self-compatible homostyles (Ch. 7), and dioecious systems may give rise to self-compatible hermaphrodites (Ch. 8). All the last three xenogamous systems are apparently secondary, having evolved from selfers, and thus if they break down they are likely to give rise to selfers again.

### *Self-pollination*

A second essential feature of self-fertilisation is self-pollination. The blocking of stigmatic sites by selfed pollen may encourage features of structure or timing in the self-incompatible flower that minimise or prevent within-flower pollination (Ch. 2). However, between-flower self-pollination (geitonogamy) will only completely disappear in those rare cases in which an individual genet has only one flower open at a time. Many partially selfing plants show little or no within-flower selfing, and thus pollen transfer between flowers is required for seed-set. Whether pollen on a stigma is predominantly selfed (from other flowers of the same genet), or predominantly outcrossed in these cases will depend on several factors, e.g. the nature or behaviour of the pollinating agent; the size of clone and the number of flowers donating or receiving pollen at the same time within a genet. These factors will vary within and between populations, and within and between flowering seasons.

It is common to find that complex, animal-pollinated flowers are incapable of within-flower selfing, but are self-compatible (most Leguminaceae, Scrophulariaceae, Lamiaceae, Orchidaceae or Araceae for instance). Such plants maintain a balanced strategy of mixed selfing and crossing, depending on pollen transfer within and between genets. In others, within-flower selfing may only be possible as the result of an animal visit. This strategy is useful for both self-incompatible and self-compatible species, in that stigmatic sites will not suffer from prior blockage by selfed pollen and such crossed pollen as may be carried by the visitor is given a chance to succeed.

### *Life style*

Outcrossing mechanisms are common, and are likely to succeed when reproductive fitness is not limited by a paucity of animal visits. Where such a limitation occurs, features that maximise within-flower selfing (autogamy) may be favoured. Fitness constraints resulting from limited visiting may occur in inimical environments in which animal visitors are few or uncertain, such as arctic, alpine or arid environments. In practice, within-flower selfing is not often conspicuously more frequent in these habitats, perhaps because most of their denizens are long-lived perennials in which constraints on reproductive fitness by seed are low.

Reproductive constraints are much greater in short-lived, and especially monocarpic, plants (flowering only once) for which a major component of fitness is the quantity and quality of the seed produced. Monocarpic plants (Ch. 10) only have one opportunity to contribute to the next generation, and thus there will be very great directional selection for this reproductive input to be maximised. Such selection pressures are likely to outweigh the disadvantages of selfing. Thus we find that almost all monocarpic plants are at least partially autogamous, and are capable of maximising reproductive output in the absence of the unreliable agents of cross-pollination.

Such is the pressure for autogamy on monocarpic species, and such can be the disadvantages to the plant brought about by repeated self-fertilisation (as we shall see later in this chapter), that it might be thought surprising that monocarpy is a successful life strategy. By restricting all reproductive energy to seed production, an annual or biennial (monocarpic) plant forgoes the chance of producing any vegetative propagule or perennating organ which would continue that genet to another flowering. Seeds have capabilities of travel, dormancy and establishment rarely found in vegetative organs. Thus, colonising species, adapted to take advantage of and rapidly cover open ground where there is low competition, will achieve greater success by maximising seed production. Energy spent on perennation, or vegetative spread would be wasted, as such open sites will show seral development. Less efficient colonisers, which allocate more reproductive energy to vegetative perennation and spread, would arrive, and by virtue of their vegetative attributes, would outcompete the *'r'* strategy colonisers by their long-term *'K'* strategy of energy allocation to vegetative organs (Ch. 10).

We find that most initial colonisers of bare ground, many of which are by their nature successful garden weeds, are monocarpic annuals, winter annuals or biennials. For these, autogamy has proved to be the most successful breeding system. In contrast, long-lived monocarpic species such as bamboos or agaves may have successful vegetative reproduction. They produce massive inflorescences and infructescences which 'satiate' flowers and seed predators. Such plants are often successful outbreeders.

## *Floral features*

Because selfing plants are not pollinated by insects or wind, they must have efficient within-flower selfing (autogamy). This is achieved by removing those features of structure and timing that have evolved in insect-visited flowers to minimise stigmatic site blocking by selfed pollen. Thus flowers that were protandrous, or protogynous (dichogamous), become homogamous, the anthers shedding pollen as the stigmas mature. Such shifts in floral timing are under polygenic control, and natural heritable variation in floral timing, which can respond to selection for increased within-flower selfing, probably occurs in all protogynous or protandrous species, as for instance in *Gilia* (see Table 9.4). Wyatt (1984) shows that *Arenaria uniflora*, widespread on granitic outcrops in the southeastern United States, shows considerable variation in protandry, which is correlated with flower size. Extreme autogamous populations, homogamous, and with many small flowers, have arisen on at least two occasions, and have been called *A. alabamica*. Many animal-visited flowers also restrict within–flower selfing by removing the anthers from the stigma spatially, so that such selfing is minimised in the absence of an insect visit. Our knowledge of homostyle *Primula vulgaris* (Ch. 7) enables us to suggest that in at least some instances a single genetic event (in this case a cross-over) can remove such spatial differences, thus maximising within-flower selfing.

In more complex flowers, special mechanisms may be required to achieve within-flower selfing. Nineteen species of the insect mimic orchids, *Ophrys*, are cross pollinated by male Hymenoptera which pseudocopulate with the flowers (Ch. 4). However, in the bee orchid, *O. apifera*, the pollinia develop abnormally long stalks and droop into the stigmatic cavities (Ch. 4, Fig. 4.34). In northern populations, in which the pollinator seems to be missing, automatic self-pollination is thus achieved. It has been suggested that more southerly populations may be cross pollinated.

In another orchid genus, *Epipactis*, allogamous species such as *E. palustris*, *E. helleborine*, *E. purpurata*, *E. atrorubens*, *E. microphylla* and *E. gigantea* have a sticky cap (viscidium) to the projection of the stigma (rostellum), above which the pollinia are situated (Ch. 4, Fig. 4.33). The wasp (*Vespa* spp.) visitors are attracted to the coloured, nectar-filled hypochile, and to reach this, brush against the viscidium, which adheres to the wasp, taking the pollinia with it. In contrast, all autogamous species (*E. leptochila*, *E. phyllanthes*, *E. youngiana*, *E. dunensis*, *E. muelleri*, *E. confusa*, *E. albensis*, *E. pontica*) lack the viscidium, an absence that may be controlled by only a single gene, and in most the rostellum is slight or absent. In these, the pollinia are released from the anther early (in bud in *E. phyllanthes*) and automatically self-pollinate (Ch. 4, Fig. 4.33b). However, despite assertions in the literature that such species are invariably selfed, these species have nectar and also receive visits from wasps, and cross pollination of loose massulae and pollen tetrads is likely to occur. Only in the scarce forms, *E. leptochila* var *cleistogama* and *E. phyllanthes* var *vectensis*, in which the flower usually fails to open at all, is self-pollination likely to be total (Richards 1982).

Not only are the autogamous *Epipactis* less variable than the allogamous species, but they tend to have smaller flowers as well. Thus, the primary cause of selfing, the absence of the viscidium, which has allowed allogamous species to evolve into autogamous species, perhaps as the result of a single mutation, is reinforced by other, less vital characters, which improve the efficiency of autogamy. The autogamous species tend to have smaller, more drooping, less open and less brightly coloured flowers with a smaller rostellum. Such features, which do not encourage insect visits, will also require a lower expenditure of energy, and may well be favoured in populations that have become largely autogamous.

It is common to find that autogamous races have smaller flowers than their allogamous relatives. The development of smaller flowers may not only augment the autogamy, but may be favoured for energetic reasons. Thus, Schoen (1982a) showed that races of the North American *Gilia achilleifolia*, which are mostly autogamous by virtue of their near homogamy and juxtaposition of their anthers and stigma, also have the smallest flowers (see Table 9.4). Likewise, Strid (1970) demonstrated that the autogamous love-in-a-mist *Nigella doerfleri* that grows on small, dry, hot Aegean islands, has much smaller flowers than its more mesic relative *N. arvensis* agg. which is also more allogamous. Indeed, derived small-flowered inbreeders may be typical of dry areas, and Moore and Lewis (1965) described the origin of a small-flowered, autogamous segregate of *Clarkia xantiana*, *C. franciscana*, after a dry period extended the desert margin near San Francisco into the range of the former, thus favouring the inbred mutant. H. Lewis (1962) has termed such phenomena 'catastrophe speciation'.

## *Cleistogamy*

The ultimate reduction in floral display is the flower that never opens, and thus shows obligate autogamy. Such closed 'cleistogamous' flowers are not uncommon, but chiefly occur in allogamous species, frequently after the 'conventional' allogamous flowers have failed to set seed. Failure of seed-set is perhaps especially common in vernal species, adapted to flower in deciduous woodland before the leaf canopy develops, which are thus subject to the vicissitudes of unreliable spring weather. The development of cleistogamous flowers later in the season, after the canopy has closed over and pollinators are few, is found in such typical herbs of the woodland floor of temperate Eurasia and North America as wood violets (*Viola* section *Rostratae*) and wood sorrel (*Oxalis acetosella* agg.). Such cleistogamous flowers may be regarded as 'fail-safe' devices, which ensure seed-set in the absence of outcrossed seed earlier on, but may also provide 'mixed strategy mating' with some variable (outbred) and some invariable (inbred) offspring. In some years, plentiful outcrossed seed is produced, thus avoiding the genetic problems resulting from repeated selfing. Other woodland herbs showing late-season cleistogamy are touch-me-not, *Impatiens noli-tangere*, and ground ivy, *Glechoma hederacea*. Annual weeds such as *Lamium amplexicaule* (hen-bit) also have some cleistogamous flowers, as do some plants from seasonally wet areas, which produce cleistogamous flowers in the dry season (*Hesperolinon* spp.). In *Commelina forskalaei*, the dry-season cleistogamous flowers are subterranean. Totally cleistogamous species are apparently very rare. Certain aquatics are usually cleistogamous, such as awl weed, *Subularia aquatica* and mud weed, *Limosella aquatica*, whereas the relatively showy flowers of water lobelia, *L. dortmanna*, are self-pollinated in bud. Casual observations can be misleading, however. Populations of *Subularia* flowering in hot weather on a dried-up lake margin do open their flowers, and are visited by small flies, although this species usually flowers under the water and is selfed.

Similarly, observations on certain annual grasses in which the stigmas and anthers are usually included within the palea can be misleading. Most British populations of the wall barley, *Hordeum murinum* appear to be cleistogamous (Fig. 9.2), yet both chromosomal and seed-protein evidence shows the origin of diversification in the group to be due to allopolyploidy resulting from between-populational crossing, and in

**Figure 9.1** Cleistogamy in the European wood violet, *Viola riviniana*. Large chasmogamous flowers in spring are often followed by small, bud-like flowers which automatically self-fertilise without opening. Cleistogamous flowers are to be seen immediately to the left of the fruit, and further left, immediately to the left of the petiole of the top left-hand leaf (×2).

warmer areas of southern Europe non-cleistogamy is frequently observed (Booth & Richards 1978, p. 122). Even in the apparently fully cleistogamous grass *Festuca microstachys* (Adams & Allard 1982), bursts of genetic variability are still observed, suggesting the occurrence of occasional outcrossing. There is an interesting review of cleistogamy by Lord (1981).

## *Outcrossing*

**Levels of outcrossing** In many inconspicuous homogamous flowers with early self-pollination, autogamy will predominate, but in others, chances for allogamy persist. The classic study by Riley (1956) on the alpine penny-cross, *Thlaspi alpestre*, is a case in point. *Thlaspi alpestre* (Fig. 9.3) is a short-lived perennial from open, metal-rich soils in montane areas, and is homogamous and self-compatible. However, it is protogynous, the stigma initially protruding through the unopened bud, and the stigma is receptive for three days before the anthers dehisce and it is self-pollinated. Riley estimated the amount of outcrossing by observing the number of pollen grains that arrived on stigmas before any flower on the plant had shed pollen (Table 9.1). This technique does not record outcrossing onto a flower after pollen has been shed on any flower on the plant, and is likely to underestimate the amount of outcrossing, which he put at 5%. It also assumes that every outcross pollen grain will fertilise a seed.

Another strategy employed by predominantly autogamous species, which may allow some outcrossing, is subgynoecy, in which some

**Table 9.1** Numbers of pollen grains on stigmas on flowers (in different phases) in inflorescences in different phases for *Thlaspi alpestre* (after Riley 1956).

| | Flower phase | | |
|---|---|---|---|
| Inflorescence phase | 1<br>Protogynous bud | 2<br>Flower open, female phase | 3<br>Flower open, male phase |
| (1) all buds | 1.48 ± 0.21 | – | – |
| (2) flowers open, all female phase | 0.81 ± 0.19 | 1.73 ± 0.29 | – |
| (3) some flowers in male phase | 10.94 ± 3.35 | 18.47 ± 5.93 | 43.37 ± 6.99 |

From the above values:

mean number of cross-pollinated grains onto female-phase inflorescences is $(0.81 + 1.73)/2 = 1.27$

mean seed-set per capsule is 7.42

proportion of stigmas exposed while inflorescence in female phase is 0.31

Thus, estimated outcrossing is $\frac{1.27}{7.42} \times 0.31 = 0.052$.

**Figure 9.2** Spikelet of the annual wall barley grass, *Hordeum murinum*. The two lateral spikelets are sterile; the central floret contains stigmas and anthers which are rarely if ever released from the palea, at least in the UK. The floret is therefore cleistogamous and automatic self-pollination almost invariably occurs. Photo by T. Booth (×4).

flowers in dense inflorescences or capitula lack stamens, and this is found in many Compositae (Asteraceae) and Umbelliferae (Apiaceae). These female florets cannot show within-flower selfing, although most pollen transfer onto their stigmas is likely to be geitonogamous, from hermaphrodite or male florets in the same inflorescence. Pollen arriving from another individual is less likely to find stigmatic sites blocked by selfed pollen. Marshall and Abbott (1984a,b) have shown that rayed morphs of the groundsel, *Senecio vulgaris*, show more outcrossing than rayless morphs, and they attribute this largely to the 10–13 female ligulate (ray) florets distributed around the capitulum. These comprise about 20% of all florets (Fig. 9.4). However, the rayed capitula are much more

**Figure 9.3** Alpine penny-cress, *Thlaspi alpestre* (Cruciferae). The small, whitish flowers are slightly protogynous in bud. Photo by author (×1).

**Table 9.2** Outcrossing frequencies of (a) the 'outer' and 'inner' floret fractions of capitula from radiate plants, and (b) the radiate and non-radiate morphs, in populations of *Senecio vulgaris* sampled from Cardiff and Leeds, together with heterogeneity $\chi^2$ comparisons (after Marshall & Abbott 1984b).

| Population | Outcrossing frequencies Cardiff | Leeds |
|---|---|---|
| (a) Floret fractions within radiate capitula | | |
| outer (ray) florets | 0.225 (436) | 0.063 (627) |
| inner (disc) florets | 0.098 (559) | 0.036 (565) |
| $\chi^2$(1) | 48.16*** | 7.62** |
| (b) Morphs† in populations | | |
| radiate | 0.137 (999) | 0.060 (1316) |
| non-radiate | 0.007 (1046) | 0.031 (1302) |
| $\chi^2$(1) | 230.6*** | 1.54 (N.S.) |

Numbers in brackets represent the progeny scored.
† Morph values are from Marshall and Abbott (1984a).
N.S., not significant.
** Significant at p = 0.01.
*** Significant at p = 0.001.

**Figure 9.4** Capitula of (from right) two unrayed groundsels, *Senecio vulgaris* (*rr*), a heterozygote rayed groundsel (*Rr*), and two homozygous rayed groundsels (*RR*). The large capitulum on the left is of the Oxford ragwort, *Senecio squalidus* from which the rayed gene in groundsel arose through introgressive hybridisation (×1.5).

conspicuous to insects than the unrayed capitula (Fig. 9.5) and Marshall and Abbott have shown that rayed capitula receive more insect visits, which may account in part for this increased level of outcrossing (Table 9.2). Nevertheless, seed resulting from the female rayed florets is more likely to be outcrossed than those from unrayed florets in the same capitulum, as estimated by the use of marker genes. Interestingly enough, such relatively high levels of outcrossing have been imposed by hybridisation of the largely autogamous groundsel, in which Abbott has shown that levels of outcrossing in wild-type unrayed populations rarely exceed 1%. Ingram *et al.* (1980) have confirmed my suggestion (Richards 1975b) that rayed morphs of the groundsel have arisen through hybridisation and introgression with the introduced Mediterranean species *Senecio squalidus* (Oxford ragwort) across a 'ploidy barrier (*S. vulgaris* $2n = 40$, *S. squalidus* $2n = 20$). *Senecio squalidus* is a longer-lived, winter annual with large, showy inflorescences. Rayed groundsels have not only inherited relatively showy inflorescences with female ray-florets which convey a higher level of outcrossing, but also a longer generation time from this parent (Richards 1975b).

**Estimation of outcrossing** There are various ways of estimating the amount of outcrossing a plant receives and none is perfect. They are summarised below:

(a) Observation of pollen received on stigmas in a protogynous species before anther dehiscence (underestimates, due to ignorance of cross-pollination after male phase commences), e.g. *Thlaspi alpestre*. A modification involves emasculation of all remaining flowers on the plant before anther dehiscence, and observing pollen on stigmas, or seed-set.

(b) Marking of pollen by external application of a fluorescent powder or blue dye to the anthers, so that outcrossed, non-marked pollen can be detected on stigmas. (This procedure is difficult to use, and some outcrossed pollen may arrive too late to effect fertilisation.)

(c) Observation of flowers that have received animal visits, or have selfed. Most relevant to the Orchidaceae in which it is usually possible to say whether in a successful pollination a pollinium has selfed or has been brought from another flower (this may nevertheless be geitonogamous).

(d) Use of dominant genetic markers on potential male parents to test the phenotype of the progeny of female parents which are homozygous for corresponding recessive alleles. This technique can be used either in experimental populations with all potential pollen donors bearing the marker, or in wild populations in which the frequency of the marker in the population is known. This technique is time-consuming, and uses a lot of space. A further draw-back is where it is suspected that the marker gene itself influences the rate of pollinator visitation (as in flower colour markers, or the radiate marker used in groundsel by Marshall and Abbott). Increasingly, isozyme morphs are used as markers, and will be more generally available than visible mutants.

(e) Inference from departures from Hardy-Weinberg equilibria (Ch. 3) of genotype frequencies in polymorphisms. In its most simple form, it is possible to hypothesise that neither genotype frequencies nor population size vary from one generation to the next. As heterozygotes selfed only give half the frequency of heterozygotes, if the number of heterozygotes remains stable, half must come from outcrossing. A comparison of this estimate of the number that have resulted from outcrossing with those expected from outcrossing in a randomly outcrossed population (Hardy–Weinberg) gives an estimate of proportion of selfing, which is, however, only very approximate.

More sophisticated models suffer from a lack of markers without dominance, in which heterozygotes can be identified (although the use of isozymes has eased this problem), and in assumptions of equal viability for all genotypes.

The variety of estimates that can be obtained for the amount of outcrossing for the same population using different techniques is exemplified by de Arroyo (1975). She compares estimates obtained from the degree of protandry observed in the flowers, from selfed seed production in insect-free conditions, and from reduction in heterozygosis in enzyme polymorphisms of natural populations of outbred and inbred species of *Limnanthes* (Table 9.3).

**Table 9.3** Variation in breeding system and enzyme polymorphism in *Limnanthes* spp. (after de Arroyo 1975).

| Taxon | Within-flower protandry (days) | Seed-set in absence of pollinators (%) | Estimated selfing from floral timing (%) | Estimated selfing from deficiency of heterozygotes (%) | PLP* (%) | $k$ | $H$ (%) |
|---|---|---|---|---|---|---|---|
| *L. alba* var *alba* | 3 | 3–7 | 0 | 0.3 | 41.7 | 1.5 | 16.3 |
| var *versicolor* | 2 | 19–32 | 0 | 2.6 | 36.1 | 1.3 | 15.1 |
| *L. flocculosa* | | | | | | | |
| ssp. *californica* | 1–2 | 100 | 50 | 26.2 | 33.3 | 1.4 | 23.5 |
| ssp. *grandiflora* | 1–2 | 93 | 50 | 21.1 | 41.7 | 1.4 | 9.7 |
| ssp. *pumila* | 1–2 | 89 | 50 | 21.2 | 41.7 | 1.4 | 5.3 |
| ssp. *flocculosa* | 0.1 | 95–100 | 100 | 79.4 | 31.2 | 1.3 | 25.6 |
| ssp. *bellingeriana* | 0 | 69–100 | 100 | 68.9 | 16.7 | 1.2 | 2.0 |

* Abbreviations: PLP, the proportion of loci polymorphic for a species or population; $k$, the mean number of alleles per locus for a species or population; $H$, the mean proportion of heterozygous loci.

**Table 9.4** Breeding system and genetic structure of *Gilia achilleifolia* (after Schoen 1982a,b).

| Population | Protandry Index | Stigma exertion (mm) | Flower weight (mg) | Seed-set in absence of pollinators (%) | $F_0$* | Number of loci | $t$† |
|---|---|---|---|---|---|---|---|
| Eagle Ridge | 0.00 | −0.6 ± 0.2 | 1.4 ± 0.3 | 91 | 0.47 | 4 | 0.15 |
| Arroyo Mocho | 0.13 | 1.7 ± 0.4 | 2.5 ± 0.2 | – | 0.05 | 6 | 0.29 |
| Metcalf Road | 0.37 | 1.2 ± 0.4 | 2.8 ± 0.3 | 84 | 0.10 | 5 | 0.42 |
| Arroyo Seco | 0.69 | 2.1 ± 0.3 | – | 42 | 0.07 | 9 | 0.58 *Adh-1*<br>0.92 *Pgi-2* |
| San Ardo | 0.69 | – | 3.5 ± 0.3 | – | 0.20 | 6 | 0.80 |
| Poly Canyon | 0.86 | 2.3 ± 0.3 | 2.7 ± 0.2 | 48 | 0.22 | 5 | 0.64 |
| Hastings | 0.88 | 1.2 ± 0.2 | 2.6 ± 0.1 | – | 0.04 | 8 | 0.85 *Adh-2*<br>1.06 *Pgi-1* |

* $F_0$, the fixation index, is based on deficiencies in heterozygote frequency from that expected by panmixis, averaged for the number of loci stated for each population, and has a maximum value (heterozygotes absent) of one.

† $t$ the estimated outcrossing rate, is based on deficiencies in heterozygote frequencies from that expected by panmixis, and is estimated from whichever of the loci *Adh-1, 2* and *Pgi-1, 2* show the highest level of polymorphism for each population.

For Arroyo Seco and Hastings, readings from two loci are given.

Modern estimates of outbreeding are derived from frequencies of alleles in offspring in relation to known maternal genotypes, and hypothetical parental allele frequencies, in open-pollinated wild populations (Schoen 1982). Such techniques (Table 9.4), which combine approaches (d) and (e) above, are unfortunately untestable, and make assumptions about allele neutrality and lack of heterozygous advantage; however, these are minimised by the use of ungerminated seed as test samples. They yield 'average' figures for a population, from which some individuals may vary. Unfortunately, pollen involved in outcross events may not be representative of the population as a whole, giving a bias towards genotypes most likely to outcross.

With such constraints in mind, Table 9.5 lists the percentage outcross-

**Table 9.5** Estimates of percentage outcrossing in different self-compatible species, using various techniques.

| Species | Estimated outcrossing (%) | References |
|---|---|---|
| *Festuca microstachys* | 0–0.01 | Kannenberg and Allard (1967) |
| *Spergula arvensis* | 0–3 | New (1959) |
| *Hordeum vulgare* | 1–2 | Allard *et al.* (1968) |
| *H. jubatum* | 1–3 | Babbel and Wain (1977) |
| *H. spontaneum* | 0–10 | Brown (1978) |
| *Avena barbata* | 1–8 | Marshall and Allard (1970) |
| *Galeopsis tetrahit* | 0–16 | Muntzing (1930) |
| *Avena fatua* | 1–12 | Imam and Allard (1965) |
| *Senecio vulgaris* (unrayed) | 0–5 | R. J. Abbott (pers. comm.) |
| *Senecio vulgaris* (rayed) | 12–26 | R. J. Abbott (pers. comm.) |
| *Thlaspi alpestre* | 5.25 | Riley (1956) |
| *Trifolium hirtum* | 1–10 | Jain (1979) |
| *Lupinus affinis* | 0–29 | Harding *et al.* (1974) |
| *L. bicolor* | 13–50 | Harding *et al.* (1974) |
| *L. nanus* | 0–100 | Harding *et al.* (1974) |
| *Gilia achilleifolia* | 15–96 | Schoen (1982b) |
| *Clarkia temborlensis* | 8–83 | Vasek and Harding (1976) |
| *C. exilis* | 43–89 | Vasek and Harding (1976) |
| *Lycopersicon pimpinellifolium* | 0–84 | Rick *et al.* (1978) |
| *Limnanthes alba* | 43–97 | Jain (1978) |
| *Plectritis congesta* | 48–80 | Ganders *et al.* (1977) |
| *Helianthus annuus* | 60–91 | Ellstrand *et al.* (1978) |
| *Eucalyptus obliqua* | 64–84 | Brown *et al.* (1975) |
| *E. pauciflora* | 62–84 | Phillips and Brown (1977) |
| *Vicia faba* | 70 | Allard *et al.* (1968) |
| *Mimulus guttatus* | 87.6 | M. McNair (pers. comm.) |
| *Cheiranthus cheiri* | 92 | Bateman (1955) |
| *Clarkia unguiculata* | 96 | Vasek (1965) |
| *Pinus ponderosa* | 96 | Mitton *et al.* (1981) |

ing estimated by a variety of techniques in 27 species of flowering plants that are self-fertile (Levin 1979). Those with levels below 1% have very small flowers which are often virtually cleistogamous, and in any case may self-pollinate in bud. Species with high levels of outcrossing such as the wallflower, *Chieranthus cheiri*, and *Clarkia unguiculata* are very strongly protogynous and protandrous respectively, to the extent that even geitonogamy is unusual, and have large attractive flowers which encourage frequent insect visits. In cases where very wide limits of estimated outcrossing are recorded, different genetic loci may give different results, or populations may differ strikingly. Thus, in *Gilia achilleifolia*, differences in estimated outcrossing rates are closely allied to the degree of protandry, the degree of stigma/anther displacement and flower size. In reality, all cases except the extreme outcrossers will be likely to show differences in the degree of outcrossing between populations, between days, and between seasons due to temperature, rainfall, wind and other external forces influencing selfing and pollen flow. This is additional to variation in outcrossing induced by genetic differences between plants in floral structure or timing, and in size. Large, vigorous plants bearing many flowers are more likely to be geitonogamously selfed than those with few flowers (Ch. 5).

## ADVANTAGES OF AUTOGAMY

There are three main reasons why a self-fertilising breeding system may succeed:

(a) Reproductive efficiency.
(b) Reduction in genetic variation and the fixation of highly adapted genotypes.
(c) Because, in a mixed population of selfers and outcrossers, the selfer is a pollen donor not only to itself, but also to outcrossers. Outcrossers may self less often than selfers outcross as males, and thus they may donate fewer outcrossing genes to the next generation than selfers will provide selfing genes.

### *Reproductive efficiency (fecundity)*

By no means all self-compatible species self-pollinate efficiently enough for all ovules to be fertilised by selfed pollen. Features of floral structure or timing ensure poor within-flower selfing and, as a rough guide, it can be said that those species normally outcrossed at more than 20% receive most selfed pollen geitonogamously by between-flower pollination. In such species, self-compatibility may increase reproductive efficiency somewhat, but only those features that lead to within-flower selfing in the absence of pollinators will maximise seed-set.

There are remarkably few examples in which reproductive efficiency in terms of percentage seed-set has been compared in related selfers and outcrossers. Nevertheless, it is clear that for species in which fitness is highly reliant on reproductive efficiency, selfing must usually be favoured. It is a common experience to discover little if any seed-set in the wild in fully self-incompatible species. In a wild population of the New Zealand alpine daisy, *Celmisia hectori*, I discovered one seed-set in 120 heads (approximately 4000 ovules; 0.025%). The best field indication of sexuality in the mostly apomictic genus *Taraxacum* (Ch. 11) is poor seed-set. Many heads in sexuals, which are mostly fully self-incompatible, set no seed at all. The average percentage seed-set in a population of the diploid sexual *T. austriacum* in Slovakia was only 20%; this coexisted with several apomictic species of *Taraxacum* in which seed-set rarely fell below 90% (author's unpublished work).

In these Asteraceae, it is easy to estimate seed-set, and to demonstrate by experimental cross pollinations that failure of cross pollination is the cause of low seed-set. However, in many plants, relatively few ovules set seed even when all have been efficiently cross pollinated. Only about 1% of the single-ovule flowers of the Australian *Banksia* usually set seed, although most or all may be cross pollinated. The fruits are very large (Fig. 9.5) and the plant is clearly adapted to develop very few in accord with energetic constraints. These are rather regularly spaced on the cone-shaped receptacle. In the related Waratah, *Telopea speciosissima*, Pyke

**Figure 9.5** A small hover-fly visits the capitulum of a rayed groundsel (×3).

(1982b) has suggested that the number of fruits set may be related to performance of the plant, rather than efficiency of pollination. Equally deceptive can be the results of disease or predation. Mitchell and Richards (1979) have shown that very poor seed-set in some seasons in the Tynemouth (UK) population of the wild cabbage, *Brassica oleracea*, is caused by a combination of seed predation by the weevil, *Ceutorhynchus assimilis*, and secondary fungal infections which set in after weevil predation. Seed-set estimated soon after flowering in this self-incompatible species is much higher than at fruit maturity.

Neither is seed-set easily estimated in some species. In the orchid genus *Epipactis*, the autogamous species develop all their capsules, and microscopic examination reveals that well over 90% of the very many seeds have embryos. In the related outbreeder *E. helleborine*, fruit-set varies from 0 to 100%, but in fruits that have set, seed which appears to be viable to the naked eye proves under the microscope to have a high proportion of testas without embryos, often over 99% (Richards 1982). On average, outbreeding species in this genus probably set less than 10% of the seed set by inbreeders.

Perhaps the most thoroughly studied example of comparative fecundity in related selfers and outcrossers is in the American cruciferous genus *Leavenworthia*. Solbrig and Rollins (1977) show that reproductive efficiency, in terms of ovule set, varies from 52 to 62% in self-incompatible populations, to 73 to 94% in self-compatible populations, thus giving approximately a 3:2 reproductive advantage to the selfers. It is noted, however, that seed-set in outcrossers varies noticeably with the weather, which presumably affects pollinator activity.

One of the few major sources relating to seed-set and reproductive capacity in populations of wild plants is Salisbury (1942). Unfortunately, this exhaustive work takes little account of the breeding system of the plants investigated, but in the poppies, *Papaver* spp., he has compared three species which are known to be self-compatible and which are liable to substantial within-flower self-pollination, with the fully self-incompatible *P. rhoeas* (Table 9.6). From these data it is by no means clear that

**Table 9.6** Reproduction data for four British species of *Papaver* (from Salisbury 1942).

| Species | Mean number of capsules per plant | Mean number of seed per capsule | Mean number of seeds per plant |
|---|---|---|---|
| *P. argemone* (s-c)* | 6.81 ± 0.38 | 313.6 ± 17.7 | 2135.6 |
| *P. hybridum* (s-c) | 7.28 ± 0.42 | 230.0 ± 14.4 | 1674.4 |
| *P. dubium* (s-c) | 6.83 ± 0.34 | 2008.0 ± 126 | 13 714 |
| *P. rhoeas* (s-i) | 12.5 ± 0.6 | 1360.0 ± 125 | 17 000 |

* Abbreviation: s-c, self-compatible; s-i, self-incompatible.

self-incompatibility is a reproductive disadvantage to *P. rhoeas*, which sets substantially more seed per plant than any of its selfing relatives. Perhaps the most meaningful comparison is with *P. dubium*, as the other two species are smaller plants. It is notable that the seed-set per capsule is less in *P. rhoeas* than in the self-compatible *P. dubium*, and the greater reproductive output is due to the greater number of capsules, which may reflect the heterotic vigour of the outbreeder. It should also be noted that all these species are opportunist annual colonisers of open ground, in which seed production is presumably very important, and unlike many such species they have large showy flowers which receive abundant insect visits. Thus even the self-compatible species probably receive much outcrossed pollen, which was estimated at between 71 and 81% by Humphreys and Gale (1974) for *P. dubium*. There is little difference in the quality of the seed between inbreeders and the outbreeders; all four species have very similar seed weights.

The primrose, *Primula vulgaris*, is a relatively long-lived species with some capacity for vegetative spread, and most populations have a dimorphic incompatibility system (Ch. 7). However, a substantial proportion of some populations in Somerset, UK, have homostyle self-compatible flowers. In view of the early flowering, paucity of pollinators and poor seed-set of this species, one might expect the homostyle forms to show a considerable advantage over the heterostyles in reproduction by seed. However, data collected by M. Wilson (personal communication) from the homostyle population at Sparkford shows that this is far from the case, at least in terms of fruit-set (Table 9.7). Piper *et al.* (1984) show that homostyles do set more seed per fruit (about 62% more, Ch. 7), and this is attributable to increased pollination efficiency (selfing) in homostyles.

The surprisingly low fruit-set of the self-pollinating, self-compatible homostyle variety may be at least partially due to the activity of mice and slugs and snails. The latter eat the anthers and stigmas, and are thus liable to make homostyles (in which both organs are at the mouth of the flower) sterile, whereas heterostyles, which have one of the organs protected in

**Table 9.7** Fruit set of homostyle and heterostyle primroses (*Primula vulgaris*) in the field, at Sparkford, Somerset (M. Wilson unpublished observations).

| Morph | Flowers | Fruits | Fruit-set (%) |
|---|---|---|---|
| pin | 342 | 59 | 17.2 |
| thrum | 156 | 31 | 19.9 |
| homostyle | 1104 | 193 | 17.4 |

the floral tube, are not so affected. As already mentioned, the ability to self-fertilise is most likely to enhance fecundity in those species whose floral structure allows efficient within-flower self-pollination, for these can maximise seed-set without requiring insect visits. Species that require insect visits for self- or cross-fertilisation will benefit less from self-fertility, but it is likely that self-fertile strains will usually set more seed (which may be of poorer quality) than self-sterile ones, especially if clones are large or individuals are well spaced or patchy in distribution. However, this does not seem to be the case in *Papaver* or *Primula*. More information on this vital aspect of breeding-system evolution is badly required; since the time of Darwin (1876) we have acquired remarkably little new information on reproductive efficiency of inbreeders in comparison to outbreeders.

## *Genetic variation and fixation of genotypes*

**Theoretical considerations** Darwin said 'nature abhors perpetual self-fertilisation' (Darwin 1862, 1876) but he was well aware of the apparently greater reproductive fitness of many selfers. Repeated self-fertilisation rapidly results in a reduction in genetic variation within populations, and in a decrease in the proportion of loci in populations and individuals that are heterozygous: most loci become fixed in the homozygous state. How this occurs is most simply examined by following the fate of a heterozygous locus *A*/*a* in a line experiencing repeated selfing. The proportion of heterozygotes is halved in each successive generation, for a selfed heterozygote yields 50% heterozygotes and 50% homozygotes, which can subsequently only form homozygotes. The proportion $f^{(n)}$ after $n$ generations selfing is thus $(\frac{1}{2})^n f^0$ where $f^0$ is the original proportion of heterozygotes (Table 9.8).

Thus within relatively few generations, often less than ten, one may

**Table 9.8** The fate of a heterozygous locus *A*/*a* during repeated self-fertilisation.

| | Fate of locus | | | Generation | Proportion of offspring of original parent heterozygous |
|---|---|---|---|---|---|
| homozygotes | | *Aa* selfed | | | |
| *AA* and *aa* | 1 *AA* | 2*Aa* | 1 *aa* | $F_1$ | 0.5 |
| can only | 1 *AA* | 2*Aa* | 1 *aa* | $F_2$ | 0.25 |
| form | 1 *AA* | 2*Aa* | 1 *aa* | $F_3$ | 0.125 |
| homozygote | 1 *AA* | 2*Aa* | 1 *aa* | $F_4$ | 0.062 |
| offspring | 1 *AA* | 2*Aa* | 1 *aa* | $F_5$ | 0.031 |
| | 1 *AA* | 2*Aa* | 1 *aa* | $F_6$ | 0.015 |
| | | etc. | | | |

presume that an obligate selfer would become homozygous (fixed) at most loci, in the absence of heterozygous advantage. It does not follow, however, that an obligate selfing population will become genetically invariable. Different lines within the population may show fixation of different homozygotes from originally heterozygous loci either by chance, if the alleles are nearly neutral with respect to each other, or by disruptive selection. If the alleles are neutral, the proportion of the two homozygotes *AA* and *aa* will depend on the initial frequency of *A* and *a* before selfing.

We have seen, however, that obligately selfing populations are probably very rare, and some outcrossing occurs onto nearly all autogams. In such a partially autogamous population, the proportion of heterozygotes $f^{(n)}$ in the population after *n* generations is calculated by:

$$(s/2)^n f^0 + 2pq[2t/2 - s][1 - (s/2)^n]$$

where *s* is the proportion of selfing, *t* is the proportion of random outcrossing $(1 - s)$, $f^0$ is the proportion of heterozygotes at generation 0, *p* is the frequency of *A* in generation 0, and *q* is the frequency of *a* in generation 0. For further reading on the modelling of the genetic structure of inbred populations, refer to Allard (1960), Gale (1980) or Cook (1971).

As *n* increases $(s/2)^n$ tends towards zero, so that at equilibrium, the frequency of heterozygotes *f* becomes $2pq(2t/2 - s)$ and is thus dependent on the initial allele frequencies, and the proportion of selfing. These equilibrium frequencies of heterozygotes will vary from 0.017 (less than 2%) where one allele is initially rare ($p = 0.05$) and selfing is high (90%, $t = 0.1$) to 0.50 where $p = 0.5$ and the populations are fully outbreeding ($t = 1$) (Table 9.9).

**Table 9.9** Expected equilibrium proportions of heterozygotes under mixed random mating and selfing (without selection) for various assumptions about *t* and *p* (initial frequency of allele $A_1$).

| Probability of outcrossing ($t$) | Initial frequency of $A_1$ ($p$) | | | | | |
|---|---|---|---|---|---|---|
| | 0.05 | 0.10 | 0.20 | 0.30 | 0.40 | 0.50 |
| 0.1 | 0.0173 | 0.0327 | 0.0582 | 0.0764 | 0.0783 | 0.0909 |
| 0.2 | 0.0317 | 0.0600 | 0.1067 | 0.1400 | 0.1600 | 0.1667 |
| 0.3 | 0.0439 | 0.0831 | 0.1477 | 0.1939 | 0.2216 | 0.2308 |
| 0.4 | 0.543 | 0.1029 | 0.1829 | 0.2610 | 0.2743 | 0.2857 |
| 0.5 | 0.0633 | 0.1200 | 0.2133 | 0.2800 | 0.3200 | 0.3333 |
| 0.6 | 0.713 | 0.1350 | 0.2400 | 0.3150 | 0.3600 | 0.3750 |
| 0.7 | 0.0783 | 0.1482 | 0.2636 | 0.3459 | 0.3953 | 0.4118 |
| 0.8 | 0.0844 | 0.1600 | 0.2844 | 0.3733 | 0.4266 | 0.4444 |
| 0.9 | 0.0900 | 0.1705 | 0.3022 | 0.3969 | 0.4548 | 0.4737 |
| 1.0 | 0.0950 | 0.1800 | 0.3200 | 0.4200 | 0.4800 | 0.5000 |

Consideration of the effect of selfing on many originally heterozygous loci is simple when they are unlinked on different chromosomes, as each locus will obey the same rules independently. The number of generations needed for an individual to become fully homozygous increases with the number of loci considered (Allard 1964, Fig. 9.7) but is still remarkably small. After 12 generations of repeated total selfing, 98% of individuals originally heterozygous for 100 loci (assuming so many could be unlinked!) will have become totally homozygous for these loci. However, even low frequencies of outcrossing, of the order of 1%, will allow few loci to fix, and fixation equilibria are rarely reached, because crossing will occur between differentially fixed homozygotes (*AA* and *aa*) in the same population.

**Linkage disequilibria** When loci are linked on the same chromosome, considerations of increased homozygosity and genetic fixation become more complex (usefully summarised in Allard *et al.* 1968). Linkage of two

**Figure 9.6** Infructescence of *Banksia quercifolia* (Proteaceae) near Sydney, Australia. Only about 1% of flowers have set fruit; the regular distribution of set fruits on the infructescence suggests that pollination was adequate, and that fruit-set is limited by resources available to female function. Fruits remain closed for years until burnt by bush fires, when seeds are released, as here (×0.25).

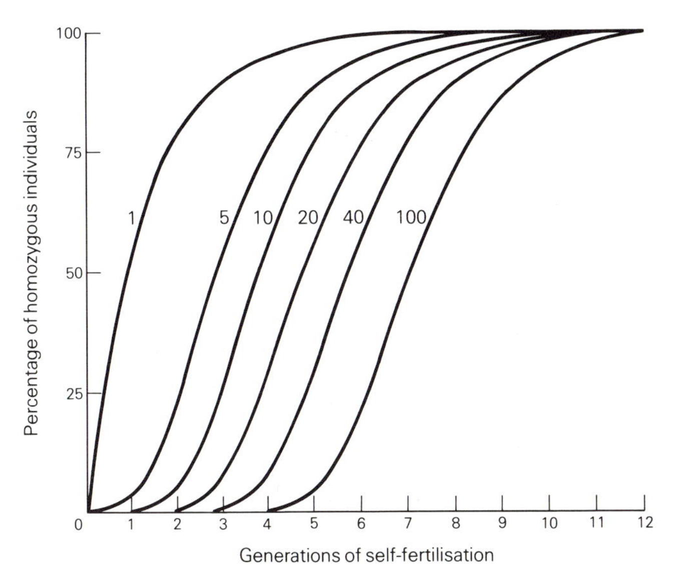


**Figure 9.7** Percentage of homozygous individuals after various generations of self-fertilisation, when the number of independently inherited heterozygous gene-pairs is 1, 5, 10, 20, 40 and 100 (after Allard 1964).

loci *A*/*a* and *B*/*b* on the same chromosome will tend to result in linkage disequilibria, in which the frequency of the four possible gametes *AB*, *Ab*, *aB* and *ab* differs from that predicted by the products of the frequencies of the alleles *A*, *a*, *B* and *b* in the population. In the absence of differential selection in an outbreeding population, linkage disequilibria will disappear through recombination in time, although the more closely linked the loci (low recombination values), the more slowly this will occur. Total selfing of a population carrying two heterozygous linked loci in disequilibrium will probably result in homozygous fixation at disequilibrium, with some genotypes (e.g. *AAbb*) being more common and some (e.g. *AABB*) less common than predicted by allele frequencies.

In a partly inbred situation, linkage can cause linkage disequilibria to be maintained, and fixation to occur very slowly. In Fig. 9.8, where $s = 0.9$ ($t = 0.1$), random segregation ($c = 0.50$) causes linkage disequilibria to almost halve in 10 generations and zygotic associations, which monitor genetic fixation, increase. However, tight linkage between two loci ($c = 0.01$) causes almost no diminution of disequilibrium, and fixation proceeds more slowly. As one allele proceeds to homozygous fixation through the force of inbreeding, it 'drags' closely linked alleles of different loci with it, irrespective of their type. Linkage, especially close linkage in which the 'normalising' effects of recombination are minimised, removes the independence of a locus proceeding to fixation.

However, in totally autogamous populations, recombination will eventually ensure that chromosomes as well as genes will become entirely homozygous (Gale 1980, pp. 22–33).

**Polyploidy** Just as linkage tends to slow down the process of homozygous fixation of polymorphic loci in inbreeders, so does polyploidy. At polysomic loci, the proportion of heterozygotes remaining in the population $H^n$ after total selfing is determined by:

$$[(4K - 3)/2(2K - 1)]^n \times H^0$$

where $H^0$ is the original frequency of heterozygotes, $n$ is the number of generations of obligate selfing, and $K$ is half the 'ploidy level (2X = 1, 4X = 2, 6X = 3, etc). After one generation of selfing in a diploid, where $H^0$ = 0.5, $H^n$ becomes 0.25, i.e. the proportion of heterozygotes is halved. Thus, in a tetraploid, $H^n$ is 0.416 and the proportion of heterozygotes is reduced by 1/6, and in a hexaploid $H^n$ is 0.45 and the proportion of heterozygotes is reduced by 1/10. At a tetrasomic locus, heterozygotes can exist in simplex (*Aaaa*), duplex (*AAaa*) and triplex (*AAAa*) forms. Thus, polyploids buffer polymorphic loci against the effects of selfing more effectively than diploids. Of course, most polyploids are allopolyploid, and are not necessarily polysomic for any one locus; however, resulting as they do from hybrids between related parents, allopolyploids may be more often effectively polysomic than appears from the homology and pairing relationships of chromosomes from the parental genomes, and tetrasomic inheritance is well known in many loci of allotetraploids such as dutch clover *Trifolium repens* or birds-foot trefoil, *Lotus corniculatus*. It has been suggested that inbreeders are more often polyploid than outbreeders; this relationship, if it exists, may be due to the failure of outbreeding systems such as dioecy (Ch. 8), heteromorphy (Ch. 7) or gametophytic self-incompatibility (Ch. 6) at the polyploid level. However, it is clear that polyploids are better able to maintain relatively high levels of variability and heterozygosity when partially selfed, or through temporary bouts of total selfing than are diploids; in high polyploids, the accumulation of homozygosity may be very slow indeed.

Inbreeders can also more successfully achieve polyploidy in the first place, by selfing. A new outcrossing tetraploid arising in a diploid population will be disadvantaged by producing sterile triploid offspring, whereas a selfer will produce fertile tetraploids (Fowler & Levin 1984).

**Disadvantageous alleles** It is likely that a high proportion of the recessive alleles carried heterozygously will be disadvantageous if expressed phenotypically in the homozygous condition. In a randomly outbred, panmictic condition, such homozygotes will occur, and be selected against, but as selection renders the allele increasingly scarce, the homozygote becomes still scarcer, at the square of the frequency of the allele. In a population in which the frequency of a recessive allele is 1 in 100 (0.01), it will be expressed in only 1 in every 10 000 individuals (0.0001). At low frequencies, selection against harmful recessive alleles proceeds very slowly indeed in panmictic populations. This allows high levels of residual variability and heterozygosity to be maintained.

The pattern in a selfing population is quite different, for heterozygotes that can protect disadvantageous recessive alleles rapidly disappear, and these alleles are exposed to selection in a homozygous condition. Instead of being protected from the rigours of selection at low frequencies, as in a panmictic condition, disadvantageous alleles will not occur in a fully selfing population, and will be unusual even in less fully selfed plants. Selfing populations carry little if any 'genetic load' of deleterious recessives, and are genetically 'clean'. However, the early phases of inbreeding may expose obvious deleterious recessives. Experimentally selfed material from naturally outcrossing populations may produce some progenies with a proportion (usually 1:3) of non-viable offspring that may be albino (chlorophyll-less), very slow growing, amino-acid requiring or which may simply fail to germinate (lethal). Continued selfing will lead to genetically 'clean' stocks in which such lethals or sublethals no longer occur, although these may then suffer from inbreeding depression, which will be discussed in a later section.

**Results from genetic structural analysis** It is currently considered that the most thorough and objective analyses of the genetic structure of populations are obtained through the techniques of gel electrophoresis of enzyme systems. Extracted enzymes that are provided with a specific substrate, and stained with a specific stain, separate as bands which, under constant conditions, have repeatable mobilities on the gel. Some enzymes produce more than one stained band and when these occur close together on the gel they are usually assumed to be, and through analytical genetic crosses should be proved to be, different alleles at the same locus. Several different loci, usually more widely separated from each other on the gel, may occur for a given enzyme. These studies of isozymes, or allozymes, allow many different loci to be examined for individuals and for populations.

A number of different statistics can be derived from the isozyme studies of wild populations, and these are listed below:

(a) PLP, the proportion of loci polymorphic for a species or population.
(b) $k$, the mean number of alleles per locus for a species or population.
(c) $H_0$, the mean proportion of loci heterozygous per individual.
(d) $H_e$, the mean proportion of loci per individual expected to be heterozygous if panmictic.

(e) $F$, fixation index, derived from $H_0$ and $H_e$, and thus a measure of departure from panmictic Hardy–Weinberg expectations of heterozygote frequencies at several loci. If $F$ is 0, panmictic expectations are met ($H_0 = H_e$). If $F$ is 1, there is full fixation, i.e. no heterozygotes occur.

(f) $t$, outcrossing rate estimate, derived from comparisons of observed and expected heterozygote frequencies in a population for a single locus.

Clearly, the statistics of the greatest interest to the student of selfing plants are $F$ and $t$. These have been calculated for mostly inbred and mostly outbred populations of *Gilia achilleifolia* by Schoen (1982a,b) (see Table 9.4).

Plants that are highly protandrous, and thus unlikely to give much within-flower selfing, have high scores on the protandry index, and are predominantly outbred, with less than 50% fruit-set in the absence of pollinators (see the bottom of the table). These populations show low levels of genetic fixation $F_0$ (scarcely varying from Hardy–Weinberg expectations), and high levels of estimated outcrossing rate, $t$, as might be predicted. In contrast, selfed populations, typified by that at Eagle Ridge, have high genetic fixation rates (0.47) and low estimated outbreeding (0.15). Schoen suggests that lowered levels of estimated outcrossing could be due to disruptive selection within the apparent population giving two differently adapted subdemes with different isozyme constitutions (the so-called Wahlund Effect, Wahlund 1928). If this were the case, it would be expected that different homozygous accessions would yield statistically different proportions of heterozygotes from seed, and this is not the case in this instance.

[Levin (1978) shows that the partially selfed *Phlox cuspidata* is less heterozygous, but has greater between-populational variability than have its outbreeding relatives *P. drummondii* and *P. roemariana*.]

Perhaps the most thorough enzymic study of autogamous populations is that of the familiar garden weed, shepherd's purse, *Capsella bursa-pastoris*, by Bosbach and Hurka (1981). Although much data is presented in this work, relatively few calculations are made, but some can be determined from the paper. These authors examined 1368 individuals from 81 populations ranging from Switzerland to northern-most Norway, a transect of some 5000 km. For the 17 loci identified in three enzyme systems, eight were heterozygous (PLP = 0.48). $k$ values were very low; in 44 populations (54%) no heterozygotes were found at all, and 13 populations (16%) were entirely monomorphic for all 17 loci; $k$ values did not exceed 1.2. $H_0$, the proportion of loci that are heterozygous, was only 2.17%. However, as many as 35% of all individuals studied may be heterozygous at at least one locus. Curiously, perhaps, they find the most geographically extreme populations tend to have the highest proportion of heterozygotes.

Clearly, much work remains to be done on the genetic structure of inbreeding populations as assessed by isozyme analysis. However, early results suggest that predictions of high levels of monomorphism within populations for some, or even all, systems are fulfilled. In addition, higher levels of homozygous fixation and much lower levels of heterozygosity than expected by panmictic equations are also found, as genetic theory would also predict. Many populations of, for instance, *Capsella bursa-pastoris*, are apparently genetically invariable. Inbreeding populations of *Leavenworthia* and *Gilia* probably experience much higher levels of outcrossing (above 10%) than *Capsella* (which may be much less than 1% outcrossed), but their genetic variability is nevertheless also much reduced.

## PHENOTYPIC (MORPHOLOGICAL) VARIATION

If selfing populations are genetically less variable than outcrossed relatives, one would expect them to be phenotypically less variable as well. However, the phenotype results from interactions between the genotype and the environment. The extent to which the same genotype can give rise to different phenotypes in different environments is here termed phenotypic plasticity, which is dealt with in a later section. Suffice it to say for the present that the degree of plasticity may itself be under genotypic control, and it might be expected that genotypically invariable populations would respond to selection pressures that encourage a high level of plastic potential to evolve.

At the simplest level, one can compare morphological variability for various metric attributes in the field between related selfing and outcrossing species, or populations, I have done this with *Epipactis* (Richards 1982) for nine populations of seven species (Table 9.10). For each of these, the coefficient of variation is compared for ten characters. Coefficient of variation is a useful statistic in such comparisons, for dividing the standard deviation of the mean by the mean, it removes scalar effects on the estimate of variability given by the standard deviation, and thus allows the variability of characters with very different means to be directly compared.

Although differences in coefficients of variation are not large when the outbreeding (allogamous) populations are compared with the selfing (autogamous) populations, there is little overlap between individual readings in the two groups, and when mean readings for each character are compared between the allogamous populations together and the autogamous populations together, in no instance does the autogamous population show greater variability in any of the ten characters.

**Table 9.10** Coefficients of variation in allogamous and autogamous populations of *Epipactis* in north-east England (from Richards 1982).

| Population | Basal leaf | | Stem leaf | | Bract | | Ovary | | Sepal | | Mean coefficient of variation |
|---|---|---|---|---|---|---|---|---|---|---|---|
| | Length | Width | Length | Width | Length | Width | Length | Width | Length | Width | |
| **Allogamous species** | | | | | | | | | | | |
| *E. atrorubens* ($n = 26$) | 0.32 | 0.25 | 0.27 | 0.27 | 0.41 | 0.34 | 0.19 | 0.18 | 0.15 | 0.18 | 0.26 |
| *E. palustris* ($n = 25$) | 0.36 | 0.25 | 0.18 | 0.22 | 0.34 | 0.23 | 0.17 | 0.20 | 0.16 | 0.15 | 0.23 |
| *E. helleborine* Dissington ($n = 16$) | 0.40 | 0.40 | 0.20 | 0.22 | 0.39 | 0.41 | 0.21 | 0.19 | 0.16 | 0.25 | 0.28 |
| *E. helleborine* Settlingstones ($n = 16$) | 0.29 | 0.28 | 0.12 | 0.20 | 0.29 | 0.37 | 0.23 | 0.31 | 0.12 | 0.12 | 0.23 |
| Mean coefficient of variation | 0.34 | 0.30 | 0.19 | 0.23 | 0.36 | 0.34 | 0.20 | 0.22 | 0.15 | 0.18 | 0.25 |
| **Autogamous species** | | | | | | | | | | | |
| *E. leptochila* ($n = 28$) | 0.26 | 0.24 | 0.11 | 0.20 | 0.45 | 0.47 | 0.14 | 0.18 | 0.11 | 0.14 | 0.23 |
| *E. dunensis* Beltingham ($n = 21$) | 0.20 | 0.19 | 0.17 | 0.20 | 0.24 | 0.29 | 0.19 | 0.23 | 0.09 | 0.16 | 0.19 |
| *E. dunensis* Holy Island ($n = 32$) | 0.25 | 0.25 | 0.23 | 0.21 | 0.31 | 0.30 | 0.14 | 0.26 | 0.11 | 0.18 | 0.22 |
| *E. phyllanthes* ($n = 17$) | 0.33 | 0.28 | 0.19 | 0.19 | 0.31 | 0.32 | 0.16 | 0.26 | 0.11 | 0.12 | 0.22 |
| *E. youngiana* ($n = 25$) | 0.22 | 0.24 | 0.15 | 0.14 | 0.23 | 0.36 | 0.17 | 0.20 | 0.12 | 0.10 | 0.19 |
| Mean coefficient of variation | 0.25 | 0.24 | 0.17 | 0.19 | 0.31 | 0.34 | 0.16 | 0.22 | 0.11 | 0.14 | 0.215 |

**Table 9.11** Coefficients of variation in fruit length and fruit width in self-compatible (s-c) and self-incompatible (s-i) populations of *Leavenworthia* (after Solbrig & Rollins 1977).

| | Parents (field) | Offspring (culture) |
|---|---|---|
| Fruit length | | |
| s-c | 0.091 | 0.118 |
| s-i | 0.133 | 0.127 |
| Fruit width | | |
| s-c | 0.066 | 0.118 |
| s-i | 0.098 | 0.119 |

Similar trends are found in Solbrig and Rollins' study of *Leavenworthia*, when the variability of fruit length and width for self-compatible and self-incompatible populations are compared for parental populations (in the field), and for their offspring raised in relatively standard conditions (Table 9.11). In all cases, self-compatible populations are on average less variable than self-incompatible populations, although these differences were not always statistically significant. Interestingly, variability in offspring was greater than in parents, and less distinct between the two breeding systems, although growing conditions were more uniform. This may be a product of the relatively relaxed stabilising selection operating in artificial growing conditions, allowing a greater range of genotypes to survive to fruiting.

As the authors predict, between-family variance is significantly greater than within-family variance in nearly all populations of the self-compatible *L. exigua*, whereas most families of self-incompatible populations gave between-family variances that were not significantly greater than within-family variances. In addition, between-family variances were greater on average in self-compatible than in self-incompatible populations. Here we see a fairly clear example of the effect of homozygous fixation in inbreeders; differential fixation of neutral or nearly neutral alleles has apparently increased between-family differences, but decreased within-family differences in inbreeders compared with related outbreeders.

The phenotypic effects of selfing can be investigated by comparing variability of the offspring of a single individual in cases where chasmogamous and cleistogamous flowers are both produced. Both Clay and Antonovics (1985) for the grass *Danthonia spicata*, and Waller (1984) for *Impatiens capensis* demonstrate that offspring from cleistogamous flowers are less variable than those from chasmogamous flowers.

## PHENOTYPIC PLASTICITY

As already discussed, most largely selfing species are opportunist annuals of open, cultivated and waste ground. The open nature of such habitats may be temporary, as in an arable field, or a garden bed, but if it is permanent, climatic or edaphic factors may prevail which preclude luxuriant plant growth. In the UK, natural annual communities may be found in such sites as sand dunes, cliff tops, very shallow soils, nutrient-poor heaths or chalk downland, etc. In most of these sites both nutrients and water availability may be very limiting to plant growth, and annuals are very small indeed. It is common to find, for instance, *Veronica arvensis* with one leaf, one flower, a total seed production of two, and an above-ground height at fruiting of 6 mm. Seeds from such plants when grown in optimum conditions can produce 150 flowers, 300 seeds and an above-ground height of 150 mm. Such plastic differentials to a factor of 100 or more are usual in autogamous annuals. They allow such plants to reproduce in the most inimical conditions. This is clearly of fundamental importance to populations that must reproduce by seed, whatever the season, often in conditions, or at a time of year, when potentially competing perennials cannot function.

Phenotypic plasticity is a fascinating subject that has been little studied and is still less understood. In particular, there is almost no work on the relationship between the breeding system, levels of genetic heterozygosity and plasticity. What little work there is, is confined to animals (*Drosophila*) and is summarised in Bradshaw (1965) which forms an early, but very useful, review of the topic. This suggests that high levels of heterozygosity are associated with phenotypic stability (low plasticity) which would conform with the high plasticity of inbred annuals. Yet Waddington (1957) associates phenotypic plasticity with the epigenetic background, and intuitively one might expect the capability for high levels of plasticity to be a product of genetic diversity: the more heterozygous, the greater number of allele-controlled enzymes that can be called upon with respect to phenotypic response. That this may well be a naïve view is suggested by the following conclusions of Bradshaw's review:

(a) Phenotypic plasticity for a character is itself under genetic control, and can respond to selection.
(b) Consequently, characters show individual responses to plasticity; one character may be much more plastic than another in one population, and vice versa in another (Sørensen 1954 for *Capsella*).
(c) Determinate, or rapidly determined characters (e.g. leaf shape or flower characters), will be much less liable to plasticity than indeterminate or slowly determined characters (e.g. size or number of organs).

(d) Whereas there is considerable evidence that the degree of plasticity varies with such environmental pressures as density, climate, soils etc., there is no clear evidence that it is related to the breeding system. Work comparing plasticity of inbred and hybrid lines of *Nicotiana* (Jinks & Mather 1955) and tomato (Williams 1960) shows no relationship between plasticity and genetic heterozygosity. It must therefore be concluded that the massive capability of many genetically invariable annuals for phenotypic plasticity has been selected for as a result of the lifestyle of these plants, and is not a direct product of the breeding system. However, such plasticity may go some way to compensating the plant for the lack of genetic variability in its offspring.

## INBREEDING DEPRESSION

It will be noted later (Ch. 10) that hybrids are usually very vigorous in comparison with their parents. This vigour, sometimes called hybrid vigour, is associated with the unusually high levels of heterozygosity found in hybrids, when it may be termed heterosis. Biometrical genetic analysis shows that heterosis is commonly the result of dispersion of dominant alleles. Thus, if one parent was *AABBccdd*, and the other was *aabbCCDD*, the F1 hybrid would be *AaBbCcDd*. If all dominants additively and equally donated vigour effects, the hybrid would be more vigorous than the parents. Increased fitness associated with dominance and heterozygosity at a single locus is called heterozygote advantage, and this may also be expressed through increased vigour in comparison with homozygotes.

The corollary of hybrid vigour, or heterosis, is inbreeding depression, for by definition inbred and relatively homozygous individuals will be less vigorous than their heterozygous counterparts. The first major investigation of inbreeding depression, and still by far the most complete, is by Darwin (1876). He compared the performance of lines that had been artificially selfed, and artificially crossed, usually for ten generations or more, in more than 40 species. This staggeringly complete piece of work has no equal, and has much to offer the modern student, although it rarely seems to be quoted today. From it, the following major conclusions may be drawn:

(a) Repeated selfing renders the majority of species lines relatively less vigorous, when measured as height, weight or reproductive capacity (and thus often less fit).
(b) This is not, however, true of all species lines investigated, and some show little if any response to repeated selfing when compared to repeated crossing (Table 9.12).

**Table 9.12** Lists of species that showed, and did not show, inbreeding depression after (usually ten) generations of artificial selfing when compared with artificially crossed lines of the same species (after Darwin 1876).

| Species showing inbreeding depression | | Species showing no inbreeding depression | |
|---|---|---|---|
| Outbreeders | Inbreeders | Outbreeders | Inbreeders |
| *Ipomaea purpurea* | *Lactuca sativa* | *Scabiosa atropurpurea* | *Borago officinalis* |
| *Primula veris* (heterostyle) | *Iberis umbellata* | *Passiflora gracilis* | *Nolana prostrata* |
| *Cyclamen persicum* | *Papaver dubium* | *Hibiscus africanus* | *Legousia hybrida* |
| *Nicotiana tabacum* | *Reseda lutea* | *Dianthus caryophylleus* | *Bartonia aurea* |
| *Lobelia fulgens* | *R. odorata* | *Primula sinensis* | *Adonis aestivalis* |
| *Nemophila insignis* | *Viola tricolor* | *Petunia violacea* | *Phaseolus multiflorus* |
| *Mimulus luteus* | *Tropaeolum minus* | | *Pisum sativum* |
| *Digitalis purpurea* | | | *Ononis minutissimum* |
| *Verbascum thapsus* | | | *Phalaris canariensis* |
| *Origanum vulgare* | | | *Primula veris* (homostyle) |
| *Brassica oleracea* | | | |
| *Escholtzia californica* | | | |
| *Pelargonium zonatum* | | | |
| *Limanthes douglasii* | | | |
| *Lupinus luteus* | | | |
| *Lathyrus odoratus* | | | |
| *Cytisus scoparius* | | | |

(c) Inbreeding depression usually manifests itself in the first generation, but may increase for ten inbred generations or more; even a single outcross will tend to nullify inbreeding depression, especially when made to another inbred line.

(d) The amount of inbreeding depression expressed after ten generations differs markedly between species, and between lines of the same species. However, it is typical in an outbreeder to find inbred lines with about 70% vigour of the outcrossed lines. Vigour in inbred lines appears to be inherited, and selection of particularly vigorous inbred lines in successive generations may eventually yield lines that are more vigorous than the most outbred (as *Petunia violacea*).

(e) On the whole, normally outbred species show more response to selfing than normally inbred species, as may be discerned from Table 9.12. This is perhaps most strikingly shown for the cowslip, *Primula veris*, for which normal heterostyle lines show marked inbreeding depression. However, homostyle lines of horticultural origin show little if any inbreeding depression when selfed (Ch. 7). It may therefore be suggested that many habitual inbreeders show little inbreeding depression because of adaptation to repeated selfing, perhaps through the selection of unusually vigorous inbred lines.

Inbreeding depression may express itself at different stages in the life cycle. In *Thlaspi alpestre*, which has been stated to be only about 5% outcrossed (Riley 1956), the percentage seed germination of artificially inbred lines is about 70% of within- and between-populational crosses, and indeed of wild collected seed. In *Spergula arvensis*, a cornfield weed in which outcrossing is estimated as below 3%, heterozygotes for the seed-coat character papillate/non-papillate germinate faster than either selfed homozygotes.

In the wild oat, *Avena fatua*, Allard *et al.* (1968) showed that survival after germination, the number of tillers produced per plant, and the time to heading were all superior in the progeny of open pollination, in comparison with the progeny of artificial selfs. This pertains despite the low level of outcrossing (less than 12%) estimated in open-pollinated progeny.

It cannot necessarily be assumed, however, that the lower vigour of inbred lines confers less fitness on those lines in comparison with outcrossed lines. For instance, slow growth rates or smaller stature might be favourable in sub-optimal conditions. Furthermore, fitness is not only an attribute of vigour, but also of reproductive capacity, and the lower production of reproductive units in a less vigorous inbred line might be more than compensated for by its greater reproductive efficiency. Regrettably, few studies comparing total fitness of inbred and outbred lines exist, and still fewer compare their performance in competition. Thus, the tentative results of Solbrig and Rollins (1977) in *Leavenworthia* are worth examination. Between-species competition experiments show that the mean dry weight of the self-incompatible outbreeder *L. alabamica* tends to be less when grown with itself than when it competes with the self-compatible *L. crassa*, *L. torulosa* and *L. exigua*. In comparison, the inbreeder *L. exigua* does better when it competes with itself than when it competes with *L. alabamica*. However, the results of *L. crassa* and *L. torulosa* are more confused, and too much should not be made of this preliminary experiment.

In a very interesting recent study, Waller (1984) showed that the offspring of allogamous flowers outcompete the offspring of cleistogamous flowers, from the same parent plant of the balsam *Impatiens capensis*. Some competitive effects were heritable, and density-dependent, and seedlings from allogamous flowers tended to be more variable than those from cleistogamous flowers for a number of attributes. This variability was considered to form an important component in competitive success.

That inbreeding depression may be allied to the most obvious genetic feature of selfing, the scarcity of heterozygotes, is perhaps not entirely self-evident. Furthermore, it is by no means clear whether inbreeding depression, if mediated by homozygosis, is the product of high levels of homozygosis at all loci, or at relatively few loci. The results of Harding and his associates, reported in Allard *et al.* (1968), on Lima beans (*Phaseolus lunatus*) suggest the latter (Fig. 9.9). The very strong inverse correlation between the fitness of heterozygotes at a single locus S/s and their frequency is most striking. Clearly, if these heterozygotes make up more than about 15% of the population, they encounter each other competitively at a level that nullifies the heterozygote advantage they possess. From this very interesting result, two important deductions may be made:

(a) Heterozygote advantage, and thus presumably heterosis, may be frequency dependent, and therefore there may be a more marked vigour response resulting from an occasional outcross in a predominantly selfed population than from habitual outcrossers.
(b) Selfers with high levels of homozygosis may be able to maintain vigour by retaining heterozygosity at few loci with marked heterozygote advantage; in a selfing condition heterozygosis at these loci will be strongly selected for, thus off-setting the tendency to homozygosity at these loci.

That there is a relationship between selfing, homozygosis and inbreeding depression is beyond dispute. However, we are still far from understanding the metabolic or molecular basis of this association. Perhaps the most useful concept at present is that of alternative metabolic pathways

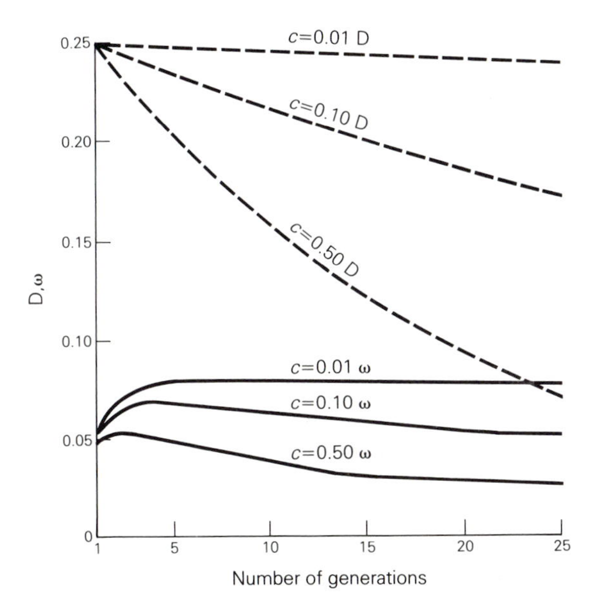


**Figure 9.8** Changes in D (gametic associations, a measure of linkage) and $\omega$ (zygotic associations, a measure of genetic fixation) in a 90% selfed ($s = 0.90$) model population for three different levels of linkage between *A*/*a* and *B*/*b*. Where $c = 0.50$ there is random segregation; where $c = 0.10$ and 0.01, *A*/*a* and *b*/*b* are recombined in 20% and 2% of meioses respectively. The experiment runs for 25 generations ($x$ axis). The original population started with the linkage groups *AB* and *ab* equally frequent. After Allard *et al.* (1968).

or 'shunts'. Presumably, the most vigorous individual is the one that is metabolically most efficient. Rates of metabolic reactions depend on each other, showing positive and negative feedback, and are thus dependent on metabolic 'bottlenecks'. Such bottlenecks may be by-passed by shunts, thus increasing metabolic rates. One may assume that plants with high levels of heterozygosis produce more types of enzyme, and may thus be capable of more metabolic shunts. This vague model may also explain why plants show vigour when heterozygous only at certain loci, for these might be providing shunts at critical points in the metabolic system.

It is also possible that inbreeding leads to reduced male fertility, perhaps as a result of homozygosity in recessive partial male-sterility genes, and that such infertility automatically selects for outbreeding events (Crawford, personal communication). This concept deserves further investigation. Certainly, Levin (1984a) shows greater levels of female sterility (as seed abortion) after selfs rather than crosses in *Phlox drummondii*. There is some indication also of sterility in near-neighbour crosses in this species.

## TAXONOMIC CONSIDERATIONS

A simple way of describing the genetic consequences of repeated selfing is that populations are likely to show less within-population variability but more between-population variability than populations with higher levels of outcrossing. Selfing leads to homozygosis within populations, and local selection pressures will favour relatively few and similar invariable lines, or even only a single line within a population. In contrast, different populations will suffer different selection pressures, and will have experienced different histories of chance fixation of neutral alleles, and thus very striking between-population differences may occur. The impact of these on the observer is emphasised by the phenotypic uniformity displayed within populations.

It is a common experience to find constant, and often very marked, differences between populations of annual garden weeds, often over short distances. Over more than a decade, transient populations of the shepherd's purse, *Capsella bursa-pastoris*, have escaped the gardener's hoe in the flower beds on the campus of the University of Newcastle upon Tyne. Year after year, I have noticed that a certain uniform phenotype is typical of a certain bed, but in other beds within 100 m, very different phenotypes occur. Such differences are so striking that they have tempted generations of taxonomists to describe them as species. Between 1907 and 1929, the Swede Almqvist (1923) described no less than 144 species in *Capsella*; many of the interspecific differences were shown by Shull (1929) to be caused by differential fixation of relatively few major genes, and these taxa are generally not employed today. Almqvist was following a tradition initiated by the Frenchman Jordan in the 1870s. In the autogamous annual *Erophila verna*, Jordan (1864) described over 200 species, and he also created lesser numbers of segregates in other largely autogamous Cruciferae (Brassicaceae) such as *Thlaspi*, *Iberis* and *Biscutella*. As a result, such segregate, or 'microspecies', in autogamous groups have sometimes been nicknamed 'Jordanons'. Unlike the segregate species described in agamospermous genera such as *Taraxacum*, *Hieracium* or *Rubus* (Ch. 11), Jordanons have largely fallen into disrepute and are rarely employed today. However, as in all taxonomy, pragmatics play a part and segregate species are still recognised in several genera in which relatively few microspecies were designated. Thus, there would appear to be no difference in principle between the segregates of *Erophila*, *Capsella*, *Thlaspi* or *Viola arvensis* which are not now used, and those of *Salicornia* (7), *Epipactis* (8), *Iberis* (8), *Arenaria serpyllifolia* s.l. (17), *Biscutella* (20), or *Euphrasia* (20) (approximate numbers of autogamous segregate species recognised in *Flora Europaea* are given in parentheses).

## ECOLOGY AND DISTRIBUTION

Many genera or related groups of species show a consistently autogamous pattern of reproduction (*Erophila, Capsella, Hordeum, Avena, Spergula*) and such genera almost invariably prove to be opportunist colonisers of open habitats which use seeds for dormancy and travel. In these, selfing can be presumed to have evolved in response to strong selection for maximum reproductive efficiency.

It is instructive to examine groups of related populations or species in which both selfing and outcrossing systems occur, to enquire whether noticeable ecological differences can be detected between plants of different breeding systems. In some cases, no such differences are apparent. *Papaver rhoeas*, the self-incompatible poppy, has a very similar ecology and a similar distribution to its self-compatible relatives (of which *P. dubium* is the commonest), and frequently coexists with them. However, it may be significant that *P. dubium* extends rather further north in the British Isles than *P. rhoeas*, and is the only species found in the windswept and pollinator-poor Hebridean islands (Fig. 9.10). The self-incompatible and self-compatible populations of *Leavenworthia* studied by Solbrig and Rollins (1977) all occur in a specialised habitat (limestone outcrops, termed 'cedar glades'); the two fully self-compatible species *L. uniflora* and *L. exigua* are more widespread, and in the case of *L. uniflora*, occur further north than the self-incompatible species. A similar

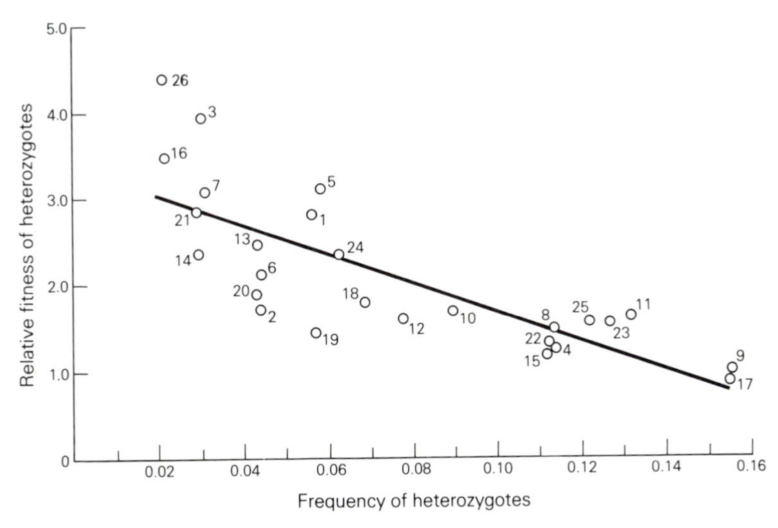


**Figure 9.9** Relationship between the frequency of heterozygotes and their fitness relative to homozygotes for the *S/s* locus in lima bean (*Phaseolus lunatus*) populations. Homozygotes were assigned a relative fitness of 1.0 (after Harding *et al*. 1966, quoted in Allard *et al*. 1968).

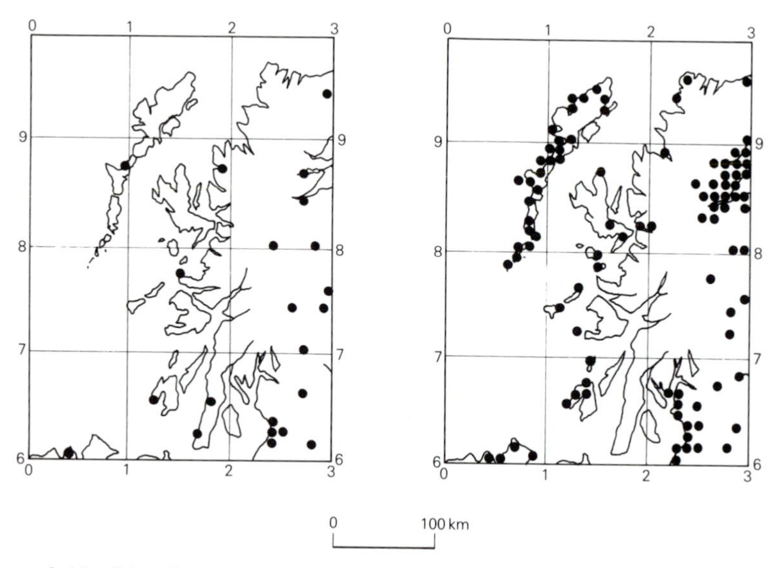


**Figure 9.10** Distribution in western Scotland of the self-incompatible outbreeding poppy *Papaver rhoeas* (left) and the self-compatible *P. dubium* (right). Unlike *P. rhoeas*, *P. dubium* has colonised the Scottish Isles. (From the Atlas of the British Flora by kind permission of the Botanical Society of the British Isles.)

pattern occurs in *Gilia achilleifolia* (Schoen 1982a), in which northern populations are more selfed than southern populations. Levin (1984b) shows that a disjunct and isolated *Phlox, P. pilosa* ssp. *sangamonensis* is electrophoretically less variable than its presumptive progenitor, ssp. *detonsa*, which he suggests may be due to founder effect and genetic drift.

It is possible to hypothesise that populations approaching the northern limit of their range in the northern hemisphere might be forced towards selfing by the paucity of insect visits, and such conditions might favour relatively few specialised genotypes in a population. Yet in *Capsella*, Bosbach and Hurka (1981) have clearly shown higher levels of heterozygosity in populations from the extreme north of Scandinavia, in comparison with the rest of Europe. However, it must not be assumed that these are necessarily more outcrossed; extreme conditions may at times favour heterozygous advantage at certain loci.

In some cases, there are clear implications that localised environmental conditions favour selfing. Self-pollinating *Epipactis* (especially *E. phyllanthes*) occur in much deeper woodland shade, where wasp visits must be more unlikely to occur, than do allogamous species. On the desert margin, small-flowered self-compatible *Clarkia* species would be less likely to receive pollinator visits than in more mesic situations (Moore & Lewis 1965). But in some cases, it is difficult to separate environmentally

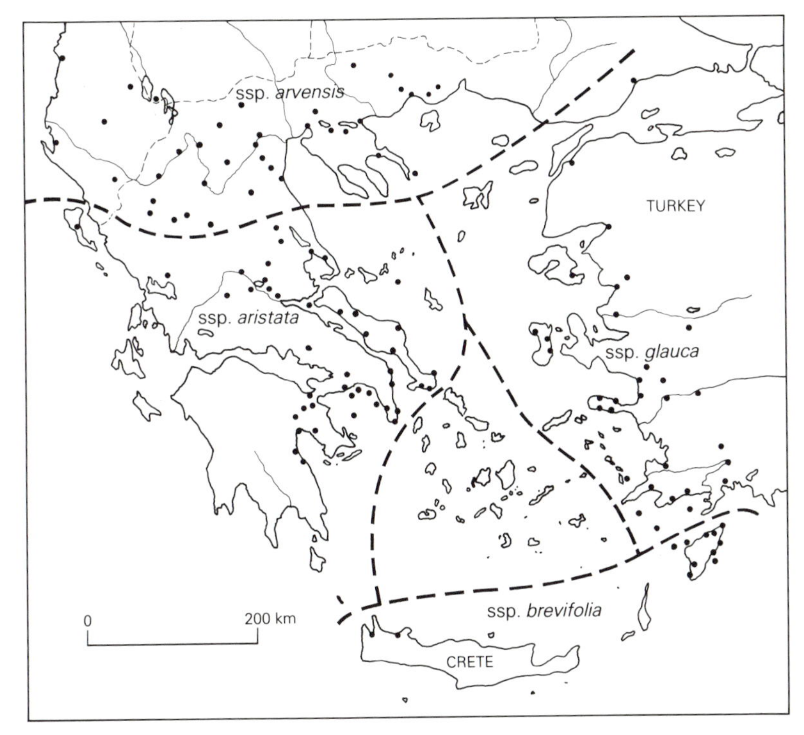


**Figure 9.11** Distribution of the self-incompatible *Nigella arvensis* (Ranunculaceae) in the Aegean region (after Strid 1970).

imposed requirements for reproductive efficiency, or genetic invariability, from the consequences of founder effect on biological islands.

It has been pointed out by Baker (1955), and expanded by Carlquist (1974) and MacArthur and Wilson (1967) among others, that the colonisation of islands occurs by a succession of rare and unlikely events; the more distant the island, the more unlikely the event. Although the islands considered are usually pieces of land in the sea, the same principal holds for any localised and distinctive habitat, for instance lakes in a land mass or isolated mountains in lowlands. As such colonising events will be rare, it is unlikely that more than a single individual will form the founder. If a species of plant relies on seed for survival and spread, only species capable of some selfing are likely to form successful colonisers. Thus, one might expect the flora of remote oceanic islands to be primarily selfed, and some results seem to show this to be true. I have argued (Ch. 8) that the high frequency of outbreeding systems of a secondary nature (especially dioecy) on some oceanic islands may be a product of selection for secondary outbreeding on mostly inbred founders. However, remote islands also tend to be small, ecologically limited

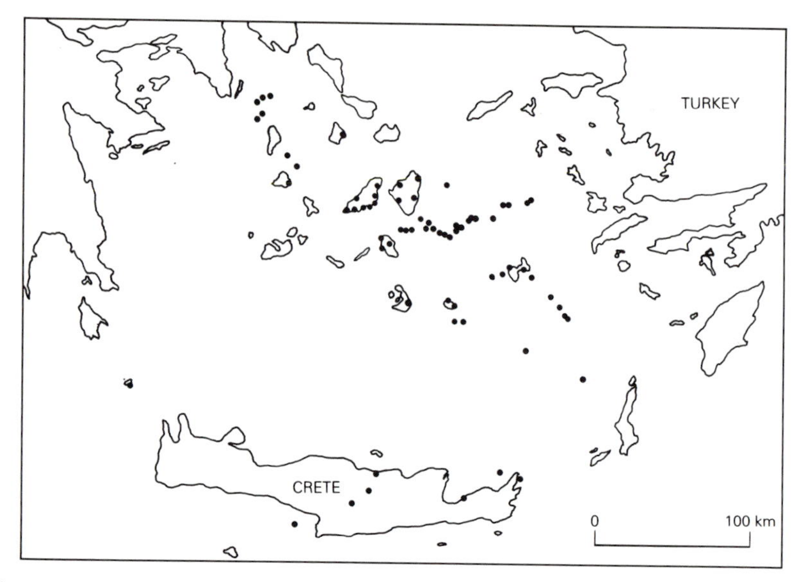


**Figure 9.12** Distribution of the self-compatible inbreeder *Nigella doefleri* in the Aegean region (after Strid 1970). Unlike *N. arvensis,* it is confined to the small and climatically extreme islands of the Cyclades, from which *N. arvensis* is largely absent.

and poor in pollinators. Thus, it can also be argued that selfing is also favoured in such situations by ecological or reproductive pressures. A very elegant example is found in the annual genus *Nigella* (love-in-a-mist). Strid (1970) has shown that whereas the self-incompatible *N. arvensis* occurs in various subspecies throughout the mainland of Greece, Turkey and the larger Aegean islands such as Rhodes and Crete, the small-flowered self-compatible inbreeder, *N. doefleri* is confined to the much smaller and climatically more extreme islands of the Cyclades (Figs. 9.11 and 9.12). But is this distinctive pattern a product of the ecology and climate of the Cyclades, or have the smaller islands only been successfully colonised by self-compatible founders? In this case, in which inter-island distances are relatively small and man has been transporting plant material between the islands for thousands of years, ecological arguments are more plausible, but the problem remains.

## CONCLUSIONS

(a) Obligate selfing is a very rare condition in the flowering plants, if it occurs at all.
(b) However, many species show high levels of selfing, which may

exceed 99% of all fertilisations. Such behaviour is particularly frequent in opportunist annuals with *'r'* strategies.

(c) Many other species are at least partially selfed. Only dioecious species, fully self-incompatible species, or the few self-compatible species that produce only one flower at a time are likely to be totally outcrossed. Even in flowers in which features of structure or timing render within-flower selfing (autogamy) impossible, between–flower selfing (geitonogamy) may be a major feature.

(d) Obligate selfing should lead to total homozygosis; however, even small amounts of outcrossing between differently fixed homozygotes will maintain some heterozygosis. This may be augmented by heterozygote advantage at some loci.

(e) Nevertheless, selfing populations should show much less within-population variability and heterozygosis, and much more between-population variability than outbreeders. In practice, this does seem to occur.

(f) The reproductive efficiency of selfers should be greater than that of outbreeders, and in practice this often seems to be the case; however, outbreeders may compensate by heterotic vigour in comparison with selfers.

(g) Artificially selfed outbreeders usually show inbreeding depression, as do some partial selfers; however, many habitual selfers show no inbreeding depression. This may be achieved by selection for particularly vigorous homozygous loci, or by heterozygous advantage at a few loci that remain heterozygous.

(h) Selfers may show unusually high capabilities for phenotypic plasticity, and this may have been the result of selection, especially in *'r'* strategy annuals.

(i) Selfing may be advantageous in some extreme, restricted or marginal habitats. However, it is difficult to separate the effects of restricted pollinator activity, restricted niches and founder effect as factors favouring selfing in such circumstances.

(j) Most plants are outbreeders, or outbreeders with some selfing. Selfing may be favoured by pressures for reproductive efficiency, especially in monocarpic plants, but these may then suffer from inbreeding depression, and lack of genetic variability. At least some selfers seem to have overcome the problem of inbreeding depression, and it is likely that the main disadvantage of high levels of selfing is the corresponding lack of genetic variability created. This will severely restrict the niches in time and space available to a population in a habitat.

## CHAPTER TEN

# *Vegetative reproduction*

---

Vegetative reproduction is the asexual multiplication of an original individual that has arisen from a sexually formed zygote (genet) into spatially separated units (ramets). It can generally be assumed that these will be genetically identical with their parent, being members of a clone, and from a genetical point of view, all sister ramets belong to one individual, the genet.

It could well be argued that vegetative reproduction has little to do with breeding systems, for breeding systems concern the release of genetic variation through the sexual process (Ch. 2), and the controls imposed upon this release. Yet, few would argue that asexual reproduction mediated through seeds (agamospermy) should be ignored as a breeding system, not least because many intermediate conditions between sexuality and agamospermy (facultative agamospermy) occur. Thus, agamospermy is the subject of the next chapter (Ch. 11). From a genetical standpoint, there is little difference between vegetative reproduction and agamospermy. Both are asexual, forming clones, through the use of different types of disseminule. Both are often described under the single term apomixis (as in Gustafsson 1946–7). Most plants that undergo vegetative reproduction can also reproduce sexually; thus vegetative reproduction can be interpreted as a facultative apomixis. As in facultative agamospermy, disruptive pressures for sexuality and asexuality have encouraged reproductive versatility.

There is another sense in which vegetative reproduction is an important component of breeding systems. Large clones take up much more space than single (non-ramifying) genets. Thus, between-individual distances may be much greater, and effective population numbers (neighbourhood sizes, Ch. 5) much smaller than is immediately apparent. Vegetative reproduction may thus restrict the gene pool, and its evolutionary opportunities. Large clones may each occupy many niches spatially within a habitat; this also reduces the amplitude of genetic variation available to a population.

## DISTRIBUTION

Asexual (somatic) reproduction is unknown in vertebrate animals, although some fish and reptiles exhibit asexual reproduction via an unreduced egg (parthenogenesis). In contrast, somatic asexual reproduction, as well as parthenogenesis, is widespread amongst invertebrate animals. Familiar examples range from the budding *Hydra*, observed by so many schoolchildren; through the immense ($10^3$–$10^4$-fold) capability for asexual reproduction exhibited by many parasites, such as the cercarian phase of liver flukes; to the great clonal colonies of polyps which form coral reefs, and transform the topography of tropical shores. Unicellular organisms, both prokaryotic (bacteria, blue-green algae), and eukaryotic (protozoans, algae) also usually have remarkable powers of asexual reproduction, which in the case of multicellular forms (for instance the Volvocales group of the Chlorophyta) is equivalent to vegetative reproduction in higher plants. All Bryophytes, and most Pteridophytes, also undergo regular vegetative multiplication, of the gametophyte and the sporophyte generations (only), respectively.

In contrast, vegetative reproduction is curiously rare in the Gymnosperms. This may be allied to the predominantly woody habit of this group. Thus, the untypically herbaceous (and systematically very distant) *Ephedra* and its ally *Gnetum*, which often forms lianes, undergo vegetative reproduction. However, in the other subclasses of the Gymnosperms, vegetative reproduction is probably confined to some alpine species of conifer with a creeping habit, the procumbent branches of which may root adventitiously. This occurs in *Podocarpus alpinus* and *P. nivalis*, from New Zealand, and in prostrate mountain forms of the juniper from northern Europe (*Juniperus communis* subsp. *nana*).

In the Angiosperms, vegetative reproduction is extremely widespread and common (Table 10.1). It is, by definition, absent from annuals and biennials. It is also lacking in many other monocarpic species (which die after flowering), for instance *Saxifraga longifolia*, the century plant *Agave americana*, *Echium* species and *Aeonium* species from the islands of the Canaries and Azores, the afroalpine *Dendrosenecios and Lobelias* etc. Yet,

**Table 10.1** Estimates of the percentage of Angiosperm species with a strong capacity for vegetative reproduction in three European floras.

| Area | Latitude (°N) | Vegetation type | Species with vegetative reproduction (%) |
|---|---|---|---|
| Petsamo, Finland (Soyrinki 1938) | 68 | birch tundra | 45 |
| Bromarv, Finland (Perttula 1941) | 62 | boreal | 80 |
| Great Britain (Salisbury 1942) | 50–59 | temperate | 46 (68% of perennials) |

bamboos (for instance *Arundinaria* spp.) have a remarkable capacity for vegetative multiplication, single clones sometimes occupying whole forests or regions, or gardens throughout the world. But bamboos are also monocarpic, and when they finally flower in approximate synchrony, their self-incompatibility may restrict seed-set disastrously if only one clone is present.

Vegetative reproduction is found in the majority of herbaceous perennials. Amongst woody plants, it is very common in dwarf or creeping shrubs, climbers and lianes, but is less common in trees. In certain temperate tree genera, for example poplars (*Populus*), elms (*Ulmus*), and cherries and plums (*Prunus*), suckering from roots is conspicuous. It may be damaging to man, as when the foundations of nearby buildings are disturbed. It may be horticulturally inconvenient, as when vigorous suckering stocks overwhelm grafted scions, for instance in *Prunus, Rosa* and *Rhododendron.* It can also provide the salvation of a species, a habitat and a landscape. The hedge elms of England, for example, have been devastated by Dutch elm disease, but many moribund trees have suckered strongly after infection. Trees and shrubs of subtropical and tropical areas undergo vegetative reproduction much more rarely.

Among land plants, vegetative cloning is perhaps most conspicuous amongst anemophilous monocotyledons, grasses (Gramineae) and sedges (Cyperaceae) (Ch. 5). Although there are a number of annual grasses, the annual habit is very probably a secondary development in this family, and most perennial grasses frequently form very large clones indeed (p. 386). Nearly all the Cyperaceae form clones of various sizes. Natural 'grasslands' range from reed beds (*Phragmites, Arundo*), marram dunes (*Ammophila*) and cord-grass marshes (*Spartina*), to steppes, prairies, savannahs, salt marshes, sea-cliff tops and alpine meadows. In all these types of habitat, throughout the world, from one to a few grass species form dominants, and are able to do so largely through a formidable capacity for vegetative reproduction. Often (*Phragmites, Ammophila*) the same species occurs in a specialised habitat throughout the world, and these are amongst the most widespread species of plants known.

The other growth form in which vegetative reproduction is remarkably successful is amongst hydrophytes (water plants Fig. 10.1). It is probably true that all bottom-rooted, submerged and free-floating water plants are excellent vegetative reproducers. Water is a more uniform and less rigorous environment for the dispersal of relatively unprotected vegetative disseminules than are terrestrial habitats; at the same time, the aquatic habitat may restrict the occurrence or efficiency of flowering, and the dispersal of seed may be less efficient than on land. The uniformity of the aquatic environment renders the polarity of plant form less evident; roots may be absent or may be produced from every stem node. Thus, the fragmentation of an individual by physical forces, which will lead to

**Figure 10.1** Vegetative reproduction in the fringed water lily *Nymphoides peltata* (Menyanthaceae) (×0.4).

death or diminution in a terrestrial habitat, may lead to rapid multiplication and dispersal in water. Roots and apical meristems can be formed from tiny stem fragments, each of which may develop into a new ramet.

The efficiency of vegetative reproduction in water plants is vividly demonstrated by ecological invaders (adventives), which may cause severe economic problems. Thus, the expensive and prestigious irrigation and hydroelectric schemes based on the Aswan and Kariba dams in Africa have both been impeded by the explosive growth of the South American water hyacinth, *Eichhornia crassipes*, which forms dense and thick mats of vegetation floating on the water surface entirely by vegetative reproduction. The flowers are tristylous (Ch. 7) and self-incompatible, so the introduced single clones set little if any seed. Most colonies are monomorphic with respect to the tristyly (Ganders 1979). This species impedes navigation, blocks conduits and pipes, and lowers water quality. The damage done by this invasive adventive is reputed to run into millions of US dollars. One of the checks to its growth in South America is said to be the large aquatic mammal, the dugong or manatee, and experiments are currently under way to introduce these as a biological control agent in Africa.

In the UK, the development of the industrial revolution during the first half of the 19th century was accompanied by the construction of a large and efficient network of canals, which could transport raw materials and manufactured goods cheaply, cleanly and safely, if rather slowly. From the years 1840 to 1870, the canals outcompeted the developing railways for industrial business, and might well have handled a substantial proportion of industrial traffic today were it not for the introduction of the Canadian pondweed, *Elodea canadensis*. This was first recorded in Britain in 1836 or earlier, having probably been accidentally introduced, perhaps from aquaria. Within a decade it had become abundant in every British canal (they are all interconnected), being fragmented and dispersed by canal traffic (Simpson, 1984). By 1870, most canals had become irretrievably clogged, and most industrial business had passed to the railways. *Elodea canadensis* is a dioecious species, and, except at one locality, the only introduced sex was female (this suggests that only one clone may have been responsible throughout, a single individual genet blocking thousands of miles of waterway). Thus, all reproduction was vegetative, sexual reproduction being impossible.

## ORGANS OF VEGETATIVE REPRODUCTION

Plants have developed many different means of mediating vegetative reproduction. These will have at least one of the following functions:

(a) Creation of new ramets.
(b) Dispersal of new ramets.
(c) Protection of new ramets, especially during dormancy (hibernation or aestivation).

In almost all cases, vegetative disseminules are modifications of stems (Table 10.2), or of axillary buds which are stem initials. Only in a few Crassulaceae, such as the familiar house plant *Bryophyllum pinnatum*, do vegetative propagules ('bulbils') form on the edges of leaves at vein ends. The suckering of certain trees from underground roots has already been mentioned.

Although underground organs such as bulbs and corms seem primarily designed for dormancy (many vernal species are above ground for remarkably short times, for instance 5–7 weeks per year for many *Galanthus*, *Erythronium* and *Trillium* species), they can be surprisingly good at dispersal. This is exemplified well by the snowdrop (*Galanthus nivalis*). In British woods, multiplication of the bulbs after die-back forms very dense clones. The smaller bulbs are squeezed upwards, to be revealed lying on the woodland floor when the protective cover of leaves of other summer-green herbs dies back in October. Various agencies can then transport them for at least 10 m (personal observation), maybe much further. These will include rodents, which eat many bulbs but carry some to store, rainstorms and the feet of larger animals including man. Seed-set is exceedingly poor in this self-incompatible species, and various marker genes including tepal doubling, tepal-mark colour, tepal length and stem length, strongly suggest that in most British populations only a single genet occurs.

In a number of genera in the Liliaceae, notably *Fritillaria*, and in some bulbous *Iris*, such as *I. danfordiae*, a system analogous to monocarpy prevails. Bulb size increases over several seasons until flowering is achieved. After flowering, the bulb divides inside the tunic into a very large number of small bulbs ('rice-grain bulblets') which may be no more than 2 mm in diameter, and up to a thousand can be produced from a single bulb (Fig. 10.2). These are squeezed onto the soil surface where they are dispersed by various agencies, including the wind, in summer-dry conditions. At the onset of the first winter rains, the roots emerge and penetrate the soil, pulling the bulblet into the soil (as also happens for *Galanthus*). They will take several seasons to reach flowering size again. This is a system that is both frustrating and useful for the growers of these beautiful plants. Dutch nurserymen have perfected techniques using hydroponics and peat, giving commercial multiplications of bulblets by the thousand-fold to saleable size in three years.

**Figure 10.2** 'Rice-grains' produced by the flowering-size bulb of the Turkish *Iris danfordiae* after flowering. The small rice-grain bulbs take several seasons to reach flowering size, a vegetative analogue to monocarpy (×1).

**Table 10.2** Attributes of vegetative reproduction organs in Angiosperms.

| Organ | Nature and origin | Example | Multiplication | Protection | Dormancy | Dispersal |
|---|---|---|---|---|---|---|
| stem | aquatic fragmentation | *Elodea* | poor | none (many aquatics also produce turions *q.v.*) | none | good |
| stem | tip rooting | *Rubus*<br>*Rosa*<br>*Hedera* | poor | none | none | poor |
| stolon | specialised above-ground stem | *Fragaria*<br>*Potentilla* spp.<br>*Ranunculus repens*<br>*Geum reptans* | poor | none | none | poor to moderate |
| rhizome | specialised below-ground stem | *Ammophila*<br>*Phalaris*<br>*Spartina* | poor | subterranean, often corky and with food reserves | especially hibernation | poor |
| corm | short, swollen below-ground stem | *Crocus*<br>*Anemone* | poor to moderate | subterranean, tunicated, food and water reserves | especially aestivation | poor to moderate |

| | | | | | | |
|---|---|---|---|---|---|---|
| bulb | as for corm, but reserves in modified leaves | *Lilium* *Allium* *Galanthus* *Narcissus* | poor to good | subterranean, tunicated, food and water reserves | especially aestivation | poor to moderate |
| bulbils | axillary buds of inflorescence | *Polygonum viviparum* *Allium* *Saxifraga cernua* | moderate | bud scales, some food and water reserves | especially hibernation (*Allium* aestivation) | moderate |
| | leaves | *Saxifraga granulata* *Lilium* | good | | aestivation | |
| leaf proliferation | | *Bryophyllum* | good | none | none | poor |
| floral proliferation (vivipary) | | *Poa alpina* | moderate | none | none | poor |
| turions | axillary buds of aquatics | *Elodea* *Hydrocharis* | good | bud scales, food reserves | hibernation | poor |
| thallus | budding | *Lemna* | good | none | none | good |
| suckers | from roots | *Ulmus* *Prunus* *Populus* | good | none | none | poor |

**Figure 10.3** Sexual flowers (above), and asexual bulbils (below) produced within the inflorescence of the pan-arctic viviparous bistort, *Polygonum viviparum* (×1).

In many species of *Lilium*, the individual bulb scales drop apart after flowering to achieve similar effects. It is curious that in the bulbous genera *Lilium* (lilies) and *Allium* (garlics and onions), many species not only show multiplication of below-ground bulbs, but also produce axillary bulbils in the leaf or bract axils. In *Allium*, these may occur intermixed with sexual flowers, or flowers may be absent. For instance, in the crow garlic, *Allium vineale*, three varieties occur in the British Isles. In the wetter western regions, forms with flowers only predominate (var. *capsuliferum*). In the drier east, forms with mixed flowers and bulbils (var. *typicum*) or forms with bulbils only (var. *compactum*) predominate. However, there is some overlap, and one not infrequently finds all three forms growing together. It is by no means certain that these varieties are even under genetic control (Salisbury 1942). However, in *Allium carinatum*, experimental crosses have shown that bulbil production (in subsp. *carinatum*) is dominant over absence of bulbils (in subsp. *pulchellum*), although other modifier genes and environmental conditions also influence the expression of bulbils. In some *Allium* complexes, there is a good deal of polyploidy, and the forms with the higher levels of 'ploidy are more likely to be bulbiliferous.

*Allium* and *Lilium* are unusual in that some species have evolved two specialised methods of vegetative dormancy and dispersal, below-ground bulbs and above-ground leaf or inflorescence bulbils.

Inflorescence bulbils also occur on a number of other plants. In the viviparous bistort, *Polygonum viviparum*, which is widespread in montane and arctic areas throughout the northern hemisphere, bulbils occur below the flowers. It is my experience that in the Scottish highlands, over

**Figure 10.4** Pseudovivipary in the inflorescence of the grass *Poa alpina*, which is widespread in alpine and arctic regions of the northern hemisphere. Some races have sexual or agamospermous florets, particularly at lower altitudes. Viviparous races such as this which proliferate vegetation plantlets in the inflorescences are more common at higher altitudes (×0.5).

750 m, most plants lack flowers, although at lower levels flowers predominate and bulbils may be absent (Fig. 10.3). However, Law *et al.* (1983) could find few convincing correlates between the degree of bulbil production and environmental variables.

Another widespread arctic–alpine plant which usually bears inflorescence bulbils is the drooping saxifrage, *S. cernua*. In Scotland, this is a very rare plant with only three known populations. At the well-known station on Ben Lawers, only bulbils are formed. At another station, nearly all plants produce one terminal flower as well as bulbils, and at the third, flowers are occasionally formed. All sites are above 1000 m, very high for Scotland, in a very harsh alpine environment. Similar variation in flower production between sites occurs elsewhere, in Norway and Greenland.

In some arctic–alpine grasses, vegetative proliferation of plantlets ('vivipary', or more correctly, pseudovivipary) occurs in place of flowers on the inflorescence. This is a condition analogous to inflorescence bulbils, but the reproductive organs usually have a shoot, leaves and even a short root. In *Poa alpina* (Fig. 10.4), Muntzing (1954) has shown that the pseudoviviparous forms (which are polyploid) tend to occur at higher levels and in more harsh conditions than the sexual and agamospermous diploids and tetraploids. In the viviparous fescue, *Festuca ovina* var *vivipara*, Harmer and Lee (1978a,b) show that plantlets have three to four times as much carbohydrate reserve and more mineral nutrients than do the seeds of *Festuca ovina*. Perhaps as a result, plantlets establish better than seedlings at low temperatures. However, they are at a disadvantage in one respect, for they have no capability for dormancy, whereas fescue seed can remain dormant for many years.

Pseudovivipary is associated with severe arctic–alpine conditions in other grasses as well (*Poa alpigena*, *P. arctica*, *Deschampsia alpina*). It may confer two benefits in such environments: reproduction and dispersal may be enhanced by the avoidance of cross pollination and by the endowment of more maternal resource per disseminule than is possible for a seed; and rigorous stabilising selection in adverse conditions may have favoured asexual rather than sexual reproduction. However, it should be noted that pseudovivipary also occurs in some lowland grasses of dry situations (*Poa bulbosa*).

In another lowland European species of summer-dry sites (the meadow saxifrage *S. granulata*), the plant is winter-green but dies back after flowering in early June (aestivates). Only the axillary buds to the rosette leaves survive the summer as bulbils. Between 10 and 40 bulbils are produced per rosette, and in a dense stand as many as 10 000 bulbils can occur in 1 $m^2$. Each is capable of germinating during the autumn to form a new rosette. Bulbils are deposited at the soil surface, which often becomes very bare in summer, for this is a species which can grow on very shallow soils. Bulbils can travel through the agencies of water, wind or small animals for 3 m or more, as has been shown by careful habitat mapping (Stevens 1985). Bulbils vary by a factor of ten or more in weight. It is likely that the ramets formed by the multitude of smaller bulbils mostly fail to survive, being outcompeted by ramets from the fewer larger bulbils. Larger bulbils can be shown experimentally to be more likely to survive, to form larger ramets, to flower, and to produce more seed than smaller bulbils. They are also more likely to form large bulbils at the next aestivation. Such differences may not be genetic, but may reflect the age and microenvironment of each bulbil line. Small bulbils may be able to disperse further than large bulbils, and if they reach a new niche, without competition, they may be able to form their own new asexual dynasty.

Although *S. granulata* flowers regularly, and forms large quantities of fertile seed, it seems that sexual reproduction is an unusual event, at least in some populations. None of the hundreds of seedlings whose fate was followed in the field by Stevens survived beyond the season in which they germinated. It may be that most populations consist of only a few widely dispersed clones, which through bulbil dispersal have become intermingled.

*Saxifraga granulata* seems to be unusual amongst plants with specialist methods of vegetative reproduction, in that no trade-off seems to occur between sexual and asexual reproduction. The most successful asexual clones (with the largest bulbils), are also potentially the most successful sexual clones, as they flower better and produce more seed. This contrasts markedly with, for instance, *Polygonum viviparum* (Law *et al.* 1983) in which sexual and asexual reproductive allocation tend to be negatively correlated (see also Harper 1977). It may be that the very successful method of vegetative dormancy and dispersal adopted by *S. granulata* has relaxed selection for sexual fertility. Many clones are aneuploid, and partially, or even completely (Ch. 8), pollen sterile, which suggest that this may be the case.

## DISPERSAL AND RATES OF MULTIPLICATION

Vegetative dispersal may be relatively effective, and of the same order of distance as seed dispersal, even for plants where new ramets remain in contact with their parents. The wild strawberry (*Fragaria vesca*) can certainly move 5 m per year in my garden using stolons, and rhizomatous weeds such as couch (*Elymus repens*) or ground elder (*Aegopodium podagraria*) can move 2–3 m per year with ease, even without the helping hand of the gardeners' spade! It is characteristic of opportunist plants of this kind that stolons and rhizomes move in a straight line, the exit and entry points onto ramets being at 180° to one another. Once the initial colonisation event has occurred, lateral colonisation from ramets proceeds, and the apical dominance is lost in subsequent events. Perhaps the outstanding performer in the UK is the sand-sedge, *Carex arenaria*, which can travel up to 10 m per year by rhizomes in perfectly straight lines over bare stable sand (Noble *et al.*, 1979). Other sand-dune species (*Ammophila arenaria*, *Elymus farctus*, *Leymus arenaria*) can also colonise bare sand very swiftly, even after substantial burying by mobile sand.

The vegetative colonisation of sand-dune species can change the shape of shorelines in only a few decades, through sand-accretion and longshore drift. An even more striking example of vegetative colonisation of maritime habitats has been provided over the last 50 years by the hybrid grasses *Spartina x townsendii* and its fertile alloploid derivative *S. anglica* on British shorelines. These cord-grasses arose from hybridisation between the rare native *S. maritima* and the American *S. alterniflora* towards the end of the 19th century, in Southampton Water. Since then they have spread rapidly through natural dispersal of rhizome fragments by the tide, and by intentional and unintentional introductions by man.

They are found today in nearly all the estuaries of the British Isles, and they are also found on the other side of the English Channel. Although *S. anglica* is seed fertile, and colonisation by seed certainly occurs, most colonisation appears to be vegetative. However, *S. anglica* is more widespread than *S. x townsendii*. Their power of colonisation of estuarine mud, where no other Angiosperm can grow, is truly awe-inspiring. In not much more than a decade, a single coloniser has covered 1.5 $km^2$ (150 ha) of mud monospecifically at Lindisfarne, Northumberland, and is severely threatening important feeding grounds for ducks, geese and waders; what was essentially intertidal mud has become land, and more than 10% of the mudflats have already disappeared. In Essex and Kent, where the cord-grasses have been present longer, coastlines change significantly by the year, and many square kilometres have been reclaimed from the sea.

Water plants are particularly suited to vegetative reproduction and dispersal, and the invasions of *Eichhornia* and *Elodea* have already been mentioned. For speed of reproduction and powers of dispersal, there is no equivalent to the highly specialised Lemnaceae (duckweeds) which cover the surface of slow-moving waters in many parts of the world. These have become reduced to tiny (1–10 mm) floating thalli, and although they flower, producing remarkably conventional stamens and gynoecia, nearly all reproduction is vegetative. They can divide vegetatively ('bud') extremely rapidly, often doubling the population of ramets in a day. This exponential population growth can result in a founder covering the surface of a pond within a month in suitable conditions. Such 'blooms', which are only equalled by certain algae and by the rather similar water fern *Azolla*, usually die back equally rapidly, perhaps because nutrients become exhausted. Duckweeds provide excellent experimental material for the investigation of models of population growth, and for experiments on the effects of pollutants, etc.

The specialised vegetative organs of water plants may account for their success, but they may also influence their distribution. The frog-bit, *Hydrocharis morsus-ranae*, is a floating plant of Europe, the axillary buds (turions) of which sink to the bottom of the water in autumn, when the rest of the plant dies back. In the spring, the turions first germinate, producing two small leaves, and then float to the surface, where they grow, produce more ramets by stolons, and flower. We have shown (Richards & Blakemore 1975) that water temperature is vitally important in controlling both germination and floating, although at least some light is also necessary. Turions germinate very poorly at 10°C, and do not float. Even at 15°C, less than half the turions have floated in three weeks, but at 20°C both germination and floating is rapid (Fig. 10.5). It is probable that this temperature control is responsible for the restriction of this species to southern England and to the more southerly regions of Europe.

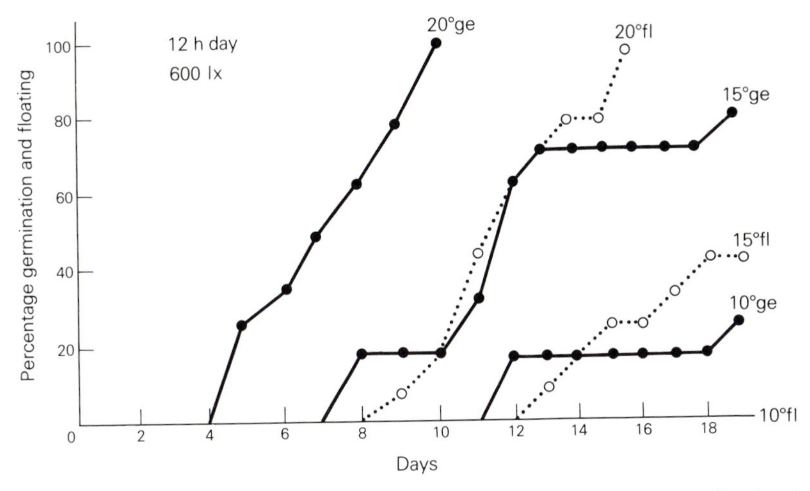


**Figure 10.5** Effect of temperature on germination (ge) and floating (fl) of turions in the frog-bit, *Hydrocharis morsus-ranae*, a free-floating European aquatic plant.

## THE AGE AND SIZE OF CLONES

It is clear from the previous sections that many plants have very considerable powers of vegetative multiplication and dispersal. Thus, it is to be expected that for many species that are good at vegetative reproduction, clones (genets) may be very large, both in number (with thousands or even many millions of ramets) and in area. In some cases, as in *Elodea canadensis*, *Spartina anglica*, *Ammophila arenaria*, *Lemna minor*, *Phragmites australis* or *Eichhornia crassipes*, the extent of single clones may extend to hectares, or even square kilometres. It follows that such clones may be of a very considerable age. Quite how old any clone is can be difficult to judge, but some may date back thousands of years. An interesting case has been studied by Y. Heslop-Harrison (1953). A small lake in the hills of Northumberland, UK (Chartners Lough) has a population of the hybrid waterlily *Nuphar* × *spennerana* (*N.* × *intermedia*). Of the parents, the commoner *N. lutea* has its nearest locality some 25 km distant. The other parent is the much rarer arctic relict *N. pumila*, which does not approach within 150 km at the present time. The hybrid, which covers the Lough, is somewhat fertile, but morphologically invariable, and may well represent a single clone. Heslop–Harrison has suggested (in Stace 1975) that this hybrid clone may well be a relict from much earlier times when both parents grew together on the Lough, and this may have been as long ago as the late-glacial period (approximately 10 000 years BP).

Another case concerns the dwarf birch, *Betula nana*, a single clone of which was discovered in Upper Teesdale, UK in 1965. Subfossil leaves of this species, some of which were dated at 10 000 years BP, occurred in peat from directly beside the survivor, but are absent from most surrounding peat deposits. Here once again is circumstantial evidence for the survival of a very old clone.

Clones may indeed be much older than is immediately evident. By sectioning dead or moribund wood of the relict shrubby cinquefoil (*Potentilla fruticosa*) growing beside the River Tees, UK, I was able to show that no above-ground stems were more than 40 years in age (Richards, 1975a). Clones are easy to delimit in this dioecious species, not only by sex, but by vegetative characteristics. Some of the biggest females were over 20 m in maximum extent, with several thousand stems. Judging by growth rates of seedlings from this population in the much less rigorous conditions of a botanic garden (about 1 m laterally in 10 years), these clones were at least hundreds of years old, much older than the stems apparent today.

Similar indirect evidence is obtained from observations of fertility. Few northern English or Scottish populations of the reed (*Phragmites australis*) are fertile today; it appears that summer temperatures are too low for seed to ripen adequately. Thus, most northern reed beds perpetuate themselves and spread through vegetative means, and many may consist of single clones only. These clones may have survived from earlier epochs with warmer summer temperatures, during the so-called 'Atlantic Maximum' from 2000 to 4000 years BP.

In order to establish the size of clones, in terms of numbers of ramets and area covered, it is necessary to use genetic markers which identify a particular genet. Mere contiguity is not enough, for as we have seen, individual genets may break up and intermingle with others. Various techniques for the identification of clones have been adopted.

## *Techniques for clone identification*

**Visual identification of morphological traits** Although of no use for many clonal species, it is surprisingly easy to delimit clones by eye for some low-growing herbs. This is especially so for clovers (*Trifolium*), where such species as dutch clover (*T. repens*) and red clover (*T. pratense*) show a great deal of variability in stature, leaflet size and shape, leaflet colour, white and anthocyanin leaflet markings and (in *T. pratense*) indumentum. Although variations might defy formal description, the 'gestalt' (or 'jizz' from American ornithology) of each clone is usually adequate for an experienced observer to delimit clones with some confidence in the field.

**Sexual incompatibility** In species with multi-allelic self-incompatibility (Ch. 6), particularly when it is under gametophytic control, many *S* alleles usually coexist in a population at roughly equal frequencies. Thus, it is usual for an individual to be cross-compatible with all others, but it will be self-incompatible. By making crosses between ramets throughout the population, it is thus possible to build up a picture of the limits of clones, either by examining seed-set, or more quickly by examining pollen-grain germination and pollen-tube growth on the stigma. Crosses made within clones will be sterile; those made between clones will nearly always be fertile. My ex-student John Faulkner (1973) has shown this to be the case for British species of *Carex* section Acutae (sedges), as part of an investigation into interspecific breeding barriers. A large proportion of the fenny basin of Otmoor, Oxfordshire, UK is dominated by *Carex nigra*, *C. acuta*, and presumptive hybrids between them, which cover many tens of hectares, with millions of ramets (these species are markedly rhizomatous). After sampling flowering spikes from different parts of the area, and examining the pollen-tube growth of a large number of experimental crosses, Faulkner found very few interfertile reactions (Fig. 10.6) and informally estimates that very few (possibly less than 10) clones occur throughout this vast area.

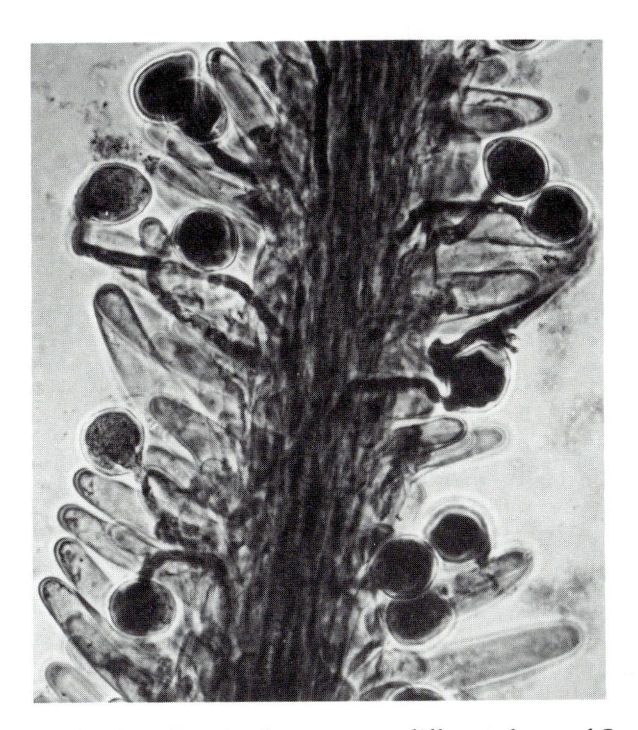

**Figure 10.6** A fertile pollination between two different clones of *Carex*; *C. elata* (male) × *C. acuta*. Microphotography by J. Faulkner (×200).

Harberd (1961, 1962) wrote two classic papers on the grasses *Festuca rubra* and *F. ovina* which are surprisingly rarely quoted. He used a combination of visual identification and sexual incompatibility to identify clones isolated from 100 square-yard quadrats (83.61 $m^2$) in Scottish hill grassland. For each species, he collected over 1000 tillers from regular points throughout the quadrat, grew these on, and classified them visually as to presumptive clonal identity. A large number of test crosses were then made within and between presumptive clones, and these very largely confirmed his provisional clonal identifications. In both species, a similar pattern of clonal distribution occurred within these quadrats. The great majority of ramet samples could be assigned to very few genets (16 were responsible for 90% of the ramets in *F. rubra*, and one genet was responsible for no less than 51% of the 1481 isolates occurring throughout the quadrat). This particular clone (appropriately labelled 'W', for it was evidently very fit!) occurred elsewhere on the moor over at least 200 m. Harberd suggests that the age of this clone must be measured at least in hundreds, and may well be measured in thousands, of years, based on current growth rates.

It is remarkable that essentially comparable results were obtained for *F. ovina*. This species, unlike *F. rubra*, is not markedly stoloniferous, but is highly tufted. Thus vegetative spread must be mediated by external forces, probably chiefly by the feet and jaws of cows and sheep, aided and abetted by frost-heave and wind. Both species set good seed, which will germinate and establish in open niches (e.g. molehills and hoofprints). Yet a great deal of vegetative competition and selection takes place, and exceptionally fit individuals can clearly survive for long times and cover large areas of ground. In a later paper, Harberd and Owen (1969) show that for *F. rubra* the lowest clone density (most 'clonal reduplication'); occurs in areas where the cover of the species, and thus presumably its optimal conditions, are highest.

**Sterility** Populations consisting of only one clone do not usually set seed. This can be due to dicliny, self-incompatibility, sterility, or hybridity. We presume that almost all British *Elodea* (p. 374) is of one clone, for it is virtually all female. Similarly, the butterbur, (*Petasites hybridus*, is only male throughout much of the UK (Fig. 10.7) with females, and hence seed, only occurring in limited areas of the midlands and northern England. It would be interesting to learn to what extent it consists of a single clone outside these areas. Many self-incompatible introduced species in the British flora rarely, if ever, set seed, presumably because introductions originate from single clones. One may list *Petasites fragrans*, *Pachyphragma macrophyllum* (Davie & Akeroyd, 1983), *Doronicum pardalianches*, several species of *Aster*, *Gaultheria shallon*, *Galanthus nivalis* (p. 374), *Iris versicolor* (p. 108) and *Lysimachia terrestris*

**Figure 10.7** Male inflorescence of the butterbur, *Petasites hybridus*, Compositae (Asteraceae); this species is only male through much of the British Isles (×0.5).

among very many other examples. A single male clone of the introduced white poplar, *Populus albus*, near Hexham, UK, has about 800 separate shoots, many of them tree-sized, and covers 1.2 ha.

Other, much more widespread, native species also rarely set seed, and these are always species with conspicuous means of vegetative multiplication. Thus fruit is rarely set on creeping jenny (*Lysimachia nummularia)*, silverweed (*Potentilla anserina*) or creeping soft-grass (*Holcus mollis*) among many other examples. Such sterility may have one of two causes; in many populations there may have been only a single founder, or only a single aggressive clone which has survived many years of asexual competition. Thus self-incompatible populations with a single genet would not set seed.

Alternatively, and this is not a self-exclusive proposition, many clones may have accumulated chromosomal abnormalities (e.g. chromosome breakages, deletions, interchanges, inversions, aneuploidy and polyploidy) somatically, so that they have defective meioses. The rôle played by meiosis in sexuality as a 'normalising' sieve that sorts out cytologically irregular somatic lines, only allowing regular lines to reproduce sexually, is argued more fully later in this chapter. In populations in which sexual reproduction is rare or missing, such chromosomal mutations may accumulate to the detriment of fertility (Müller's ratchet, Ch. 2).

*Potentilla anserina* is an interesting case. It is self-incompatible, and occurs as tetraploid and hexaploid clones. The tetraploids are variably, but reasonably, fertile, whereas the hexaploids (which may be autotriploids) show high levels of pollen sterility (2–28% fertile), and some clones never apparently set seed (Rousi 1965, Ockendon & Walters 1970, Cobon & Matfield 1976).

For *Holcus mollis*, Harberd (1967) detected only four genets in 58 isolates from sampling sites ranging over 4 km. One clone occurred over at least 1 km. Three of the genets were pentaploid, and the fourth was triploid; all were sterile.

In some genera, sterile but vigorous hybrids form large and often widespread clones. In cases where clones can be transported between sites (many water plants, for instance hybrid *Potamogetons*; or hybrids of garden origin), it can be difficult to determine whether all such hybrids belong to a single genet. That is, has the hybrid arisen on more than one occasion? In many cases, for instance hybrid mints of garden origin (*Mentha* × *piperita*, the peppermint), a single clone of considerable age and distribution may indeed be involved. However, in this instance, different varieties of the hybrid are known (var *citrata*), and clones with different chromosome numbers have been recorded ($2n = 66, 72$) (Harley, 1972). It is not clear whether these have arisen somatically after the original hybridisation event, or if they represent different hybridisations. Nevertheless, it is clear that in some cases single sterile hybrid clones can cover large amounts of ground within localities, and disperse to many different localities, even throughout an area such as the British Isles.

A good example is one of the monkey flowers, *Mimulus luteus* × *guttatus*, the parents of which are native to Chile and California respectively. Although the seed-fertile *M. guttatus* ($2n = 28$) has become common beside the rivers of upland Britain, *M. luteus* ($2n = 60, 62, 64$) is much rarer, and most plants given this name are in fact the hybrid, which is very pollen sterile and completely seed sterile (Roberts cited in Stace, 1975). All chromosome counts of the wild British hybrid are $2n = 45$, despite the fact that synthesised hybrids can be $2n = 44, 45$ and 60. This suggests that only one hybrid clone, perhaps of garden origin, is involved. It is now widespread on upland rivers of northern Britain. These *Mimulus* are highly stoloniferous, and even tiny fragments root with ease. Carry between river systems may have occurred unintentionally by man (on feet) or by wildfowl.

**Genetic markers** The use of visual markers and incompatibility markers in identifying clones has been covered under visual identification and sexual incompatibility. Single genetic markers are unsatisfactory, as they probably occur polymorphically in a population and do not allow a clone to be identified with certainty. The most satisfactory method is to use isozymes to identify clones; if four or five systems with 10 to 20 variable loci are employed, the chances of two genets giving the same zymograms become minimal. Unfortunately, this method is both costly and time consuming, and requires the provision of an electrophoretic laboratory, with technical expertise. [Wu *et al.* (1975) claim, however, that it is less costly in time and material than Harberd's 'cloned-clones, incompatibility' method.] However, populational work using isozymes is fashionable at present, and some interesting results on the genetic structure of populations with large clones is emerging.

The first report in the literature seems to have been Wu *et al.* (1975) on populations of *Agrostis stolonifera* which have colonised copper-polluted soils in the Liverpool region of England. Although electrophoretically crude by more recent standards, this study was able to distinguish clearly between old grasslands at Freshfield and Sefton Park, in which only one and two genotypes respectively were detected, and three lawns that had been established on copper soils for less than a decade. In these, 9, 10 and 15 genotypes were discovered; altogether the study identified 34 genotypes ('zymograms') of which only one occurred at more than one site. Generally, not more than two samples had the same genotype on the new lawns. It was calculated that about 100 genets occurred per 100 $m^2$ ($10 \times 10$ m) of new lawn, a density of one genet per $m^2$. The authors considered this to be a high diversity of genets, particularly in a species that can colonise by stolons at the rate of 1 m a year. However, considering the youth of the populations, and the many thousands of seeds that must have been sown in the area, selection of genets had proceeded apace; in the polluted soil, relatively few would have shown high levels of tolerance in any case.

In contrast to grasses such as *Festuca* and *Agrostis*, the vegetative spread of dandelions (*Taraxacum*) is very limited, as this is a rosette-forming tap-rooted herb. However, as most members of the genus are obligate agamosperms (reproduce asexually by seed, Richards, 1973) seed-clones are formed, which have a great capacity for dispersal, but a limited capacity for generating genetic variability (Ch. 11). Ford and Richards (1985) showed that in a 100 $m^2$ area of Northumberland sand dunes, 18 different genotypes occurred, which could be classified into 10 agamospecies. Five agamospecies, morphologically uniform, displayed more than one zymogram genotype (Ch. 11, p. 439, Fig. 11.13). Thus genet diversities were lower than in recently established swards of the stoloniferous *A. stolonifera*, but greater than in long-established populations at Freshfield and Sefton Park. The ready dispersal of genets by seed in dandelions, and their poor vegetative spread, is likely to result in a greater diversity of asexual genets per unit area of long-established ground, than in grasses in which asexual reproduction is limited to vegetative spread.

## GENET ESTABLISHMENT AND SURVIVAL

Grime (1973) has produced a model that describes relationships in grassland between species diversity, seral colonisation and stress. He suggests that in areas with low stress, colonisation proceeds by vigorous species, which by virtue of their height, shading attributes, persistent litter and pervasive roots discourage coexistence by competing species. After this initial colonisation phase, one or more edaphic requirements for growth are used up, and becomes scarce and limiting, that is the colonisers invoke their own stress. Other less vigorous species are now able to coexist, and although the original colonisers may become rare, diversity increases. Eventually, intraspecific vegetative competition between coexisting clones results in the survival and success of relatively few, widely dispersed genets, and diversity will drop again as the community reaches its subclimax equilibrium. This distribution of diversity with time has resulted in this being termed the 'hump-backed' model. Grubb (1977) suggests that diversity in a stable community is dependent on the number of reproductive niches, both sexual and asexual, that are available. This has been more thoroughly discussed in Chapter 4.

Soane and Watkinson (1979) have modelled the relative effects of vegetative and sexual reproduction on genet diversity in a grassland, in the absence of differential selection between genets. They show that from initial establishment by seed, genet diversity will drop to an equilibrium point where loss of genets by death is replaced by the birth of new genets from seed. The proportion of ramets born sexually (from seed) will be extremely low compared with the asexual (vegetative) birth rate, but nevertheless plays a vital part in maintaining genet diversity in the population.

Thus, as a grassland matures, the specific diversity should increase and then decrease to an equilibrium dependent on reproductive niches. The intraspecific diversity should decrease to an equilibrium dependent on the proportion of ramet replacement that occurs from seed. This general model seems to be borne out by Wu *et al.* (1975) for the Liverpool lawns, in so far as genet diversity apparently decreases with the age of the community. Some interesting work by Gray (1982) concerns recruitment after fire of enzymically identified genets of the local grass *Agrostis setacea* in heathlands of southwestern England. He found no further recruitment of genets three years after the fire. Seedlings recruited in the third year after fire did not flower, being outcompeted by first and second year seedlings. This is a strongly tufted species with very limited means of vegetative spread. Yet clearly the viable seed population after fire, and the rate of seedling growth, is such that continuous seedling replacement is a rare event.

Further confirmation of the model comes from recent work by T. McNeilly (unpublished observations) on *Lolium perenne*. Zymogram identification of clones, using $\gamma$ ornithine aminotransferase and phosphoenolpyruvate, showed that two-year-old sown leys had the following numbers of genets or per 0.25 $m^{-2}$ under different grazing regimes: cattle, 45; sheep, 42; ungrazed, 36. In contrast, a permanent pasture known to be 45 years in age averaged only five genets or per 0.25 $m^{-2}$, and only six genets $m^{-2}$.

More evidence of seedling recruitment into a stable population at equilibrium comes from Lovett Doust (1981). For the stoloniferous buttercup *Ranunculus repens* in woodland, she found that only 0.7% of new ramets arose from seed, and that after seedling mortality, only 0.3% of surviving ramets came from seedlings. This was not due to a shortage of seed. Over 1000 viable seeds occurred per $m^2$, and the percentage of seed available that was recruited may have been only about 0.3% per year, of which about 0.1% survived. Similar results were obtained by Culwick (1982) for the introduced rhizomatous pirri-pirri bur *Acaena novae-zelandiae* (Ch. 5) which sets abundant viable and well-dispersed seed. Yet Culwick found very few seedlings, and of the 4000 marked seedlings whose fate she followed, only one survived more than a year. In comparable areas, many thousands of new vegetative ramets are established each year.

### *Clonal strategies*

Harper (1977) has written a very full account of demographic processes in plants, which has become a key work in the understanding of plant populations. Part of this is concerned with vegetative growth by clonal species, and he introduces the concept of viewing a sward 'from the plant's eye view'. This is done by assessing within and between genet, and within and between species plant contacts. It is through these contacts that competition, and hence the structure of vegetatively reproducing plant communities, is regulated. Some of these ideas were propounded earlier by Bradshaw (1972). Lovett Doust (1981) developed these concepts to describe two contrasting clonal strategies which she terms 'phalanx' and 'guerilla'. She considered *Ranunculus repens*, in which interspecific contacts are maximised by isolated advancing stolons, to be a guerilla species. In contrast, rhizomatous or tufted grasses, such as *Lolium perenne*, *Agrostis setacea* or the *Festucas*, are phalanx species that advance on a broad front and minimise interspecific contact. Lovett Doust and Lovett Doust (1982) have suggested that such phalanx species grow in a geometrical pattern that maximises 'packing' (space filling). Such packing maximises within-genet contacts and minimises between-genet and between-species contacts. In contrast, guerilla species grow asymmetrically, thus minimising within-genet contacts

**Table 10.3** Growth strategies of vegetatively reproducing clones (after Lovett Doust & Lovett Doust 1982).

| Strategy | Guerilla | Phalanx |
|---|---|---|
| mode of growth | asymmetrical | geometrical |
| organ | stolon | rhizome |
| within-clone packing | poor | good |
| within-clone contacts | few | many |
| between-clone contacts | many | few |
| predominant subclass | dicotyledons | monocotyledons |
| predominant seral stage | pioneer | stable |

but maximising between-genet contacts. They claim that most dicotyledons are guerilla types, and most monocotyledons are phalanx species. They also suggest that guerilla species tend to predominate in pioneer habitats, while phalanx species are more typical of stable habitats (Table 10.3). Lovett Doust (1981) points out that this variation of strategy with habitat can even occur within a species. *Ranunculus repens* is more 'guerilla' in woodland and arable, and more 'phalanx' in stable grassland. Casual observation would seem to confirm this.

Apart from within-clone architecture, another factor that will influence clonal strategies and patterns of clonal growth is density-dependent regulation of ramet recruitment and death. This has been discussed by Lovett Doust (1981) and by Noble *et al.* (1979); the demonstration of genuine density dependence, rather than non-causal correlation, is difficult to establish. However, there is some indication of density-dependent regulation of both ramet recruitment and death in *R. repens;* interestingly there is no such relationship with ramet birth, suggesting that in dense clones of this species, within-clone between-ramet competition is an important phenomenon.

The suggestion that within-species variation in pattern of clonal growth may occur in *R. repens* is echoed by the interesting work of Turkington and Harper (1979). This concerned competitive relationships within a pasture, and showed that two of the most important species in the pasture, *Lolium perenne* and *Trifolium repens,* showed a positive response to each other in mixed culture. In contrast, they each showed a poorer growth response to coexistence with all other species occurring in the pasture. These two species are apparently well suited to growing together, as is very well known agriculturally, for these are the two main species sown in permanent leys in Britain. There was a strong indication that individual genets of each species were coadapted to growing together. When ramets were dug up, multiplied in pure culture and then transplanted back into the pasture, they performed significantly better in the sites from which they had originally been removed, than in other sites in the pasture.

This demonstration of the rôle played by interspecific competition in selecting for certain clonal genotypes may help to explain the predominance of certain genets of *Festuca* (Harberd 1961, 1962). It is also a powerful argument for the maintenance of sexual reproduction in predominantly vegetatively reproducing species (Ch. 2). However rare sexual reproduction may be in stable grassland communities, its continuance is vital for the plant to maintain genet diversity, with respect to interspecific competition. Here is a clear example of the 'Red Queen effect' (Ch. 2) in operation.

It is notable, however, that Ford (1981) has shown that different *Taraxacum* agamospecies which coexist in the absence of sexuality, show differential adaptation to successful coexistence with competing grass species. Some competitive niche diversification is possible in the absence of sex.

## *Resource allocation*

Gadgil and Solbrig (1972) have adapted the concepts of MacArthur and Wilson (1967) to plants. Taking constants from Malthusian equations of population growth, they have characterised populations as being relatively '*r*' or '*K*'. The symbol '*r*' concerns the innate reproductive growth rate of a population when there are no checks to population growth. The symbol '*K*' describes the carrying capacity of the environment (as density of ramets) for that population. Thus '*r*' strategy plants will be adapted to the rapid colonisation of pioneer habitats, whereas '*K*' strategy plants will be adapted to maintain stable population densities in stable, subclimax or climax communities. Although it has been usual to discuss the attributes of sexual reproduction with respect to '*r*-ness' or '*K*-ness', it is just as valid to examine attributes of asexual reproduction, including vegetative growth, in this context (Table 10.4).

Thus, a typical '*r*' strategy plant would be a groundsel (*Senecio vulgaris*) which grows rapidly, flowering in a few weeks from germination. It is monocarpic, flowering only once before dying, and does not reproduce vegetatively but produces large numbers of small well-dispersed seeds, using about half of its total resource in sexual reproduction. It is largely autogamous (Ch. 9), genetically invariable, and highly plastic, and rapidly colonises fertile, open ground with large, short-lived populations of small plants that cannot compete with larger and later colonisers.

In contrast, a typical '*K*' strategy plant would be an English oak (*Quercus robur*). This grows slowly from seed, and only starts to flower after 30 years; however, it may then flower annually for 500 years. Although it does not usually reproduce vegetatively, it may form a massive individual plant dominating many square metres of ground, and if felled before senescence, it will produce large numbers of new

**Table 10.4** Attributes of alternative resource-allocation strategies.

| Strategy | '*r*' | '*K*' |
|---|---|---|
| life span of genet | short | long |
| vegetative growth-rate of ramet | fast | slow |
| total biomass of ramet | small (to large) | large |
| total biomass of genet | small (to large) | large |
| number of ramets per genet | 1 to few | many |
| number of reproductive events in time | 1 to few | many |
| allocation of total resource to reproduction | large | small |
| allocation of reproductive resource to sex | large | small |
| size of propagule | small | large |
| proportion of reproductive resource invested in a single propagule | small | large |
| number of propagules produced in unit time | large | small |
| propagule dispersal | good | poor (to good) |
| habitat | pioneer | stable |
| length of habitat occupancy | short | long |
| genetic variation within a population | low | high |
| potential for phenotypic plasticity | high | low |

ramets (suckers) from the roots. It sets seed poorly and irregularly, and most seeds are eaten before maturity, as it presents a very obvious 'target' for predators. Reproduction by seed is thus a rare event, and usually only succeeds when the canopy is opened; however, the seeds ('acorns') are large (10 g) and well provided with resource, although their dispersal is very poor, unless dropped or stored by predators such as jays and squirrels. It is wind pollinated and self-incompatible, genetically very variable, and shows little plasticity. It rarely occurs in pioneer communities, except as a relict, but it dominates climax lowland vegetation in north-western Europe (Morris & Perring 1974).

These are two extreme examples, but it is also possible to consider two species in a genus relatively '*r*' and '*K*' with respect to each other (*Senecio vulgaris* and *S. cineraria*, or *Quercus coccifera* and *Q. robur* respectively), or even two ecotypes of a species. Ibrahim (1979) studied demographic and reproductive processes in two populations of the cowslip (*Primula veris*) in Northumberland, UK; one growing in stable, mown grassland 130 years old, and the other in the open and recent habitat of a disused limestone quarry which had ceased working 20 years previously. She showed that for seedling density, recruitment of adults from seedlings, adult mortality, and predicted and actual increases in adult densities, the quarry site gave values approximately three times those of the grassland site. Seedling mortality and adult density were about the same in both sites at the time of the study. The adult half-life and the population turnover time were about three times greater in the grassland. It is possible to suggest that the quarry population showed more '*r*' type characteristics than the grassland population. However, the grassland site produced a five times greater density of seed than the quarry (and a seed was thus some 15 times less likely to produce a detectable seedling). In this attribute the grassland site could be said to be more '*r*'. Linhart *et al.* (1979) demonstrated '*r*' and '*K*' selection within a Colorado population of *Pinus ponderosa*. Genetically slow-growing individuals showed greater sexual fertility ('*r*') than fast growing genets ('*K*').

Ford (1985) has shown that two coexisting, and morphologically very similar, English sand-dune asexual dandelions, *Taraxacum brachyglossum* and *T. lacistophyllum*, differ markedly in reproductive output and population turnover time, the former being markedly more '*r*' in its characteristics. He suggests that this allows each to succeed in the presence of the other, by filling different time or space niches in the habitat.

One may predict from '*r*' to '*K*' theory that different habitats would encourage species with different patterns of reproductive resource allocation. This has been shown to be true by Abrahamson (1979) in a study of dry weight allocation to vegetative and floral organs in 50 North American species from woodland and grassland habitats. The woodland species tended to put more resource into vegetative organs ('*K*') and the grassland species put more resource into floral organs and seed production ('*r*'). It is therefore not surprising to learn (Sarukhan 1977) that the stoloniferous *Ranunculus repens* puts more resource into vegetative reproduction than do the other two common western European buttercups, *R. acris* and *R. bulbosus*, which are tufted. *Ranunculus repens* occurs in more shaded habitats than the other species.

We have already learnt (Lovett Doust 1981) that *R. repens* adopts a guerilla strategy of vegetative reproduction, with isolated advancing stolons. She considers, however, that woodland populations are more 'guerilla' than are grassland populations. As far as I know, the resource allocation attributes of 'guerilla' and 'phalanx' strategies have yet to be compared, and this would form an interesting project. One might predict that guerilla strategies will prove more '*r*', i.e. more of their resource is devoted to creating new ramets and less in maintaining old ramets. After all, guerilla strategies have been associated with pioneer habitats in which sexual '*r*' strategies predominate. It is noteworthy that the total reproductive resource allocation to the three species of buttercups, is identical; it is the partition of resources between vegetative and sexual reproduction that differs.

## SOMATIC VARIABILITY

It is a genetic truism that no genetic variation occurs within a genet. The genet has arisen from a single zygote, which possesses an unique genotype conferred upon it by the sexual processes of recombination, segregation and fusion (Ch. 2). All ramets of that genet should be exactly duplicated by mitosis so that they share an identical genotype, which should thus differ from any other genet.

But to what extent is this truism in fact true? The original source of genetic variability is mutation, caused by the substitution, deletion, duplication or rearrangement of the nucleotide bases on the DNA molecule which form the genetic code. Mutations occur somatically during vegetative growth; whether they survive within a genet will depend on where they occur, and their ability to persist in competition with non-mutant cell lines. Thus a viable mutant cell line with normal growth characteristics which arises early in the formation of a meristem initial should be able to form part of the resultant organ, which will therefore be a genetic chimaera. However, the majority of somatic mutations will be lost, being non-viable or out-competed, or through not occurring in a meristem.

Whether a new mutant can become established in a ramet will depend on the nature of the meristem in which it arises. If it arises in a floral initial, it may be carried by pollen or egg cells, and may thus be incorporated into new genets arising from seed. If it arises in an initial that forms a stolon, rhizome, corm, bulbil or suckering root, it may give rise to a new ramet that is partly or wholly mutant. A great deal of somatic competition within genets, between genetically different cell lines that have arisen by mutation, must occur in every plant.

This is an important topic about which we know very little indeed. We sample the plant chromosomes in most cases in two organs only; the root tip, which is rarely involved in the creation of new ramets, and the anther, which helps to form new genets, sexually. We have for the most part very little idea of what is happening in between. We have even less idea about other genetical chimaeras, which are not visible through the chromosomes. Occasionally they are strikingly visible, as in pigment chimaeras such as chlorophyll mutants (variegations) which are propagated for decorative purposes (Fig. 10.8). Anthocyanin chimaeras may be equally conspicuous (Fig. 10.9) and can lead to flower-colour polymorphisms in the following generation (Ch. 5)

Electrophoretic work on isozymes occasionally leads one to suspect that isozymal chimaeras occur. It is good practice to subsample different organs, and organs of different ages for isozymes, to ensure that one's technique is not biased by choice of tissue. In most cases, within-plant

**Figure 10.8** Leaf chimaera caused by mutant plastids lacking chlorophyll in Norway maple, *Acer platanoides* (×0.5).

**Figure 10.9** Anthocyanin chimaera in petal of red campion, *Silene dioica*, which is half pink (left) and half white. Photo by M. Wilson (×2).

samples are reassuringly uniform. Just occasionally an anomaly occurs, as Mogie (1982) found for one individual of accession 20 of the dandelion *Taraxacum pseudohamatum*. When tested for esterase, for one leaf, this individual clearly lacked a band present in all other individuals of this accession. Repeated tests of other leaves of this individual were normal, the band in question always being present. In such tests, a 'Catch 22' operates, for they are by definition unrepeatable. If the chimaera occurs in only one organ, which is destructively harvested, it is impossible to be sure of one's result (Ch. 11).

That such chimaeras may lead to genetic variation between plants is suggested by a result in another species, *T. unguilobum*. In this case, one sibling of accession 14 displayed an extra esterase band in comparison with all its siblings, and in comparison with four other uniform accessions of that species. Repeated tests of different leaves showed that this plant appeared to be homogeneous for this mutant; it should be emphasised that this mutant had also arisen asexually, for this is an obligate agamosperm, and all siblings are members of the same seed clone (Ch. 11). There is clearly much more scope for within-genet electrophoretic sampling to establish levels of somatic genetic variation.

### *Cytoplasmic variation*

By no means all heritable material that enters a new cell is DNA. Organelles of the cytoplasm (plastids, mitochondria, ribosomes and endoplasmic reticulum) also do so, and they are for the most part self-replicating, with a history independent of the nuclear DNA. Many cytoplasmic mutants are well known (e.g. Michaelis 1954), but their study has been neglected in higher plants, although they are currently much researched in lower phyla.

Cytoplasmic variation is well known to hybridists, for reciprocal crosses between the same interfertile individuals of different species, or even races, often produce very different hybrid phenotypes. The nuclear contribution from both parents is approximately the same whatever the direction of the cross. It is the cytoplasmic contribution that differs, for that originates almost entirely from the mother. I will take a single example from my own laboratory with which I have been made familiar on the day I am writing this. When the closely related sexual dandelions *Taraxacum alacre* and *T. brevifloroides* are crossed, the seeds germinate all together four days after sowing when *T. alacre* is the female plant. When it is the male parent, no germination at all is achieved after two weeks, and the seeds may well be sterile. This result is obtained repeatedly for crosses between the same individuals, and between different individuals of these two species. Presumably the cytoplasm (mitochondria?) of *T. brevifloroides* disagrees with the nuclear component from *T. alacre* (if endosperm/embryo interactions were involved, as discussed in Ch. 3, both reciprocal crosses would probably have been at least partially sterile).

One fascinating aspect of such plasmagenes is that they may be exempt from the 'central dogma' of molecular genetics. This states that the interaction between the phenotype and the genotype operates in one direction only. The genotype, composed of the DNA code, governs the phenotype. However, interactions between the environment and the phenotype cannot change the genotype. The DNA cannot mutate in any directional way, to order; but only randomly, without order. This stems from the nature of the DNA molecule. Thus, any heritable factors that are coded by the DNA cannot be subject to direct environmental influence. There can be no 'inheritance of acquired characteristics', as promoted by Lamarck (1809) and others. Evolution can only proceed by selection for randomly generated mutations of the DNA.

DNA occurs not only within the nucleus, in chromosomes, but also cytoplasmically, in mitochondria and plastids. Cytoplasmic DNA, being non-chromosomal, does not undergo meiotic forms of recombination and segregation, and at fusion it is inherited almost entirely maternally, through the egg. Yet, it still must obey the 'central dogma'. It may play a central rôle in coding for part of a major metabolic enzyme. Ribulose diphosphate carboxyase is a dimer, the regulatory monomer of which is inherited through the nucleus, and the catalytic monomer of which is coded by chloroplast DNA (Kung 1977). Techniques that allow the direct sequencing of mitochondrial DNA in plants are currently becoming available, and promise to tell us much about phylogeny, distant relationships and rates of evolution. They may also throw light on forms of maternal inheritance and reciprocal differences in hybrids.

However, it is likely that many forms of plastid inheritance have nothing to do with cytoplasmic DNA, but depend on morphological or chemical changes in the cytoplasmic organelles themselves. This is as yet a very little understood area, which deserves further study. However, if such organelle 'mutants' do occur, they will not be subject to the 'central dogma', for they are independent of the DNA. They may indeed respond directly to environmental pressures, and the new types of organelle that result may be inherited maternally through the cytoplasm.

Thus, it has been known for many years that exposure to quinacrine mustards will cause many micro-organisms to form small-celled, slow-growing phenotypes, which are inherited. Only mating with normal strains will cause this 'petite' strain to be lost. These 'petite' phenotypes, resulting from a change in the mitochondria, have been rendered small with few cristae and a poorly operating terminal oxidase systems. This is the inheritance of acquired characteristics; an environmental influence, however artificial, directly results in an inherited trait.

In a most significant paper, which has not been pursued as thoroughly as might have been the case, Breese *et al.* (1965) demonstrated somatic selection within clones of rye grass, *Lolium perenne*. They were able to show that it was possible to select for different growth rates between ramets within genets, and that these differences were inherited by generations of daughter ramets. It would be interesting to learn whether such differences could also be inherited sexually from these selected ramets. However, the authors did establish two significant tendencies with respect to the likelihood of somatic variation occurring:

(a) Somatic variation is more frequent in genets that have undergone long periods of asexual reproduction.
(b) Somatic variation is more frequent in the shoots of young rather than old ramets.

Breese *et al.* (1965) suggested that this variation is arising through changes in the 'plasmon', and that natural selection is occurring within clones for attributes that are not controlled by the DNA. They did not show that such attributes arise directionally, that they are 'acquired', but clearly, this could be a possibility.

## POTENTIAL DISADVANTAGES OF VEGETATIVE REPRODUCTION

It is commonly observed in the horticultural world that well-known clones become weaker with age. Continual vegetative propagation may be considered desirable to perpetuate a 'good form', but it leads to reduced vigour. Renewed vigour will usually result from sexual propagation by seed. It is unclear to what extent such effects operate in the field. The very considerable age and size of many clones in nature, discussed earlier in this chapter, suggest that deterioration is relatively unimportant. Perhaps species that habitually reproduce vegetatively have evolved some unexplained immunity to such clonal weakness; many garden clones come from species that regularly undergo sexual reproduction in nature. There are two main reasons for loss of vigour in old clones as detailed below.

### *Virus-induced losses*

As a genet ages, it will build up an increased viral load through viral multiplication and reinfection. An early infection can later cause simultaneous death in ramets that become separated by great distances. A classic case has recently occurred in a single hybrid clone between two Himalayan primulas, *P. scapigera* and *P. bracteosa* (*P.* × *'scapeosa'*). This clone was raised from an intentional hybridisation by R. B. Cooke in 1949 at Kilbryde Garden, Corbridge, Northumberland, UK. It proved to be an attractive and extremely vigorous clone with a high potential for vegetative multiplication, and by 1972 it was offered by many nurserymen, and grown by specialists as far away as Canada, Australia and New Zealand. At this date, the Royal Botanic Garden, Edinburgh, alone grew almost a thousand separate 'clumps' (ramets), almost as a bedding plant. It was highly pollen sterile, and never set seed, doubtless as a result of its hybridity. By then, it was, however, showing unmistakable signs of infection by cucumber mosaic virus, with distorted leaves and petals, the latter with colour streaking. As late as 1978, it was still a vigorous clone, but was showing signs of weakness. By 1982 it had completely disappeared from all gardens, even in New Zealand and Australia. Evidence from my own plants, and those at Edinburgh, showed that it suddenly became very weak, slow-growing and difficult. Many other *Primula* clones have a similar history, which is readily overcome by growing new genets from seed.

### *Somatic mutation*

It has already been argued that 'Müller's Ratchet' (Ch. 2) causes many clones to accumulate disadvantageous mutants. As the accumulation of mutations in a cell line shows a linear relationship with time, it might be presumed that the older the clone, the more disadvantageous mutants it will possess. These may lead to a loss in vigour. This concept disregards the possibility of somatic selection against weakening mutants, which we presume must operate, although there is very little hard evidence for it. It seems likely that some selection will occur between cell lines within a ramet (so that highly disadvantageous mutants probably never establish) and between ramets within a genet.

It seems possible that such somatic selection will not only 'weed out' many somatic mutants, but may also increase the 'niche width' of a clone, by allowing genetically different cell lines of a genet with different strategies to coexist. Thus, the same clone might evolve differentially to allow the coexistence of different ramets with different competitors (Turkington & Harper 1979, Ford 1981). Once again, it must be emphasised that such considerations are purely speculative, except for the seed clones of *Taraxacum* (Ch. 11).

However, it is clear that at least some species with habitual vegetative reproduction accumulate chromosomal mutants that are harmful at least with respect to sexual reproduction. As has been pointed out above, many successful clones are partially or completely sterile, due to high polyploidy (*Potentilla anserina*; Ockendon & Walters 1970), uneven polyploidy (*Holcus mollis*, Harberd 1967), aneuploidy (*Saxifraga granulata*, Stevens 1985), chromosomal rearrangement (*Crocus speciosus*, from *C. cancellatus*, Brighton *et al.* 1983) and high levels of B chromosomes.

Thus, when opportunities for sexual reproduction occur, there will be selection for ramets with higher levels of fertility; this selection will occur between genets, or more rarely within genets, when chromosomal mutant chimaeras have accumulated within a genet. This is the 'meiotic sieve' in operation; meiosis will have a normalising function by excluding accumulated mutants that reduce sexual fertility, and thus the sexual function will tend to dissipate the effects of 'Müller's Ratchet' (Ch. 2). It will also help to segregate and recombine mutants. The high level of chromosomal abnormality encountered in some habitual vegetative reproducers testifies to the rarity of sexual reproduction in their populations. Vegetatively, such mutants may be very successful.

Repeated tissue culture of previously sexually propagated sexual strains, may lead to a build-up of sterility mutants to the extent that the cultivar clone cannot be reproduced by seed. As tissue culture is becoming increasingly important in the horticultural market, such mutants, which are already evident in tissue-cultured dwarf roses, may prove an embarassment to micropropagators in the future.

### *The 'clean egg' hypothesis*

Thus, a study of vegetatively reproducing plants is able to throw a new light on the functions of sexual reproduction (Ch. 2). It is clear that prime functions of sexuality are to generate variability, to dissipate Müller's Ratchet and break up linkage disequilibrium. However, the sexual process also engenders:

(a) Gametes that have passed through the meiotic process and are thus chromosomally fairly 'normal' ('meiotic sieve').
(b) Zygotes that are free from virus (plant viruses seem unable in most cases to survive the sexual process in their hosts, although the reason for this appears to be obscure).

This concept of the 'cleansing' function of sex, resulting in offspring genets that are chromosomally and virally 'pure' has been called the 'clean egg' hypothesis. The function and importance of the clean egg in considerations of the rôle of sex in higher plants with vegetative reproduction has rarely been noted. This may be because most sexual theoreticians concern themselves with animals or with annual plants. The latter provide the better model systems, which may constitute a warning!

It is not even certain that sex alone provides the genetic variation in a population that can fully exploit the diverse niches engendered by any habitat. The rôle played by somatic variability in fulfilling this function for vegetative reproducers has still to be explored. In Chapter 11, I shall report how seed-cloned (agamospermous) plants generate genetic variability that can lead to evolution. Although asexual, they have the additional bonus of reproducing via an egg, which is virally, but not chromosomally, clean.

CHAPTER ELEVEN

# *Agamospermy*

---

Asexual reproduction in seed plants (apomixis) can be divided into two main classes; vegetative reproduction (Ch. 10) and agamospermy. Agamospermy can be defined as the production of fertile seeds in the absence of sexual fusion between gametes or 'seeds without sex'. Sexual fusion presupposes a reductional meiosis if the 'ploidy level is to remain stable. In agamospermy, a full reductional meiosis is usually absent, and thus chromosomes do not segregate.

In some cases (*Citrus*), the gametophytic generation is avoided entirely, and the new sporophyte embryo is budded directly from the old sporophyte ovular tissue, usually the nucellus (adventitious embryony) (Fig. 11.1).

More commonly, a female gametophyte (embryo-sac) is produced with the sporophytic chromosome number. This non-reduction of chromosome number is achieved either by a complete avoidance of female meiosis (apospory and mitotic diplospory), in which case the embryo-sac is usually budded directly from the nucellus; or by failure and non-reduction (restitution) of the female meiosis (meiotic diplospory).

If a diplosporous female meiosis occurs, there is usually an absence of chiasmata, and hence no recombination. Thus, genes remain inextricably linked to chromosomes and indeed to the whole genotype; even if recombination occurs, it can only do so between chromosomes that are unable to segregate. Genetic variation can only result from 'position effects' of genes entering a new chromosomal environment. In the absence of segregation and fusion, the agamospermous cell line forms one gigantic linkage group in which advantageous genes are unable to escape from an accumulation of unsuccessful mutants (Müller's ratchet) by sexual processes (Ch. 2).

Thus, it has become customary (Janzen 1977) to consider the agamospermous cell line to be a single, invariable linkage group, unmodified by time or space. One vivid conceptualisation is for all the living individuals (ramets) of an agamospermous cell line to be envisaged as being equivalent to shoots of a single, gargantuan (and presumably buried!) tree (for instance a dandelion agamospecies, which in some successful cases may through introduction be almost worldwide). In other words, a genetically invariable agamospecies should consist of a single seed clone, or genet, through time and space. As we shall see, such a traditional interpretation is probably far from the truth in most cases. If, however, we follow

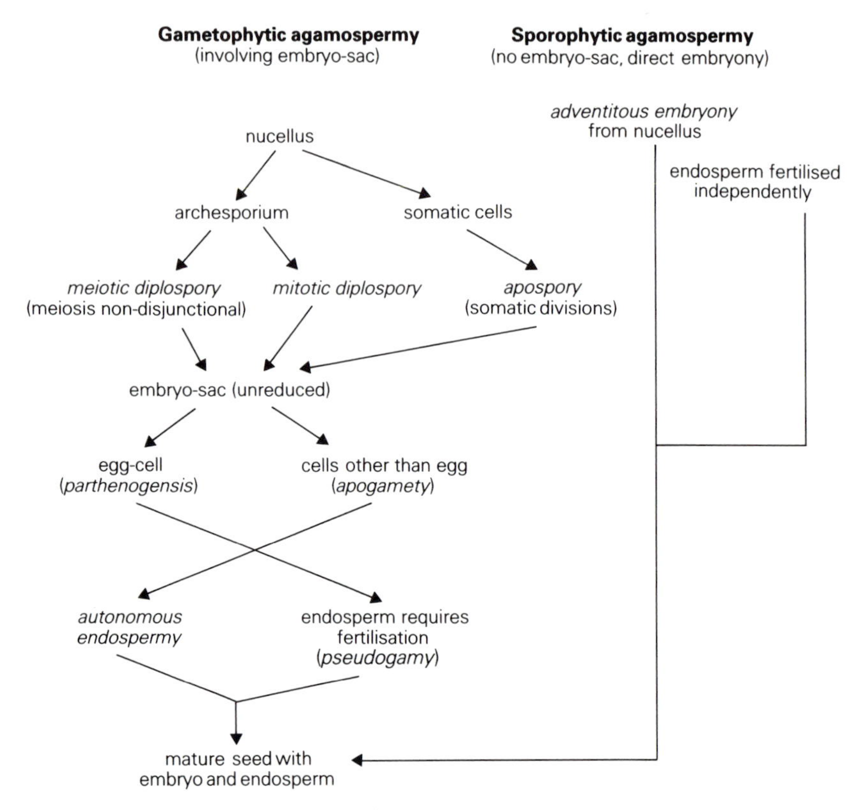


**Figure 11.1** A scheme illustrating the main types of agamospermy (after Gustafsson 1946–7).

Darlington (1939) or Stebbins (1950) in assuming its veracity for the time being, we can make certain assumptions about the probable advantages and disadvantages inherent in such an agamospermous cell line.

## POTENTIAL ADVANTAGES AND DISADVANTAGES OF AGAMOSPERMY

### *Advantages*

There are many apparent advantages to agamospermy:

(a) Assured reproduction, even in the absence of pollination. This should favour agamospermy in extreme climatic conditions. However, most species with apospory or adventitious embryony are pseudogamous, requiring pollination for asexual seed development.

(b) Clonal reproduction by seed, combining all the advantages of a seed (e.g. 'clean egg' hypothesis, dispersal and dormancy; Ch. 2) with the genetic equivalent of vegetative reproduction.
(c) The 'cost of meiosis' (Williams 1975, Lloyd 1980b) (Ch. 2) is avoided; i.e. in the absence of recombination and segregation, maternal energy is not wasted on the aftercare of unfit zygotes, for all offspring are as fit as the mother. It is of theoretical advantage to the mother that her genetic contribution to her offspring is 100%, not 50% as in a sexual outbreeder.
(d) Some agamosperms are able to avoid male costs inherent in a sexual hermaphrodite by producing no pollen (agamospermy rarely develops from dioecious plants); however, male sterility is not widespread in agamosperms because male-sterile genes are unable to spread between clones (this is discussed more fully in Ch. 2). Thus, many agamosperms are still able to act as male parents to related sexuals, which may be a counter-advantage, balancing the energetic advantage of losing pollen.
(e) Agamosperms fix and disseminate an extremely fit genotype (the '$W_{max.}$' genotype for a given niche), other less fit agamic genotypes having disappeared from that niche by the processes of natural selection. Darlington (1939) has emphasised that most agamosperms have morphological and chromosomal features that strongly suggest that they arose from sexually sterile hybrid polyploids ('apomixis is an escape from sterility'). Being hybrid and polyploid, agamosperms can be expected to be highly heterozygous, and thus heterotic and vigorous (Ch. 2), the heterosis being 'fixed' by agamospermy (see Table 11.5, p. 435).

### *Disadvantages*

There are four main disadvantages to agamospermy:

(a) Inability to escape from accumulating disadvantageous (but non-lethal) mutants ('Müller's ratchet'), for the fittest genotypes would be inevitably 'saddled' with the mutants that became linked to it, unable to escape by the processes of recombination and segregation.
(b) Inability to recombine novel advantageous mutants that allow current evolution in the face of environmental change, such as evolution by sexual competitors (the 'Red Queen hypothesis', Ch. 2).
(c) A very narrow population niche width; I have assumed (Ch. 2) that one of the main advantages of panmictic sexuality is that genetic variation between individuals in the gene pool remains in intermittent contact with itself in time and space. This maximises the fitness and abundance of DNA by allowing it to fill the maximum number of niches in the environment (Dawkins 1976).

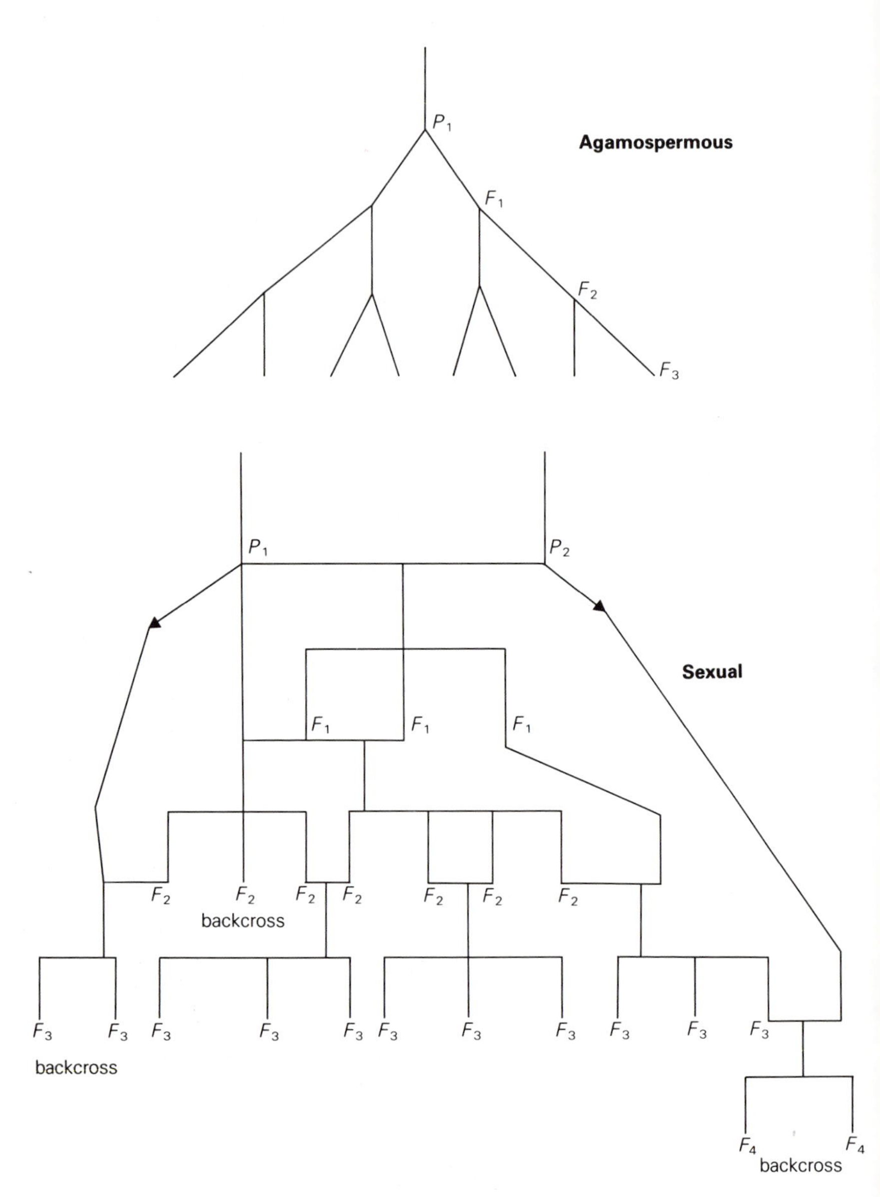


**Figure 11.2** Genetic contact in sexual and agamospermous lineages.

(d) As hybrids, most agamosperms might be expected to lack the adaptive 'fine-tuning' to a particular environment niche that would be expected of its parental sexual species. In general, an empirical assessment of many agamospecies would agree with this expectation. Many are weeds, or plants of open, transitory or unstable habitats.

However, if point (c) is considered further, the agamospermous gene pool should consist of only a single genotype, and genetic contacts should only be in direct, vertical lineage from mother to daughter in time, rather than being reticulate (Fig. 11.2). As a result, the niche width of the gene pool, and thus the number of opportunities that an evolutionary unit of DNA has to exploit the environment, should be severely limited. This also severely limits the fitness of agamospermous mothers.

In practice, it seems that at least some of this potential niche limitation is offset by unusually high levels of phenotypic plasticity in many agamospecies (as in dandelions, *Taraxacum*, Richards 1972). This stratagem is also typical of many relatively invariable inbreeders (Ch. 9).

It should also be pointed out that for so-called 'populations' of related agamosperms, for instance belonging to the same genus, many genotypes (agamospecies) may coexist. To take one example, H. Øllgaard (personal communication) has identified over 100 *Taraxacum* agamospecies inhabiting one hectare of waste ground near Viborg, Denmark. In the UK, it is commonplace to find 20–30 *Taraxacum* agamospecies coexisting. Such levels of diversity can also be found in some other agamic complexes (e.g. *Rubus* and *Ranunculus auricomus*), but in others (e.g. *Alchemilla* and *Sorbus*) it is unusual to find more than two agamospecies coexisting.

The definition of a population in an agamic complex cannot be the same as in a sexual gene pool. One could point out that 10 species of grass can coexist, entirely by apomictic (vegetative) reproduction, in a lawn that is regularly mown. We would not consider these to belong to the same population, for the species are not in genetic contact with each other. Similarly, coexisting *Taraxacum* agamospecies are not in genetic contact with each other. Thus, although an agamic complex can show considerable diversity within a location, the diversity within an evolutionary unit (gene pool of DNA) should be very low.

## MECHANISM OF AGAMOSPERMY

For obligate agamospermy to function correctly, it is necessary for three or four changes to be made to the sexual process; these may be independent of each other, or they may be connected.

(a) Avoidance of reductional meiosis.
(b) Avoidance of sexual fusion.
(c) Spontaneous embryony.
(d) Spontaneous development of the endosperm (not necessary in pseudogamous species), and adjustment to different genetic balances between the embryo and endosperm.

There are many and complex mechanisms by which agamosperms have bypassed the basic sexual functions of reductional meiosis and sexual fusion. These mechanisms have suffered detailed, and at times confusing nomenclatures which have been comprehensively reviewed by Gustafsson (1946–7) and Battaglia (1963). There is little point in covering all these, with their manifold variations, in the present volume. The main mechanisms are presented as a flow diagram (Fig. 11.3), and are discussed briefly in the following sections. For references to source material, the reader is referred to Gustafsson (1946–7).

### *Adventitious embryony (sporophytic agamospermy)*

In *Citrus* and in most other genera showing adventitious embryony, agamospermy only seems to occur in the presence of normal sexual reproduction. Typically, pollination is followed by the double fertilisation of a conventional reduced sexual embryo-sac, and the embryo and endosperm start to develop. The stimulus of embryo development results in the growth of further apomictic embryos in the nucellus (pollination by itself may also act as a stimulus). Various fates of the sexual embryo and the apomictic embryos are now possible, but typically one of the apomictic embryos invades the embryo-sac, outcompeting the other apomictic and sexual embryos, and commandeers the sexually produced endosperm. Alternatively, the sexual embryo and one of the apomictic embryos coexist, sharing the same endosperm. Thus, the mature seed may have one sexual embryo, one apomictic embryo, or two (or rarely, more) embryos (polyembryony), one of which is sexual and the other apomictic in origin.

It will be seen that in these cases, agamospermy is almost always facultative (that is agamospermous and sexual reproduction coexist, in the same population, individual mother, or even seed). Thus genetic fixation is tempered by sexual variability. Also, unlike other agamospermous mechanisms, only one mutation is necessary for adventitious embryony to arise. Thus, agamospermy may respond readily to selection, and it is perhaps curious that adventitious embryony is not a more common mechanism. Although sexual fusion and a reductional meiosis are avoided by the spontaneously arising embryo, adventitious embryony nevertheless depends vicariously on the sexual process. Thus,

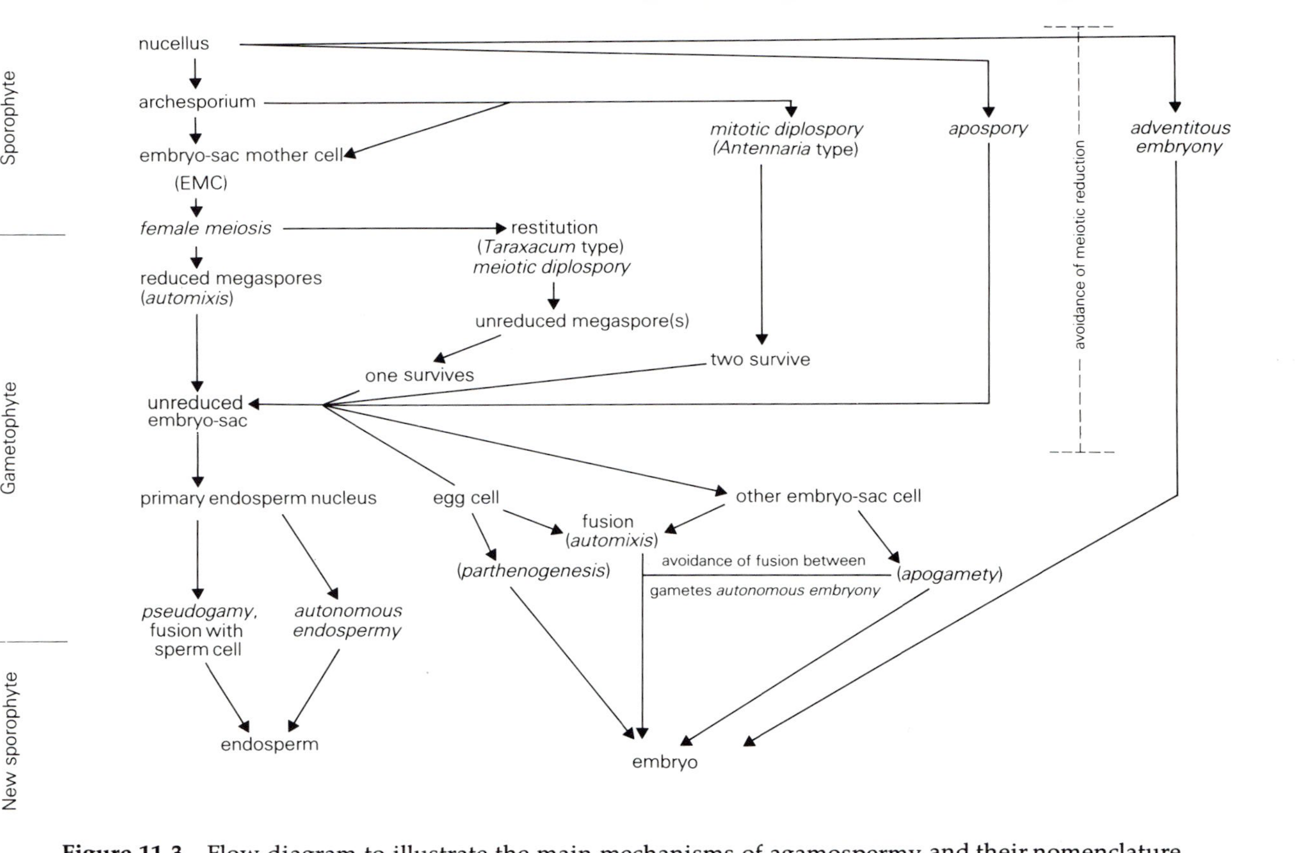


**Figure 11.3** Flow diagram to illustrate the main mechanisms of agamospermy and their nomenclature.

pollination and sexual fertility (favouring diploidy and non-hybridity) are required. In common with aposporous agamosperms, there remains a dependence on the production of an endosperm by conventional sexual means (pseudogamy).

### *Gametophytic agamospermy*

Unlike adventitious embryony, all other forms of agamospermy involve an embryo-sac (female gametophyte) (see Fig. 11.1). A distinction is generally made between apospory, in which the gametophyte originates from the nucellus, and diplospory, where the gametophyte originates in the archesporium. However, this is not always a very clear demarcation, as pointed out by Gustafsson (1946–7). The archesporium itself originates in the nucellus. In the case of so-called mitotic diplospory, a single distinctive embryo-sac mother call (EMC), which undergoes at least the earlier stages of female meiosis, is replaced by multiple cells which resemble an archesporium and several EMCs (which do not undergo meiosis) to different extents. It is important to distinguish between mechanisms that undergo meiotic diplospory, whether complete and reductional as in *Rubus* automixis, or incomplete and restitutional as in *Taraxacum*, and those that do not. The former have at least the potential for recombination (through crossing over) and segregation so that off-spring may differ, whereas this is impossible in the products of apospory and mitotic diplospory.

The distinction betwen apospory and diplospory is also genetically important. Aposporous agamospermy is usually facultative, and diplo-sporous agamospermy is usually obligate. Just as the nucellar embryos of *Citrus* allow the sexual embryo-sac to undergo regular sexual fertilisation and embryony, so the nucellar embryo-sacs of apospory allow the archesporium to undergo a reductional meiosis which can result in a sexual embryo-sac as well. If this is fertilised, embryos of sexual and agamospermous origin may coexist in the ovule; the mature seed may have both (as is often the case in *Poa*), or one only (most frequently the agamospermous one) may survive. The point is, that in aposporous mechanisms the archesporium, by definition, remains intact, and thus there is always the potential for sexuality. In such plants, agamospermy must always be considered at least potentially facultative rather than obligate.

In the three aposporous groups most thoroughly investigated, the *Poa pratensis* and *P. alpina* complexes, the *Potentilla verna* and *P. argentea* complexes, and the *Ranunculus auricomus* aggregate, agamospermy is very often facultative. Polyembryony is well known in all three groups, and nucellar agamospermy coexists with archesporial sexuality within the population, the individual, or even the ovule. However, individuals which by virtue of their hybridity, triploidy or other genetic disfunction have highly irregular female meioses may show poor or quite sterile sexual function, and these will approach obligate agamospermy.

## *Endosperm development and pseudogamy*

For an embryo to develop normally, it is also necessary that the endosperm should develop. In sexual plants the endosperm only develops after the fusion of the diploid polar nucleus of the embryo-sac with a sperm cell to form a triploid primary endosperm cell (Ch. 3). The balance of 'ploidy between the diploid embryo and the triploid endosperm, and the balance in genome contributions between the embryo and endosperm are probably critical for the successful development of the sexual endosperm, and embryo.

In adventitious embryony, diploid, maternal apomictic embryos commandeer triploid, sexual endosperm. This is a form of pseudogamy, in which pollination and fertilisation of the endosperm is required for the successful agamospermous development of the embryo. Pseudogamy is also usual in aposporous agamosperms; in these the balance between the embryo and the endosperm in 'ploidy, and in genome constitution, will differ from that in sexual relatives, and will need to be accommodated by the new agamosperm. The successful agamospermous embryo is accompanied by an endosperm from the same asexual and unreduced embryo-sac. This endosperm is achieved in most cases by the pseudogamous fertilisation of the tetraploid polar nucleus by a haploid sperm cell. (There are exceptions, for instance four-celled embryo-sacs in which the polar nucleus is diploid, and not the product of autofusion; and diploid sperm cells from unreduced pollen.)

As a result, the 'ploidy level of the endosperm will normally differ from that of the embryo in an aposporous pseudogam by 5:2 (3:2 in sexuals) and the male genome contribution balance between the endosperm and embryo will be 0.2:0 (0.33:0.5 in sexuals). The evaluation of pseudogamy has been modelled in a recent paper (Strenseth *et al.* 1985).

Apospory is closely associated with pseudogamy; nearly all aposporous agamosperms are pseudogamous, requiring fusion of the sperm cell with the polar nucleus for endospermy, and hence successful seed-set to occur. This association may reflect the facultative nature of aposporous agamospermy. Because the archesporium remains to undertake sexual reproduction, which often occurs, there is no benefit arising from the loss of pollen or pollination. Indeed, there will be a continuing pressure for the regular presence of fertile pollen tubes at the ovule in order that sexual fusion can occur. Thus, mutants that confer autonomous development of the endosperm in the absence of polar nucleus fertilisation will not be at a selective advantage, and male sterile mutations will be disadvantageous.

Gustafsson (1946–7) shows that when the polar nucleus remains unfertilised in pseudogams, the endosperm fails to develop. Whereas the embryo of *Poa alpina* or *Potentilla collina* may develop as far as the 256-cell stage in the absence of the endosperm, its form is abnormal, and abortion eventually results. In most pseudogams, fertilisation of the endosperm seems to be the only function of pollination, for embryo development precedes pollination (it is precocious). Only in *Ranunculus auricomus* does it seem that pollination (but not pollen-tube growth or fertilisation) is also necessary for embryony to proceed. We have already seen (Ch. 3), that embryony in a conventional sexual plant is often dependent on the stimulus of pollination, in addition to fertilisation, perhaps through a hormonal message. Thus, most pseudogams seem to have escaped the need for the stimulus of pollination for successful embryony. In this way, they also escape fertilisation of the egg cell, by developing their embryos early. Ironically perhaps, they have not escaped the need for the fertilisation of the primary endosperm cell (fused polar nucleus).

In contrast to agamosperms with adventitious embryony and apospory, nearly all those that are diplosporous are non-pseudogamous; i.e. they have autonomous endosperm development. Little seems to be known about the physiological or genetic control of autonomous endospermy. It is possible that it can derive from the same processes that result in autonomous embryony. In any case, neither the endosperm nor the embryo will have a paternal genome contribution. In all known cases, the endosperm usually has twice the 'ploidy level of the embryo, having resulted from the fusion of unreduced polar nuclei. Occasionally, high polyploid endosperm cells, which presumably result from endoduplication, are found as well. The adjustment of non-pseudogamous diplosporous agamosperms to an endosperm : embryo 'ploidy ratio of 2 : 1 (3 : 2 in sexuals) may be less difficult than the adjustment of pseudogamous aposporous agamosperms to a 5 : 2 ratio. In so far as the male : female genome balance is critical in controlling relative rates of endosperm and embryo growth (Ch. 3) this should not trouble the non-pseudogam, with its entirely maternal endosperm and embryo.

## *Diplospory*

As they are derived from the archesporium, diplosporous unreduced embryo-sacs have forfeited the use of the archesporium for reduced sexual embryo-sacs. Any sexuality in diplosporous agamosperms must therefore depend on inconstancy in the avoidance of sexual fusion, which may be accompanied by inconstancy in spontaneous embryony and by some reduction in the female meiosis. Most commonly, diplosporous plants show obligate agamospermy, and are thus non-pseudogamous, for reasons argued above.

In diplosporous species, the archesporium either undergoes a female meiosis, which fails in some way so that it is restitutional ('*Taraxacum*-type', Gustafsson 1946–7) or it divides to form the embryo-sac mitotically ('*Antennaria*-type').

A regular reductional meiosis in sexuals seems to be dependent on regular chromosome pairing (synapsis) and chiasma formation between homologues; synaptic bivalents auto-orientate on the spindle in such a manner that regular reduction and segregation of half-bivalents to the poles of anaphase I is achieved. In the absence of synapsis, the univalents fail to line up on the spindle at metaphase I, and in the absence of bivalent auto-orientation, disjunction fails to occur. Thus, in *Taraxacum* at least (Richards 1973), restitution (non-reduction) appears to be a function of asynapsis, through the failure of chromosome homologues to form chiasmata (and thus bivalents at diakinesis and metaphase I) (Fig. 11.4). Asynapsis can be shown to be under simple Mendelian dominant control, unlinked to the control for precocious embryony (also a simple Mendelian dominant).

However, it is not necessary for a female meiosis to be totally asynaptic for it to be restitutional. In certain synaptic triploid species of *Taraxacum*, many meioses are so disturbed (apparently as a result of the allotriploidy

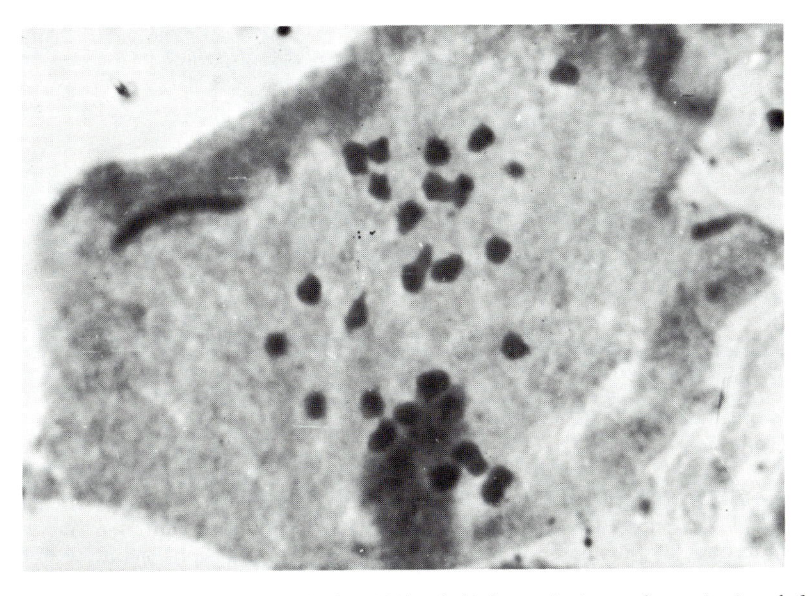

**Figure 11.4** Microphotograph (×2000) of diakenesis in male meiosis of the obligate agamosperm, *Taraxacum hamatum*, with 24 univalents. This meiosis is fully restitutional.

itself) that disjunction fails, although neighbouring meioses may, by chance, be sufficiently regular for disjunction and reduction to proceed (Figs. 11.5 and 11.6) (Richards 1970a, Müller 1972) (p. 425). Malecka (1965, 1967, 1971, 1973) shows that in some obligately agamospermous *Taraxaca*, asynapsis in female meiosis may be only partial, several bivalents forming in addition to many univalents. These meioses are restitutional.

## *Embryony*

Most forms of agamospermy pass through an embryo-sac, and this is most usually eight celled (but is four celled in panicoid grasses). These cells show much less constancy in form than is usual in a sexual embryo-sac. Multiple egg cells, an absence of synergids, poor differentiation between egg cells and antipodals, etc. are commonplace. However, in some genera (e.g. *Taraxacum*) the embryo-sac is conventional in appearance, and can apparently function sexually on occasions. Variation in embryo-sac cells can mean that distinctions between development of the egg cell (parthenogenesis) and another cell (apogamety) into the embryo, apparently straightforward, can in practice be difficult to draw.

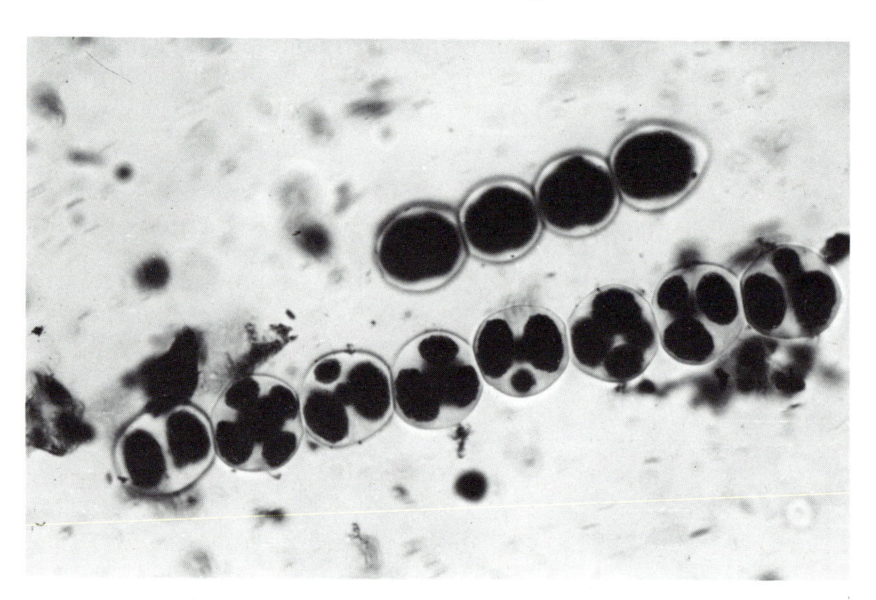

**Figure 11.5** Microphotograph (×200) of pollen 'tetrad' formation in *Taraxacum subcyanolepis*. From the left is a diad (meiosis restitutional), a pentad, a diad with two micronuclei (partial reduction), a tetrad (meiosis reduction), a diad with two micronuclei, a tetrad, and two unequal tetrads caused by partial reduction. The parent has $2n = 24$, and pollen grains will contain from 2 to 24 chromosomes, with most intermediate counts. Only grains with $n = 8$–$12$ usually penetrate the stigma.

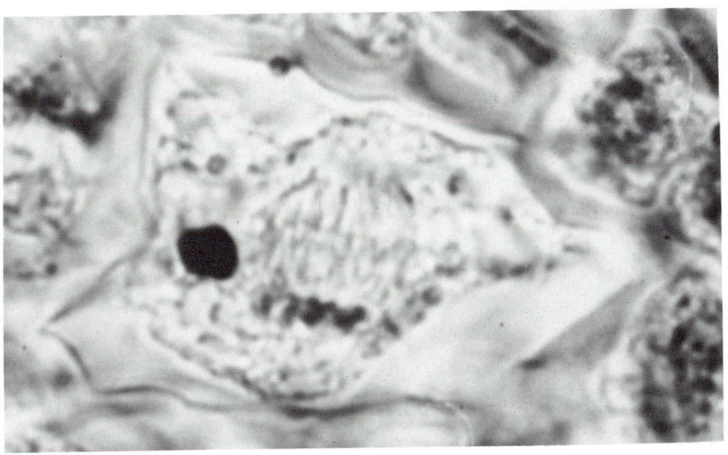

**Figure 11.6** Microphotograph (×1500) of anaphase I at female meiosis in *Taraxacum brachyglossum* ($2n = 24$). This is a reductional meiosis (the spindle and one segregating set of chromosomes only are visible in this plane of focus) in a triploid facultative agamosperm. It will give rise to a sexual embryo-sac. In the same plant, restitutional meioses giving rise to asexual embryo-sacs are also found.

However, in the vast majority of cases, it is a single egg cell that forms the embryo.

The polar nuclei, which usually fuse in the conventional fashion with each other, may develop autonomously to give an endosperm with twice the number of chromosomes of the embryo and a maternal genotype. Alternatively, they may fuse with a sperm cell to give a tissue with 5/2 the embryo chromosome count, 1/5 male genetically (pseudogamy).

The agamospermous attributes of failure of fertilisation, and autonomous embryony, are theoretically quite separate, but may in practice be closely correlated. In very many cases, both aposporous and diplosporous (but *not* in adventitious embryony) the egg cell develops into the embryo before the flower opens (precocious embryony), so that fertilisation of the agamospermous egg is impossible. Thus, in *Taraxacum* (Murbeck 1904, Richards 1973) embryony proceeds 48 hours before anthesis. Even in pseudogamous aposporous species, it is commonly found that the embryo has developed before the flower opens, so that later fertilisation of the polar nuclei takes place in the presence of a sizeable embryo. An exception appears to be in *Ranunculus auricomus* (goldilocks), in which autonomous embryony only proceeds after the fertilisation of the polar nuclei (Nogler 1972). In such a case, pseudogamous agamospermy may be inferred by a maternal inheritance of chromosome numbers and of other genetic characters (and an avoidance of female meiosis). Only very careful and painstaking cytology can prove that fusion of a sperm cell with the polar nuclei occurs in the absence of fusion with an egg cell. Non-precocious embryony is also recorded in the diplosporous, non-pseudogamous *Antennaria*. In such a plant, agamospermy is readily proven by emasculation followed by bagging of the flowers to prevent cross pollination, or by a removal of the stigmas (not possible for pseudogamy in which pollination is required).

The well-known 'emasculation' procedure for showing agamospermy in *Taraxacum* (Fig. 11.7) removes not only the anthers before anthesis, but also the stigmas, thus making cross pollination quite impossible in any circumstances. In agamospecies, this traumatic procedure is generally followed by perfect fruit-set.

It is clear that in *Taraxacum*, and other species with precocious embryony, failure of fertilisation is primarily mediated by precocious development of the embryo before the flower opens and pollination becomes possible (Fig. 11.8). Other mechanisms that prevent fertilisation might include full male sterility, failure of the pollen grain to germinate on the stigma or grow down the style, or failure of the sperm cell to fuse with the egg cell. All of these, if they occur (and it is only certain that the last mechanism operates in *Ranunculus* and *Antennaria*) may result from modifications to a self-incompatibility system. However, the physiologi-

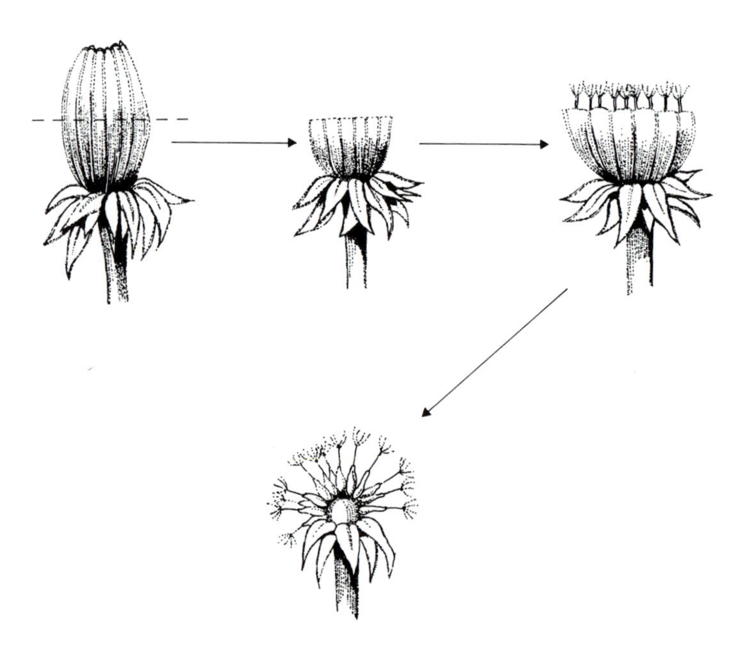

**Figure 11.7** Emasculation of an agamospermous *Taraxacum*. The top portion of the bud, containing anthers and stigma, are sliced off with a razor, but fertile seed is set asexually (×1).

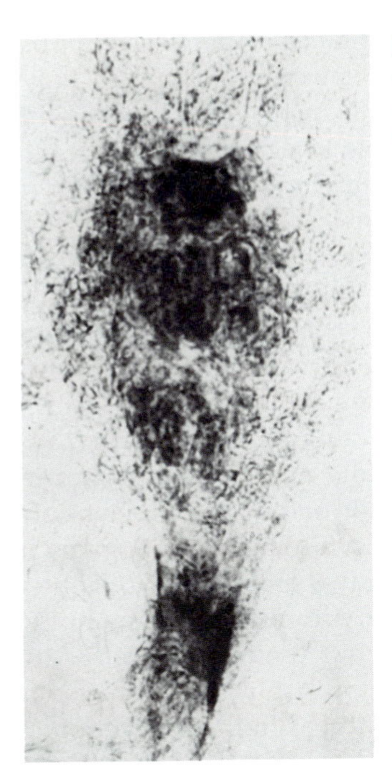

**Figure 11.8** Microphotograph (×60) of an eight-cell embryo (centre), with endosperm (above), surrounding nucellus, and neck and basal cell (below) dissected from the bud of *Taraxacum hamatum*, an obligate agamosperm, 24 hours before the inflorescence opens (precocious embryony).

cal basis of failure of fertilisation in *Ranunculus* and *Antennaria* seem to be unknown.

As yet, we also seem to be largely ignorant of the physiological processes that lead to autonomous embryony and, in non-pseudogams, autonomous endospermy. Many complex mysteries remain to be discovered by modern physiological and biochemical techniques. Much of the basic work on agamospermy was completed before 1950, and the study of agamospermous processes has not been fashionable in the last three decades, although theoretical and populational studies have grown in popularity. Thus, we have no idea how, for instance, in facultatively agamospermous *Taraxaca*, with irregular partially reductional meioses, the unreduced egg cells develop into embryos precociously in the agamospermous manner, but the unreduced eggs in the same capitulum do not so develop, but wait until anthesis when they are available for sexual fusion (Richards 1970a). Apparently the message that controls spontaneous embryony is in this case dependent on the chromosome number of the embryo-sac, which is very difficult to credit. However, Asker (1979) quotes many other examples in which this also appears to operate (see Fig. 11.10).

Before proceeding with the next section, it is necessary to refer briefly to a curious situation, by no means fully resolved, which seems to occur in *Rubus* (brambles). In this genus, relatively few, widespread species are wholly sexual diploids such as the raspberry (*R. idaeus*), and the bramble *R. ulmifolius*. In other somewhat variable, relatively widespread polyploid species (which have been termed 'circle-species'), apospory with pseudogamy prevails. These species are able to act as female parents when the sexual embryo-sac resists suppression by aposporous embryo-sacs. In these aposporous species, agamospermy is sufficiently prevalent for species integrity to persist in most areas (Haskell 1966). Some 30 circle species are recognised in the UK.

There is, however, a third situation, which seems to be typical of very localised tetraploid ($2n = 28$) *Rubus* agamospecies of which there are very many, hundreds in certain areas, for instance northern France, and the southern midlands of England. These undergo a fairly regular, reductional female meiosis to give rise to a diploid embryo-sac. Although pollination is needed for seed-set, sexual fusion of the egg-cell seems not to take place, and genetic inheritance is maternal. Thus, pseudogamy is suspected, and Thomas (1940) suggests that two embryo-sac cells fuse (automixis) to give rise to a tetraploid embryo. Although such an embryo might be considered to be totally homozygous (being the product of fusion between two identical cells), it would probably be heterozygous for many homoeologous loci which are duplicated in the two genomes that make up the allotetraploid. The endosperm would be the product of conventional fusion, and should be hexaploid, with two maternal and one paternal contribution.

Such automixis results from a reductional diplospory, and appears to be unique among plants (but compare *Rosa*, p. 426). However, similar conditions are widespread amongst parthenogenetic animals (White 1970, Vepsalainen & Jarvinen 1979). They are interesting, because both recombination and segregation occur, in the absence of sexual fusion. It must be expected that some genetic variation in automictic offspring will result. It is also likely that autofusion in the *Rubus* embryo-sac will frequently fail, and that reductional eggs (as well as automictic tetraploid gametes) will at times be able to undergo sexual fusion. The resultant capability for genetic variation, both within-clonal and hybrid, renders the taxonomy of such localised automictic brambles extremely problematical (p. 452).

## ORIGINS OF AGAMOSPERMY

Modern thinking about the origin of agamospermy, and its control, is reviewed in two influential papers by Asker (1979, 1980). The theme of Asker's work is that genes that control successful agamospermy when they occur in combination, can be found singly as aberrations in many conventionally sexual populations, where they are likely to be unsuccessful. Examples of such aberrations are listed in Table 11.1. It will be seen

**Table 11.1** Occurrence of independent elements typical of agamospermy amongst habitually sexual species.

| Autonomous embryony | Endosperm development | Avoidance of reductional meiosis |
|---|---|---|
| haploid parthenogenesis<br>e.g. *Zea* (Chase 1969) | autonomous endospermy<br>*Anemone nemorosa* (Trela 1963)<br>*Triticum aestivum* (Kandelaki 1976) | total inhibition of meiosis<br>*Zea* (Palmer 1971) |
| progressive subhaploidy<br>4X *Solanum tuberosum* (Breukelen *et al.* 1975) ↓<br>2X (dihaploid) → (autodiploid)<br>↓ (chromosome doubling)<br>1X (haploid) | lack of fusion of polar nuclei in four-cell aposporous embryo-sac followed by sperm-cell (pseudogamous) fusion<br><br>Panicoid grasses | total asynapsis (rare)<br>*Zea* (Beadle 1930) |
| synergid apogamety (*Zea*) | | partial asynapsis (desynapsis)<br><br>*Datura*, *Zea* (Catcheside 1977) |
| diploid parthenogenesis<br>(from endomitotic tetraploid embryo-sac mother cell followed by reductional female meiosis)<br>*Brassica* (Eeinnk 1974) | | non-disjunction at meiosis I or II<br>e.g. *Zea*, *Datura* (Rhoades 1956) |
| | | aposporous development of embryo-sac<br>× *Raphanobrassica* (Ellerstrom & Zagorcheva 1977)<br>*Sanguisorba* (Nordborg 1967) |

that all the functional features of various forms of agamospermy can be found exceptionally in sexual populations (adventitious embryony is excluded as it leads to facultative agamospermy *per se*). Thus, Asker argues that agamospermy is best viewed as an extension of the sexual process. This is supported by his contention that truly obligate agamospermy, in which all possibility of sexuality has been lost, is a rare phenomenon. He argues that some sexuality continues to be a feature of most agamospermous plants. This thesis, which supports that of Gustafsson (1946–7) and de Wet and Stalker (1974) is discussed below.

Clones that carry only one agamospermous mutant are likely to be unsuccessful, since they will reproduce in one of the following ways:

(a) Production of parthenogenetic haploids (autonomous embryony).
(b) Production of unbalanced high polyploids (avoidance of reductional meiosis).
(c) Non-formation of embryo (avoidance of sexual fusion).
(d) Non-formation of endosperm, leading to seed sterility.

However, if two particular mutants come together in a single genet, agamospermy may be successfully achieved at a stroke. For example, an aposporous mutant would generate unreduced embryo-sacs which, after fertilisation, would give rise to sterile autotriploids. A separate parthenogenetic mutant in the same species would give rise to sterile, and very possibly non-viable, haploids. If, however, the parthenogenetic mutant (male) is crossed with the aposporous mutant, some offspring might well prove to be aposporous, parthenogenetic triploids which were fertile agamosperms.

If the original parthenogenetic mutant is hybrid and polyploid, with a disturbed meiosis that is partially non-disjunctional, it may be able to reproduce agamospermously for non-reduced ovules. This seems to be the case for some triploid dandelions (*Taraxacum*) which still have synaptic meioses (Richards 1970a, 1973). Obligate agamospermy in *Taraxacum* only develops after a second mutation for asynapsis (resulting in a totally non-reductional female meiosis). This would presumably be favoured in such facultative triploids with relatively poor seed-set.

Thus, at least two, and often three or more, independent mutations are required before most forms of agamospermy can function. Because reductional meiosis is bypassed by most primary agamospermous mutants, the power to recombine different mutations onto a single chromosome (linkage group) is missing. Segregation of agamospermous alleles will occur in agamosperm × sexual backcrosses. Therefore, it is not possible for agamosperms to form a single coadapted linkage group in the way that dioecious or heteromorphic (Chs. 7, 8) plants can. Agamospermy is only likely to become stable if obligate agamospermy arises, and the whole genotype becomes a single linkage group. Because the evolution of obligate agamospermy, and the loss of recombinational variation, may not be favoured in many subsexual populations, the maintenance of subsexuality is reinforced by the independent segregation of different alleles for agamospermy.

We find that agamospermy is closely associated with certain life styles and genetic conditions in flowering plants. We can assume that such attributes predispose plants to the establishment of agamospermy. These attributes were:

(a) Perenniality, with limited vegetative spread, and good vegetative persistence. (Note, however, that very efficient vegetative dispersers have their own form of apomictic dissemination, and in these agamospermy might not be successful.)
(b) Hybridity; hybrids may favour agamospermy by being partially sterile and such hybrids will be vigorous, thus promoting good vegetative persistence until mutants come together. Hybrids may have disturbed metabolic systems which promote abnormal functions such as apospory or autonomous embryony and may also be more likely to combine different mutants in independent genetic lines. The disturbed female meioses of hybrids are likely to be non-disjunctional (restitutional).
(c) Polyploidy; polyploidy may encourage a higher level of parthenogenesis *per se* as in maize (Yudin 1970) and it may also lead to partial sterility, which should promote agamospermy; polyploids (especially triploids which will be common when meiosis fails) will have disturbed, and at least partially unreduced, meioses. Additionally, mutants that bypass meiosis (e.g. apospory) will produce unreduced embryo-sacs, so that new agamosperms may originate from crosses between $2X$ and $X$ gametes and thus be triploid ($3X$). [Fowler and Levin (1984) have discussed the 'minority type disadvantage' problems which a newly formed polyploid encounters. Most of its outcrossed matings will be with diploids, giving rise to sterile triploids.]

Parents of agamosperms carrying mutants that avoid meiosis or promote parthenogenesis, will be sexually unsuccessful, and thus the combination of these mutants in a single genet through sexual reproduction will occur very infrequently. It is not surprising that agamospermy is a relatively uncommon phenomenon. It has been suggested (Stebbins 1950) that the rarity of agamospermy is a measure of its long-term failure. Conversely, there is little doubt that agamospermy is a great success in genera such as *Taraxacum*, *Hieracium* or *Rubus*, and some agamospecies may be of a great age, dating back 10 000 years (Wendelbo 1959), or even to the early Pleistocene (Richards 1973). Modern work tends to emphasis the fitness and evolutionary potential in most agamic complexes (p. 432) suggesting that the scarcity of agamospermy is rather the product of the low likelihood of agamospermy arising in a group, or spreading amongst a genus. The continuing sexuality of many groups of agamosperms, espoused by Gustafsson (1946–7) and emphasised by Asker (1980), suggests that active evolution is still continuing in agamic complexes. It now seems likely that even obligately agamospermous groups are capable of genetic variation and evolution (p. 432). This contrasts markedly with the earlier 'gloom and doom' view of the evolutionary future of obligate agamospermy propounded by Darlington (1939) and Stebbins (1950).

Although Asker suggests that most agamospermous mutants are recessive, this is by no means always the case, as in *Taraxacum* where they are dominant in polyploids (Richards 1973). In a theoretical study, Marshall and Brown (1981) show that single mutants for agamospermy should always succeed in competition with sexuals, be they recessive or dominant. As they concede, although the facultative mechanisms of adventitious embryony and apospory may be under the control of a single locus, diplospory and obligate agamospermy never are. We await the production of more complex and realistic models on the evolution of agamospermy.

## GENETIC CONTROL OF AGAMOSPERMY

This is also comprehensively reviewed by Asker (1980). For aposporous systems, genetic analysis by studying the inheritance of agamospermous characteristics in conventional crossing programmes is readily achieved. A few well-known cases are simplified in Table 11.2. It will be seen that agamospermy is most commonly controlled by two unlinked loci, between which there may or may not be epistatic reactions, and that the Mendelian alleles controlling agamospermy are most commonly recessive.

Diplosporous systems and some aposporous systems tend to be complicated by the rôle played by polyploidy, especially triploidy, in influencing the fate of the meiosis. Thus in *Potentilla*, tetraploids tend to arise by sexual apospory from *aaBB* or *aaBb* diploid genotypes, which are usually the product of sexual and agamosperm crosses. These tetraploids are usually sexual or parthenogenetic. In the latter case, they may generate agamospermous dihaploids *aabb*. Thus, there is a tendency for a 2X agamosperm/4X sexual/2X agamosperm cycle to arise, although the results are in fact much more complex.

Similar cycles are also reported in the African grasses *Panicum*, *Bothriochloa* and *Dichanthium*, which form dominant tussocks over much of the savannah and probably in the southern hemisphere *Cordaterias*, including some of the 'pampas grasses' (Asker 1979, Philipson 1978, Connor

**Table 11.2** Some examples of the genetic control of agamospermy (after Asker 1980).

| | |
|---|---|
| *Sorbus* (aposporous) | *A* genomes for *S. aria* (Whitebeam), *B* genomes from *S. aucuparia (Rowan)*. *A* genomes apparently influence apospory in polyploids, thus *AA* (*S. aria*), *BB* (*S. aucuparia*) are sexual and *AAAA* (*S. rupicola*), *AAAB*, *AAB* (*S. minima*, *S. arranensis*) are obligate agamosperms. *AABB* (*S. intermedia*, *S. anglica*) are facultative agamosperms with *ABB* (*S.* × *pinnatifida*) being mostly sexual (Liljefors 1955, Richards 1975c) |
| *Pennisetum* (aposporous) | two loci *A*/*a* and *B*/*b*, *A* causes agamospermy, *B* overrides it to give sexuality, e.g. *Aabb* is agamospermous and *AaBb* is sexual (Taliaferro & Bashaw 1966) |
| *Panicum* (aposporous) | two loci control agamospermy recessively with complementation; thus either recessive homozygote with a heterozygote is agamospermous; *aabb*, *aaBb*, *Aabb* are agamospermous and *AABB AABb*, *AaBB*, *AaBb*, *aaBB*, *AAbb* are sexual (Hanna *et al.* 1973) |
| *Potentilla* (aposporous) | complex, and influenced by 'ploidy, but in some, two-locus recessive control is apparent: *AABB*, *AaBb*, *AABb*, *AaBB* are sexual; *aaBB*, *aaBb* are aposporous and sexual, giving tetraploid sexuals; *AAbb*, *Aabb* are parthenogenetic and meiotic, giving haploids or diploids; *aabb* is agamospermous (Muntzing 1945) |
| *Taraxacum* (diplosporous) | diploids always sexual: parthenogenesis and precocious embryony dominant for unreduced egg cells (on chromosome *H*?) in polyploids with irregular meiosis (facultative); asynapsis, leading to totally unreduced egg-cell dominant (on chromosome *D*?) (obligate) |

1979). These cycles depend on sexual apospory in diploids giving rise to sexual tetraploids which, when crossed with agamospermous tetraploids, generate hybrids with parthenogenesis but no apospory, which give sexual dihaploids (Fig. 11.9).

The great majority of polyploid *Taraxacum* species are obligate agamosperms. However, some sexuality, which can lead to sexual/agamosperm cycles depending on chromosome number, has been reported to arise from two distinct phenomena.

First, in a much quoted piece of work, Sørensen and Gudjonsson (1946) and Sørensen (1958) apparently show that certain disomic aberrants ($2n = 23$) from triploid species ($2n = 24$) which are obligate agamosperms, demonstrate some sexuality. They claim that the eutriploid progenitors are autotriploid, having the eight distinct chromosomes of the genome

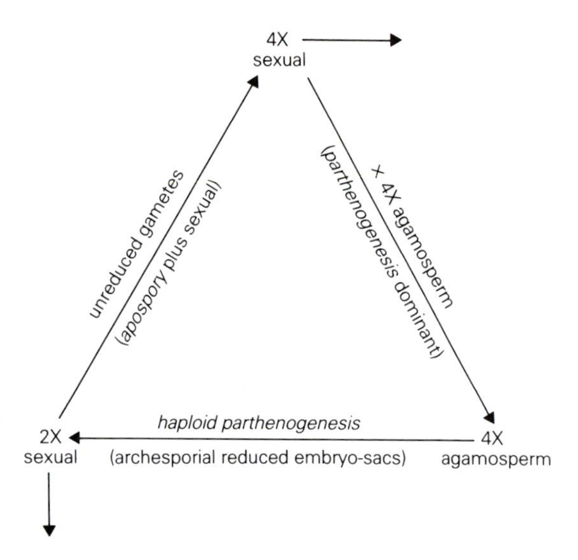


**Figure 11.9** Diploid/tetraploid cycle in *Panicum* and *Dichanthium* (after Asker 1979).

each present three times. Disomic aberrants lacking a chromosome *D* give rise to sexual diploid offspring, whereas aberrants lacking a chromosome *H* give rise to higher polyploids. The inference is (Richards 1973) that the dominant control for synapsis (giving unreduced eggs) is on one of the *D* chromosomes, whereas the dominant control for parthenogenesis and autonomous embryony is on one of the *H* chromosomes. Thus, if the asynapsis gene is lost by loss of chromosome *D*, a reductional meiosis (leading to some reduced sexual eggs) takes place. If the parthenogenetic gene is lost by loss of a chromosome *H*, parthenogenesis is missing, and unreduced eggs are fertilised.

Unfortunately, it has to be said that the karyology of triploid *Taraxacum* species (including the species investigated by these authors) is nothing like they state it to be. Chromosome types are fairly readily distinguishable, but never occur in regular sets of three (several authors, summarised in Richards 1972, Mogie 1982). Nor would one expect them to, for all triploid dandelions are almost certainly hybrid allotriploids arising from species with very diverse karyology. I have occasionally encountered weakly sexual forms with $2n = 23$, which are accompanied by apparent hybrids, and I have little doubt of the basic truth of Sørensen and Gudjonsson's findings. However, we must treat the claimed identification of eight distinct morphological aberrants, each resulting from a different missing chromosome, and the chromosomal location of the agamospermous genes with grave suspicion.

Second, both I (1970) and Müller (1972) have noted that facultative agamospermy is found in a few *Taraxacum* species from central Europe, and in the more widespread *T. brachyglossum* (which occurs in the UK, and adventively in North America). In such plants, emasculation of triploids leads to 40–70% seed-set, and gives rise to triploid offspring agamospermously. However, cross pollination of these heads gives a higher percentage seed-set, and some of the offspring are diploid or subdiploid ($2n$ = 16–18) sexuals. The female meiosis appears to be synaptic, and it is presumed that some female meioses fail, leading to unreduced, agamospermous eggs (which develop parthenogenetically). Other meioses are apparently reductional, leading to haploid, or near-haploid, sexual eggs (Fig. 11.10). Diploid sexuals may cross with reductional pollen, to give rise to more sexual diploids. However, some of the pollen from triploids is non-reductional, or partially reductional, and triploid or near triploid ($2n$ = 22–29) agamospermous offspring can arise from agamosperm (male) and sexual diploid crosses (parthenogenesis being dominant and operating only on unreduced eggs). This remarkable mechanism poses more questions than it solves, but it may account for the high level of sexual diploidy found among the dandelions of some parts of Europe (Fürnkranz 1960, 1966, den Nijs & Sterk, 1980, 1984). I have suggested (Richards 1973) that such plants represent an intermediate condition in the evolution of agamospermy in *Taraxacum*, in which the gene for the parthenogenetic precocious development of unreduced eggs is present, and functions in triploids with an irregular

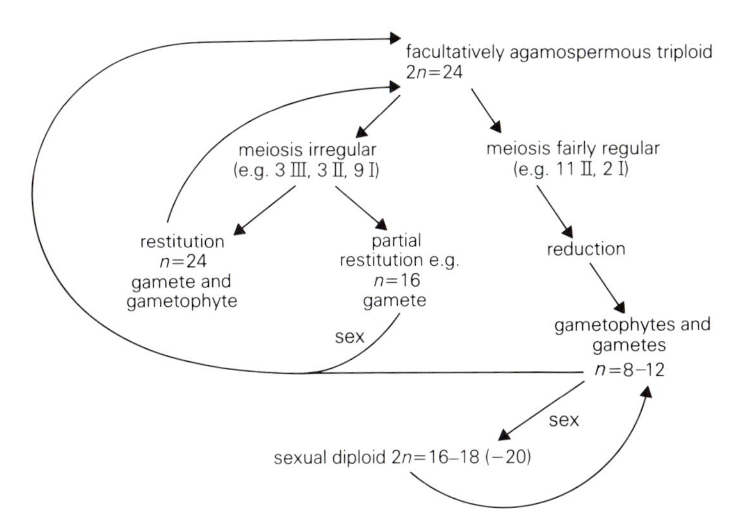


**Figure 11.10** Diploid/triploid cycle in facultatively agamospermous *Taraxacum* (after Richards 1970a).

meiosis. However, meiotic asynapsis, leading to total non-reduction of female meiosis and hence obligate agamospermy (as is found in most *Taraxaca*), is not present.

Before leaving this section, I feel bound to remark on a very interesting mechanism that is apparently unique in its nature and operates in most polyploid roses such as the dog roses (*Rosa canina* agg.) (Gustafsson 1944, Gustafsson & Håkansson 1942, Fagerlind 1944). These may be tetraploid ($2n = 28$) to hexaploid ($2n = 42$); most have $2n = 35$. In all, there is unequal synapsis in both male and female meiosis. Seven bivalents form and disjoin regularly. The remaining chromosomes (from 14 to 28 depending on the 'ploidy level) remain as univalents. They all become included in one of the daughter nuclei at telophase I of the female meiosis. Thus, one daughter nucleus has $n = 7$, and the other from $n = 21$ to $n = 35$ depending of the 'ploidy level.

After male meiosis, only the $n = 7$ microspores survive meiosis II, the others aborting. Thus, all pollen is $n = 7$, those chromosomes having passed through a regular disjunctional meiosis with recombination and segregation. In contrast, after female meiosis, the $n = 7$ megaspores abort, and one of the others develops an embryo-sac, which will have an egg cell with from $n = 21$ to $n = 35$. Although seven of the chromosomes in this egg have undergone a disjunctional meiosis with recombination and segregation, the remainder have not. Instead, they have been inherited maternally as an invariable, quasi-agamospermous package.

Regular sexual fusion now occurs between the $n = 7$ sperm cell and the $n = 21$ to 35 egg cell to give an embryo of $2n = 28$ to 42. There is also fusion between the other sperm cell and the fused polar nucleus. As a result, the endosperm has a female : male ratio of between 6 and 12 to one, much higher than is normal.

In a sense, the *Rosa* system is equivalent to facultative agamospermy, in that sexual and asexual reproduction by seed coexist. What is unique is that features of sexual and asexual reproduction coexist *within* every individual meiosis. Thus, those genes that are carried on the seven pairs of homologous, synaptic chromosomes undergo the regular sexual cycles of recombination, segregation and random fusion. The remaining genes are protected from the sexual process by asynaptic maternal inheritance. Thus, it should pay a rose to translocate genes that donate fitness and vigour onto the asexual chromosomes, and to transfer disadvantageous mutants onto the sexual chromosomes, where they can be recombined successfully, or lost by segregation. Genes that convey variability in niche specificity might also be most successfully linked to sexual chromosomes.

To my mind, there is a theoretical elegance about the rose system which is totally delightful. If one was to 'play God' and design an ideal breeding system, evolutionarily, this would be it. However, the likelihood of such a system evolving by the processes of mutation and natural selection would seem to be very small, which may explain why *Rosa* seems to be the only example of this mechanism. In particular, the regular inclusion of the asynaptic chromosomes in one daughter nucleus, and the differential abortion of high and low chromosome number spores after male and female meioses, must be very rare phenomena in regular sexual populations from which the rose system must have evolved.

## DISTRIBUTION OF AGAMOSPERMY

Although agamospory is widespread in Pteridophytes (Manton 1950), agamospermy is limited to Angiosperms amongst seed plants, and does not occur in the Gymnosperms. Perhaps the most comprehensive review on the occurrence of agamospermy in the Angiosperms is to be found in the appendix of Nygren (1967). Excluding the viviparous species which he includes (but which I prefer to treat as a form of vegetative reproduction, Ch. 10), and cases of haploid pathenogenesis (as in *Dactylorhiza*, Hagerup 1944, 1947 and *Epipactis*, Hagerup 1945), he lists about 300 taxa which have been shown to be agamospermous. However, this figure is particularly difficult to establish, and must be treated with great caution, due to taxonomic difficulties as to the nature of a species in agamic complexes. Thus Nygren has been, understandably perhaps, very inconsistent as to the taxonomic rank of his units. For instance, many *Sorbus* and *Alchemilla* agamospecies are treated as units, whereas the genera *Crataegus* (over 100 agamospecies) and *Rubus* and *Hieracium* (well over 2000 agamospecies each) are each treated as a single unit. *Taraxacum* is given an intermediate position, for 9 of the 25 or so sections of the genus which contain agamospecies (Richards 1973) are individually treated, although the 2000 odd agamospecies are ignored. With these major reservations in mind, it is interesting to compare numbers of taxonomic units which possess each of the three main classes of agamospermy (Table 11.3). [Note that in deriving the numbers in this table the categories queried by Nygren are not queried here, and species which have not been determined but occur in genera with a consistent system are 'rounded up' to that system for the sake of simplicity; a further problem arises in distinguishing between plants showing apospory and those showing mitotic diplospory (some *Potentilla, Hieracium*); it can be difficult to decide whether initials in the nucellus form an archesporium and an EMC, or not (Gustafsson 1946–7); these figures must not be treated as more than rough estimates.]

Some interesting points arise from a consideration of Table 11.3. Whereas agamospermy is recorded from about 15% of plant families (34 in all), no less than 75% of agamospermous taxa belong to only three

**Table 11.3** Numbers of agamospermous taxonomic units showing various types of agamospermy (after Nygren 1967).

| Family | Adventitious embryony | Apospory | Diplospory | Total |
|---|---|---|---|---|
| Compositae | | 18 | 51 | 69 |
| Gramineae | | 68 | 27 | 95 |
| Liliaceae | 6 | | 1 | 7 |
| Rosaceae | | 65 | 3 | 68 |
| Rutaceae | 5 | 2 | | 7 |
| Urticaceae | | 2 | 7 | 9 |
| 28 other families | 33 | 5 | 15 | 53 |
| | 44 | 160 | 104 | 308 |

families, the Compositae, Gramineae and Rosaceae. Although the Compositae and Gramineae comprise two of the three largest plant families; nevertheless, these three families only contain about 10% of Angiosperm species. Thus, it might be supposed that plants in these families are in some sense 'preadapted' to agamospermy, showing for instance unusually high levels of polyploidy, hybridisation or haploid parthenogenesis amongst their sexual members (p. 421).

An alternative hypothesis, that agamospermy, having arisen in a family, leads to the extensive genetic migration of agamospermous traits, or extensive radiation of agamospermous genera through that family, must be discounted. Although facultative agamospermy may encourage the spread of agamospermy within part of a genus, there is no indication that a single origin of agamospermy has ever given rise to more than one genus. Every agamospermous genus has some sexual taxa, which must be considered primitive to it, and agamospermy must be thought of as being essentially highly polyphyletic in nature, arising *de novo* on many occasions, even in closely related genera. Thus in the family tribe Cichorieae (Compositae) which have yellow, dandelion-like flowers, three massive agamic complexes have arisen in the hawksbeards (*Crepis*), hawkweeds (*Hieracium*) and dandelions (*Taraxacum*), as well as in smaller genera (*Chondrilla, Ixeris*). Yet each genus has its primitive sexual progenitors, and in each the mechanism of agamospermy differs (*Crepis*, apospory with parthenogenesis; *Hieracium* subgenus *Hieracium*, non-meiotic diplospory with apogamety; *Hieracium* subgenus *Pilosella*, apospory with apogamety; *Taraxacum*, meiotic diplospory with parthenogenesis; Table 11.4).

An examination of Table 11.3 lends some credence to the 'preadaptation' hypothesis. Thus, apospory is a much commoner method of circumventing meiosis in the Gramineae, and especially in the Rosaceae, than it is in the Compositae, in which diplospory prevails. The pattern of

**Table 11.4** Some of the more important agamospermous genera, and their major agamospermous mechanisms (after Nygren 1967).

**Adventitious embryony** (with some sexual fusion and with pseudogamy)
*Citrus* (most species and hybrids to varying extents), *Mangifera* (Mango), *Spathiphyllum*, *Opuntia*, *Capparis* (caper), some *Euonymus* spp. (spindles), *Pachira*, *Garcinia mangostana* (Mangosteen), some *Eugenia* spp. (rose-apples), *Ochna serratula*, *Nigritella*

**Apospory** (with pseudogamy and occasional sexual polyembryony)
*Poa pratensis/alpina*, *Panicum maximum* agg., *Potentilla argentea* agg., *P. verna* agg., *Dichanthium/Bothriochloa*, *Cortaderia* sp., *Paspalum* spp., *Pennisetum* spp., *Urochloa* spp., *Rubus fruticosus* agg., *Malus* spp., *Alchemilla* spp., *Cotoneaster* spp.,? *Sorbus* spp.,? *Crataegus* spp., *Ranunculus auricomus* agg., *Crepis* spp., *Hieracium* subgenus *Pilosella*, *Skimmia* spp.

**Mitotic diplospory** (sexuality very rare, no pseudogamy)
*Antennaria*, *Ixeris*, *Hieracium* subgenus *Hieracium*, *Arnica* spp., *Erigeron* spp., *Elatostema* spp., *Balanophora* spp., *Calamagrostis* spp.

**Meiotic diplospory** (sexuality usually absent except for *Rubus*, no pseudogamy)
*Taraxacum*, *Chondrilla* spp., *Rudbeckia* spp., *Nardus stricta*,? *Limonium* spp., automictic *Rubus*.

distribution is certainly consistent with a hypothesis that suggests some 'preadaptation' to a certain sort of agamospermous mechanism within a family.

Predispositions to agamospermy, if they exist, are not necessarily confined to chromosomal or physiological phenomena. The predominant growth habits and ecology of the various families are similar. Thus the Gramineae and Compositae are mostly comprised of perennial herbs of temperate areas, and indeed are the dominant families in these biomes. The Rosaceae, although including some agamospermous perennial herbs (*Alchemilla*, *Potentilla*) or even annuals (*Aphanes*), also has many agamospermous trees (*Crataegus*, *Sorbus*). Common factors in these three great families are non-specific pollination mechanisms by wind (Gramineae) or insects, temperate or grassland distributions, and single-seeded fruits. Non-specific pollination may be associated with hybridisation; and temperate distributions seem to have encouraged polyploidy (associated with glacial conditions which may also have favoured agamospermy).

In a single-seeded fruit, the selective fitness of an asexual embryo (of a maternal genotype) may be optimised with respect to sexually arising half-siblings. In comparison, asexually and sexually arising half-siblings which compete for maternal resource in a multi-seeded fruit may be compelled to share that resource more equally. There may also be greater competition between half-sibling seedlings after dispersal from certain sorts of multi-seeded fruits. As yet there seems to be no information to support these very tentative suggestions.

We do not seem to understand what has apparently predisposed the Compositae to diplospory, and the Gramineae and Rosaceae to apospory. If there is a logical basis to these apparent preadaptations, we may have to seek it in currently little-understood areas of developmental control and differentiation. However, our confidence that preadaptive mechanisms for the origin of agamospermy repay consideration is strengthened by an examination of the distribution of adventitious embryony. This is totally lacking in the temperate, often herbaceous, predominantly polyploid and single-seed fruited Compositae, Gramineae and Rosaceae. Yet 58% of the remaining 76 agamospermous taxa have adventitious embryony. These are typically subtropical or tropical, woody, diploid and with fleshy, multi-seeded fruits (Table 11.4), and thus quite different from aposporous and diplosporous species.

The control of adventitious embryony is at least partially hormonal, and labile, so that the predominance of asexual or sexual embryony varies with the environment and can be modified experimentally. As no embryo-sac is involved, the mechanism is entirely sporophytic. Also, it may be able to arise through a single mutation. As the archesporium is not involved in the agamospermy, mechanisms that avoid fusion and circumvent reductional meiosis are irrelevant; indeed most if not all agamosperms with adventitious embryony can undertake perfectly regular sexual reproduction. Thus, the correlates of polyploidy, hybridisation, temperate climate and vegetative persistence which characterise apospory and diplospory tend to be absent (p. 408). Because each of the seeds emanating from a population at one time tend to share a similar type of genetic origin (all sexual, or all agamospermous) there is no competition between sexual and non-sexual half-siblings before or after seed dispersal. This may explain why animal-dispersed multi-seeded fruits are commonly associated with this mechanism (this is a frequent form of dispersal among tropical trees, Ch. 8).

It has frequently been pointed out (e.g. Stebbins 1950), that agamospermy becomes more frequent in the flora as one proceeds further north or further up mountains, as does polyploidy and the frequency of perennial herbs. As far as I am aware, this relationship has not been investigated with increasing latitude southwards. Various explanations for increasing levels of agamospermy with increased latitude northwards can be presented:

(a) Arctic, boreal and northern temperate floras have been more thoroughly examined, thus the relationship with latitude and altitude is an artefact.
(b) For non-pseudogamous agamosperms, the lack of a requirement for pollination is an advantage in areas with cold summers.

(c) The heterotic vigour of agamospermous hybrids is an advantage, and the lack of precise adaptation to a complex habitat is not a disadvantage in open arctic, montane or boreal habitats.

(d) The onset of glacial epochs encouraged migration, resulting in hybridisation; and encouraged rapid evolution, optimised by the genetic isolation and variability of polyploids. Both polyploidy and hybridisation are associated with agamospermy, as discussed above, thus agamospermy in high latitudes and altitudes is an accidental correlate of hybridisation and polyploidy.

(e) Agamospermy is chiefly found in a few families which show a preadaptation towards it; these families predominate at high latitudes and altitudes for reasons which do not, or only partially, coincide with the preadaptive attributes.

It is likely that all these explanations contribute to the correlation between agamospermy and latitude/altitude, and most are to some extent interdependent or interrelated. However, there still remains a considerable doubt about explanation (a). Recent reports (Kaur *et al.* 1978) suggest that many tropical trees of south-east Asia may be agamospermous, but it has been long known that many tropical grasses such as *Panicum* (Usberti & Jain 1978) can be agamospermous. There seems little doubt, though, that gametophytic agamospermy such as apospory, and especially diplospory, becomes commoner in northern areas. This may not be the case for adventitious embryony (p. 430).

It has frequently been noted that agamospermous complexes have a wider distribution geographically and ecologically than their sexual counterparts and progenitors, and this has been explained by the greater levels of heterozygosity, capacity for plasticity, reproductive efficiency, and short-term fitness of agamospermous genotypes (e.g. Babcock & Stebbins 1938, Stebbins 1950 for North American *Crepis*). However, this apparent relationship appears to be the product of some confused and possibly wishful thinking, and requires a closer examination. For instance, agamospermous dandelions (*Taraxacum*) undoubtedly cover a wider area of the earth's surface than do sexual dandelions; probably about twice the area. Yet, there are about 10 times as many agamospermous taxa as sexual taxa in *Taraxacum*. Very many of the agamospecies are extremely localised, whereas sexual species such as *T. serotinum*, *T. brevirostre* or *T. bicolor* range for thousands of kilometres (Richards 1973, van Soest 1963). Typically, sexual species are palaeoendemic and may have wide, interrupted distributions, whereas agamospecies are neoendemic with narrow, but expanding, distributions, particularly as a result of introduction by man.

There is also a danger of comparing sexual species, as a group or individually, with agamospecies, as a group or individually. The philosophy surrounding specific limits is unavoidably very different for diverse breeding systems (p. 448), and it is difficult to make direct comparisons between the two. However, it is possible to point out that for *Crepis*, *Taraxacum*, or other partially agamospermous genera, the agamospermous mechanism appears to be currently more successful, than sexuality for it is found in more taxa and over a wider area in these genera. Such a statement infers nothing about the ecological versatility of the agamospermous genotype, but does highlight its greater fitness in some genera. In others (for instance *Sorbus*), the agamospecies are numerically and distributionally much more limited than the sexual species, and some are verging on extinction in the British Isles (e.g. Clapham *et al.* 1962, Perring & Sell 1968).

## GENETIC DIVERSITY IN AGAMOSPERMS

I have already contrasted the old-fashioned 'gloom and doom' view of the evolutionary potential of agamosperms with the more 'optimistic' modern approach. It was presupposed by Darlington (1939) and Stebbins (1950) that many agamosperms never underwent sexual reproduction and consequently generated no genetic variation amongst their offspring. As evolution depends on a supply of genetic variation, and it was expected that plants had to change evolutionarily in order to survive (the 'Red Queen' hypothesis, Ch. 2), the scarcity of agamospermy was explained by the frequent extinction of agamospermous lines. More recently, Maynard Smith (1978) has suggested another reason for the lack of success of asexual reproduction. Asexual (agamospermous) lines do indeed generate variability, as they accumulate mutations. However, the majority of these mutations will be harmful, and as the agamosperm is unable to recombine them, or lose them by segregation of chromosomes, each line is 'stuck' with them (Ch. 2, p. 17, and p. 405).

However, Gustafsson (1946–7) and Asker (1979, 1980) show that most agamosperms retain some sexuality; obligate agamospermy is rare, and is probably confined to diplosporous species. Even in these, it is by no means certain that the generation of useful genetic variation ceases. Genetic variation amongst the offspring of parthenogenetic animals is well known in snails (Stoddart 1983), sea anemones (Schick *et al.* 1979), moths (Mitter *et al.* 1979), earthworms (Jaenike & Selander 1979) and aphids (Blackman 1979) (reviewed by Vepsalainen & Jarvinen 1979). Such variation seems to occur as a result of somatic recombination within the genome by processes of chromosome breakage and fusion which may be mediated by transposable genetic elements (TGEs or so-called 'jumping genes'). These are pieces of DNA that promote chromosome breakage next to them and thus roam around the genome recombining fragments of chromosome as they go. Their activity is known to be under genetic control (Fincham & Sastry 1974). Thus, it is possible that high levels of TGE activity, which promote high levels of somatic recombination, might permit the success of obligate agamosperms which have otherwise lost the power of genetic recombination. Evidence is presented below that this may be the case in *Taraxacum*.

As yet, there seem to have been few studies that compare levels of genetic variation in sexual and agamospermous progeny. Thus, although we can expect that facultative agamosperms will retain some genetic variation, there is little indication of the quantitative levels expected. The only study of which I am aware in which such comparisons have been made is for the guinea-grass, *Panicum maximum* (Usberti & Jain 1978). Variation is assessed at seven electrophoretic loci in the alcohol dehydrogenase (ADH) and esterase systems, and for several morphological characters in 28 populations, 25 of which are 'predominantly asexual' and three '50% or more sexual'. For both morphological and electrophoretic characters, a consistent pattern emerges, in which the 'sexual' populations are on average more variable than the 'apomictic' populations. However, substantial variation for most characters is also noted in the asexual populations, and in some characters there was more variability in the asexual than the sexual populations. Thus, for one asexual population, the seven esterase bands did not vary ($H' = 0$), but for another $H' = 0.42$ (for three sexual populations $H' = 0.60$, 0.65 and 0.68 where $H'$ is the proportion of individuals that are heterozygous for one or more loci). Yet the population which was invariable for esterase showed substantial variation for alcohol dehydrogenase (ADH). In fact the mean $H'$ estimate for asexual populations was 0.45 and that for sexual populations 0.66.

Most unfortunately, direct assessment of the breeding systems of the experimental material by embryological investigations was not made. Instead, the breeding system was characterised by the segregation of offspring at esterase loci (Marshall & Brown 1974). This technique is justified by the latter authors in cases where apomixis has already been cytologically characterised, in that large samples can be scored. Unfortunately, a number of other phenomena can influence the variability of electrophoresis bands in progeny; these are not mentioned in Marshall and Brown:

(a) Variation due to technical artefacts of pH, temperature, etc.
(b) Variation due to phenotypic effects, ageing, etc.
(c) Dissociation of polymeric enzymes into multiple bands, and other forms of non-Mendelian segregation (often untestable genetically in apomicts).
(d) Differential fitness of isozyme morphs.
(e) Non-random mating, and other abnormalities of a sexual breeding system such as selfing, etc.

As a result, the genetics of an electrophoretic locus, and the breeding system of a plant, have to be very well understood if that locus is to be accurately employed in the characterisation of a breeding system. It is probably safer to use cytological or, for non-pseudogams, emasculation procedures to establish the breeding system. In particular, circularity can be involved when electrophoretic variation in progeny is used to characterise a breeding system, and variation is then compared between presumptively sexual and agamospermous strains, using the same electrophoretic system!

In the genus *Taraxacum*, Jane Hughes (unpublished) has compared levels of heterozygosity in an unidentified population of self-incompatible diploid sexual plants from the Auvergne, France, and in an obligately agamospermous population of *T. pseudohamatum*. The breeding system of each plant was checked. For eight electrophoretic loci that were polymorphic in the sexual population, five were monomorphically heterozygous in the agamospermous population, and a sixth almost so. Of 16 loci that were examined in total, eight (50%) were polymorphic in the sexuals, and only one (6%) in the agamosperm. However, it is, noteworthy that 7/16 (43%) of the agamosperms had fixed heterozygosity (Table 11.5). The lack of variability, but high level of heterozygosity, theoretically typical of an agamosperm, is well illustrated here.

Morphological variation in the asexual progeny of guayule, *Parthenium argentatum*, and mariola, *P. incanum*, was demonstrated non-quantitatively by Rollins (1945). However, genetic variation in apparently obligately agamospermous populations of plants seems to have been first demonstrated by Solbrig and Simpson (1974) in American *Taraxacum*. A more widespread study was made by Lyman and Ellstrand (1984). In the first study, four different isozyme patterns were detected from a limited area. In the second, 21 isozyme phenotypes were discovered in 22 populations from various regions of North America.

Unfortunately, these studies suffer from the same shortcomings as those of Usberti and Jain (see above). In neither case was the breeding system checked, although this is very simply and rapidly done for *Taraxacum* by emasculation (see Fig. 11.7). As we have seen, triploid *Taraxaca* can at times show some sexuality.

Although it was claimed that the isozyme phenotypes represented distinct genotypes, this was only tested by Lyman and Ellstrand (1984) by reference to the very distantly related goatsbeard, *Tragopogon*. To establish the genetic basis of isozyme bands in a partially asexual genus such as *Taraxacum*, it is necessary to make test crosses for the same systems between related sexual species in the same genus, as we have done (p. 436).

The third problem has been discussed earlier in this chapter (p. 407). We can assume that different genotypes do coexist in the North American

**Table 11.5** Comparison of variation between an asexual and a sexual population of Taraxacum.

| | Phenotype | | | |
|---|---|---|---|---|
| | Sexual | | Asexual | |
| Enzyme locus | Ho | He | Ho | He |
| MDH-1 | 0.84 | 0.16 | – | 1.00 |
| ME | 0.98 | 0.02 | 1.00 | – |
| 6PGDH-1 | 0.98 | 0.02 | 1.00 | – |
| 6PGDH-2 | 0.70 | 0.30 | 0.06 | 0.94 |
| TYR | 0.74 | 0.26 | – | 1.00 |
| PER | 0.61 | 0.39 | – | 1.00 |
| SOD-1 | 0.78 | 0.22 | – | 1.00 |
| ACPH | 0.53 | 0.47 | – | 1.00 |
| MDH-2 | 1.00 | – | 1.00 | – |
| IDH | 1.00 | – | 1.00 | – |
| GDH | 1.00 | – | 1.00 | – |
| CAT | 1.00 | – | – | 1.00 |
| SOD-2 | 1.00 | – | 1.00 | – |
| SOD-3 | 1.00 | – | 1.00 | – |
| GOT-2 | 1.00 | – | 1.00 | – |
| GOT-3 | 1.00 | – | 1.00 | – |

Ho, homozygous; He, heterozygous.
Mean heterozygosity per individual: sexual = 0.12; asexual = 0.43.

dandelion populations of Solbrig and Simpson, and Lyman and Ellstrand (but in the latter case we must ignore the fruit-colour phenotypes, for these are insufficiently characterised in what is a very plastic character). However, we do not know if they belong to the same or different agamospecies. The North American 'weedy' dandelions are introduced from Eurasia, and as yet nobody has identified them to the agamospecies, preferring the 'coverall' name *Taraxacum officinale*. It is not of any great interest to show that different agamospecies coexist in North America, as they do in Europe. Any *Taraxacum* taxonomist could demonstrate that in seconds. It would, however, be of great interest to show that one agamospecies displays genetic variation, even between individuals of a single progeny. Although this has yet to be done for North American *Taraxacum*, we have demonstrated such variability in European populations.

*Taraxacum pseudohamatum* is invariably triploid ($2n = 24$) and sets good seed after emasculation in every one of hundreds of tests. It is an obligate agamosperm, at least as a female parent (it produces viable pollen that can hybridise with sexual forms). Mogie (1982) grew 12

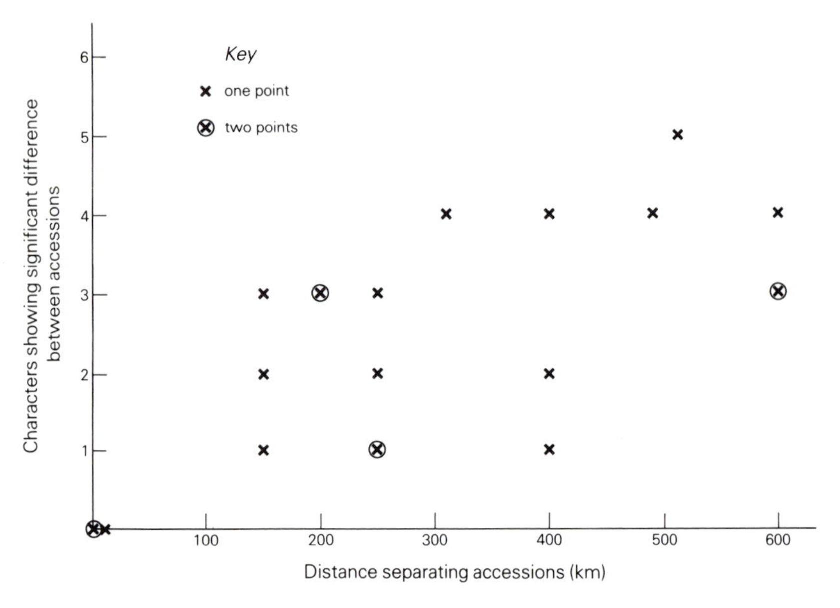


**Figure 11.11** Relationship between geographic distance between the origin of the accessions and the number of characters by which they differ.

seedlings from each of seven different accessions of this species (from origins up to 600 km apart) in standard conditions and in randomised blocks. He then compared within-accession (between-block) variation with between-accession (within-block) variation by the analysis of variance (anovar) of six quantitative characters that are known to be highly heritable in this species. There was in total no significant variation within accessions for any character (6% of anovars showed significant variation at the 5% level). However, 38% of anovars showed significant between-accession variation at the 5% level, and these included all six characters; thus within the taxonomic agamospecies *T. pseudohamatum* there were marked phenotypic differences which were probably genetic as they were heritable and did not occur between sibling replicates. Even more interestingly, there was a geographical pattern to this variation. Three accessions obtained from within a few kilometres of each other showed no significant variation between each other in any character. Of the others, there was a clear relationship between geographic distance between the origin of the accessions, and the number of characters by which they differed (Fig. 11.11).

As yet no clear information is available on whether phenotypic variation occurs within families, that is between siblings. This is best tested by comparing variation between siblings with variation between ramets of a single genet, propagated by root cuttings. This experiment is being undertaken as this book is being written. In an earlier limited experiment with *T. pseudohamatum*, Mogie (1985) showed that for one accession there was no significant variation between seed-grown and root cutting-grown material from a single accession. However, in a second accession, root-grown material differed significantly from seed-grown material of the same accession (which did not vary significantly within itself) for two out of six characters. It remains to be seen whether this tentative indication of the creation of somatic genetic-based variation in an obligately agamospermous *Taraxacum* is confirmed by further studies.

Certainly, there are strong indications that electrophoretic variation occurs within *Taraxacum* agamospecies, and even between siblings, and that this variation is heritable, and represents genetic variation. Mogie (1985) identified eight different zymograms (electrophoretic banding patterns) for esterase among 60 accessions of *Taraxacum*, which are placed in eight closely related agamospecies in section Hamata (Table 11.6). Although *T. pseudohamatum* showed the greatest variability, with four zymograms in 18 accessions, only three of the agamospecies failed to show variability in this small sample. However, there are some worrying features of these results. Although there is little doubt that the zymogram did not vary for an individual (most samples were examined on a number of occasions), the pattern of variation shows that one zymogram is the most common for all species, as if the others are minor variants'. Furthermore, work on related sexual species (Hughes & Richards 1985), shows that the inheritance of esterase bands in *Taraxacum* is very complex, and can only be explained on the basis of a number of loci, and the widespread occurrence of null alleles. Similar types of esterase inheritance are known in other plants, for instance barley (Hvid and Nielsen 1977).

Mogie (1985) found little variation in esterase zymograms between siblings. He analysed 135 seedlings from nine accessions of *T. pseudohamatum*, and 98 seedlings from three accessions of another

**Table 11.6** The esterase zymograms of 60 accessions in eight species of *Taraxacum* section Hamata (after Mogie 1982).

| | Zymograms | | | | | | | |
|---|---|---|---|---|---|---|---|---|
| *Taraxacum* spp. | a | b | c | d | e | f | g | h |
| *T. pseudohamatum* | 7 | 1 | 5 | | | | 5 | |
| *T. quadrans* | 12 | | | | | 1 | 1 | |
| *T. atactum* | 5 | | | | | | | |
| *T. hamatiforme* | 4 | 1 | 1 | | | 1 | | 1 |
| *T. hamatum* | 5 | 1 | | 1 | 1 | | | |
| *T. boekmanii* | 3 | | | | | | | |
| *T. kernianum* | 2 | | | | | | | |
| *T. hamiferum* | 1 | | | | | | | 1 |

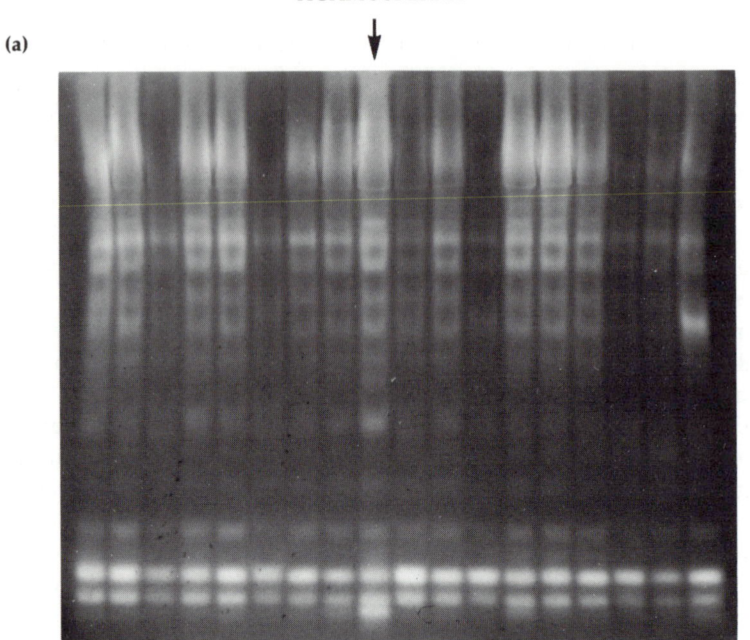


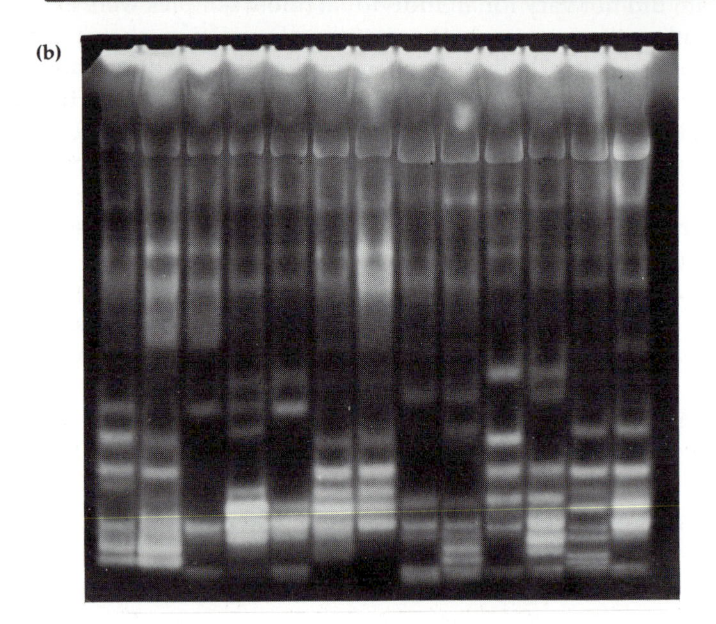


**Figure 11.12** Gel-electrophoretic bands representing esterase isozymes in the leaves of 18 siblings from a single agamospermous mother of *Taraxacum pseudohamatum* (a) and from a single population of sexual diploids from the Auvergne, France (b). The sexual population shows great variability. One of the agamosperm offspring in (a) (arrowed) has an extra band at the base, indicating a change in genotype in this asexual plant.

agamospecies (which lacks pollen) *T. unguilobum*. He found no variation within families in *T. pseudohamatum*, although one leaf gave an aberrant zymogram lacking a major band that could not be repeated on other leaves of the same individual. In later work, Jane Hughes (personal communication) identified one seedling with an extra esterase band (type c, Table 11.6) amongst 18 seedlings from three parents from the same population (type b) which were otherwise identical (Fig. 11.12).

One seedling of *T. unguilobum* (out of 98) also clearly differed from all the rest in possessing an extra esterase band on every occasion it was examined. Unfortunately, we do not know that this extra esterase band is heritable, so that it is not proven that a pollen-less obligate agamosperm can create genetic variation in this case (but see p. 441).

Ford and Richards (1985) have examined electrophoretic patterns for esterase in a number of coexisting *Taraxacum* agamospecies from a Northumberland sand dune (Fig. 11.13). In a study area of 100 m$^2$, 97 individuals were randomly sampled, and grown on in uniform conditions. They were identified as belonging to 10 agamospecies. In the sample, five agamospecies were found to have no intraspecific isozyme variability. Two esterase morphs were discovered in *T. unguilobum*, *T. fulviforme* and *T. nordstedtii*, three in *T. brachyglossum*, and four in *T. lacistophyllum*, the dominant species in the area. All plants were tested for sexuality by emasculation, and found to be obligate agamosperms. They included triploids ($2n = 24$), tetraploids ($2n = 32$) and a hexaploid species ($2n = 48$). We can conclude that isozyme variation can occur within agamospecies populations, as well as between populations. Within an agamospecies, there is a similar frequency of different isozyme morphs between populations (as in Table 11.6), at 0.22, to that within a population (Fig. 11.13) at 0.19.

No variation in these populations was detected for tyrosinase, or acid phosphatase. The tyrosinase system is of particular interest, as Hughes and Richards (1985) show that tyrosinase has a simple one-locus two-allele Mendelian inheritance in sexual dandelions. Homozygotes each have one band, and heterozygotes have both bands. Three families were grown up from parents, all of which were heterozygotes $T^1T^2$, and the seedlings were examined for tyrosinase:

| | Parents | | |
|---|---|---|---|
| Offspring | brachyglossum | lacistophyllum 1 | lacistophyllum 2 |
| $T^1T^2$ | 8 | 18 | 5 |
| $T^1T^1$ | 12 | 3 | 5 |
| $T^2T^2$ | 1 | 2 | 3 |
| | 21 | 23 | 13 |

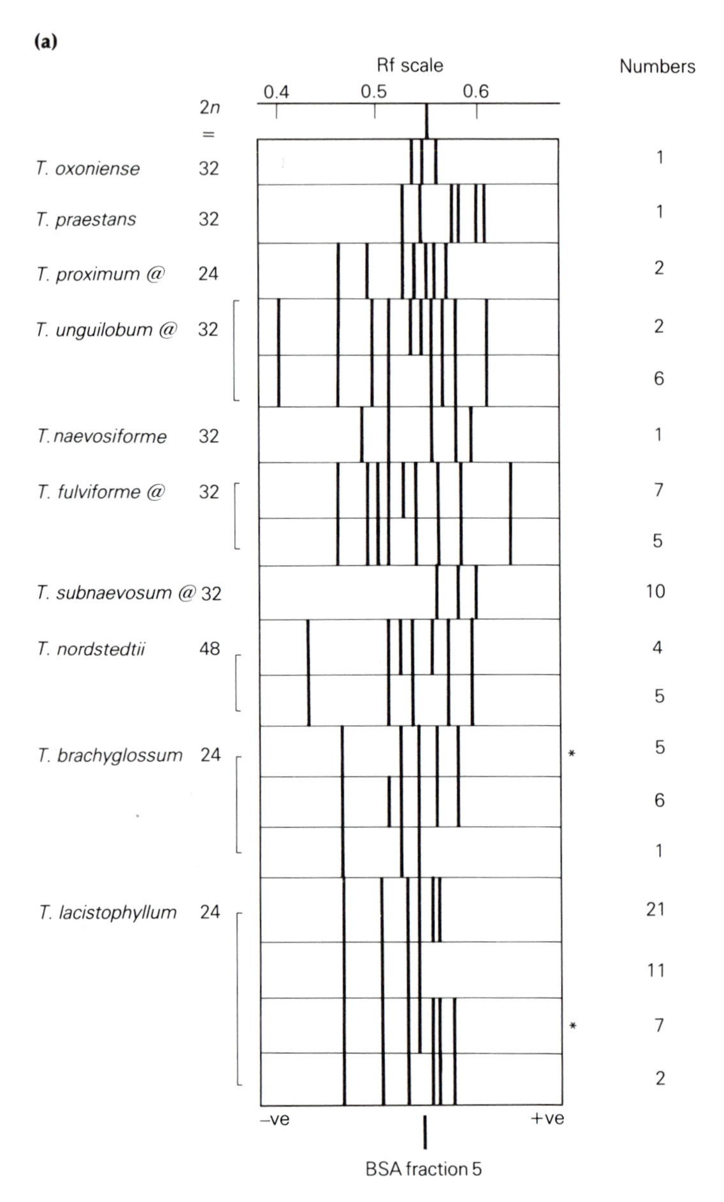


*Taraxacum species*, chromosome number, zymogram, number in population sample (total = 97), @ indicates those species in which pollen is absent

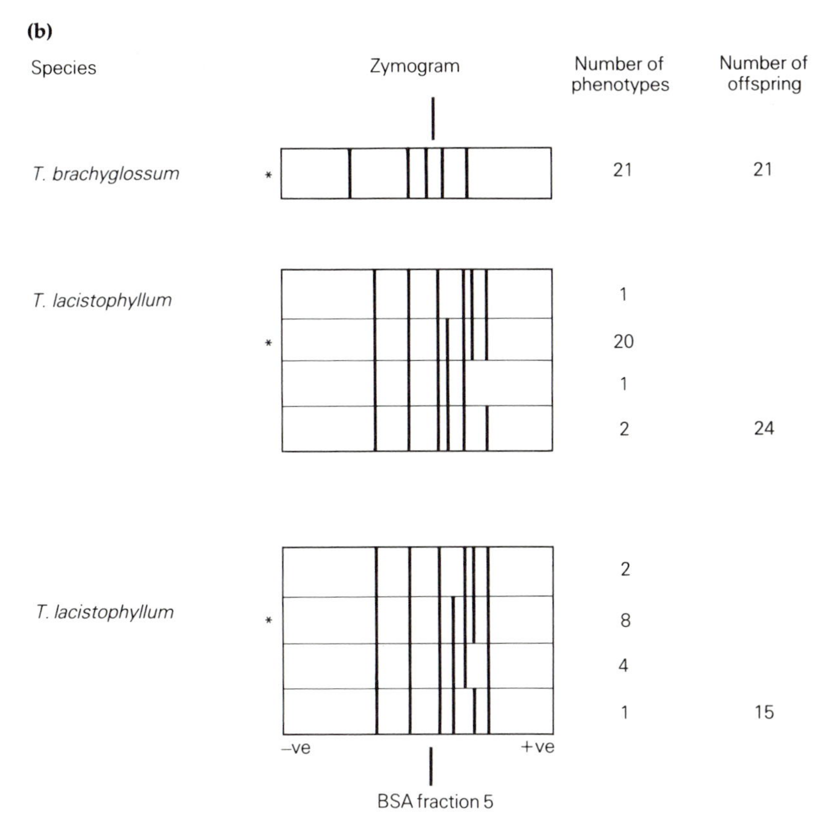


Zymograms of the offspring of apomictic *Taraxacum* parents (parental type asterisked)

**Figure 11.13** Representation of esterase gel-electrophoretic band phenotypes in 97 randomly sampled individuals of *Taraxacum* from 100 $m^2$ of sand-dune grassland in Northumberland, UK (left). Ten agamospecies were represented, with a total of 18 esterase phenotypes. Three offspring families were also examined, from two parents (*). BSA is the bovine serum albumen standard run with each gel.

We can conclude that variation occurs between agamospermous seedlings of a single parent at a surprisingly high frequency in this system, 0.45 being variants. As only heterozygotes $T^1T^2$ were found in the parental population, it seems that there may be a very strong heterozygous advantage, for the homozygotes $T^1T^1$ and $T^2T^2$ which arise so commonly in offspring, are apparently outselected in the wild. A much lower rate of variation (0.15) was also detected in the same seedlings for esterase. They did not vary for acid phosphatase.

Mogie and Richards (1983) have suggested that the morphological and genetic variation that occurs within agamospecies can result in the origin of new agamospecies. In a cytological survey of the 22 species placed in section Hamata (a morphologically discrete section of western Europe dandelions), they show that all species invariably show a cytological abnormality, with two rather than three chromosomes having satellites (associated with the nucleolar organiser). Most other triploid *Taraxaca* have three satellite chromosomes. Sexuality is likely to result in the segregation of satellited chromosomes, resulting in clones with different numbers of such chromosomes. Thus, they suggest that all the agamospecies have arisen since the evolution of obligate agamospermy, without the help of any sexuality (all are obligate agamosperms today).

It is not yet entirely clear what mechanism leads to the creation of genetic variation amongst obligately agamospermous *Taraxaca*. Mogie (1982) has shown that obligate agamosperms in section Hamata although usually eutriploid ($2n = 24$) with two satellited chromosomes, show great karyological diversity. He shows a clear association between karyological similarity and the degree of family relationship between cells (Table 11.7), using an index of karyotype similarity (KSI).

Unfortunately, insufficient comparisons were made between siblings to establish whether variation within individuals leads to variation within families. However, the level of variation observed between cells was astonishing, and on average each cell differed from the next by the effects of one chromosomal interchange. If this actually occurs, and we are not viewing an artefact, very high levels of somatic recombination must be taking place in meristems. Our confidence of the validity of these results is strengthened by two observations:

(a) As expected, a proportion of anaphases show chromosome 'bridges' resulting from the disjunction of the centromeres of a single dicentric chromosome which has arisen from an interchange; the daughter cells of such bridge anaphases may be non-viable or aneuploid (aneuploid cell lines are rather frequent).

**Table 11.7** Percentage mean similarity of karyotypes (KSI) between cells within and between individuals in *Taraxacum* section Hamata (after Mogie 1984).

| Species | Within individuals | Between accessions | Between species |
|---|---|---|---|
| *T. pseudohamatum* | 87.5 | 82.5 | 83.0 |
| *T. hamatum* | 87.9 | 84.1 | 82.2 |
| *T. boekmanii* | 92.1 | 85.8 | 83.0 |
| *T. atactum* | 90.8 | 82.1 | 82.2 |
| mean | 89.6 | 83.6 | 82.6 |

**Table 11.8** Percentage mean similarity of karyotypes (KSI) between cells within individuals and percentage of anaphase cells showing bridges, in *Taraxacum* of different breeding systems (after Mogie 1982).

| Species | Breeding system | Mean KSI (%) | Cells with bridges (%) |
|---|---|---|---|
| *T. pseudohamatum* | obligate agamosperm | 87.5 | 10.5 |
| *T. lacistophyllum* | obligate agamosperm | 89.6 | 3.8 |
| *T. unguilobum* | obligate agamosperm | 90.9 | 6.6 |
| *T. breviforoides* × *oliganthum* | wide sexual hybrid | 85.6 | 8.1 |
| *T. alacre* | sexual outbreeder | 97.5 | 0 |
| *T. brevifloroides* | sexual outbreeder | 99.4 | – |
| *T. bessarabicum* | sexual inbreeder | 100 | 1.0 |

(b) Sexual dandelions show much higher levels of karyotype similarity within and between individuals, and correspondingly low proportions of chromosome bridges at anaphase in comparison with agamosperms (Table 11.8). The low levels of karyotype similarity in an artificially created broad hybrid, which are similar to those of the agamosperms, strongly suggests that the karyotypic instability of agamospermous *Taraxaca* (also shown here for the unrelated species *T. lacistophyllum* and *T. unguilobum*) is a product of its hybridity.

As mentioned above, it is possible that the very high levels of chromosome breakage and refusion cycles (somatic recombination) which are apparently being observed in agamospermous *Taraxacum* are a product of transposable genetic elements, or TGEs. In any case, we may suppose that such phenomena will generate genetic variation by position effects. It is thoroughly established that genetic products are a function of the genetic environment, and the expression of a gene will depend on the genes to which it is linked on the chromosome. For instance, operators may allow a gene to be expressed (e.g. Ch. 6, Fig. 6.4), or enzymes may be produced as monomers or polymers. Thus, somatic recombination should greatly affect the phenotypic expression of a genotype, either morphologically or as isozyme bands.

It must not be forgotten that genetic variability may also arise amongst obligately agamospermous *Taraxacum* (and other genera such as *Erigeron* and *Hieracium*) as a result of events taking place during female meiosis, loosely classed together by Gustafsson (1946–7) as 'autosegregation'. Under this heading he includes (a) the results of faulty non-disjunction, resulting in the loss or duplication of a chromosome in the embryo-sac and embryo, as in Sørensen's disomic ($2n = 23$) aberrants which differ from their mothers phenotypically and (b) the results of synapsis (chiasma formation) between some homologous chromosomes in a subsequently non-disjunctional meiosis, as observed in a few *Taraxacum* species by Malecka (1965, 1967, 1971, 1973). In the latter case, meiotic recombination within the genome would result in position effects just as somatic recombination would. However, meiotic recombination would be more limited, only involving segments of homologous or homoeologous chromosomes.

Autosegregation can only occur in agamosperms with a meiotic diplospory. Presumably, somatic recombination or autosegregation are in any case most likely to be favoured in diplosporous species in which agamospermy is most likely to be obligate. Facultative agamosperms, with apospory or with adventitious embryony, will generate some genetic variability by sexual means.

## *Obligate agamospermy*

It seems likely (from the above section) that obligately agamospermous *Taraxacum* can generate heritable genetic variation by several mechanisms:

(a) Somatic recombination (chromosome breakage and refusion).
(b) Meiotic recombination (probably not widespread).
(c) Chromosome loss and gain.
(d) Accumulation of mutants.

It also seems likely that this variation is subject to natural selection so that it can result in asexual speciation (Mogie & Richards 1983). Although similar mechanisms have not yet been demonstrated in other obligate agamosperms, they may well occur, for instance in *Hieracium*, in which many very localised obligately agamospermous endemics are found.

Having established that obligate agamosperms can vary and evolve, contrary to earlier ideas, there remains the question as to how common obligate agamospermy is. As discussed above, it is probably confined to diplosporous species, but in these, various forms of facultative agamospermy may also occur, as in *Taraxaca* with chromosomal loss or partially reductional meioses (p. 425). Even when plants are obligate agamosperms as mothers, they may be able to act as fathers. Very few obligate agamosperms lack pollen, for male-sterile mutants, even if advantageous, cannot spread to other genomes (Ch. 2). Only 17% of British dandelion agamospecies lack pollen.

It is a remarkable feature of agamospermous *Taraxaca*, emphasised by Gustafsson (1946–7) that *male* meioses are usually relatively regular and reductional. This is in strict contrast to the restitutional, and usually totally asynaptic, *female* meiosis of the vast majority of obligately agamospermous species. A number of workers such as Richards (1970b) have shown that pollen from *Taraxacum* agamosperms can fertilise sexuals or

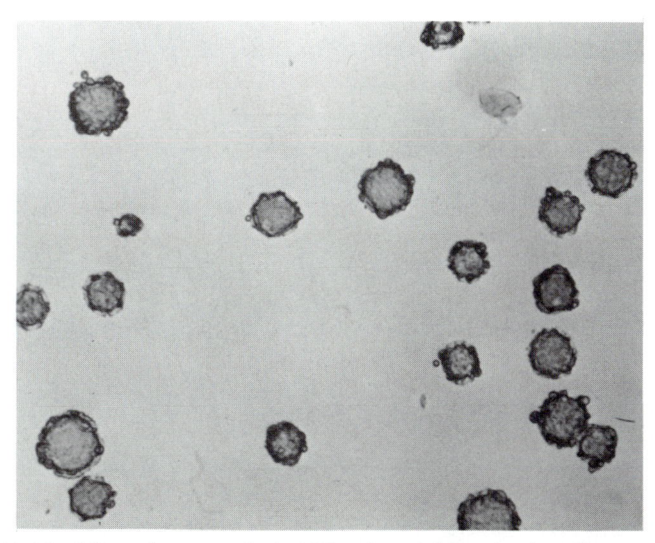

**Figure 11.14** Microphotograph (×100) of variably sized pollen typical of a triploid ($2n = 24$) obligately agamospermous *Taraxacum* with irregular male meiosis (see also Fig. 11.5).

facultative agamosperms, often creating hybrids, which vary in their breeding system depending on their chromosome number (p. 425). C. Brazier (personal communication) has shown that pollen from agamosperms is less fertile than that from sexuals. Agamospermous pollen is very variable in size (Figs 11.5 and 11.14) and only a fraction of grains of median size (probably with $n = 8$ from a triploid $2n = 24$) are fertile (Fig. 11.15).

It seems not to have been suggested before that the fertility, regularity, and the more or less reductional nature of the pollen of many agamospermous dandelions may be adaptive. Mutants that led to asynapsis in the female meiosis alone may have been more successful than those that also gave asynapsis in male meiosis. Agamosperms with reductional pollen can not only act as female parents with a full maternal genetic control (p. 405), but can also act as male parents to coexisting sexuals. In many parts of Europe and Asia, such as Japan (Morita 1976, 1980), sexuals and agamosperms regularly occur together. Even in areas such as the UK and the USA, where sexuals are rare or absent, extant agamosperms must have originally evolved agamospermy sympatrically with sexuals. Thus, obligate agamosperms may also be able to generate genetic variability (and new agamospermous biotypes) by hybridisation with sexuals. Sexual/agamosperm hybrid swarms of this kind have been described by Fürnkranz (1961, 1966) and Richards (1970c).

Although it often seems to have been assumed that certain taxa are

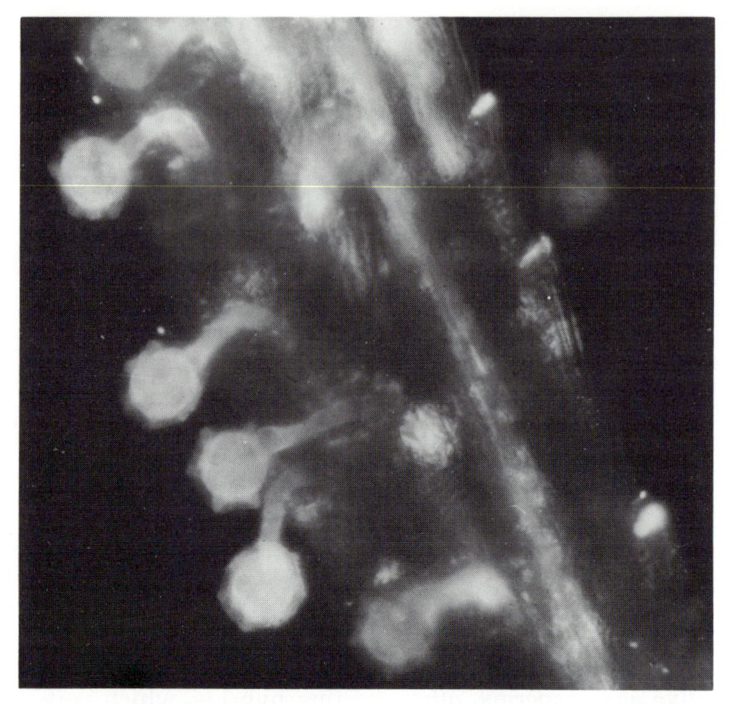

**Figure 11.15** Pollen germinating and penetrating stigmatic papillae in a sexual *Taraxacum*. Microphotograph under ultraviolet illumination (×400).

obligately agamospermous, with no means of sexual reproduction of any kind, these assumptions seem rarely to have been critically examined. For obligate agamospermy to be verified, the following conditions have to be met:

(a) Diplospory (in apospory and adventitious embryony the archesporium is available for the production of sexual embryo-sacs, and pseudogamy prevails).
(b) Male sterility, or the total absence of coexisting related sexuals.
(c) No difference in seed-set between emasculated flowers and pollinated flowers.
(d) Invariably restitutional female meioses, or no female meioses.
(e) Absence of regularly occurring sexual 'mutants' caused by chromosome loss, etc.
(f) Only maternal chromosome numbers amongst offspring.
(g) Absence of paternal characters amongst offspring. As we have seen, non-variability of the offspring is not a critical condition in itself.

An examination of Table 11.4 suggests that there are relatively few diplosporous candidates for obligate agamospermy. Unfortunately, none of these are sufficiently well known to be sure that obligate agamospermy operates, apart from some *Taraxacum* species. However, certain *Hieracium* species (in, for instance, sections Alpina, Subalpina and Alpestria), some *Limonium* spp. (with only one heteromorph; Ch. 7), some *Antennaria* (which are sexually dioecious with only females) and *Ixeris* and *Chondrilla* are strong candidates.

With respect to *Taraxacum*, we can be certain that sexuality is a very rare phenomenon in northern Europe (i.e. UK, Iceland and all of Scandinavia apart from the Gothenburg region where the sexual endemic *T. obtusilobum*, which may now be extinct, used to occur). In North America, all the lowland adventive species are probably obligate agamosperms, although native sexuals are known from Greenland (*T. pumilum*), the high Californian Sierras and South America. However, even in these regions, the occurrence of very occasional sexuality by disomic aberrants (Sørensen 1958) remains a possibility. In practice, such aberrants seem usually to be weak, and probably rarely survive long in the wild. In much of the rest of the range of *Taraxacum* some sexuality probably remains, either by facultative agamospermy, or through hybridisation by obligate agamosperms of coexisting sexuals. This detailed example warns against the facile allocation of obligate agamospermy to any taxon without detailed study.

## *Variability in agamospermous behaviour*

There are several reported cases in which the reproductive behaviour of facultative agamosperms (usually pseudogamous and aposporous) varies with environmental conditions. Perhaps the best known is in the grass *Bothriochloa* (Saran & de Wet 1979) which behaves sexually during the summer, but as a facultative agamosperm during the Oklahoma winter. Embryological studies of plants kept experimentally under different day-length regimes demonstrated that the relevant environmental trigger was only two hours difference in photoperiod (Table 11.9). There

**Table 11.9** The effect of day length (photoperiod) on the production of sexual (S) or aposporous (A) ovules in *Bothriochloa* spp. (after Saran & de Wet 1970)

| | Ovules | | | |
|---|---|---|---|---|
| Day length (photoperiod) (hours) | % S embryo-sacs only | % S embryo-sacs with non-functional A sacs | % with functional S and A embryo sacs | % with functional A sacs only |
| 14 | 78.4 | 21.6 | | |
| 12 | 36.6 | | 20.0 | 43.4 |

are several other examples of related Andropogonoid grasses, in which apospory is related to photoperiod, and in all cases, short days favour high levels of agamospermy. Thus, Knox (1967) and Knox and Heslop-Harrison (1963) show that in northern Australian populations (9°S) of *Dichanthium aristatum*, with less than 14 hour days at flowering, 91% of embryo-sacs are aposporous. Further south (to 36°S), the longer day lengths result in only 60% of the embryo-sacs being aposporous. These responses are plastic, without genetic differentiation between areas. As the summer progresses and day lengths decline, the proportion of apospory increases, and this can be repeated in the photoperiod experimental chamber. Somewhat unexpectedly, short days also resulted in high levels of pollen sterility. It is difficult to see how the pseudogamous apospory can operate in such conditions, without fertile pollen to fertilise the endosperm nucleus.

We may presume that these plastic responses to photoperiod are adaptive. Curiously, short days may be related to very different adverse conditions, which favour agamospermy, in the two areas. Short days in Oklahoma presage the cold plains winter. Short days in Australia are associated with the searing summer heat of the northern tropics. In a pseudogamous species, there is no advantage for agamospermy in having reproductive efficiency in hostile conditions. Conceivably, harsh conditions for seedling establishment favour a low genetic variability in the offspring, which is associated with the agamospermous mode of reproduction. If the mother is adapted to resist harsh conditions, seedlings of a maternal type may be better suited to establish in these conditions. Seedlings establishing at a more favourable time of year, or a more favourable locality might be able to colonise a wider range of niches in the environment if they displayed the genetic variability expected to result from sexual reproduction.

## TAXONOMY OF AGAMOSPERMS

This book is not about taxonomy, and the taxonomy of agamospermous groups has been thoroughly reviewed by Gustafsson (1946–7) and Grant (1981) among others. The latter account has interesting discussions concerning the agamic complexes in *Crepis, Citrus, Rubus* and *Bouteloua* in particular.

However, agamic complexes present the most daunting problems that face the plant taxonomist and, as these are the product of the breeding system, they merit a short discussion here. In particular, I wish to emphasise the point that agamospermous breeding systems vary widely, and are never exactly the same in any two complexes. The nature of the taxonomic problems also varies widely, and it is a mistake to expect that the best taxonomic solution for one agamospermous genus is applicable to another. The philosophical basis to the taxonomy of an agamic complex should be based on a thorough understanding of the breeding system, as well as on the pattern of morphological, geographical and cytological variation displayed.

I have been closely involved in the taxonomy of one agamic complex, *Taraxacum*, for 20 years, and I am struck by the lack of sympathy and understanding of agamospecies taxonomy displayed by authors who discuss the subject, but have not themselves worked on the taxonomy of such a group. In some agamic complexes, very large numbers of agamospecies have been described (notably in *Taraxacum*, *Hieracium* and *Rubus*, in each of which between 2000 and 3000 agamospecies are in use). These have understandably dismayed many authorities, who doubt whether such classifications are useful (without attempting to use them). Yet these are morphologically, ecologically and geographically very diverse genera, and comparable species diversity is accepted in totally sexual genera such as *Carex* (2000 species) *Astragalus* (2000 species) or *Senecio* (3000 species) without comment.

In a genus such as *Taraxacum*, new agamospecies are published to clarify taxonomic problems, when it seems to the taxonomist that their recognition is unavoidable. Thus, Grant (1981, p. 443) states 'Øllgaard (1978) has recently described 26 new "species" (read microspecies) from Denmark alone' (in *Taraxacum*). I was associated with the work that led to the publication of this paper, and I know that it was the product of years of painstaking experimental work by the foremost authority on the genus. Many of the agamospecies described in this paper also occur elsewhere, as in the British Isles, and their recognition has substantially clarified rather than confused a number of taxonomic problems in the British *Taraxacum* flora.

It is unavoidable that the complexities of such large agamic genera should render their taxonomy available only to a smaller number of specialists. This is also true to some extent for large sexual genera such as *Carex*. This does not invalidate such classifications, but it does make them inaccessible to most non-specialists. Within the framework of a conventional taxonomy there exist taxa of a wider scope, such as sections and subsections, which can be employed in an agamic complex by the non-specialist. However, he must be prepared to encounter difficulties, for the evolutionary unit is the agamospecies, and not all of these will conveniently fit into a taxon of wider scope in a highly reticulate, multidimensional pattern of morphological relationships between agamospecies.

Unfortunately, specialists and non-specialists alike have been tempted to invoke taxonomic categories to deal with the particular taxonomic problems of certain agamic complexes that are not recognised by the

International Code for Botanical Nomenclature. Although I write of 'agamospecies' in this account, this category merely recognises the special nature of the breeding system, and hence of the pattern of species delimitation, and the species are named by a conventional binomial, e.g. *Taraxacum pseudohamatum*, placed in section Hamata. In the quote from Grant above, it is clear that he cannot accept such a taxon as a species ('read microspecies') yet the International Code has no category of 'microspecies', and it cannot be employed. Further, as a description of taxonomic philosophy it is also too vague, for so-called 'microspecies' have also been invalidly employed for autogamous taxa (Ch. 9), fixed interchange heterozygotes such as *Oenothera*, and unique breeding systems such as that in *Rosa* (p. 426).

In *Crepis*, the correspondence between morphologically well-separated sexual diploids, and 'pillars' of agamospermous polyploids, each related to one sexual species, has tempted Babcock and Stebbins (1938) to adopt a trinomial category *forma apomictica* for the agamic taxa which are recognised thus, e.g. *Crepis acuminata* apm. *sierrae*, where *C. acuminata* is a sexual diploid. This category, however desirable, has no recognition in taxonomic law, and must be disregarded, unless the mandarins of the International Association of Plant Taxonomists can be persuaded to modify the 'Code' accordingly. One might also mention the '*species principae collectivae*' described for *Hieracium* by Zahn (1921–3), the 'Gruppe' of Dahlstedt (1921) for *Taraxacum*, and the 'circle-species' employed for *Rubus*. None of these has any validity, and any attempt to create species at more than one level in an agamic complex should be disregarded (including the 'species groups' of Richards & Sell 1976!). It is also probably inappropriate to distinguish agamic taxa as subspecies, as has traditionally been the case for the segregates of goldilocks, *Ranunculus auricomus* aggregate, in Scandinavia and central Europe. Although subspecies are perfectly valid taxa, they are used in a particular way for sexual plants. Until the Code is modified to create special taxonomic categories for agamic taxa, it is simplest if only one taxonomic category, the species, is used to cover segregates of genera with unconventional breeding systems.

As mentioned above the number of taxa described in agamic groups shows a huge variation. In many cases (*Aphanes arvensis*, *Nigritella nigra*), only one taxon is employed. In others, a few (less than 20) agamospecies are recognised (*Poa pratensis* agg., *Antennaria*, *Rudbeckia*). In *Ranunculus auricomus*, *Sorbus*, *Crataegus* and *Alchemilla*, between 100 and 300 taxa have been described. There follows a large numerical gap until we reach the three genera, *Taraxacum*, *Rubus* and *Hieracium* in which the taxa are numbered in thousands. There seems to be some correspondence between the number of taxa described in a group and its breeding system. Many of the cases of adventitious embryony are monotypic. In a case like *Citrus*, a number of diploid sexual species show variable capabilities for agamospermy. Because sexuality remains in each of these species, their taxonomy is straight forward. Through the activities of man, a complex series of diploid and polyploid hybrids has arisen, and in some of these sexual sterility has increased the rôle played by agamospermy. As a result, we see today a confusing reticulate pattern of variation, in which many man-produced hybrids maintain themselves by agamospermy, but also occasionally backcross to sexuals. The taxonomic confusion is a product of the activities of man.

Among aposporous and diplosporous species, the number of agamospecies recognised in a complex is dependent on three parameters:

(a) The amount of taxonomic attention that a group has received. It is perhaps no accident that most of the great agamic complexes are from Europe, for it is in this continent that the indigenous flora has been studied for the longest time, is best known, and leisured people have had the greatest opportunity to describe complex variation in detail. (However, exceptions occur in the North American *Crataegus*, and the Japanese *Taraxacum*, each of which have over 100 agamospecies.)
(b) The innate potential of a plant to engender complex morphological variation which can be used by taxonomists for agamospecies delimitation. Thus, the *Taraxacum* leaf shape, the nature and cover of the hairs of *Alchemilla* and *Hieracium*, the diversity of prickle types in *Rubus* have (together with many other characters) encouraged taxonomists to describe many variants. In comparison, grasses such as *Poa* or *Dichanthium* have few characters that vary in a way that encourages taxonomists to describe species.
(c) The ability of the breeding system to create and stabilise morphological variation in a way that will encourage taxonomists to create new species. Thus, although new variation will certainly arise in aposporous groups, the residual sexuality which is typical of many will blur the morphological limits of such variants and tend to discourage taxonomists from describing minor variants. A broader view of the agamospecies will be suggested to the taxonomist by the results of the breeding system.

In contrast, we can expect that diplosporous systems with obligate agamospermy will create many discrete, but relatively minor, variants which the taxonomist can recognise. It is important to the taxonomist that his taxa are discrete, and repeatedly recognisable in time and space. They must also be predictable. Thus, an agamospecies will be expected to have a coherent ecology and geographical distribution, and to be relatively invariable for biological characters such as presence of pollen, chromosome number, or secondary chemical products. We can expect that when sexuality is rare or absent, but new variation continues to arise asexually on which natural selection can act, many 'useful' agamospecies of a narrow morphological scope can be recognised. This is the case in diplosporous *Taraxacum*, and *Hieracium* subgenus *Hieracium*. Interestingly, in the aposporous *Hieracium* subgenus *Pilosella* (see Table 11.4), far fewer, and more variable agamospecies are recognised, and most of these are considered to be *de novo* hybrids.

The exception to this general pattern is in *Rubus*, the brambles, which have already been discussed. Our understanding of the reproductive processes of brambles are far from complete, but we do know that they are very complex. Thus, it is not surprising that patterns of morphological variation are equally complex. More than in any other genus, large numbers of describable variants occur in localised concentrations, and many of these variants have very restricted distributions. Thus, in many regions of the UK there are very few distinct taxa, which are readily learnt by expert tuition in a few days. However, in some 'hot spots', hundreds of variants coexist, which may be very limited in space (only one hedge), or time (they cannot be refound in later years). It may be that this breeding system should dictate that an unusually conservative view of *Rubus* agamospecies limits is adopted, at least until we understand the breeding system better.

*Rosa* is another genus that is taxonomically extremely difficult, as we would expect from our understanding of its unique breeding system. A very broad concept of taxonomic limits in the Caninae roses has been adopted by nearly all taxonomists in recent years, which is, however, quite inadequate to deal with the patterns of variation encountered. This is another situation in which the Linnean binomial system seems quite inappropriate and inadequate, and a radical rethink of nomenclatural law may be beneficial.

In summary, it is narrow minded to adopt a bigoted approach to the question of agamospecies philosophy, in other words to take sides in the old 'splitters and lumpers' controversy. A good taxonomy conveys the maximum amount of useful information on variation and evolution in a group. In some agamic complexes, the nature of the breeding system, and the resulting pattern of morphological variation, may justify a conservative ('lumping') taxonomy, and this may be especially true where a considerable amount of sexuality remains in most genets (adventitious embryony and apospory). In other complexes, especially diplosporous groups in which agamospermy is mostly obligate, a much narrower concept of the species may be more useful ('splitting'). A thorough knowledge of the breeding system may even justify the use of different approaches within a complex. Thus, whereas a narrow agamospecies concept seems to be justified for the dandelions of northern Europe, in which sexuality is absent, in many parts of central and southern Europe this may not be possible. I have studied dandelion populations in the

Auvergne, France in which sexuals outnumber agamosperms, and in which any agamospecies concept appears to be quite unworkable. It is impossible to treat this situation usefully by the use of the Linnean binomial system, but until the Code offers us other possibilities, we must continue to use the species rank for both sexuals and agamosperms, delimiting species where it seems most practicable.

Above all, each complex must be considered separately, and judged on the merits of our knowledge of its breeding system.

## AGAMOSPERMY AND PLANT BREEDING

For a plant breeder, the 'ideal' plant is highly heterozygous, with heterotic vigour and the potential for future variability, but breeds true. The basic problem that the plant breeder has to face with most plant material is that the concepts of heterozygosity and true breeding are antagonistic, for heterozygotes release variability. Thus, it would seem that agamospermy is potentially an ideal breeding system from a crop-breeding viewpoint, for an agamosperm is both heterozygous and true breeding.

Unfortunately, agamospermy is very rare in crop plants, the best-known cases being in *Citrus* (oranges, lemons, grapefruits, etc), and *Mangifera* (mango), although it also occurs in some minority crops such as prickly pear, caper, mangosteen, brambles, species apples, and in some blue grasses (*Poa* spp.).

A great deal of money and effort has been expended in trying to induce agamospermy in sexual crops (reviewed in Asker 1979), and currently this subject has a renewed popularity, for instance in the USSR. As will be seen from Table 11.1, some agamospermous traits have arisen in certain crops, notably maize (*Zea*). As yet, there seems to be no case in which these traits have been successfully combined to give a useful agamospermous plant in a group in which agamospermy was not hitherto known.

It seems likely that the advent of tissue culture, in which a single desirable cultivar can be micropropagated by the thousand, is likely to reduce the demand for the development of agamospermous crops, although it may be accompanied by some mutational problems.

## CONCLUSION

For agamospermy to succeed, it is necessary for the independent functions of avoidance of meiosis, and failure of fertilisation/autonomous embryony to occur together. Except in the case of adventitious embryony, at least two separate mutants need to be combined. Although such mutations arise in sexual populations, they are not widespread, and the reduced sexual function that they engender singly reduces the likelihood of their co-occurrence in the same genet. As a result, agamospermy is an uncommon phenomenon, which is most frequent in a few families of plants that may show preadaptation to the origin of certain types of agamospermy.

Obligate agamospermy is rare, and is probably limited to a few diplosporous genera in which pollen is missing (unusual as male sterile mutants cannot be recombined), or sexuals are absent. Most agamosperms retain some sexuality (facultative) which allows them to maintain some genetic variability. Nearly all agamosperms with apospory or adventitious embryony can be sexual, and retain the requirement for fertilisation of the endosperm nucleus (pseudogamy). They thus retain good pollen function , which can also be used in sexual reproduction. At least some obligate agamosperms are also able to generate some genetic variability by somatic recombination, and/or various forms of autosegregation. Thus, it is probably untrue that agamosperms generate no genetical variability, and have no evolutionary potential. In any case, it seems that some agamospecies are able to persist for thousands of years.

Nevertheless, agamosperms may suffer reduced fitness in comparison with sexuals through their lower generation of variability, which reduces the number of niches in the environment that they are able to fill. They are also likely to suffer from the accumulation of disadvantageous mutants in the agamospermous linkage group, and poor recombination and migration of advantageous mutants as a result of reduced sexual function.

However, agamosperms possess a number of theoretical advantages over sexuals which should result in their automatic success in competition with sexuals, at least in the unreal condition of a single environmental niche. These are:

(a) As agamospermous offspring are of a maternal genotype, the agamospermous mother wastes no energy succouring offspring (seeds) which through the variability engendered by sexuality are relatively maladapted and unfit; this advantage is difficult to quantify.
(b) Agamosperms donate genes to sexuals as male parents, but receive no genes from sexuals; this advantage is frequency dependent (Lloyd 1980b), and decreases with the frequency of agamosperms in the population.
(c) Agamosperms have an advantage over dioecious and androdioecious sexuals, in that they produce no males; this advantage is independent of the sex ratio of the diclinous population (Charlesworth 1980), and should be 2 : 1.

(d) Male-sterile agamosperms, which are uncommon, have an advantage over hermaphrodite sexuals; this is *not* a function of the relative resource used in male and female reproduction, but depends on the fraction of the total energy output of the plant used for male reproduction (Ch. 2); however, male steriles suffer from not donating genes to sexuals.

(e) Non-pseudogamous agamosperms have a reproductive advantage over outbreeding sexuals in conditions of limited cross pollination.

It should also be noted that male-fertile agamosperms, with pseudogamy, possess no advantage over sexual hermaphrodites when they are unable to act as male parents (for instance when sexuals are absent), except that of avoiding the 'cost of meiosis'. Some of these points are expressed in Table 11.10. It is difficult to quantify advantages of different types together.

One of the features of agamosperms is that nearly all have close sexual relatives, so that it is possible to assess the breeding system of plants that have given rise to agamosperms. Nearly all such sexual progenitors are perennial hermaphrodite outbreeders with limited powers of vegetative spread. Nearly all diplosporous agamosperms, and many aposporous agamosperms, differ from their sexual progenitors in being polyploid (often of uneven 'ploidy') and hybrid (this is not true of agamosperms with adventitious embryony which can arise by a single mutation).

It is supposed that polyploidy and hybridity encourage the evolution of agamospermy by leading to the failure of meiosis, by recombining the separate functions of agamospermy from different parents, and by being seed sterile, thus giving an immediate advantage to agamospermous seeds. Perenniality and vegetative persistence should allow sterile

**Table 11.10** Energetic advantages of agamosperms over coexisting sexuals.

| | Male function of agamosperm with equal male and female expenditure | | |
|---|---|---|---|
| | No pollen | Non-functional pollen (e.g. sexuals absent) | Functional pollen |
| dioecious sexual with equal sex ratios and equal male and female expenditure | 2:1 | 1><2:1 | 2>:1 (but unlikely) |
| hermaphrodite sexual with equal male and female expenditure | 1>⩽2:1 | 1:1 | 3:2 to 2:1 (depending on agamosperm frequency) |

clones, which possess one of the functions of agamospermy, to persist until a second function is incorporated into that clone (linkage group) by mutation or hybridisation. Thus, it is in those types of plants that agamospermy has arisen and established (rather than those features causing the origin of agamospermy *per se*).

Agamospermy seems to have arisen very rarely from other types of breeding system, although there are some examples:

(a) Autogamy; autogamous species will have a low potential for combining the different functions of agamospermy. They also possess some of the advantages of agamospermy. These include: partial avoidance of the cost of meiosis' (offspring of a mostly maternal type depending on the level of homozygosity and autogamy); reproductive efficiency; a low male energetic load; and the donation of pollen to outbreeders in the absence of receipt of pollen from outbreeders. Thus, there will be a relatively small advantage for agamospermous mutants in an autogam. The parsley-piert, *Aphanes arvensis,* which has apparently arisen from the sexual autogam *A. microcarpa* is the only example of which I am aware.

(b) Dioecy; a dioecious agamosperm will only produce females. This should be a serious disadvantage as pseudogamy is not possible (autonomous endospermy as well as embryony must arise) and agamosperms cannot act as male parents. Also, female agamosperms have to be obligate, without any chance of sexuality. The only dioecious agamosperm of which I am aware is *Antennaria*.

(c) Heteromorphy; a heteromorphous agamosperm is limited in its power of pollen donation to sexuals, and pollen reception if pseudogamous or facultatively agamospermous. Only one morph will be fixed as an agamosperm. These may not be major disadvantages. The scarcity of agamospermy among heteromorphs may reflect the scarcity of heteromorphy as a phenomenon. *Limonium* seems to be the only example of a heteromorphous agamosperm.

We can conclude that agamosperms have both advantages and disadvantages in comparison with sexuals. It is very difficult to balance these in any model system to predict the final evolutionary fate of agamosperms, particularly if a changing environment is presupposed. As yet, mathematical models that assess the success of agamospermy are hopelessly naïve, and take account of too few factors. An examination of the real world suggests that agamospermy has proved to be a successful form of reproduction for a number of genera in which new variation can be generated and old variation fixed.

CHAPTER TWELVE

# Conclusion

This book has attempted to show that:

(a) Seed plants have a wide variety of breeding systems. Mate choice is controlled by a diversity of mechanisms. Many of these mechanisms are interrelated. They occur together within one plant or population, and influence one another.

(b) Plant breeding systems are not static, or fixed, but are variable and flexible. Genetically controlled and heritable variation for all breeding-system attributes occurs between plants in many populations. Further variation, or novel attributes can also arise by mutation. As a result, breeding systems are subject to natural selection and adaptive evolution in nearly all plants in every generation.

(c) Breeding-system attributes have profound effects on the reproductive fitness of individual genets in all plants.

(d) Breeding-system attributes influence the genetic structure of individuals. They determine patterns of gene linkage and levels of heterozygosity in vitally important ways. As a result, they also influence the genetic structure of populations – the ways that plants within a population differ from one another genetically.

(e) As a result, breeding systems control the nature of evolution in seed plants. Because breeding systems themselves evolve constantly, they are an integral part of this evolution, and are subject to feedback effects from the results of the evolution they control.

(f) The taxonomic and classificatory systems that we impose on plant variation are influenced by the pattern of variation that we perceive. No two plant populations have exactly similar breeding systems and exactly similar patterns of variation. Our taxonomic philosophy should take account of this. At present it conspicuously fails to do so, resulting in tedious and unnecessary controversies of the 'splitter'/'lumper' type.

In this brief concluding chapter I shall examine some of the effects that different plant breeding systems have on the genetic variability of plant populations. This account owes much to the review by Hamrick *et al.* (1979). Our knowledge of genetic variation in plants has been greatly expanded in recent years through the use of gel electrophoresis to examine the distribution of isozymes, the direct products of genes.

Although electrophoretic techniques have frequently been open to criticism, and the interpretation of electrophoretic results has often been less than rigorous, we now have a much clearer idea than we did 15 years ago as to how genetic variation within and between plants is organised.

Hamrick *et al.* (1979) compared isozyme attributes with types of breeding system and other life-history characters for 113 taxa of seed plants. They examine three features of genetic variability as revealed by isozyme electrophoresis (Ch. 9): PLP (P) the percentage of loci polymorphic per population (or species); *k* (*A*) the mean number of alleles per locus per population (or species); PI the mean proportion of loci expected to be heterozygous if the population is panmictic (Hardy–Weinberg frequencies, Ch. 2). (Symbols in parentheses are those used by the authors, which differ from those used in this book; note that these figures relate to variation in populations, rather than *H* figures which refer to heterozygosity in *individuals*.]

If mean values are calculated for all 113 taxa, an estimate of values typical of seed plants can be obtained. Interestingly, seed plants have average levels of genetic variability considerably higher than those estimated for vertebrate animals, which have few reproductive options (Ch. 1), but are roughly similar to estimates for invertebrates which have more reproductive options (Table 12.1).

In a comparison of genetic variability between major plant taxa, and different plant life strategies, some interesting features emerge (Table 12.2). We can expect Gymnosperms on average to be more variable than Angiosperms, for most Gymnosperms are long-lived woody perennials with efficient seed and pollen dispersal (usually by wind), and they are either monoecious or dioecious. When monoecious, they are often markedly protandrous, so they are usually outcrossed. It is perhaps rather more surprising that monocotyledons tend to be slightly more genetically variable than are dicotyledons. However, there are fewer monocotyledonous than dicotyledonous annuals, and many of the

**Table 12.1** Mean values for estimates of genetic variation in seed plants, invertebrates and vertebrates (see the text for explanation of symbols) (after Hamrick *et al.* 1979).

| Group | PLP (%) | *k* | PI |
|---|---|---|---|
| seed plants | 36.8 | 1.69 | 0.141 |
| invertebrates | 46.9 | – | 0.135 |
| | 39.7 | – | – (different estimates) |
| vertebrates | 24.7 | – | 0.061 |
| | 17.3 | – | – (different estimates) |

**Table 12.2** Mean values and standard errors for estimates of genetic variation in populations of seed plants with different life strategies, and belonging to different major taxa (see text for explanation of the symbols) (after Hamrick *et al.* 1979).

| Group | PLP (%) | *k* | PI | Mean number of loci examined | Number of taxa examined |
|---|---|---|---|---|---|
| Gymnosperms | 67.0 ± 8.0 | 2.1 ± 0.2 | 0.27 ± 0.04 | 9.2 | 11 |
| Dicotyledons | 31.3 ± 3.3 | 1.5 ± 0.1 | 0.11 ± 0.01 | 11.4 | 74 |
| Monoctyledons | 39.7 ± 6.0 | 2.1 ± 0.2 | 0.16 ± 0.3 | 11.6 | 28 |
| | | | | | |
| annual | 39.5 ± 4.3 | 1.7 ± 0.1 | 0.13 ± 0.02 | 11.2 | 42 |
| biennial | 15.8 ± 5.1 | 1.3 ± 0.1 | 0.06 ± 0.02 | 17.2 | 13 |
| short-lived perennial | 28.1 ± 5.1 | 1.5 ± 0.1 | 0.12 ± 0.02 | 12.0 | 31 |
| long-lived perennial | 65.7 ± 5.1 | 2.0 ± 0.1 | 0.27 ± 0.03 | 7.6 | 27 |
| | | | | | |
| coloniser | 29.7 ± 3.8 | 1.6 ± 0.1 | 0.12 ± 0.01 | 12.5 | 54 |
| mid-successional | 37.9 ± 4.4 | 1.6 ± 0.1 | 0.14 ± 0.02 | 9.7 | 49 |
| climax vegetation | 62.8 ± 5.3 | 2.1 ± 0.2 | 0.27 ± 0.04 | 12.0 | 10 |
| | | | | | |
| xerophyte | 15.4 ± 8.2 | 1.1 ± 0.1 | 0.05 ± 0.04 | 8.8 | 4 |
| hydrophyte | 27.7 ± 10.3 | 1.6 ± 0.2 | 0.14 ± 0.05 | 13.0 | 8 |
| mesophyte | 36.0 ± 3.6 | 1.6 ± 0.1 | 0.15 ± 0.02 | 11.4 | 82 |
| | | | | | |
| haploid chromosome number | | | | | |
| $n$ = 5–10 | 35.5 ± 3.6 | 1.5 ± 0.1 | 0.11 ± 0.01 | 13.1 | 50 |
| $n$ = 11–15 | 37.4 ± 5.3 | 1.7 ± 0.1 | 0.17 ± 0.02 | 10.0 | 44 |
| $n$ = 16+ | 41.6 ± 7.2 | 2.1 ± 0.1 | 0.22 ± 0.03 | 8.9 | 16 |

monocotyledons in the sample are grasses that have wind pollination and fairly efficient seed dispersal.

As expected, long-lived perennials are genetically far more variable than plants with shorter life histories. In long-lived perennials, reproductive fitness is most often optimised by vegetative rather than sexual reproduction (*'K'* rather than *'r'* strategy), and so selfing and agamospermous mechanisms that respond to pressures for the optimisation of seed-set and reduce genetic variability are relatively unusual. It is interesting that annuals are more variable than biennials or short-lived perennials, at least for the percentage of loci that are polymorphic per population (PLP). This measure does not differentiate between homozygotes, e.g. *AA* and *aa* from heterozygotes *Aa* in a population, but merely states whether both *A* and *a* are present in a population. Annuals, which are mostly selfing plants, may show high levels of differential homozygous fixation at a locus within a population in comparison with outbreeders, and this may account for the relatively high PLP reading. PI readings, which determine *potential* levels of heterozygosity, do not differ between annuals and short-lived perennials, but annuals are likely to be less heterozygous.

The correspondence of patterns of genetic variation with ecological attributes are very much as one might predict. Colonising species, and species occurring in very dry habitats and in water, are less variable than plants that occur in stable climax and mesic habitats. Extreme habitats impose severe stabilising selection, and encourage breeding systems through which the release of genetic variation is minimised. Colonising species can best initiate a new population, and optimise reproduction by seed, when they are self-fertilising, and self-fertilisation will reduce genetic variability. Mesic and stable habitats will provide more niches and will encourage breeding systems that maximise the release of genetic variability.

The rôle played by the chromosome number of the plant in controlling the release of genetic variation is well illustrated. Chromosomes are linkage groups, and the more chromosomes that are present, the greater the potential for segregational variability (Ch. 2). Polyploids, with high haploid numbers, also buffer the effects of selfing with respect to loss of variability (Ch. 9). High chromosome numbers will be advantageous for plants with outbreeding strategies, whereas plants that are favoured by a low release of genetic variability may select for low chromosome numbers through translocations.

Hamrick *et al.* (1979) also compared the genetic structure of seed plant populations with breeding system attributes, and these results are central to the theme of this book (Table 12.3). Once again, they fulfil expectations well. Results from self-pollinating and self-fertilising species are very similar, probably because they refer to essentially the same samples. They show very much less genetic variability than do insect- and wind-pollinated species, and species that are predominantly cross fertilised.

There is an interesting contrast between animal- and wind-dispersed pollination. As discussed in Chapter 5, wind-dispersed pollen travels much further than animal-dispersed pollen, although as a pollination mechanism it is much less efficient, and thus more expensive. We must attribute the notably higher genetic variation of populations that are wind pollinated to this difference. Wind-pollinated plants will tend to have much larger neighbourhood areas *A* and sizes $N_e$ than insect-pollinated plants. As plant populations are frequently very small (Ch. 2 and

**Table 12.3** Mean values for estimates of genetic variation in populations of seed plants with different breeding system attributes (see the text for explanation of symbols) (after Hamrick *et al.* 1979).

| Breeding system | PLP (%) | *k* | PI | Mean number of loci examined | Mean number of taxa examined |
|---|---|---|---|---|---|
| autogamous | 19.0 ± 3.5 | 1.3 ± 0.1 | 0.06 ± 0.02 | 14.2 | 33 |
| zoophilous | 38.8 ± 3.9 | 1.5 ± 0.1 | 0.13 ± 0.01 | 9.5 | 55 |
| anemophilous | 57.4 ± 6.3 | 2.3 ± 0.2 | 0.26 ± 0.03 | 10.7 | 23 |
| fruit dispersal | | | | | |
| epizoochorous | 28.8 ± 5.5 | 1.5 ± 0.1 | 0.09 ± 0.02 | 11.1 | 16 |
| endo-zoochorous | 33.0 ± 8.2 | 1.4 ± 0.1 | 0.13 ± 0.04 | 7.0 | 20 |
| anemochorous | 44.9 ± 7.2 | 1.9 ± 0.1 | 0.19 ± 0.03 | 12.2 | 21 |
| self-fertilising | 17.9 ± 3.2 | 1.3 ± 0.1 | 0.06 ± 0.01 | 14.2 | 33 |
| mixed selfing and crossing | 14.2 ± 4.9 | 1.8 ± 0.1 | 0.18 ± 0.02 | 8.6 | 42 |
| cross fertilising | 51.1 ± 4.9 | 1.8 ± 0.1 | 0.18 ± 0.02 | 11.3 | 36 |

Ch. 9), wind pollination may often serve to overcome inbreeding effects (with consequent loss of genetic variability) engendered by small populations.

The effectiveness of wind as a dispersal agent is also illustrated by the data for fruit dispersal. The genetic variability of populations that are dispersed on the outside of animals is less than that of populations for which the fruits are eaten by animals. Populations with wind-dispersed fruit show most variation. These differences in genetic variability are probably attributable to differences in seed dispersal variance, and hence in neighbourhood size (Ch. 5). Fruits that are dispersed the furthest give rise to the largest, and hence the most variable, populations. There is no significant difference in the genetic variability of populations with fruits of different sizes (results not given).

The comparison between the three sets of data for types of fertilisation is particularly interesting. For the percentage of all loci that vary (PLP), mixed strategy plants do not differ from selfers; relatively few loci are polymorphic within a population in comparison with outcrossers. However, the mean number of alleles per locus ($k$) in mixed strategy plants does not differ from outcrossers, and is higher than in selfers. This can only mean that mixed strategy plants have more than two alleles at a locus in a population more frequently than do selfers. A combination of genetic fixation through selfing, and gene migration through bouts of outcrossing, may allow this to occur. Also, mixed strategy plants do not differ from outcrossers in the proportion of loci that are heterozygous (PI), and this proportion is much higher than in selfers. We can conclude that a mixture of outcrossing and selfing tends to promote gene migration, recombination and heterozygosity, and is as effective at this as is outcrossing. However, it does not allow so many disadvantageous recessives to be maintained, as these will be exposed to selection by bouts of selfing. Mixed strategy mechanisms are revealed to be evolutionarily rather efficient, which may help to explain why so many species of plants possess them.

Unfortunately, the present set of data is not large enough to allow generalisations to be made about more specific features of breeding systems with respect to the genetic variation of populations. We have very little information about asexual strategies, or about the effects that dioecy, gynodioecy, heteromorphy, ramet size, genet size, floral structure, floral timing, monoecy, resource allocation to sexes, variation in reward or a host of other features that control breeding systems have on genetic variation. What information that we do have has been presented in the relevant chapters.

# *Glossary*

**abiotic** Not involving living organisms.
**abscission** Dropping off of organs, e.g. leaves, flowers, fruits.
**actinomorphic** Radially symmetrical (rotate), of a flower.
**adventitious embryony** A form of agamospermy in which embryos are budded directly from the nucellus without an intervening embryo-sac.
**agamic** Without sex, asexual.
**agamospecies** 'Microspecies' or 'segregate species' of low taxonomic amplitude used in **agamic** complexes.
**agamospermy** The formation of seeds without sexual reproduction.
**agglutination** The conglomeration and precipitation of molecules of a protein.
**allele** One of two or more forms that a gene may take.
**allogamy** Fertilisation between pollen and ovules of different flowers.
**allopolyploid** A plant with more than two sets of chromosomes which originate from two or more parents, at least some chromosomes of which are non-homologous with respect to each other.
**allozyme** A form of an enzyme which behaves allelically with respect to other allozymes of that enzyme.
**anaphase** The stage in chromosome division **meiosis** or mitosis at which the chromosomes or chromatids pull apart.
**androdioecy** Where male and **hermaphrodite** genets coexist.
**androecy** Maleness.
**andromonoecy** Where a **hermaphrodite** bears both male and hermaphrodite flowers.
**anemochory** Dispersal by wind.
**anemophily** Pollination by wind.
**aneuploid** A plant with a chromosome number which is not a direct multiple of the base chromosome number of the group.
**Angiosperm** A flowering plant, i.e. belonging to the class Angiospermae.
**anisogamy** Sexual fusion between gametes of unequal size.
**anisoplethy** The departure of ratios of **genet** morphs in population from the expectation of unity (one to one, e.g. of males and females, **pins** and **thrums**).
**annulus** Constricting ring at the top of a floral tube.
**anther** The part of the stamen that contains pollen (a microsporangium).
**antherozoid** A motile male gamete, as in the Pteridophyta and Cycadopsida.
**anthesis** The opening of a flower; usually used to denote the stage when the flower first donates pollen, or is receptive to pollen, whichever is the earlier.
**anthocyanin** Group of pigments that most commonly give flowers pinkish, purplish or blueish tints.
**antigen** Substance that stimulates the formation of an antibody, or seroprotein.
**antipodal** Nucleus (usually one of three) at the chalazal end of the embryo-sac.
**apogamety** Autonomous development of a nucleus apart from the egg nucleus into an embryo in an **agamosperm**.
**apomixis** Asexual reproduction; includes both **agamospermy** and vegetative reproduction.

**apospory** Development of the embryo-sac from a tissue apart from the **archesporium**, usually the nucellus, in **agamospermy**.

**archegonium** Female sex organ, containing the egg, in the gametophyte generation of the Bryophytes, Pteridophytes and Gymnosperms.

**archesporium** The tissue within the nucellus of a young ovule which usually gives rise to the embryo-sac mother cell, female **meiosis** and the embryo-sac.

**assortive (assortative)** Mating, where mating occurs between gametes of the same morph more frequently than would be expected by a random mating system (**panmixis**).

**asynapsis** The failure of **homologous** or **homoeologous** chromosomes to form chromosome associations and thus chiasmata at **meiosis**.

**autogamy** Within-flower fertilisation.

**automixis** Fusion of nuclei within the embryo-sac.

**auto-orientation** The positioning of multivalents at diakenesis and metaphase I of **meiosis** in a polyploid in such a way that a regular segregation of centromeres is achieved and fertile spores result.

**autopolyploid** A plant with more than two sets of chromosomes which originate from a single parent, or from different parents which are **homologous** in their chromosomes (*see* **allopolyploid**).

**autosegregation** The regular, or somewhat regular, disjunction of chromosomes at anaphase I of **meiosis** between which there has been poor synapsis or asynapsis, as in agamosperms or hybrids.

**autosome** A chromosome that is not a sex chromosome.

**axile (placentation)** Arrangement of the ovules down the central walls of septae in a fused ovary.

**Batesian mimicry** The adaptive mimicry of an organism or an organ to a model (*see* **Müllerian mimicry**).

**bract** Leafy organ subtending an inflorescence.

**Bryophyte** Member of the subphylum Bryophyta, a moss (Musci) or liverwort (Hepaticae).

**bulbil** Vegetative organ of dispersal or perennation, usually formed from an axillary bud.

**callose** A complex carbohydrate $\beta$1,3,glucan, usually formed as a wounding reaction.

**calyx** The sepals of a flower.

**campylotropous** Form of an ovule in which the micropile and **chalaza** are placed laterally to the placenta (fig. 3.11).

**cantharophily** Pollination by beetles.

**capsule** A dry, dehiscent fruit.

**carpel** The segment of an ovary.

**caudicle** The sticky, elastic base to an orchid pollinium, formed from the modified connective.

**centric** Containing the centromere (of a chromosome).

**certation** Differential growth of pollen grains of different *S* allele constitutions when grown on a legitimate stigma; can cause cross-compatible genotypes to appear cross incompatible.

**chalaza** The opposite end of the ovule to the **micropile**.

**chiasma** Where chromatids of different **homologous** chromosomes pairing at **meiosis** I break and reciprocally join with each other, resulting in genetic crossing over, and hence in recombination within linkage groups.

**chimaera** The coexistence of cells of more than one **genotype** within a **genet**.

**chirepterophily** Pollination by bats.

**chorology** The study of the geographical distribution of organisms.

**chromatid** The product of longitudinal division of a chromosome; at **meiosis** in a diploid, two **homologous** chromosomes produce four chromatids which may form chiasmata with each other resulting in recombination; each spore resulting from a **meiosis** will receive only one **chromatid** from each pair of **homologous** chromosomes (bivalent).

**chromosome** A structural unit, or 'package' of genetic information contained within the nucleus and consisting of DNA and histones; chromosomes segregate independently from one another at **meiosis**, and so each chromosome forms a linkage group, which may be broken by chiasmata (**recombination**). The number of chromosomes is usually characteristic for a species.

**cistron** A unit of genetic information coding for a polypeptide which cannot usually be broken by **recombination**.

**cleistogamy** Where flowers do not open, and are thus inevitably autogamous.

**clone** A number of ramets which belong to the same **genet**.

**cob** A heteromorphic condition in the family Plumbaginaceae characterised by short stigmatic papillae.

**cohort** 1. Individuals of similar age in a population. 2. Pollinators of different species exhibiting similar flower choice.

**column** Part of the **gynoecium** of orchids, homologous with the style, which is a **gynostegium**, carrying the pollinia as well as a stigmatic cavity.

**complementation** Interaction of two or more genetic loci in the expression of a single phenotypic character.

**corolla** The petals of a flower.

**crypsis** Where an organism or organ is difficult to perceive or see.

**cuticle** The superficial proteinaceous and waxy layer secreted by the epidermis which covers aerial plant organs (*see* **pellicle**).

**cutinase** Esterase enzymes that break down the waxy cuticle, especially of stigmatic papillae.

**cypsela** The one-seeded fruit of Compositae (Asteraceae) which commonly bears a feathery pappus which is derived from the **calyx** and promotes dispersal by wind.

**deletion** The loss of a gene or part of a chromosome.

**deme** A population of potentially interbreeding individuals.

**demography** The study of populations.

**diakenesis** The stage during **meiosis** I at which chromosomes become fully contracted, and centromeres start to disjoin.

**di-allelic** An incompatibility system containing only two *S* **alleles**, *S* and *s*, usually as the diploid genotypes *Ss* and *ss* which are between-morph compatible and within-morph incompatible; usually associated with **heteromorphy**.

**dialysis** Technique for separating molecules of different sizes through a membrane of known pore diameter, usually across an electropotential gradient.

**dichogamy** separation of anther dehiscence and stigma receptivity within a flower in time so that **autogamy** cannot occur (*see* **protandry** and **protogyny**).

**dicliny** Where not all genets in a population are regularly **hermaphrodite**; i.e. males, or females, or both occur.

**dicotyledon** Member of the class Dicotyledones which includes the majority of Angiosperms, and is characterised by such features as two cotyledons in a seedling, vascular bundles in a ring, secondary thickening, broad, net-veined leaves and floral parts in fours and fives.

**dihaploid** the result of doubling the chromosomes in a **haploid**, usually resulting in a plant that is totally **homozygous**.

**dikaryon** A cell containing two nuclei.

**dimer** A protein (e.g. an enzyme) containing two subunits.

**dimorphism** (y) The coexistence of two genetically controlled floral types in a population, e.g. **pins** and **thrums** (**heterostyly**), **cobs** and **papillates**, or males and females.

**dioecy** Where all **genets** in a population are either male or female.

**diploid** A **genet** with two homologous sets of chromosomes (**genomes**).

**diplospory** Development of the embryo-sac from the **archesporium**, often by way of an irregular **meiosis**, in **agamosperms**.

**disjunction** The pulling apart of the chromosomes at **anaphase** in **meiosis** or mitosis.

**disomic** Where a chromosome or gene is present twice in a **genet**.

**disseminule** An organ of reproductive dispersal (*see also* **propagule**).

**dissortive (dissortative)** Mating, where mating occurs between gametes of different morphs more frequently than would be expected by a random mating system (**panmixis**).

**duftmale** An area of a petal that secretes scent.

**duplex** *AAaa* heterozygote in **tetrasomic** inheritance (*see* **simplex, triplex**).

**duplication** Where a gene, supergene or part of a chromosome occurs twice within a **genome**.

**EMC (the embryo-sac mother cell)** The archesporial cell which undergoes female **meiosis**.

**ecodeme** A population or gene pool adapted to a particular environment.

**elaiophore** An oil-secreting gland on a flower.

**elaiosome** A gelatinous projection of the testa of the seed, attractive to ants.

**electrophoresis** Technique for separating molecules of different sizes and charges (especially **isozymes**) by diffusing eluates through gels of acrylamide or starch along an electropotential gradient, and staining the products produced by suitable substrates to give rise to visible bands (*see* **zymogram**).

**embryo** The young stage of a new **genet** within the seed, usually developed from an egg cell, which on germination will give rise to a seedling.

**embryony** The development of an embryo.

**embryo-sac** The female gametophyte of flowering plants (**Angiosperms**), contained within the ovule, developing from the surviving **megaspore** after female **meiosis**, and containing eight nuclei (Fig. 3.10).

**endemic** A plant **taxon** (usually species) which is confined to a small geographical area.

**endoduplication** The multiplication of **genomes** within cells of certain tissues.

**endosperm** The nurse tissue of the embryo within the seed, formed from fusion of a sperm cell with the fused polar nucleus, and usually triploid.

**endozoochorous** Dispersed within animals, i.e. eaten (of a seed or fruit).

**entomophily** Pollination by insects.

**epidermis** The outermost layer of cells of plant organs.

**epiphyty** Of plants which live rooted to the aerial parts of other plants, e.g. on tree trunks.

**epizoochorous** Dispersed on the outside of animals (of a seed or fruit).

**esterase** Enzymes that lyse lipids and waxes.

**ethology** The study of animal behaviour.

**eukaryote** A higher organism with clearly differentiated cell organelles such as nucleus, **plastids**, mitochondria (*see* **mitochondrion**) etc.; includes algae (except Cyanophyta), Bryophyta, Pteridophyta and Spermatophyta (Gymnospermae and Angiospermae).

**eutroploid** With exactly three sets of **homologous** or **homoelogous** chromosomes (**genomes**).

**exine** The outer wall of an **Angiosperm** pollen grain, columnar in structure, perforated with baculae, with an outer **tectum** and an inner **nexine**, and constructed from **lipoprotein (sporopollenin)** (Fig. 3.8).

**extrorse** Of an anther which dehisces away from the centre of the flower.

**facultative** Partial, not complete, as in facultative agamospermy where some sexuality is also found.

**fecundity** Ability to reproduce, particularly as the production of seeds per **genet**.

**filiform apparatus** Outgrowths of one or both synergid nuclei, which engulf the pollen tube apex just prior to double fertilisation in the **Angiosperms**.

**fixation index** ($F_0$) A measure of the genetic effects of inbreeding, based on the deficiency of the frequency of heterozygotes compared with those predicted by the Hardy–Weinberg Law.

**frameshift** A form of DNA mutation in which deletion of one or two base-pairs of the DNA next to a 'stop' code causes a whole sequence of DNA triplets to be translated differently.

**fused polar nucleus** (*See also* **primary endosperm nucleus**). The **dikaryon** in the **Angiosperm** embryo-sac which after fusion with a sperm cell forms the **endosperm**.

**gamete** Specialised reproductive cells (usually **haploid**) which fuse in sexual reproduction to give the **zygote**, and thus the embryo; in seed plants the male gametes are sperm cells or **antherozoids**, and the female gamete is the egg cell nucleus.

**gametophytic** Of the gametophyte (usually **haploid**) generation; in the seed plants the contents of the pollen grain and the pollen tube, and the embryo-sac; in **gametophytic** incompatibility it is the **genotype** of the pollen grain contents which is significant.

**geitonogamy** Fertilisation between pollen and **ovules** of different flowers on the same **genet**.

**gene pool** Concerning **genets** which are in genetic contact through sexual reproduction in time and space.

**genet** A genetical individual, resulting from a single sexual fusion (**zygote**), consisting of one to many **ramets**, and usually genetically distinct from all other genets.

**genetic drift** Non-adaptive change in gene frequencies in a population caused by chance (stochastic) fluctuations in small populations.

**genome** A set of **homologous** chromosomes; a **diploid** has two **genomes**.

**genotype** The genetical constitution of a **genet**; its phenotype will be the result of interaction between the **genet** and the environment.

**gibberellin** A series of plant growth substances (gibberellic acids) which influence seed germination, growth, flowering and seed development.

**glycoprotein** Complex moieties of protein and carbohydrate.

**grex** A group of hybrids of different origins but similar parentage.

**guerilla** a vegetative growth strategy in which some **ramets** advance ahead of others (*see* **phalanx**).

**Gymnosperm** A seed plant (Spermatophyta) belonging to the class Gymnospermae, characterised by naked ovules and the absence of vessels in xylem among many other features; includes conifers, cycads, yews, ginkgos and gnetums.

**gynodioecy** Where female and **hermaphrodite** genets coexist.

**gynoecium (pistil)** The female parts of the flower, including **stigma, style** and **ovary**.

**gynoecy** Femaleness.

**gynomonoecy** Where a **hermaphrodite** bears both female and hermaphrodite flowers.

**gynostegium** A complex **gynoecium** on which are borne the anthers (as in Orchidaceae and Asclepiadaceae).

**half-siblings** Offspring possessing one parent (usually the mother) in common.

**haploid** A **genet** with one set of **homologous** chromosomes (**genome**).

**haustoria** Organs which filamentously penetrate other organs or **genets** for nourishment.

**hemigamy** Where gametes fuse, but form a **dikaryon** without the nuclei fusing.

**hemizygous** The condition of X-linked genes in the **heterogametic** (XY) sex, in which recessive X-linked **alleles** may be expressed.

**herkogamy** Separation of **anthers** and **stigma** in space within a flower in such a way that **autogamy** cannot occur in the absence of an insect visit (*see* **dichogamy**).

**hermaphrodite** A **genet** with both male and female function; it may have either monoecious (single sex) or **hermaphrodite** (both sexes) flowers.

**heterochromatin** Parts of chromosomes which remain more condensed at interphase and less condensed at metaphase and which tend to have replicated sequences of DNA which are non-functional, or service the functioning of the chromosome (centromere, telomeres, nucleolar organiser regions).

**heterogametic** The sex of a unisexual organism which has the number of X chromosomes half the number of genomes, e.g. XY, XO; it is usually, but not always, male.

**heteromorphy** The coexistence of two or three genetically controlled **hermaphrodite** floral types in a population, e.g. pins and thrums (**heterostyly**), cobs and papillates, and tristylous conditions (*see* **di-allelic, dimorphism**).

**heterosis** Vigour associated with outbreeding, hybridity, or levels of heterozygosity; the corollary of inbreeding depression.

**heterosporous** The production of two types of spore, **microspores** which give rise to male **gametophytes**, and **megaspores** which give rise to female gametophytes.

**heterostyly** The coexistence of genetically controlled **hermaphrodite** floral types with different style lengths, and usually with reciprocal anther positions, e.g. **pins** and **thrums (distyly)**, tristylous conditions; here usually used in place of **distyly**, with **tristyly** treated separately (Fig. 7.1).

**heterozygous** With more than one **allele** at a locus in a **genet**.

**hexaploid** A **genet** with six sets of homologous or **homoeologous** chromosomes (**genomes**).

**homoeologous** (Of chromosomes) from different parents, having partial, but not complete homology (ability to pair and recombine in a hybrid).

**homogametic** The sex of a unisexual organism which has the number of X chromosomes equal to the number of genomes, e.g. XX (*see* **heterogametic**); it is usually, but not always, female.

**homogamy** Coincidence of anther dehiscence and stigma receptivity within a flower, so that autogamy is possible in the absence of **herkogamy** (*see* **dichogamy**).

**homoiothermic** (Of animals) 'warm-blooded', i.e. with constant body temperature, as in birds and mammals.

**homologous** (Of chromosomes) having the ability to pair and recombine freely at **meiosis**.

**homomorphy** With only one floral type in the population; used in contrast with **heteromorphy** with respect to derivatives of usually heteromorphic species.

**homostyly** **Homomorphy** specifically derived from **heterostyly** (as in *Primula*).

**homozygous** With only one **allele** at a locus (*see* **heterozygous**).

**hydrophily** Pollination by water.

**hypogynous** Flower with a superior ovary, i.e. with ovary borne distal to the **receptacle**.

**illegitimate** Pollination of a flower of the same self-incompatibility **genotype** as the pollen parent (*see* **legitimate**).

**inbreeding** A breeding system which is non-panmictic to the extent that the frequency of heterozygotes falls below that expected by the Hardy–Weinberg equilibrium; may be caused by selfing, assortive mating or small populations.

**inbreeding depression** Lack of vigour associated with inbreeding and low levels of heterozygosity; the corollary of **heterosis**.

**inflorescence** A group of flowers borne on a stem in proximity to one another.

**integuments** The outer tissues of the ovule which form the **testa** in the seed.

**interchange** Where two non-homologous chromosomes break and exchange segments.

**intine** The inner cellulose wall of an **Angiosperm** pollen grain (*see* **exine**) (Fig. 3.8).

**introgression** Where one species gains some genes from another species as a result of occasional hybridisation across a partial breeding barrier.

**introrse** Of an anther which dehisces towards the centre of a flower (*see* **extrorse**).

**inversion** Where a chromosome breaks twice, and the intervening segment turns round.

**isozyme** A form of an enzyme with a particular electrophoretic mobility (compare **allozyme**).

**Jordanon** Segregate species of a taxonomic complex of largely inbred plants.

**'*K*' strategy** **Ramets** or **genets** which assign low energy allocations to reproduction, having approached the biomass carrying capacity of the environment.

**karyology** The study of chromosome morphology.

**karyotype** The morphology and number of chromosomes commonly found within a plant.

**labellum** The modified, usually lower, petal of the orchids.

**lamella** The central layer of the cell-wall, made of pectate and pectin.

**lectin** A chemical that binds specifically to another chemical.

**legitimate** Pollination of a flower of a different self-incompatibility **genotype** to the pollen parent, likely to result in fertile seed-set (*see* **illegitimate**).

**leptokurtosis** Shape of a graphical curve such that the greater the value on the $x$-axis, the lesser the reduction of the value of the $y$-axis becomes per $x$-axis unit.

**liane** A woody climbing plant.

**linkage disequilibrium** Where the frequency of chromosome genotypes for two or more linked polymorphic genes differs from that predicted by **allele** frequencies.

**linkage group** Genes borne on the same chromosome, or which for other reasons do not segregate from each other at **meiosis**; the closer together genes occur on a chromosome, the less often they will be recombined, and the 'tighter' the linkage group will be.

**lipoprotein** An intimate association between a lipid and a protein.

**locus** The position of a gene on a chromosome; each gene consists of an **allele** at a locus.

**majoring** The 'learning' of a rewarding floral patch by a bee; to 'major' is the North American word for to 'graduate'.

**megaspore** Gives rise to a female **gametophyte**.

**meiosis** Reduction division of chromosomes, involving recombination and segregation of the chromosomes, and resulting in (usually **haploid**) spores.

**meiotic sieve** The tendency of **meiosis** to only function regularly in tissues without major chromosomal abnormalities, hence meiotic products tend to be chromosomally normal.

**melittophily** Pollination by bees.

**micropile** The pore by which the pollen tube gains access to the interior of the ovule.

**microspore** Gives rise to a male **gametophyte** (includes pollen).

**miss-sense** A form of DNA mutation (frequently a deletion) in which the resulting coded polypeptide is non-functional.

**mitochondrion** The cell organelle within which the terminal oxidase system of respiration resides.

**monocarpic** Of a **genet** (or more rarely **ramet**) which only flowers once.

**monocotyledon** Member of the class Monocotyledones which includes a minority of Angiosperms (*see* **dicotyledon**) and is characterised by such features as one cotyledon in a seedling, scattered vascular bundles, an absence of secondary thickening, narrow, parallel-veined leaves and floral parts in threes and sixes. Probably derived from the dicotyledons.

**monoecy** **Hermaphrodite** genets in which anthers and gynoecia occur in different flowers (male and female function are separated).

**monomer** A protein (e.g. an enzyme) with only one subunit (*see* **dimer**).

**monomorphic** Of only one form or **genotype**, as in a population that is not polymorphic for a gene.

**monophylesis** Having a single origin in evolutionary history.

**Müllerian mimicry** The sharing of a feature or signal by a number of different species to the mutual benefit of all (*see* **Batesian mimicry**).

**multi-allelic** A gene with more than two alleles at a locus.

**multilocular** Of a carpel with more than one ovule.

**mutagenesis** Increase in the rate of mutation through the use of chemicals, radiation etc.

**myophily** Pollination by flies.

**neighbourhood** Theoretical concept of an area containing a number of genets between which there is **panmixis** (the effective population number, or neighbourhood size); see Chapter 5, p. 175 for further explanation.

**neoendemic** A species whose restricted geographical distribution is a function of its recent origin.

**neoteny** The persistence of a juvenile condition into an adult stage.

**nexine** The inner layer of the **exine**.

**non-disjunction** The failure of homologous chromosomes to pull apart at anaphase in mitosis or **meiosis**.

**nonsense** A form of DNA mutation in which one or more base-pairs of the DNA code change to read as the 'stop' code (ATG) thus causing a polypeptide translation to end prematurely.

**nucellus** Tissue of the ovule, inside the integuments, which acts as a nurse to the **archesporium** and embryo-sac (or **prothallus**).

**nucleolus** Body within the nucleus containing RNA; the number of nucleoli per nucleus is usually the same as the number of nuclear organiser regions in the **karyotype**, which is often the same as the number of genomes present in the **karyotype** (i.e. a diploid has two).

**nucleotide** The nucleotide bases adenine, guanine, cytosine and thymidine are the material from which the triplet code of the DNA is formed.

**oligophilic** A flower that is visited by few species of pollinator (*see* **polyphilic**).

**oligotropic** A pollinator that visits few species of flower (*see* **polytropic**).

**ornithophily** Pollination by birds.

**orthotropous** Form of an ovule in which the **chalaza** is next to the placenta and micropile is distal to the placenta.

**osmophore** Floral organ, often formed from a petal, which is long, narrow, drooping, often dark, and usually foul-smelling, usually as part of a myophilous syndrome.

**outbreeding** A breeding system which is panmictic, giving rise to heterozygote frequencies as predicted by the Hardy–Weinberg equilibrium (*see* **inbreeding**).

**ovary** Part of the **gynoecium**, containing one or more carpels, which houses the ovules and forms the fruit.

**overdominance** Condition in which a heterozygote is superior to either homozygote.

**oviposition** Egg laying (of insects).

**ovule** The female sporangium of seed plants, containing the integuments, nucellus and embryo-sac or **prothallus** when mature, and after fertilisation forming the seed.

**PMC (the pollen mother cell)** The cell which undergoes male **meiosis** to give rise to microspores (pollen).

**palaeoendemic** A species whose restricted geographical distribution is considered to be a relict from an earlier wider distribution.

**panicle** An inflorescence which is a determinate branched **raceme**.

**panmixis** A theoretical concept of a breeding system which has an infinitely large number of genets which are equally likely to mate with each other; *see* **outbreeding, inbreeding, neighbourhood**.

**papilionoid** The typical flower form of the family Leguminosae (Fabaceae) subfamily Papilionaceae, with a dorsal standard, lateral wings, and two petals fused to form a keel, within which the stamens and style are held under tension (a 'pea flower').

**papillae** The specialised often extruded epidermal cells of the stigma which receive the pollen grains, and which the pollen tubes penetrate.

**papillate** A heteromorphic condition in the family Plumbaginaceae characterised by long stigma papillae (*see* **cob**).

**parietal (placentation)** Arrangement of the ovules around the outer walls of a carpel.

**parthenogenesis** The autonomous development of an egg into an embryo in the absence of fertilisation.

**patch** A group of flowers, often of the same species, presenting a similar syndrome to pollinators, which 'major' on it.

**pectins** Complex polysaccharides forming the middle lamella of the cell wall; also occur as salts or pectates.

**pectinases** Enzymes that lyse pectin and pectinates.

**pellicle** The proteinaceous outer layer of the cuticle of stigma papillae.

**pentaploid** With five sets of **homologous** or homoeologous chromosomes (**genomes**).

**perianth** The petals and sepals, or tepals, of a flower.
**phalaenophily** Pollination by moths.
**phalanx** A vegetative growth strategy in which ramets advance together (*see* **guerilla**).
**phenotype** The expression of a **genotype**, i.e. the result of interaction between the **genotype** and the environment.
**phenotypic plasticity** The capability of a **genotype** to assume different phenotypes.
**pheromone** Volatile hormone used by insects as a mating attractant.
**phytosociology** The classification of plant ecological communities.
**pin** The long-styled morph of a distylous heteromorph.
**placenta** The organ that attaches an ovule to the carpel wall.
**plasmagene** A hereditary factor that is borne on an organelle outside of the nucleus and chromosomes, is (usually) maternally inherited, and is not subject to recombination and segregation.
**plasmalemma** The reticulum of membranes that permeates the contents of a cell.
**plastids** Cell organelles that contain (for instance) photosynthetic pigments.
**pleiotropy** The ability of a gene to influence more than one phenotypic attribute.
**pollen** Microspores, which germinate to give pollen tubes (male gametophytes).
**pollenkitt** Coating of lipoprotein on the outside of pollen grains, derived from the **tapetum**, often coloured orange or yellow by caroteins, and involved in sporophytic incompatibility recognition mechanisms (*see also* **tryphine**).
**pollen tube** The male gametophyte produced by the pollen grain which has three nuclei (in **Angiosperms**), penetrates the stigma papilla, and grows down the style to the ovule where it releases two sperm cells.
**pollinium** The male dispersal unit in the orchids, derived from an anther locule, containing a sticky mass of pollen tetrads, and with a flexible sticky stalk (**caudicle**).
**polyembryony** The occurrence of more than one embryo in an ovule, usually caused by the coincidence of **apospory** and sexuality.
**polygamy** Where genets with male only, female only, **hermaphrodite** only, and gynomonoecious and/or andromonoecious flowers coexist.
**polymer** A molecule made of repetitive subunits.
**polymorphism** The occurrence of more than one **allele** at a locus in a population at a frequency greater than that assignable to mutation or migration.
**polyphilic** A flower that is visited by many species of pollinator (*see* **oligophilic**).
**polyphylesis** Having more than one origin in evolutionary history (*see* **monophylesis**).
**polyploid** A **genet** with more than two **homologous** or **homoeologous** sets of chromosomes.
**polysaccharide** Polymer of sugar molecules.
**polysomic** Where a chromosome or gene is present more than twice in a **genet**.
**polytropic** A pollinator that visits many species of flower (*see* **oligotropic**).
**poricidal** Anther dehiscence through apical pores (as in Ericaceae).

**precocious embryony** Development of the embryo asexually before the flower opens and anther dehiscence occurs.

**primary endosperm nucleus** *see* **fused polar nucleus.**

**proboscis** The tubular sucking mouthparts of insects such as lepidoptera.

**propagule** An organ of reproduction (*see* **disseminule**), e.g. seed, bulbil, stolon.

**protandrous** Dichogamous condition in which anther dehiscence precedes stigma receptivity within a flower (*see* **protogynous**).

**prothallus** The **gametophyte** generation of Pteridophytes, and the female gametophyte of Gymnosperms (equivalent to the Angiosperm embryo-sac).

**protogynous** Dichogamous condition in which stigma receptivity precedes anther dehiscence within a flower (*see* **protandry**).

**pseudocompatibility** Loss of self-incompatibility due to environmental stimuli such as heat, electrical stimulation, physical damage to the stigma, or bud pollination.

**pseudogamy** Agamospermy in which pollination and fertilisation of the primary endosperm nucleus is required for the successful development of the asexual embryo.

**pseudoviviparous** Vegetative proliferation of plantlets in the inflorescence axes of (usually) grasses.

**psychophily** Pollination by butterflies.

**Pteridophyte** Member of the subphylum Pteridophyta, a vascular plant without flowers or seeds, and with a free-living **gametophyte** generation, including ferns, horsetails, quillworts and clubmosses.

**quadrivalent** An association caused by the chiasmatic pairing of four chromosomes at **meiosis**.

**'*r*' strategy** ramets or genets which assign high energy allocations to reproduction and hence potentially show exponential population growth rates.

**raceme** An unbranched inflorescence.

**ramet** A physiologically independent individual; from one to many may make up a **genet** (*see also* **clone**).

**receptacle** Part of a flower on which the **perianth**, stamens, and in **hypogynous** flowers, the **gynoecium** are borne.

**reciprocal** Of experimental crosses between the same two genets in different directions, i.e. $A$ male × $B$ and $B$ male × $A$.

**recombination** The reassociation of genes and parts of chromosomes between **homologous** or **homoeologous** chromosomes after **meiosis** by the action of chiasmata, thus breaking up linkage groups.

**restitution** Where most or all the chromosomes of a meiotic or mitotic division remain together in a single daughter nucleus, usually as the result of non-disjunction.

**rugose** Roughly sculptured, as of the **exine** of a pollen grain.

***S* alleles** **Alleles** that control the self-incompatibility response.

**sapromyophily** Pollination of flowers by flies, where the flower mimics rotting meat.

**saprophyty** Where plants lack chlorophyll, and use decomposing vegetation as an energy source.

**satellite** The euchromatic region of a chromosome distal to the heterochromatic nucleolar organiser region.

**section** Taxonomic rank at a level between species and genus.

**self-compatible** A **genet** capable of self-fertilisation if self-pollinated.

**self-incompatible** A **genet** incapable of self-fertilisation, but capable of cross fertilisation with another genet.

**selfing** Fertilisation of an ovule by a pollen grain of the same **genet**.

**semi-compatibility** Where two genets share some but not all gametophytic *S* alleles, and thus in crosses between them some pollen grains can effect fertilisation and others cannot.

**septae** Walls between carpels in a fused ovary.

**sere** A succession of plant communities in a given habitat leading to a climax association.

**seroprotein** Protein produced in response to an antigen in an immunological reaction.

**siblings** Offspring of the same parents.

**simplex** *Aaaa* heterozygote in tetrasomic inheritance (*see* **duplex, triplex**).

**solar furnace** Flower that focuses solar radiation so that the temperature at the centre of the flower is above the ambient temperature.

**somatic** Non-meiotic or non-sexual; thus **somatic** recombination occurs by chromosome breakage/refusion events in non-sexual tissues.

**sphingid** Hawk-moth (family Sphingidae).

**sporangium** Organ that contains spores.

**sporophyll** Organ that bears sporangia.

**sporophytic** Of the sporophyte (usually diploid) generation; in the seed plants the dominant generation, the **genet** that develops from a **zygote**, is the sporophyte; in sporophytic incompatibility it is the **genotype** of the anther that produces the pollen grain which is significant, the anther being of the sporophyte generation.

**sporopollenin** The lipoprotein which is the main constituent of the **exine** of the Angiosperm pollen grain.

**staminode** Floral organ derived from a stamen, but no longer bearing an anther.

**stigma** Part of the gynoecium that receives pollen grains on stigma papillae, often borne on the end of a style.

**strategy** A teleological but useful word used to describe a group of attributes which result in a particular life-style or (in animals) behaviour pattern; compare syndrome.

**style** Part of the **gynoecium** that links the stigma to the ovary, down which the pollen tubes grow; usually slender.

**subandroecy** Where wholly female genets coexist with andromonoecious genets.

**subgynoecious** Where wholly male genets coexist with gynomonoecious genets.

**supergene** Linkage group of coadapted loci.

**sympatric** Having overlapping geographical distributions.

**synapsis** The close association of regions of **homologous** or homoeologous chromosomes during zygotene of **meiosis** preparatory to the formation of chiasmata.

**syndrome** A group of coadapted attributes, or attributes which occur together in a particular condition; a syndrome can contribute to a strategy.

**synergid** Nuclei (usually two) positioned at the micropilar end of the **Angiosperm** embryo-sac, and involved in the process of double fertilisation (*see* **filiform apparatus**); sometimes known as the egg apparatus.

**tapetum** The nurse tissue to the developing pollen grains, within the anther.

**taxon** A taxonomic category, e.g. species, section, genus, family.

**tectum** The outer layer of the pollen grain **exine** in Angiosperms.

**tepal** A perianth organ not differentiated into a petal or sepal.

**testa** The seed coat, derived from the ovule integuments.

**tetrads** Groups of spore nuclei in fours, the immediate product of **meiosis**.

**tetrasomic** Where a chromosome or gene is present four times in a **genet**; **tetrasomic** inheritance involves three types of heterozygote (*see* **simplex, duplex, triplex**).

**thrum** The short-styled morph of a distylous heteromorph.

**transcription** The formation of messenger RNA from the coded sequences of DNA genes on chromosomes prior to the formation of polypeptides at ribosomes (**translation**).

**translocation** Where a chromosome breaks, and part joins to a non-homologous chromosome (*see* **interchange**).

**transposable genetic elements (TGE)** Short sequences of DNA which promote chromosome breakage probably by the formation of endonucleases near to the TGE, which is this enabled to 'roam' around the **karyotype** in association with linked genes by a succession of breakage/refusion events.

**trap-lining** Where certain tropical bees are **oligotropic** with respect to distantly spaced individuals of flowering plant, and thus travel long distances between bouts of flower visiting.

**tricolporate** A pollen grain with three pores, each contained within a groove (sulca).

**trioecy** Where genets with male only, female only, and **hermaphrodite** only flowers coexist; compare **polygamy**.

**triplex** *AAAa* heterozygote in **tetrasomic** inheritance (*see* **simplex, duplex**).

**tripping** The explosive release of the stamens and style held in tension in the keel of a papilioniod flower, by a pollinator.

**tristyly** A heteromorphic heterostylous condition in which short-styled, mid-styled and long-styled morphs with reciprocal anther positions coexist.

**trivalent** An association caused by the chiasmatic pairing of three chromosomes at **meiosis**.

**tryphine** A coating of lipoprotein on the outside of pollen grains derived from the **tapetum**; more or less identical with **pollenkitt**, but not necessarily pigmented.

**turgor** Where a cell has absorbed water through osmosis until the cell wall is rigid.

**unilocular** Of a carpel with only one ovule.

**unisexual** Of a **genet** which is either male or female, equivalent to dioecy in seed plants.

**vascular plant** Plant containing vascular tissue, xylem and phloem; i.e. Pteridophytes, Gymnosperms, Angiosperms.
**viscidium** The sticky cap to the rostellum on the end of the column to which the pollinia adhere in some species of orchid.

**xenogamy** Fertilisation between pollen and ovules of different genets.
**xeromorphic** With phenotypic attributes enabling a plant to survive dry conditions.

**zoophily** Pollination by animals.
**zygomorphic** Bilaterally symmetrical (two-lipped), of a flower.
**zygote** The result of sexual fusion between two gametes.
**zymogram** Configuration of isozyme bands in the electrophoretic analysis of an enzyme system, typical of a **genet**.

# References


Abrahamson, W. G. 1979. Patterns of resource allocation in wildflower populations of field and wood. *Am. J. Bot.* **66**, 71–9.

Adams, W. T. and R. W. Allard 1982. Mating system variation in *Festuca microstachys*. *Evolution* **36**, 591–5.

Adey, M. E. 1982. *Taxonomic aspects of plant-pollinator relationships in the Genistinae (Leguminosae)*. PhD thesis, University of Southampton, UK.

Allard, R. W. 1960. *Principles of plant breeding*. New York: Wiley.

Allard, R. W., S. K. Jain and P. L. Workman 1968. The genetics of inbreeding populations. *Adv. Gen.* **14**, 55–131.

Almqvist, E. 1923. Studien über *Capsella bursa-pastoris* (L.). II. *Acta Hort. Berg.* **7**, 41–95.

Altman, P. L. and D. S. Dittmer 1964. *Biology data book*, 428–31. Washington, DC: Federation of American Societies for Experimental Biology.

Anderson, E. 1924. Studies on self-sterility. VI. The genetic basis of cross-sterility in *Nicotiana*. *Genetics* **9**, 13–40.

Anderson, M. K., N. L. Taylor and J. F. Duncan 1974. Self-incompatibility, genotype identification and stability as influenced by inbreeding in red clover (*Trifolium pratense* L.). *Euphytica* **23**, 140–8.

Annerstedt, I. and A. Lundquist 1967. Genetics of self-incompatibility in *Tradescantia paludosa* (Commelinaceae). *Hereditas* **58**, 13.

Arasu, N. N. 1968. Self-incompatibility in angiosperms: a review. *Genetica* **39**, 1–24.

Arroyo, M. T. K. de 1975. Electrophoretic studies of genetic variation in natural populations of allogamous *Limnanthes alba* and autogamous *Limnanthes floccosa* (Limnanthaceae). *Heredity* **35**, 153–64.

Arroyo, M. T. K. de and P. H. Raven 1975. The evolution of subdioecy in morphologically gynodioecious species of *Fuchsia* sect. Encliandra (Onagraceae). *Evolution* **29**, 500–11.

Ascher, P. D. 1966. A gene action model to explain gametophytic self-incompatibility. *Euphytica* **15**, 179–83.

Ascher, P. D. and S. J. Peloquin 1966. Effect of floral ageing on the growth of compatible and incompatible pollen tubes in *Lilium longiflorum*. *Amer. J. Bot.* **53**, 99–102.

Asker, S. 1979. Progress in apomixis research. *Hereditas* **91**, 231–40.

Asker, S. 1980. Gametophytic apomixis: elements and genetic regulation. *Hereditas* **93**, 277–93.

Assouad, M. W., B. Dommée, R. Lumaret and G. Valdeyron 1978. Reproductive capacities in the sexual forms of the gynodioecious species *Thymus vulgaris* L. *Bot. J. Linn. Soc.* **77**, 29–39.

Attia, M. S. 1950. The nature of incompatibility in cabbage. *Proc. Am. Soc. Hort. Sci.* **56**, 369–71.

Babbel, G. R. and R. P. Wain 1977. Genetic structure of *Hordeum jubatum*. I. Outcrossing rates and heterozygosity levels. *Can. J. Gen. Cyt.* **19**, 143–52.

Babcock, E. B. and G. L. Stebbins 1938. The American species of *Crepis*. *Publ. Carnegie Inst.*, Wash. **504**, 1–199.

Baker, H. G. 1948. Corolla-size in gynodioecious and gynomonoecious species of flowering plants. *Proc. Leeds Phil. Soc.* (Scientific section) for 1948, 136–9.
Baker, H. G. 1953a. Dimorphism and monomorphism in the Plumbaginaceae. II. Pollen and stigmata in the genus *Limonium. Ann. Bot.* II **17**, 433–45.
Baker, H. G. 1953b. Diomorphism and monomorphism in the Plumbaginaceae. III. Correlation of geographical distribution patterns with dimorphism and monomorphism in *Limonium. Ann. Bot.* II **17**, 615–27.
Baker, H. G. 1955. Self-compatibility and establishment after 'long-distance' dispersal. *Evolution* **9**, 347–9.
Baker, H. G. 1966. The evolution, functioning and breakdown of heteromorphic incompatibility systems. 1. The Plumbaginaceae. *Evolution* **20**, 349–68.
Baker, H. G. 1975. Sporophyte-gametophyte interaction in *Linum* and other genera with heteromorphic self-incompatibility. In *Gamete competition in plants and animals,* D. L. Mulcahy (ed.). Amsterdam: North Holland Publ. Co.
Baker, H. G. and I. Baker 1973a. Amino acids in nectar and their evolutionary significance. *Nature* **241**, 543–5.
Baker, H. G. and I. Baker 1973b. Some anthecological aspects of the evolution of nectar-producing flowers, particularly amino acid production in nectar. In *Taxonomy and ecology,* V. H. Heywood (ed.), 243–64. London: Academic Press.
Baker, H. G. and I. Baker 1975. Studies of nectar-constitution and pollinator-plant coevolution. In *Coevolution of animals and plants,* L. E. Gilbert and P. H. Raven (eds). Austin: University of Texas Press.
Baker, H. G. and I. Baker 1977. Intraspecific constancy of floral nectar amino acid complements. *Bot. Gaz.* **138**, 183–91.
Baker, H. G. and I. Baker 1983. Floral nectar sugar constituents in relation to pollinator type. In *Handbook of experimental pollination,* C. E. Jones and J. Little (eds). New York: Von Nostrand.
Baker, R. R. 1969. The evolution of the migratory habit in butterflies. *J. Anim. Ecol.* **38**, 703–46.
Bannister, M. H. 1965. Variation in the breeding system of *Pinus radiata. The genetics of colonizing species,* H. G. Baker and G. L. Stebbins (eds), 353–72. New York: Academic Press.
Barons, K. 1938. *Proc. Am. Soc. Hort. Sci.* **36**, 637.
Bateman, A. J. 1947. Contamination in seed crops. III. Relation with isolation distance. *Heredity* **1**, 303–36.
Bateman, A. J. 1952. Self-incompatibility systems in angiosperms. I. Theory. *Heredity* **6**, 285–310.
Bateman, A. J. 1954. Self-incompatibility systems in angiosperms. II. *Iberis amara. Heredity* **8**, 305–32.
Bateman, A. J. 1955. Self-incompatibility systems in angiosperms. III. Cruciferae. *Heredity* **9**, 53–68.
Battaglia, E. 1963. Apomixis. In *Recent advances in the embryology of angiosperms,* P. Maheshwari (ed.), 221–64. Delhi: University of Delhi.
Batygina, T. B. 1974. Fertilization process of cereals. In *Fertilization in higher plants,* H. F. Linskens (ed.), 205–20. Amsterdam: North Holland.
Baur, E. 1919. Ueber Selbststerilität und uber Kreuzungsversuche einer selbstfertilen und einer selbststerilen Art in der Gattung *Antirrhinum. Z. Indukt. Abst.* **21**, 48–52.
Bawa, K. S. 1979. Breeding systems of trees in a tropical wet forest. *N.Z. J. Bot.* **17**, 521–4.
Bawa, K. S. 1980. Evolution of dioecy in flowering plants. *Ann. Rev. Ecol. Syst.* **11**, 15–39.

Bawa, K. S. 1981. Modes of pollination, sexual systems and community structure in a tropical lowland rainforest. *Abstracts, XIII International Botanical Congress, Sydney*, **103**.

Bawa, K. S. 1983. Patterns of flowering in tropical plants. In *Handbook of experimental pollination biology*, C. E. Jones and R. J. Little (eds), 394–410. New York: Scientific and Academic Editions.

Bawa, K. S. and J. H. Beach 1983. Self-incompatibility systems in the Rubiaceae of a tropical lowland wet forest. *Am. J. Bot.* **70**, 1281–8.

Bawa, K. S. and P. A. Opler 1975. Dioecism in tropical forest trees. *Evolution* **29**, 167–79.

Beach, J. H. and K. S. Bawa 1980. Role of pollinators in the evolution of dioecy from distyly. *Evolution* **34**, 1138–42.

Beadle, G. W. 1930. Genetical and cytological studies of Mendelian asynapsis in *Zea mays*. *Cornell Univ. Agric. Expt. St. Mem.* **129**.

Beattie, A. J. 1976. Plant dispersion, pollination and gene flow in *Viola*. *Oecologia* **25**, 291–300.

Beattie, A. 1978. Plant–animal interactions affecting gene flow in *Viola*. In *The pollination of flowers by insects*, A. J. Richards (ed.), 151–64. London: Academic Press.

Beattie, A. J. and D. C. Culver 1979. Neighbourhood size in *Viola*. *Evolution* **33**, 1226–9.

Berry, R. J. 1977. *Inheritance and natural history*. London: Collins New Naturalist **61**.

Best, L. S. and P. Bierzychudek 1982. Pollinator foraging on foxglove (*Digitalis purpurea*): a test of a new model. *Evolution* **36**, 70–9.

Bino, R. J. and A. D. J. Meeuse 1981. Entomophily in dioecious species of *Ephedra*: a preliminary report. *Acta Bot. Neerl.* **30**, 151–3.

Blackman, R. L. 1979. Stability and variation in aphid clonal lineages. *Biol. J. Linn. Soc.* **11**, 259–77.

Bodmer, W. F. 1958. Natural crossing between homostyle plants of *Primula vulgaris*. *Heredity* **12**, 363–70.

Booth, T. A. and A. J. Richards 1978. Studies in the *Hordeum murinum* aggregate: disc electrophoresis of seed proteins. *Bot. J. Linn. Soc.* **76**, 115–25.

Bosbach, K. and H. Hurka 1981. Biosystematic studies on *Capsella bursapa storis* (Brassicaceae): enzyme polymorphism in natural populations. *Pl. Syst. Evol.* **137**, 73–94.

Bradshaw, A. D. 1965. Evolutionary significance of phenotypic plasticity in plants. *Adv. Gen.* **13**, 115–55.

Bradshaw, A. D. 1972. Some of the evolutionary consequences of being a plant. *Evol. Biol.* **5**, 25–47.

Brantjes, N. B. M. 1978. Sensory responses to flowers in night-flying moths. In *The pollination of flowers by insects*, A. J. Richards (ed.), 13–19. London: Academic Press.

Bredemeyer, G. M. M. 1973. Peroxidase activities and peroxidase isoenzyme patterns during growth and senescence of the unpollinated style and corolla of tobacco plants. *Act. Bot. Neerl.* **22**, 40–8.

Bredemeyer, G. M. M. 1975. The effect of peroxidase on pollen germination and pollen tube growth *in vitro*. *Incomp. Newslett. Ass. EURATOM-ITAL* Wageningen **5**, 34–9.

Bredemeyer, G. M. M. and J. Blaas 1975. A possible role of a stylar peroxidase gradient in the rejection of incompatible growing pollen tubes. *Act. Bot. Neerl.* **24**, 37–48.

Breese, E. L., M. D. Hayward and A. C. Thomas 1965. Somatic selection in perennial rye-grass. *Heredity* **20**, 367–79.
Breukelen, E. W. M. van, M. S. Ramanna and J. G. T. Hermsen 1975. Monohaploids ($n = X = 12$) from autotetraploid *Solanum tuberosum* ($Zn = 4X = 48$) through two successive cycles of female parthenogenesis. *Euphytica* **24**, 567–74.
Brewbaker, J. L. 1957. Pollen cytology and incompatibility systems in plants. *J. Hered.* **48**, 217–77.
Brewbaker, J. L. 1959. Biology of the angiosperm pollen grain. *Ind. J. Genet. Pl. Breed.* **19**, 121–33.
Brewbaker, J. L. and B. H. Kwack 1963. The essential role of calcium ions in pollen germination and pollen tube growth. *Am. J. Bot.* **50**, 859–65.
Brighton, C. A., B. Mathew and P. Rudall 1983. A detailed study of *Crocus speciosus* and its ally *C. pulchellus* (Iridiceae). *Pl. Syst. Evol.* **142**, 187–206.
Broker, W. 1963. Genetisch-physiologische Untersuchungen über die ainkvertraglichkeit von *Silene inflata* Sm. *Flora*, Jena, B **153**, 122–56.
Brown, A. H. D. 1978. Isozymes, plant population genetic structure and genetic conservation. *Theoret. Appl. Genet.* **52**, 145–57.
Brown, A. H. D., A. C. Matheson and K. G. Eldridge 1975. Estimation of the mating system of *Eucalyptus obliqua* L'Heret by using allozyme polymorphisms. *Aust. J. Bot.* **23**, 931–49.
Bruun, H. G. 1930. The cytology of the genus *Primula*. *Sv. Bot. Tidskr.* **24**, 468–75.
Bruun, H. G. 1932. Studien an heterostyler pflanzen. I. Versuch einer verknüpfung von chromosomenzahl und heterostylie. *Sv. Bot. Tidskr.* **26**, 163–74.

Cahalan, C. and C. Gliddon 1985. Genetic neighbourhood sizes in *Primula vulgaris*. *Heredity* **54**, 65–70.
Calos, M. P. and J. H. Miller 1980. Transposable elements. *Cell* **20**, 579–95.
Carlquist, S. 1974. *Island biology*. New York: Columbia University Press.
Casper, B. B. and E. L. Charnov 1982. Sex allocation in heterostylous plants. *J. Theoret. Biol.* **96**, 143.
Catcheside, D. G. 1977. The genetics of recombination. *Genetics – principles and perspectives*. 2. London: Edward Arnold.
Charlesworth, B. 1980. The cost of sex in relation to the mating system. *J. Theoret. Biol.* **84**, 655–71.
Charlesworth, B. and D. Charlesworth, 1978. A model for the evolution of dioecy and gynodioecy. *Am. Nat.* **112**, 975–97.
Charlesworth, D. 1979. The evolution and breakdown of tristyly. *Evolution* **33**, 489–98.
Charlesworth, D. and B. Charlesworth, 1979. A model for the evolution of heterostyly. *Am. Nat.* **114**, 467–98.
Charnov, E. L. 1982. *The Theory of Sex Allocation*. Princeton: Princeton University Press.
Charnov, E. L. 1984. Behavioural ecology of plants. In *Behavioural ecology, an evolutionary approach*, J. J. Krebs and N. B. Davies (eds), 362–79. Oxford: Blackwell Scientific.
Charnov, E. L., Maynard Smith and J. J. Bull 1976. Why be hermaphrodite? *Nature* **263**, 125–6.
Chase, S. 1969. Monoploids, and monoploid-derivatives of maize (*Zea mays* L.). *Bot. Rev.* **35**, 117–67.
Chu, C. and S. Y. Hu 1981. The development and ultrastructure of wheat sperm cell. *Abstracts, XIII International Botanical Congress, Sydney*, 61.

Clapham, A. R., T. G. Tutin and E. F. Warburg 1962. *Flora of the British Isles*, 2nd edn. Cambridge: Cambridge University Press.

Clay, K. and J. Antonovics 1985. Quantitative variation of progeny from chasmogamous and cleistogamous flowers in the grass *Danthonia spicata*. *Evolution* **39**, 335–48.

Clayberg, C. D., L. Butler, E. A. Kerr, C. M. Rick and R. W. Robinson 1966. Third list of known genes in the tomato: with revised linkage map and additional rules. *J. Hered.* **57**, 188–96.

Cobon, A. M. and B. Matfield 1976. Morphological and cytological studies on a hexaploid clone of *Potentilla anserina* L. *Watsonia* **11**, 125–9.

Connor, H. E. 1973. Breeding systems in *Cortaderia* (Gramineae). *Evolution* **27**, 663–78.

Connor, H. E. 1979. Breeding systems in the grasses: a survey. *NZ. J. Bot.* **17**, 547–74.

Cook, L. M. 1971. *Coefficients of natural selection*. London: Hutchinson.

Corbett, S. A. 1978. Bees and the nectar of *Echium vulgare*. In *The Pollination of Flowers by Insects*, A. J. Richards (ed.), 21–30. London: Academic Press.

Correns, C. 1912. Selbststerilität und Individualstoffe. *Festschr. d. mat. nat. Gesell*, zur. 84. Versamml. Deut. Naturforsch. Artze, Munster i W., 1–32.

Correns, C. 1913. Selbststerilität und Individualstoffe. *Biol. Centr.* **33**, 389–423.

Correns, C. 1916a. Individuen und Individualstoffe. *Die Naturwissensch.* **4**, 183–7, 193–8, 210–13.

Correns, C. 1916b. Untersuchungen über geschlechtsbestimmung bei distelarten. *Sitz. Konigl. Preuss. Akad. Wiss.* **20**, 448–77.

Correns, C. 1928. *Bestimmung, Vererbung und Verteilung des Geschlechtes bei den hoheren Pflanzen*. Berlin: Borntraeger.

Crane, M. B. and W. J. C. Lawrence 1929. Genetical and cytological aspects of incompatibility and sterility in cultivated fruits. *J. Pomol. Hort. Sci.* **7**, 276–301.

Crane, M. B. and D. Lewis 1942. Genetical studies in pears. III. Incompatibility and sterility. *J. Genet.* **43**, 31.

Crane, M. B. and K. Mather 1943. The natural cross-pollination of crop plants with particular reference to the radish. *Ann. Appl. Biol.* **30**, 301–8.

Crawford, R. M. M. and J. Balfour 1983. Female predominant sex-ratios and physiological differentiation in arctic willows. *J. Ecol.* **71**, 149–60.

Crawford, T. J. 1984a. The estimation of neighbourhood parameters for plant populations. *Heredity* **52**, 273–83.

Crawford, T. J. 1984b. What is a population? In *Evolutionary ecology*, B. Shorrocks (ed.), 135–73. Oxford: Blackwell Scientific.

Crosby, J. L. 1949. Selection of an unfavourable gene-complex. *Evolution* **3**, 212–30.

Crosby, J. L. 1959. Outcrossing on homostyle primrose. *Heredity* **13**, 127–31.

Crosby, J. L. 1960. The use of electronic computation in the study of random fluctuations in rapidly evolving populations. *Phil. Trans. R. Soc. Lond.* **242**, 551–72.

Crowe, L. K. 1954. Incompatibility in *Cosmos bipinnatus*. *Heredity* **8**, 1–11.

Crowe, L. K. 1971. The polygenic control of outbreeding in *Borago officinalis*. *Heredity* **27**, 111–18.

Cruden, R. W. 1977. Pollen-ovule ratios: a conservative indicator of breeding systems in flowering plants. *Evolution* **31**, 32–6.

Cruden, R. W. and S. Miller-Ward 1981. Pollen-ovule ratio, pollen size, and the ratio of stigmatic area to the pollen-bearing area of the pollinator: an hypothesis. *Evolution* **35**, 964–74.

Cruden, R. W., S. Kinsman, R. E. Stockhouse and Y. B. Linhart 1976. Pollination, fecundity, and the distribution of moth-flowered plants. *Biotropica* **8**, 204–10.

Culwick, E. G. 1982. *The biology of* Acaena novae-zelandiae *Kirk on Lindisfarne.* Ph.D thesis, University of Newcastle upon Tyne.

Currah, L. 1981. Pollen competition in onion (*Allium cepa* L.). *Euphytica* **30**, 687–96.

Currah, L. and D. J. Ockendon 1978. Protandry and the sequence of flower opening in the onion (*Allium cepa* L.). *New Phytol.* **81**, 419–28.

Dahlstedt, H. 1921. Die Svenska arterna av slaktet *Taraxacum*. I, II, III. *Act. Fl. Suec.* **I**, 1–160.

Damme, J. M. M. van 1984. Gynodioecy in *Plantago lanceolata*. III. Sexual reproduction and maintenance of male steriles. *Heredity* **52**, 77–94.

Darlington, C. D. 1939. *The evolution of genetic systems.* Cambridge: Cambridge University Press.

Darlington, C. D. 1971. The evolution of polymorphic systems. In *Ecological genetics and evolution*, E. R. Creed (ed.), 1–19. Oxford: Blackwell Scientific.

Darlington, C. D. and K. Mather 1949. *The elements of genetics.* London: Allen and Unwin.

Darwin, C. 1862. *The various contrivances by which orchids are fertilised.* London: Murray.

Darwin, C. 1871. *The descent of man and selection in relation to sex*, 2 vols. London: Murray.

Darwin, C. 1876. *The effects of cross and self fertilisation in the vegetable kingdom.* London: Murray.

Darwin, C. 1877. *The different forms of flowers on plants of the same species.* London: Murray.

Davey, A. J. C. and C. M. Gibson 1917. Note on the distribution of the sexes in *Myrica gale*. *New Phytol.* **16**, 147–51.

David, R. W. 1977–82. (The British distribution of uncommon Carices). *Watsonia* **11**, 377–8; **12**, 47–9, 158–60, 257–8, 335–7; **13**, 53–4, 124–5, 225–6, 318–21; **14**, 68–70.

Davie, J. H. and J. R. Akeroyd 1983. *Pachyphragma macrophyllum* (Hoffm.) Busch (Cruciferae), a Caucasian species naturalised in Co. Avon, England. *Bot. J. Linn. Soc.* **87**, 77–82.

Dawkins, R. 1976. *The selfish gene.* Oxford: Oxford University Press.

Delannay, X. 1978. La gynodioecie chez les angiospermes. *Natur. Belges* **59**, 223–37.

Denward, T. 1963. The function of the incompatibility alleles in red clover (*Trifolium pratense* L.). *Hereditas* **49**, 289–334.

Dickinson, H. G. and D. Lewis 1974. Changes in the pollen grain wall of *Linum grandiflorum* following compatible and incompatible intra-specific pollinations. *Ann. Bot.* **38**, 23–9.

Dickinson, H. G. and D. Lewis 1975. Interaction between the pollen grain coating and the stigmatic surface during compatible and incompatible interspecific pollinations in *Raphanus*. In *The biology of male gamete*, J. G. Duckett and P. A. Racey (eds). *Biol. J. Linn. Soc.* **7** (suppl.), 165–75.

Dickinson, H. G., J. Moriarty and J. Lawson 1982. Pollen-pistil interaction in *Lilium longiflorum*: the role of the pistil in controlling pollen tube growth following cross and self-pollination. *Proc. R. Soc. Lond. B* **215**, 45–62.

Dodson, C. H. 1962. Pollination and variation in the subtribe Catasetinae (Orchidaceae). *Ann. Missouri Bot. Gard.* **49**, 35–56.

Dodson, C. H. and G. P. Frymire 1961. Natural pollination of orchids. *Bull. Missouri Bot. Gard.* **49**, 133–52.

Dommée, B. 1976. La stérilité mâle chez *Thymus vulgaris* L.: répartition écologique dans la région mediterranéene française. *Compt. Rend. Hebd. Seances Acad. Sci.* **282** D, 65–8.

Donk, J. A. W. van der 1974. Synthesis of RNA and protein as a function of time and type of pollen tube-style interaction in *Petunia hybrida* L. *Molec. Gen. Genet.* **134**, 93–8.

Dowrick, V. P. J. 1956. Heterostyly and homostyly in *Primula obconica*. *Heredity* **10**, 219–36.

Dressler, R. L. 1968. Pollination by euglossine bees. *Evolution* **22**, 202–10.

Dressler, R. L. 1980. *The orchids, natural history and classification.* Cambridge, Mass.: Harvard University Press.

Dulberger, R. 1964. Flower dimorphism and self-incompatibility in *Narcissus tazzetta* L. *Evolution* **18**, 361–3.

Dulberger, R. 1970. Floral dimorphism in *Anchusa hybrida* Ten. *Israel J. Bot.* **19**, 37–41.

Dulberger, R. 1974. Structural dimorphism of stigmatic papillae in distylous *Linum* species. *Am. J. Bot.* **61**, 238–43.

Dulberger, R. 1975a. *S* gene action and the significance of characters in the heterostylous syndrome. *Heredity* **35**, 407–15.

Dulberger, R. 1975b. Intermorph structural differences between stigmatic papillae and pollen grains in relation to incompatibility in Plumbaginaceae. *Proc. R. Soc. Lond. B* **188**, 257–74.

Durand, B. 1963. Le complex *Mercurialis annua* L. *s.l.* Une étude biosystematique. *Ann. Sci. Nat., Bot.* IV, **4**, 625–736.

East, E. M. 1915a. An interpretation of self-sterility. *Proc. Nat. Acad. Sci.* **1**, 95–100.

East, E. M. 1915b. The phenomenon of self-sterility. *Am. Nat.* **49**, 77–88.

East, E. M. 1917a. The behaviour of self-sterile plants. *Science* n.s. **46**, 221–2.

East, E. M. 1917b. The explanation of self-sterility. *J. Heredity* **8**, 382–3.

East, E. M. 1918. Intercrosses between self-sterile plants. *Mem. Brooklyn Bot. Gdn.* **1**, 141–53.

East, E. M. 1919a. Studies on self-sterility. III. The relation between self-fertile and self-sterile plants. *Genetics* **4**, 341–45.

East, E. M. 1919b. Studies on self-sterility. IV. Selective fertilization. *Genetics* **4**, 346–55.

East, E. M. 1919c. Studies on self-sterility. V. A family of self-sterile plants, wholly cross-sterile *inter se*. *Genetics* **4**, 356–63.

East, E. M. 1940. The distribution of self-sterility in flowering plants. *Proc. Am. Phil. Soc.* **82**, 449–518.

East, E. M. and A. J. Mangelsdorf 1925. A new interpretation of the hereditary behaviour of self-sterile plants. *Proc. Nat. Acad. Sci. (Wash)* **11**, 166–71.

East, E. M. and J. B. Park 1917. Studies on self-sterility. 1. The behaviour of self-sterile plants. *Genetics* **2**, 505–609.

Eenink, A. H. 1974. Matromorphy in *Brassica oleracea* L. II. Differences in parthenogenetic ability and parthenogenesis inducing ability. *Euphytica* **23**, 435–45.

Eisikowitch, D. 1978. Insect visiting of two subspecies of *Nigella arvensis* under adverse seaside conditions. In *pollination of flowers by insects*, A. J. Richards (ed.), 125–32. London: Academic Press.

Eisikowitch, D. and S. R. J. Woodell 1975. The effect of water on pollen germination in two species of *Primula*. *Evolution* **28**, 692–4.

Elkington, T. T. 1969. Cytotaxonomic variation in *Potentilla fruticosa*. *New Phytol.* **68**, 151.

Elkington, T. T. and S. R. J. Woodell 1963. *Potentilla fruticosa* L. Biological flora of the British Isles. *J. Ecol.* **51**, 769.
Ellerstrom, S. and L. Zagorchera 1977. Sterility and apomictic embryo-sac formation in *Raphanobrassica*. *Hereditas* **87**, 107–120.
Ellstrand, N. C. 1984. Multiple paternity within the fruits of the wild radish, *Raphanus sativus*. *Am. Nat.* **123**, 819–28.
Ellstrand, N. C., A. M. Torres and D. A. Levin 1978. Density and rate of apparent outcrossing in *Helianthus annuus* (Asteraceae). *Syst. Bot.* **3**, 403–7.
Emerson, S. 1939. A preliminary survey of the *Oenothera organensis* population. *Genetics* **24**, 524–37.
Epling, C. and Th. Dobzhansky 1942. Genetics of natural populations. VI. Microgeographic races in *Linanthus parryae*. *Genetics* **27**, 317–32.
Ernst, A. 1933. Weitere Untersuchungen zur Phänanalyse, zum Fertilitätsproblem and zur Genetik heterostyler Primeln. 1. *Primula viscosa*. *Arch. J. K.-Stift. Ver., Soc. Rass.* **8**, 1–215.
Ernst, A. 1936. Weitere Untersuchungen zur Phänanalyse, zum Fertilitätsproblem and zur Genetik heterostyler Primeln. II. *Primula hortensis* Wettst. *Arch. J. Kl.-Stift. Ver., Soc. Rass.* **11**, 1–280.
Ernst, A. 1955. Self-fertility in monomorphic primulas. *Genetica* **27**, 91–148.

Faegri, K. and L. van der Pijl 1979. *The principles of pollination ecology*, 3rd edn. Oxford: Pergamon.
Fagerlind, F. 1945. Die bastarde der *Canina*-Rosen, ihre Syndese und Formbildungsverhältnisse. *Act. Hort. Berg.* **14**, 9–37.
Faulkner, J. S. 1973. Experimental hybridisation of north-west European species in *Carex* section Acutae (Cyperaceae). *Bot. J. Linn. Soc.* **67**, 233–53.
Favre-Duchatre, M. 1974. Phylogenetic aspects of the spermatophytes double fertilization. In *Fertilization in higher plants*, H. F. Linskens (ed.), 243–52. Amsterdam: North Holland.
Feinsinger, P. and L. A. Swarm 1982. 'Ecological release', seasonal variation in food supply, and the hummingbird *Amazilia tobaci* on Trinidad and Tobago. *Ecology* **63**, 1574–87.
Feinsinger, P., J. A. Wolfe and L. A. Swarm 1982. Island ecology: reduced hummingbird diversity and the pollination biology of plants, Trinidad and Tobago, West Indies. *Ecology* **63**, 494–506.
Felsenstein, J. 1974. The evolutionary advantage of recombination. *Genetics* **78**, 737–56.
Ferrari, T. E., S. S. Lee and D. H. Wallace 1981. Biochemistry and physiology of recognition in pollen-stigma interactions. *Phytopathology* **71**, 752–5.
Filzer, P. 1926. Selbststerilität von *Veronica syriaca*. *Z. Indukt. Abstamm. Vererbl.* **41**, 137–97.
Fincham, J. R. S. and G. R. K. Sastry 1974. Controlling elements in maize. *Ann. Rev. Genet.* **8**, 15–50.
Fisher, R. A. 1961. A model for the generation of self-sterility alleles. *J. Theoret. Biol.* **1**, 411–14.
Ford, E. B. 1964. *Ecological genetics*. London: Methuen.
Ford, H. 1981. Competitive relationships amongst apomictic dandelions. *Biol. J. Linn. Soc.* **15**, 355–68.
Ford, H. 1985. Life history strategies in two coexisting agamospecies. *Biol. J. Linn. Soc.* in press.
Ford, H. and A. J. Richards 1985. Isozyme variation within and between *Taraxacum* agamospecies in a single locality. *Heredity* **55**, 289–91.

Fowler, N. L. and D. A. Levin 1984. Ecological constraints on the establishment of a novel polyploid in competition with its diploid progenitor. *Am. Nat.* **124**, 703–11.

Frankel, R. and E. Galun 1977. *Pollination mechanisms, reproduction and plant breeding.* Monogr. Theoret. Appl. Genetics 2. Berlin: Springer-Verlag.

Frankie, G. W. 1976. Pollination of widely dispersed trees by animals in Central America, with emphasis on bee pollination systems. In *Tropical trees: variation, breeding and conservation*, J. Burley and B. T. Styles (eds), 151–9. New York: Academic Press.

Frankie, G. W., P. A. Opler and K. S. Bawa 1976. Foraging behaviour of solitary bees: Implications for outcrossing of a neotropical forest tree species. *J. Ecol.* **64**, 1049–58.

Free, J. B. 1962. Studies on the pollination of fruit trees by Honey-bees. *J. R. Hort. Soc.* **87**, 302–9.

Freeman, D. C., E. D. McArthur, K. T. Harper and A. C. Blauer 1981. Influence of environment on the floral sex ratio of monoecious plants. *Evolution* **35**, 194–7.

Fürnkranz, D. 1960. Cytogenetische Untersuchungen an *Taraxacum* im raume von Wien. *Öst. Bot. Z.* **107**, 310–50.

Fürnkranz, D. 1961. Cytogenetische Untersuchungen an *Taraxacum* im raume von Wien. II. Hybriden zwischen *T. officinale* und *T. palustre*. *Öst. Bot. Z.* **108**, 408–15.

Fürnkranz, D. 1965. Untersuchungen an Populationen des *Taraxacum officinale* – Komplexes im Kontaktgebiet der diploiden und polyploiden Biotypen. *Öst. Bot. Z.* **113**, 427–47.

Gabe, D. R. 1939. Inheritance of sex in *Mercurialis annua*. *Compt. Rend. Acad. Sci. URSS* **23**, 478–81.

Gadgil, M. and O. T. Solbrig 1972. The concept of r- and K-selection: evidence from wildflowers and some theoretical considerations. *Amer. Nat.* **106**, 14–31.

Gale, J. S. 1980. *Population genetics.* London: L. Blackie.

Galen, C. and P. G. Kevan 1980. Scent and color, floral polymorphisms and pollination biology in *Polemonium viscosum* Nutt. *Am. Midl. Nat.* **104**, 281–89.

Galen, C. and P. G. Kevan 1983. Bumblebee foraging and floral scent dimorphisms: *Bombus Kirbyellis* Curtis (Hymenoptera: Apidae) and *Polemonium viscosum* Nutt. (Polemoniaceae). *Can. J. Zool.* **61**, 1207–13.

Galil, J. and D. Eisikowitch 1968. On the pollination ecology of *Ficus sycamorus* in East Africa. *Ecology* **49**, 259–69.

Galil, J. and D. Eisikowitch 1969. Further studies on the pollination ecology of *Ficus sycamorus* L. *Tijdschr. Ent.* **112**, 1–13.

Ganders, F. R. 1974. Disassortative pollination in the distylous plant *Jepsonia heterandra*. *Can. J. Bot.* **52**, 2401–6.

Ganders, F. R. 1975. Mating patterns in self-incompatible distylous populations of *Amsinckia* (Boraginaceae). *Can. J. Bot.* **53**, 773–9.

Ganders, F. R. 1976. Pollen flow in distylous populations of *Amsinckia* (Boraginaceae). *Can. J. Bot.* **54**, 2530–5.

Ganders, F. R. 1979. The biology of heterostyly. *NZ. J. Bot.* **17**, 607–35.

Ganders, F. R., K. Carey and A. J. F. Griffiths 1977. Natural selection for a fruit dimorphism in *Plectritis congesta* (Valerianaceae). *Evolution* **31**, 873–81.

Gastel, A. J. G. van 1972. Spontaneous stylar part mutations in *Nicotiana alata* Link and Otto. *Incomp. Newslett. Assoc. EURATOM-ITAL, Wageningen* **1**, 12–13.

Gastel, A. J. G. van 1974. Radiogenetics of self-incompatibility. *Ann. Rep. Comm. Europ. Comm., Progr. Biol. Health Protection* (1974).

Gastel, A. J. G. van and D. de Nettancourt 1974. The effects of different mutagens on self-incompatibility in *Nicotiana alata* Link and Otto. 1. Chronic gamma radiation. *Radiation Bot.* **14**, 43–50.
Gastel, A. J. G. van and D. de Nettancourt 1975. The effects of different mutagens on self-incompatibility in *Nicotiana alata* Link and Otto. II. Acute irradiations with X-rays and fast neutrons. *Heredity* **34**, 381–92.
Gentry, A. H. 1974. Flowering phenology and diversity in tropical Bignoniaceae. *Biotropica* **6**, 64–8.
Gentry, A. H. 1976. Bignoniaceae of southern Central America: distribution and ecological specificity. *Biotropica* **8**, 117–31.
Gerstel, D. U. 1950. Self-incompatibility studies in Guayule. II. Inheritance. *Genetics* **35**, 482–506.
Gerwitz, A. and G. J. Faulkner 1972. *National Vegetable Research Station 22nd Annual Report 1971*, **32**. Warwick: Wellesbourne.
Ghosh, S. and K. R. Shivanna 1980. Pollen-pistil interaction in *Linum grandiflorum*. Scanning electron microscopic observations and proteins of the stigma surface. *Planta* **149**, 257–61.
Gianordoli, M. 1974. A cytological investigation on gametes and fecundation among *Cephalotaxus drupacea*. In. *Fertilization in higher plants*, H. F. Linskens (ed.), 221–32. Amsterdam: North Holland.
Givnish, T. J. 1980. Ecological constraints on the evolution of breeding systems in seed plants: dioecy and dispersal in gymnosperms. *Evolution* **34**, 959–72.
Gleaves, J. T. 1973. Gene flow mediated by wind borne pollen. *Heredity* **31**, 355–66.
Godley, E. J. 1964. Breeding systems in New Zealand plants. 3. Sex ratios in some natural populations. *NZ. J. Bot.* **2**, 205–12.
Godley, E. J. 1975. Flora and vegetation. In *Biogeography and ecology in New Zealand*, G. Kuschel (ed.), 177–229. The Hague: W. Junk.
Godley, E. J. 1979. Flower biology in New Zealand. *NZ. J. Bot.* **17**, 441–66.
Golynskaya, E. L., N. V. Bashkirova and N. N. Tomchuk 1976. Phytohaemagglutanins of the pistil in *Primula* as possible proteins of generative incompatibility. *Sov. Pl. Physiol.* **23**, 69–77.
Grant, V. 1952. Isolation and hybridisation between *Aquilegia formosa* and *A. pubescens*. *Aliso* **2**, 341–60.
Grant, V. 1963. *The origin of adaptations*. New York: Columbia University Press.
Grant, V. 1981. *Plant speciation*, 2nd edn. New York: Columbia University Press.
Grant, V. and K. A. Grant 1965. *Flower pollination in the Phlox family*. New York: Columbia University Press.
Gray, A. J. 1982. In *Institute of Terrestrial Ecology Annual Report for 1981* (no page numbers).
Grewal, M. S. and J. R. Ellis 1972. Sex determination in *Potentilla fruticosa*. *Heredity* **29**, 359–62.
Griffiths, D. J. 1950. The liability of seed-crops of Perennial Rye Grass (*Lolium perenne*) to contamination by wind-borne pollen. *J. Agric. Sci.* **40**, 19–38.
Grime, J. P. 1973. Competition and diversity in herbaceous vegetation. *Nature* **244**, 311.
Grubb, P. J. 1977. The maintenance of species-richness in plant communities: the importance of regeneration niche. *Biol. Rev.* **52**, 107–45.
Gustafsson, A. 1944. The constitution of the *Rosa canina* complex. *Hereditas* **30**, 405–28.
Gustafsson, A. 1946–7. *Apomixis in higher plants*. I–III. *Lunds Univ. Arsskr.* **42**, 1–67, 43, 69–179, 183–370.
Gustafsson, A. and A. Håkansson 1942. Meiosis in some rose hybrids. *Bot. Not.* (1942), 331–42.

Hagerup, O. 1944. On fertilization, polyploidy, and haploidy in *Orchis maculatus*. *Dansk. Bot. Ark.* **11**, 1–26.

Hagerup, O. 1945. Facultative parthenogenesis and haploidy in *Epipactis latifolia*. *Kl. Danske vidensk Selsk.* **19**, 1–13.

Hagerup, O. 1947. The spontaneous formation of haploid, polyploid and aneuploid embryos in some orchids. *Kl. Danske Vidensk Selsk.* **20**, 1–22.

Hainsworth, F. R. and L. L. Wolf 1972. Energetics of nectar extraction in a small, high altitude, tropical hummingbird, *Selasphorus flammula*. *J. Comp. Physiol.* **80**, 377–87.

Haldane, J. B. S. 1922. Sex-ratio and unisexual sterility in hybrid animals. *J. Genet.* **12**, 101–9.

Hamrick, J. L., Y. B. Linhart and J. B. Mitton 1979. Relationships between life history characteristics and electrophoretically detectable genetic variation in plants. *Ann. Rev. Ecol. Syst.* **10**, 173–200.

Handel, S. N. 1982. Dynamics of gene flow in an experimental garden of *Cucumis melo* (Cucurbitaceae). *Am. J. Bot.* **69**, 1538–46.

Handel, S. N. 1983a. Contrasting gene flow patterns and genetic subdivision in adjacent populations of *Cucumis sativus* (Cucurbitaceae). *Evolution* **37**, 760–71.

Handel, S. N. 1983b. Pollination ecology, plant population structure, and gene flow. In *Pollination biology*, L. Real (ed.), 163–211. New York: Academic Press.

Handel, S. N. and J. Le Vie Mishkin 1985. Temporal shifts in gene flow and seed set: evidence from an experimental population of *Cucumis sativus*. *Evolution* **38** (in press).

Hanna, W., J. Powell, J. Millot and G. Burton 1973. Cytology of obligate sexual plants in *Panicum maximum* Jacq. and their use in controlled hybrids. *Crop. Sci.* **13**, 695–7.

Harberd, D. J. 1961. Observations on population structure and longevity of *Festuca rubra* L. *New Phytol.* **60**, 184–206.

Harberd, D. J. 1962. Some observations on natural clones in *Festuca ovina*. *New Phytol.* **61**, 85–100.

Harberd, D. J. 1967. Observations on natural clones of *Holcus mollis*. *New Phytol.* **66**, 401–8.

Harberd, D. J. and M. Owen 1969. Some experimental observations on the clone structure of a natural population of *Festuca rubra* L. *New Phytol.* **68**, 93–104.

Harding, J., C. B. Mankinen and M. H. Elliott 1974. Genetics of *Lupinus*. VII. Outcrossing, autofertility, and variability in natural populations of the *nanus* group. *Taxon* **23**, 729–38.

Harley, R. M. 1972. *Mentha* L. in *Flora Europaea* **3**, T. G. Tutin *et al.* (eds), 183–6. Cambridge: Cambridge University Press.

Harmer, R. and J. A. Lee 1978a. The growth and nutrient content of *Festuca vivipara* (L.) Sm. plantlets. *New Phytol.* **80**, 99–106.

Harmer, R. and J. A. Lee 1978b. The germination and viability of *Festuca vivipara* (L.) Sm. plantlets. *New Phytol.* **81**, 745–51.

Harper, J. L. 1977. *The population biology of plants*. London: Academic Press.

Haskell, G. 1966. The history, taxonomy and breeding system of apomictic British *Rubi*. In *Reproductive biology and taxonomy of vascular plants*, J. G. Hawkes (ed.), 141–51. Oxford: Pergamon.

Heinrich, B. 1972a. Temperature regulation in the bumblebee *Bombus vagans*: a field study. *Science* **175**, 185–7.

Heinrich, B. 1972b. Energetics of temperature regulation and foraging in a bumblebee, *Bombus terricola* Kirby. *J. Comp. Physiol.* **77**, 49–64.

Heinrich, B. 1975. Energetics of pollination. *Ann. Rev. Ecol. Syst.* **6**, 139–70.

Heinrich, B. 1976. The foraging specializations of individual bumblebees. *Ecol. Monogr.* **46**, 105–28.
Heinrich, B. 1979a. Resource heterogeneity and patterns of movement in foraging bumblebees. *Oecologia* **40**, 235–45.
Heinrich, B. 1979b. *Bumblebee economics.* Cambridge, Mass: Harvard University Press.
Heinrich, B. 1979c. "Majoring" and "minoring" by foraging bumblebees, *Bombus vagans*: An experimental analysis. *Ecology* **60**, 245–55.
Heinrich, B. and P. H. Raven 1972. Energetics and pollination ecology. *Science* **176**, 597–602.
Heitz, B. 1973. Hétérostylie et spéciation dans le groupe *Linum perenne*. *Ann. Sci. Nat. Bot. Biol. Veg.* **14**, 385–405.
Heslop-Hárrison, J. 1957. The experimental modification of sex expression in flowering plants. *Biol. Rev. Camb. Phil. Soc.* **32**, 38–90.
Heslop-Harrison, J. 1975a. The physiology of the pollen grain surface. *Proc. R. Soc. Lond. B* **190**, 275–99.
Heslop-Harrison, J. 1975b. Incompatibility and the pollen stigma interaction. *Ann. Rev. Pl. Physiol.* **26**, 403–25.
Heslop-Harrison, J. 1978. Genetics and physiology of angiosperm incompatibility systems. *Proc. R. Soc. Lond. B* **202**, 73–92.
Heslop-Harrison, J. 1979a. Aspects of the structure, cytochemistry and germination of the pollen of rye (*Secale cereale* L.). *Ann. Bot.* **44** (Suppl.), 1–47.
Heslop-Harrison, J. 1979b. An interpretation of the hydrodynamics of pollen. *Am. J. Bot.* **66**, 737–43.
Heslop-Harrison, J. 1982. Pollen-stigma interaction and cross-incompatibility in the grasses. *Science* **215**, 1358–64.
Heslop-Harrison, J. 1983. Self-incompatibility: phenonemology and physiology. *Proc. R. Soc. Lond. B* **218**, 371–95.
Heslop-Harrison, J. and Y. Heslop-Harrison 1982a. Pollen-stigma interaction in the Leguminosae: constituents of the stylar fluid and stigma secretion of *Trifolium pratense* L. *Ann. Bot.* **49**, 729–35.
Heslop-Harrison, J. and Y. Heslop-Harrison 1982b. The pollen-stigma interaction in the grasses. 4. An interpretation of the self-incompatibility response. *Act. Bot. Neerl.* **31**, 429–39.
Heslop-Harrison, J., R. B. Knox and Y. Heslop-Harrison 1974. Pollen-wall proteins: exine-held fractions associated with the incompatibility response in Cruciferae. *Theoret. Appl. Genet.* **44**, 133–7.
Heslop-Harrison, Y. 1953. *Nuphar intermedia* Ledeb., a presumed relict hybrid, in Britain. *Watsonia* **3**, 7–25.
Heslop-Harrison, Y., J. Heslop-Harrison and K. R. Shivanna 1981. Heterostyly in *Primula*. 1. Fine-structural and cytochemical features of the stigma and style in *Primula vulgaris* Huds. *Protoplasma* **107**, 171–87.
Hickey, L. J. and J. A. Doyle 1977. Early Cretaceous fossil evidence for Angiosperm evolution. *Bot. Rev.* **43** (1), 3–93.
Hildebrand, F. 1863. *De la variation des animaux et des plantes a l'état domestique.* Paris: C. Reinwald.
Ho, T. Y. and M. D. Ross 1973. Maintenance of male sterility in plant populations. II. Heterotic models. *Heredity* **31**, 282–6.
Hogenboom, N. G. 1972. Breaking breeding barriers in *Lycopersicon*. 1, 2, 3, 4, 5. *Euphytica* **21**, 221–7, 228–43, 244–56, 397–404, 405–14.
Hooper, J. E. and S. J. Peloquin 1968. X-ray inactivation of the stylar component of the self-incompatibility reaction in *Lilium longiflorum*. *Can. J. Gen. Cytol.* **10**, 941–4.

Horovitz, A. and A. Beiles 1980. Gynodioecy as a possible population strategy for increasing reproductive output. *Theoret. Appl. Genet.* **57**, 11–15.

Howlett, B. M., R. B. Knox, J. H. Paxton and J. Heslop-Harrison 1975. Pollen-wall proteins: physicochemical characterisation and rôle in self-incompatibility in *Cosmos bipinnatus*. *Proc. R. Soc. Lond. B* **188**, 167–82.

Hughes, J. and A. J. Richards 1985. The inheritance of isozymes in sexual *Taraxacum*. *Heredity* **54**, 245–49.

Hughes, M. B. and E. B. Babcock 1950. Self-incompatibility in *Crepis foetida* L. subsp. *rhoedaifolia*. *Genetics* **35**, 570–88.

Humphreys, M. O. and J. S. Gale 1974. Variation in wild populations of *Papaver dubium*. VIII. The mating system. *Heredity* **33**, 33–42.

Hvid, S. and G. Nielsen 1977. Esterase isozyme variants in barley. *Hereditas* **87**, 155–62.

Ibrahim, H. 1979. *Population studies in Primula veris L. and P. vulgaris Huds.* Ph.D. thesis, University of Newcastle upon Tyne.

Imam, A. G. and R. W. Allard 1965. Population studies in predominantly self-pollinated species. VI. Genetic variability between and within natural populations of wild oats *Avena fatua* L. from different habitats in California. *Genetics* **51**, 49–62.

Ingram, R., J. Weir and R. J. Abbott 1980. New evidence concerning the origin of inland radiate groundsel, *S. vulgaris* L. var. *hibernicus* Syme. *New Phytol.* **84**, 543–6.

Inouye, D. W. 1980. The effect of proboscis and corolla tube lengths on patterns and rates of flower visitation by bumblebees. *Oecologia* **45**, 197–201.

Jacob, F. and J. Monod 1961. Genetic regulatory mechanisms in the synthesis of proteins. *J. Mol. Biol.* **3**, 318–56.

Jaenicke, J. and R. K. Selander 1979. Evolution and ecology of parthenogenesis in earthworms. *Am. Zool.* **19**, 729–37.

Jain, S. K. 1978. Breeding system in *Limnanthes alba*: several alternative procedures. *Am. J. Bot.* **65**, 272–5.

Jain, S. K. 1979. Estimation of outcrossing rates: some alternative procedures. *Crop. Sci.* **19**, 23–6.

Janick, J. and E. C. Stevenson 1955. Genetics of the monoecious character in spinach. *Genetics* **40**, 429–37.

Janzen, D. H. 1971. Euglossine bees as long-distance pollinators of tropical plants. *Science* **171**, 203–5.

Janzen, D. H. 1977. What are dandelions and aphids? *Am. Nat.* **111**, 586–9.

Jinks, J. L. and K. Mather 1955. Stability in development of heterozygotes and homozygotes. *Proc. R. Soc. Lond. B* **143**, 561–78.

Johnson, G. B. 1977. Assessing electrophoretic similarity: the problem of hidden heterogeneity. *Ann. Rev. Ecol. Syst.* **8**, 309–28.

Johri, B. M. 1981. Transfer cells: their rôle in reproductive structures of Angiosperms. *Abstracts, XIII International Botanical Congress, Sydney* 61.

Jordan, A. 1864. *Diagnoses d'espèces nouvelles ou méconnues pour servir de matériaux a une flore réformée de la France et des Contrées voisines.* Paris: Savy.

Kakizaki, Y. 1930. Studies on the genetics and physiology of self- and cross-incompatibility in the common cabbage. *Jap. J. Bot.* **5**, 133–208.

Kandeaki, G. V. 1976. Remote hybridization and the phenomenon of pseudogamy. In *Apomixis and breeding*, S. S. Khokhlov (ed.), 179–89. New Delhi: Amerind.

Kannenberg, L. W. and R. W. Allard 1967. Population studies in predominantly self-pollinated species. VIII. Genetic variability in the *Festuca microstachys* complex. *Evolution* **21**, 227–40.
Kanno, T. and K. Hinata 1969. An electron microscope study of the barrier against pollen tube growth in self-incompatible *Cruciferae*. *Pl. Cell Physiol.* **10**, 213–16.
Kaur, A., C. D. Ha, K. Jong, V. E. Sands, H. Chan, E. Soepadmo and P. S. Ashton 1978. Apomixis may be widespread among trees of the climax rain forest. *Nature* **271**, 440–1.
Kay, Q. O. N. 1978. The role of preferential and assortative pollination in the maintenance of flower colour polymorphisms. In *The pollination of flowers by insects*, A. J. Richards (ed.), 175–90. London: Academic Press.
Kay, Q. O. N. 1982. Intraspecific discrimination by pollinators and its role in evolution. In *Pollination and evolution*, J. A. Armstrong, J. M. Powell and A. J. Richards (eds), 9–28. Publ. Royal Botanic Gardens, Sydney.
Kay, Q. O. N., A. J. Lack, F. C. Bamber and C. R. Davies 1984. Differences in floral morphology, nectar production and insect visits in a dioecious species, *Silene dioica*. *New Phytol.* **98**, 515–29.
Kerster, H. W. 1964. Neighbourhood size in the rusty lizard, *Sceloporus olivaceus*. *Evolution* **18**, 445–57.
Kerster, H. W. and D. A. Levin 1968. Neighbourhood size in *Lithospermum carolinense*. *Genetics* **60**, 577–87.
Kevan, P. G. 1978. Floral coloration, its colorimetric analysis and significance in anthecology. In *The pollination of flowers by insects*, A. J. Richards (ed.), 51–78. London: Academic Press.
Kevan, P. G. 1984. Pollination by Animals and Angiosperm Biosystematics. In *Plant biosystematics*, W. F. Grant (ed.), 271–92. Toronto: Academic Press.
Kevan, P. G. and H. G. Baker 1983. Insects as flower visitors and pollinators. *Ann. Rev. Entomol.* **28**, 407–53.
Kevan, P. G. and A. J. Lack (in press). Pollination in a cryptically dioecious plant *Decaspermum parvifolium* (Lam.) Indonesia. *Biol. J. Linn. Soc.*
Kimura, M. 1979. The neutral theory of molecular evolution. *Scientific American* **251** (5), 94–105.
Knox, R. B. 1967. Apomixis: seasonal and population differences in a grass. *Science* **157**, 325–6.
Knox, R. B. and J. Heslop-Harrison 1963. Experimental control of aposporous apomixis in a grass of the Andropogoneae. *Bot. Not.* **116**, 127–41.
Knox, R. B., R. Willing and A. E. Ashford 1972. Role of pollen-wall proteins as recognition substances in interspecific incompatibility in poplars. *Nature* **237**, 381–3.
Knox, R. B., A. E. Clarke, S. Harrison, P. Smith and J. J. Marchalonis 1976. Cell recognition in plants: determinants of the stigma surface and their pollen interactions. *Proc. Nat. Acad. Sci. USA* **73**, 2788–92.
Knuth, P. 1906–9. *Handbook of flower pollination*. Transl. J. R. Ainsworth Davis (3 vols., I, 1906, II, 1908, III, 1909). Oxford: Oxford University Press.
Kolreuter, J. G. 1763. Vorläufige Nachricht von einigen das Geschlecht der Pflanzen betreffenden Versuchen und Beobachtungen, nebst Fortsetzungen 1, 2 v. 3, 266. *Ostwald's Klassiker*, 41. Leipzig: Engelmann.
Krebs, J. R. 1978. Optimal foraging. In *Behavioural ecology: an evolutionary approach*, J. R. Krebs and N. B. Davies (eds). Oxford: Blackwell Scientific.
Krohne, D. T., I. Baker and H. G. Baker 1980. The maintenance of the gynodioecious breeding system in *Plantago lanceolata* L. *Am. Midl. Nat.* **103**, 269–79.
Kuhn, E. 1939. Selbstbestäubungen subdiöcischer Blütenpflanzen, ein neuer

beweis für die genetische theorie der geschlechtsbestimmung. *Planta* **30**, 457–70.

Kullenberg, B. 1956. On the scents and colours of *Ophrys* flowers and their specific pollinators among the Aculeate Hymenoptera. *Svensk Bot. Tidskr.* **50**, 25–46.

Kung, S. 1977. Expression of chloroplast genomes in higher plants. *Ann. Rev. Pl. Physiol.* **28**, 401–37.

Lamarck, J. B. 1809. *Philosophie Zoologique*. Transl. H. Elliot as *Zoological Philosophy* (1914), London: Macmillan.

Law, R., R. E. D. Cook and R. J. Manlove 1983. The ecology of flower and bulbil production in *Polygonum viviparum*. *Nord. J. Bot.* **3**, 559–65.

Lawrence, M. J., C. H. Fearon, M. A. Cornish and M. D. Hayward 1983. The genetical control of self-incompatibility in rye grasses. *Heredity* **51**, 461–6.

Lawrence, M. J., D. F. Marshall, V. E. Curtis and C. H. Fearon 1985. Gametophytic self-incompatibility re-examined: a reply. *Heredity* **54**, 131–8.

Lawton, J. H. 1973. The energy cost of "food-gathering". In *Resources and population*, B. Benjamin, P. R. Cox and J. Peel (eds), 59–76. London: Academic Press.

Lehmann, E. 1926. The heredity of self-sterility in *Veronica syriaca*. *Mem. Hort. Soc. NY* **3**, 313–20.

Levin, D. A. 1969. The effect of corolla color and outline on interspecific pollen flow in *Phlox*. *Evolution* **23**, 444–5.

Levin, D. A. 1972. Low frequency disadvantage in the exploitation of pollinators by corolla variants in *Phlox*. *Am. Nat.* **104**, 455–67.

Levin, D. A. 1975. Gametophytic selection in *Phlox*. In *Gamete competition in plants and animals*, D. L. Mulcahy (ed.), 207–17. Amsterdam: North Holland.

Levin, D. A. 1978a. Pollinator behaviour and the breeding structure of plant populations. In *The pollination of flowers by insects*, A. J. Richards (ed.), 133–50. London: Academic Press.

Levin, D. A. 1978b. Genetic variation in annual *Phlox*: self-compatible versus self-incompatible species. *Evolution* **32**, 245–63.

Levin, D. A. 1979. Pollinator foraging behaviour: Genetic implications for plants. In *Topics in plant population biology*, O. T. Solbrig, S. Jain, G. B. Johnson and P. H. Raven (eds), 131–53. New York: Columbia University Press.

Levin, D. A. 1984a. Inbreeding depression and proximity-dependent crossing success in *Phlox drummondii*. *Evolution* **38**, 116–27.

Levin, D. A. 1984b. Genetic variation and divergence in a disjunct *Phlox*. *Evolution* **38**, 223–5.

Levin, D. A. and D. E. Berube 1972. *Phlox* and *Colias*: the efficiency of a pollination system. *Evolution* **26**, 242–50.

Levin, D. A. and H. W. Kerster 1968. Local gene dispersal in *Phlox*. *Evolution* **22**, 130–9.

Levin, D. A. and H. W. Kerster 1969a. The dependence of bee-mediated pollen and gene dispersal upon plant density. *Evolution* **23**, 560–71.

Levin, D. A. and H. W. Kerster 1969b. Density-dependent gene dispersal in *Liatris*. *Am. Nat.* **103**, 61–74.

Levin, D. A. and H. W. Kerster 1971. Neighbourhood structure in plants under diverse reproductive methods. *Am. Nat.* **105**, 345–54.

Levin, D. A. and H. W. Kerster 1973. Assortative pollination for stature in *Lythrum salicaria*. *Evolution* **27**, 144–52.

Levin, D. A. and H. W. Kerster 1974. Gene flow in seed plants. *Evol. Biol.* **7**, 139–220.

Levin, D. A. and L. Watkins 1984. Assortative mating in *Phlox. Heredity* **53**, 595–602.

Lewis, D. 1941. Male sterility in natural populations of hermaphrodite plants. *New Phytol.* **40**, 56–63.

Lewis, D. 1942a. The evolution of sex in flowering plants. *Biol. Rev.* **17**, 46–67.

Lewis, D. 1942b. The physiology of incompatibility in plants. I. The effect of temperature. *Proc. R. Soc. Lond. B* **131**, 13–26.

Lewis, D. 1943. The physiology of incompatibility in plants. II. *Linum grandiflorum. Ann. Bot.* II **7**, 115–22.

Lewis, D. 1947. Competition and dominance of incompatibility alleles in diploid pollen. *Heredity* **1**, 85–108.

Lewis, D. 1949a. Incompatibility in flowering plants. *Biol. Rev.* **24**, 427–69.

Lewis, D. 1949b. Structure of the incompatibility gene. II. Induced mutation rate. *Heredity* **3**, 339–55.

Lewis, D. 1952. Serological reactions of pollen incompatibility substances. *Proc. R. Soc. Lond. B* **140**, 127–35.

Lewis, D. 1954. Comparative incompatibility in Angiosperm and Fungi. *Adv. Gen.* **6**, 235–85.

Lewis, D. 1955. Sexual incompatibility. *Sci. Progress* **172**, 593–605.

Lewis, D. 1960. Genetic control of specificity and activity of the S antigen in plants. *Proc. R. Soc. Lond. B* **151**, 468–77.

Lewis, D. 1964. A protein dimer hypothesis on incompatibility. *Proc. 11th Int. Congr. Genet.* The Hague 1963. In *Genetics today*, S. J. Geerts (ed.) **3**, 656–63.

Lewis, D. and L. K. Crowe 1953. Theory of revertible mutation. *Nature* **171**, 501.

Lewis, D. and L. K. Crowe 1954. Structure of the incompatibility gene IV. Types of mutation in *Prunus avium* L. *Heredity* **8**, 357–63.

Lewis, D. and L. K. Crowe 1956. The genetics and evolution of gynodioecy. *Evolution* **10**, 115–25.

Lewis, D., S. Burrace and D. Walls 1967. Immunological reactions of single pollen grains, electrophoresis and enzymology of pollen protein exudates. *J. Exp. Bot.* **18**, 371–8.

Lewis, H. 1962. Catastrophic selection as a factor in speciation. *Evolution* **16**, 257–71.

Liljefors, A. 1955. Cytological studies in *Sorbus. Act. Hort. Berg.* **17**, 47–113.

Linder, R. and H. F. Linskens 1972. Evolution des acides aminés dans le style d'*Oenothera missouriensis* vierge, autopollinisé et xénopolinisé. *Theoret. Appl. Genet.* **42**, 125–9.

Linhart, Y. B. 1973. Ecological and behavioural determinants of pollen dispersal in hummingbird-pollinated *Heliconia. Am. Nat.* **107**, 511–23.

Linhart, Y. B., J. B. Mitton, D. M. Bowman, K. B. Sturgeon and J. L. Mamrick 1979. Genetic aspects of fertility differentials in ponderosa pine. *Genet. Res.* **23**, 237–42.

Linskens, H. F. 1960. Zurfrage der Entstehung der Abwehrkörper bei der Inkompatilibätsreaktion von *Petunia.* III. Mitteilung: Serologische teste mit Leitgewebs – und der Pollen extrakten. *Z. Bot.* **48**, 126–35.

Linskens, H. F. 1967. Pollen. In *Encyclopedia of plant physiology*, W. Ruhland (ed.), vol. XXVIII. Berlin: Springer.

Linskens, H. F. 1975. Incompatibility in *Petunia. Proc. R. Soc. Lond. B* **188**, 299–311.

Linskens, H. F., J. A. M. Schrauwen and M. van der Donk 1960. Überwindung der Selbstinkompatibilität durch Röntgenbestrahlung des Griffels. *Naturwiss.* **46**, 547.

Lloyd, D. G. 1969. Petal colour polymorphism in *Leavenworthia* (Cruciferae). *Contrib. Gray Herbarium Harvard* **198**, 9–40.

Lloyd, D. G. 1972a. Breeding systems in *Cotula* L. (Compositae, Anthemideae). I. The array of monoclinous and diclinous systems. *New Phytol.* **71**, 1181–94.
Lloyd, D. G. 1972b. Breeding systems in *Cotula* L. (Compositae, Anthemideae). 2. Monoecious populations. *New Phytol.* **71**, 1195–202.
Lloyd, D. G. 1974a. Female-predominant sex ratios in angiosperms. *Heredity* **32**, 35–44.
Lloyd, D. G. 1974b. Theoretical sex ratios of dioecious and gynodioecious angiosperms. *Heredity* **32**, 11–34.
Lloyd, D. G. 1975a. The maintenance of gynodioecy and androdioecy in Angiosperms. *Genetica* **45**, 325–39.
Lloyd, D. G. 1975b. Breeding systems in *Cotula*. III. Dioecious populations. *New Phytol.* **74**, 109–23.
Lloyd, D. G. 1976. The transmission of genes via pollen and ovules in gynodioecious angiosperms. *Theoret. Pop. Biol.* **9**, 299–316.
Lloyd, D. G. 1979. Evolution towards dioecy in heterostylous populations. *Pl. Syst. Evol.* **131**, 71–80.
Lloyd, D. G. 1980a. Sexual strategies in plants. I. An hypothesis of serial adjustment of maternal investment during one reproductive session. *New Phytol.* **85**, 265–73.
Lloyd, D. G. 1980b. Benefits and handicaps of sexual reproduction. In *Evolutionary biology*, M. K. Hecht, W. C. Steere and B. Wallace (eds), 69–110. New York: Plenum Press.
Lloyd, D. G. 1983. Evolutionary stable sex ratios and sex allocations. *J. Theoret. Biol.* **105**, 525–39.
Lloyd, D. G. and C. J. Webb 1977. Secondary sex characters in plants. *Bot. Rev.* **43**, 177–216.
Lord, E. M. 1981. Cleistogamy: a tool for the study of floral morphogenesis, function and evolution. *Bot. Rev.* **47**, 421–49.
Löve, A. 1944. Cytogenetic studies on *Rumex* subgenus Acetosella. *Hereditas* **30**, 1–136.
Löve, A. and N. Sarker 1956. Cytotaxonomy and sex determination of *Rumex paucifolius*. *Can. J. Bot.* **34**, 261–8.
Lovett Doust, L. 1981. Population dynamics and local specialization in a clonal perennial (*Ranunculus repens*). 1. The dynamics of ramets in contrasting habitats. *J. Ecol.* **69**, 743–55.
Lovett Doust, J. and P. B. Cavers 1982. Sex and gender dynamics in jack-in-the-pulpit *Arisaema triphyllum* (L.) Schott (Araceae). *Ecology* **63**, 797–808.
Lovett Doust, L. and J. Lovett Doust 1982. The battle strategies of plants. *New Scientist*, 8 July 1982, 81–4.
Lundqvist, A. 1956. Self-incompatibility in rye. I. Genetic control in the diploid. *Hereditas* **42**, 293–348.
Lundqvist, A. 1960. The origin of self-compatibility in rye. *Hereditas* **46**, 1–19.
Lundqvist, A. 1961. A rapid method for the analysis of incompatibilities in grasses. *Hereditas* **47**, 705–7.
Lundqvist, A. 1962. The nature of the two-loci incompatibility system in grasses. I. The hypothesis of a duplicative origin. *Hereditas* **48**, 153–68.
Lundqvist, A. 1964. The nature of the two-loci incompatibility system in grasses. IV. Interaction between the loci in relation to pseudocompatibility in *Festuca pratensis* Huds. *Hereditas* **52**, 221–34.
Lundqvist, A. 1968. The mode of origin of self-fertility in grasses. *Hereditas* **59**, 413–26.
Lundqvist, A., U. Østerbye, K. Larsen and I. Linde-Laursen 1973. Complex

self-incompatibility systems in *Ranunculus acris* L. and *Beta vulgaris* L. *Hereditas* **74**, 161–8.

Lyman, J. C. and N. C. Ellstrand 1984. Clonal diversity in *Taraxacum officinale* (Compositae), an apomict. *Heredity* **50**, 1–10.

MacArthur, R. H. and E. O. Wilson 1967. *The theory of island biogeography*. Princeton: Princeton University Press.

McCraw, J. M. and W. Spoor 1983a. Self-incompatibility in *Lolium* species. I. *Lolium rigidum* Gaud. and *L. multiflorum* L. *Heredity* **50**, 21–7.

McCraw, J. M. and W. Spoor 1983b. Self-incompatibility in *Lolium* species. 2. *Lolium perenne* L. *Heredity* **50**, 29–33.

McCusker, A. 1962. Gynodioecism in *Leucopogon melaleucoides*. *Proc. Linn. Soc. NSW* **87**, 286–9.

Macior, L. W. 1966. Foraging behaviour of *Bombus* (Hymenoptera: Apidae) in relation to *Aquilegia* pollination. *Am. J. Bot.* **53**, 302–9.

Macior, L. W. 1982. Plant community and pollinator dynamics in the evolution of pollination mechanisms in *Pedicularis* (Scrophulariaceae). In *Pollination and evolution*, J. A. Armstrong, J. M. Powell and A. J. Richards (eds), 29–45. Sydney: Royal Botanic Gardens.

Macior, L. W. 1983. The pollination dynamics of sympatric species of *Pedicularis* (Scrophulariaceae). *Am. J. Bot.* **70**, 844–53.

McLean, R. C. and W. R. Ivimey-Cook 1956. *Textbook of theoretical botany*, vol. 2. London: Longman.

McNeilly, T. 1968. Evolution in closely adjacent plant populations. III. *Agrostis tenuis* on a small copper mine. *Heredity* **23**, 99–108.

McNeilly, T. and J. A. Antonovics 1968. Evolution in closely adjacent plant populations. IV. Barriers to gene flow. *Heredity* **23**, 205–18.

Maheshwari, P. 1949. The male gametophyte of Angiosperms. *Bot. Rev.* **15** (1), 1–75.

Maheshwari, P. 1950. *An introduction to the embryology of angiosperms*. New York: McGraw-Hill.

Maheshwari, P. and N. S. Rangaswamy 1965. Embryology in relation to genetics. In *Advances in botanical research*, vol. 2, R. D. Preston (ed.), 219–321. London: Academic Press.

Malecka, J. 1965. Embryological studies in *Taraxacum palustre*. *Act. Biol. Crac.* **8**, 223–35.

Malecka, J. 1967. Cytoembryological studies in *Taraxacum scanicum* Dt. *Act. Biol. Crac.* **10**, 195–206.

Malecka, J. 1971. Cytotaxonomic and embryological investigations on a natural hybrid between *Taraxacum kok-saghyz* Rodin and *T. officinale* Web. and their putative parent species. *Act. Biol. Crac.* **14**, 179–96.

Malecka, J. 1973. Problems in the mode of reproduction in microspecies of *Taraxacum* section Palustria Dahlstedt. *Act. Biol. Crac. ser. bot.* **16**, 37–84.

Manton, I. 1950. *Problems of cytology and evolution in the pteridophyta*. Cambridge: Cambridge University Press.

Marshall, D. F. and R. J. Abbott 1984a. Polymorphism for outcrossing frequency at the ray floret locus in *Senecio vulgaris* L. II. Confirmation. *Heredity* **52**, 331–6.

Marshall, D. F. and R. J. Abbott 1984b. Polymorphism for outcrossing frequency at the ray floret locus in *Senecio vulgaris* L. III. Causes. *Heredity* **53**, 145–50.

Marshall, D. R. and R. W. Allard 1970. Maintenance of isozyme polymorphism in natural populations of *Avena barbata*. *Genetics* **66**, 393–99.

Marshall, D. R. and A. H. D. Brown 1974. Estimation of the level of apomixis in plant populations. *Heredity* **32**, 321–33.

Marshall, D. R. and A. H. D. Brown 1981. The evolution of apomixis. *Heredity* **47**, 1–15.

Mather, K. 1950. The genetical architecture of heterostyly in *Primula sinensis. Evolution* **4**, 340–52.

Mather, K. and D. De Winton 1941. Adaptation and counteradaptation of the breeding system in *Primula. Ann. Bot.* II **5**, 297–311.

Mattson, O. 1983. The significance of exine oils in the initial interaction between pollen and stigma in *Armeria maritima*. In *Pollen biology and applications for plant breeding*, D. L. Mulcahy and E. Ottaviano (eds), 257–67. New York: Elsevier.

Maynard Smith, J. 1971. The origin and maintenance of sex. In *Group selection*, G. C. Williams (ed.), 163–75. Chicago: Aldine Atherton.

Maynard Smith, J. 1978. *The evolution of sex*. Cambridge: Cambridge University Press.

Mayo, O. and D. L. Hayman 1968. The maintenance of two loci systems of gametophytically determined self-incompatibility. *Proc. 12th Int. Congr. Genet.*, C. Oshima (ed.), 331.

Meagher, T. R. 1980. Population biology of *Chamaelirium luteum*, a dioecious lily. I. Spatial distributions of males and females. *Evolution* **34**, 1127–37.

Meagher, T. R. 1981. Population biology of *Chamaelirium luteum*, a dioecious lily. II. Mechanisms governing sex ratios. *Evolution* **35**, 557–67.

Meeuse, A. D. J. 1973. Anthecology and Angiosperm evolution. In *Taxonomy and ecology*, V. H. Heywood (ed.), 189–200. London: Academic Press.

Meeuse, A. D. J. 1978. Entomophily in *Salix*: theoretical considerations. In *The pollination of flowers by insects*, A. J. Richards (ed.), 47–50. London: Academic Press.

Meeuse, A. D. J. 1978. The physiology of some sapromyophilous flowers. In *The pollination of flowers by insects*, A. J. Richards (ed.), 97–104. London: Academic Press.

Menzel, M. Y. 1964. Meiotic chromosomes of monoecious Kentucky hemp (*Cannabis sativa*). *Bull. Torrey Bot. Club* **91**, 193–205.

Michaelis, P. 1954. Cytoplasmic inheritance in *Epilobium* and its theoretical significance. *Adv. Gen.* **6**, 287–401.

Miri, R. K. and J. B. Bubar 1966. Self-incompatibility as an outcrossing mechanism in birdsfoot trefoil (*Lotus corniculatus*). *Can. J. Pl. Sci.* **46**, 411–18.

Mitchell, N. D. and A. J. Richards 1979. Biological flora of the British Isles. *Brassica oleracea* ssp. *oleracea. J. Ecol.* **67**, 1087–96.

Mitter, C., D. J. Futuyama, J. C. Schneider and J. D. Hare 1979. Genetic variation and host plant relations in a parthenogenetic moth. *Evolution* **33**, 770–90.

Modilbowska, I. 1942. Bimodality of crowded pollen tubes in *Primula obconica. J. Heredity* **33**, 187–90.

Mogford, D. J. 1974. Flower colour polymorphism in *Cirsium palustre*. 2. Pollination. *Heredity* **33**, 257–63.

Mogford, D. J. 1978. Pollination and flower colour polymorphism, with special reference to *Cirsium palustre*. In *The pollination of flowers by insects*, A. J. Richards (ed.), 191–9. London: Academic Press.

Mogie, M. 1982. *The status of Taraxacum agamospecies*. Ph.D. thesis, University of Newcastle upon Tyne.

Mogie, M. 1985. Morphological, developmental and electrophoretic variation within and between obligately apomictic *Taraxacum* species. *Biol. J. Linn. Soc.* **24**, 207–16.

Mogie, M. and A. J. Richards 1983. Satellited chromosomes, systematics and phylogeny in *Taraxacum*. *Pl. Syst. Evol.* **141**, 219–229.
Mohl, H. von 1863. Einige Beobachtungen über dimorphe Blüten. *Bot. Z. Berl.* **21**, 309.
Moldenke, A. R. 1975. Niche specialization and species diversity along a Californian transect. *Oecologia* **21**, 219–42.
Moore, D. M. and H. Lewis 1965. The evolution of self-pollination in *Clarkia xantiana*. *Evolution* **19**, 104–14.
Morita, T. 1976. Geographical distribution of diploid and polyploid *Taraxacum* in Japan. *Bull. Nat. Sci. Mus. Tokyo, ser B* **2**, 23–38.
Morita, T. 1980. A search for diploid *Taraxacum* in Korea and eastern China, by means of pollen observations on herbarium specimens. *J. Jap. Bot.* **55**, 33–44.
Morris, M. G. and F. H. Perring 1974. (eds). *The British oak: its history and natural history*. Publ. Bot. Soc. Br. Is., Faringdon: E. W. Classey.
Mukerji, S. K. 1936. Contributions to the autecology of *Mercurialis perennis*. *J. Ecol.* **24**, 38–91, 317–39.
Mulcahy, D. L. 1967. Optimal sex ratio in *Silene alba*. *Heredity* **22**, 411–23.
Mulcahy, D. L. 1968. The significance of delayed pistillate anthesis in *Silene alba*. *Bull. Torrey Bot. Club.* **95**, 135–9.
Mulcahy, D. L. and G. B. Mulcahy 1983. Gametophytic self-incompatibility reexamined. *Science* **220**, 1247–51.
Müller, H. 1883. *The fertilisation of flowers*. Transl. W. D'Arcy. London: Thompson.
Müller, U. 1972. Zytologisch-embryologische Beobachtungen an *Taraxacum*-arten aus der Sektion *Vulgaria* Dahlst. in der Schweiz. *Ber. Geobot. Inst. Eth. Stif. Rübel* **41**, 48–55.
Muntzing, A. 1930. Outlines to a genetic monograph of the genus *Galeopsis* with special reference to the nature and inheritance of partial sterility. *Heredity* **13**, 185–341.
Muntzing, A. 1945. The mode of reproduction of hybrids between sexual and apomictic *Potentilla argentea*. *Bot. Not.* **107**, 49–71.
Muntzing, A. 1954. The cytological basis of polymorphism in *Poa alpina*. *Hereditas* **40**, 459–516.
Murbeck, S. 1904. Parthenogenese bei den Gattungen *Taraxacum* und *Hieracium*. *Bot. Not.* **57**, 285–96.
Murray, B. G. 1979. The genetics of self-incompatibility in *Briza spicata*. *Incomp. Newslett.* **11**, 42–5.

Nakanishi, T. and K. Hinata 1973. An effective time for $CO_2$ gas treatment in overcoming self-incompatibility in *Brassica*. *Plant Cell Physiol.* **14**, 873–9.
Nasrallah, M. E. and D. H. Wallace 1967. Immunogenetics of self-compatibility in *Brassica oleracea* L. *Heredity* **22**, 519–27.
Nasrallah, M. E., J. T. Burber and D. H. Wallace 1969. Self-incompatibility proteins in plants: detection, genetics and possible mode of action. *Heredity* **24**, 23–7.
Nettancourt, D. de 1972. Self-incompatibility in basic and applied researches with higher plants. *Genetica Agraria* **26**, 163–216.
Nettancourt, D. de 1975. Facts and hypotheses on the origin and on the function of the *S* gene in *N. alata* and *L. peruvianum*. *Proc. R. Soc. Lond. B* **188**, 345–60.
Nettancourt, D. de 1977. *Incompatibility in Angiosperms*. Berlin: Springer-Verlag.
Nettancourt, D. de, R. Ecochard, M. D. G. Perquin, T. van der Drift and M. Westerhof 1971. The generation of new *S* alleles at the incompatibility locus of *L. peruvianum* Mill. *Theoret. Appl. Gen.* **41**, 120–9.

Nettancourt, D. de, M. Devreux, U. Laneri, M. Cresti, E. Pacini, G. Sarfatti 1974. Genetical and ultrastructural aspects of self- and cross-incompatibility in interspecific hybrids between self-compatible *Lycopensicon esculentum* and self-incompatible *L. peruvianum*. *Theoret. Appl. Genet.* **44**, 278–88.
New, J. K. 1959. A population study of *Spergula arvensis*. II. Genetics and breeding behaviour. *Ann. Bot. n.s.* **23**, 23–33.
Nijs, J. C. M. den and A. A. Sterk 1980. Cytogeographical studies of *Taraxacum* Sect. Taraxacum (=sect. Vulgaria) in central Europe. *Bot. Jahrb. Syst.* **101**, 527–54.
Nijs, J. C. M. den and A. A. Sterk 1984. Cytogeography of *Taraxacum* sectio Taraxacum and sectio Alpestria in France and adjacent parts of Italy and Switzerland, including some taxonomic remarks. *Act. Bot. Neerl.* **33**, 1–24.
Nitsch, J., E. B. Kurtz, J. L. Livermann and F. W. Went 1952. The development of sex expression in *Cucurbit* flowers. *Am. J. Bot.* **39**, 32–43.
Noble, J. C., A. D. Bell and J. L. Harper 1979. The population biology of plants with clonal growth. I. The morphology and structural demography of *Carex arenaria*. *J. Ecol.* **67**, 983–1008.
Nogler, G. A. 1972. Genetik der aposporie bei *Ranunculus auricomus*. II. Endospermzytologie. *Ber. Schw. Bot. Ges.* **82**, 54–63.
Nordborg, G. 1967. Embryologic studies in the *Sanguisorba minor* complex (Rosaceae). *Bot. Not.* **129**, 109–20.
Nygren, A. 1967. Apomixis in the angiosperms. *Handb. der Pflanzenphys.* **18**, 551–96.

Ockendon, D. J. 1968. Biosystematic studies in the *Linum perenne* group. *New Phytol.* **67**, 787–813.
Ockendon, D. J. 1974. Distribution of self-incompatibility alleles and breeding structure of open-pollinated cultivars of Brussels sprouts. *Heredity* **33**, 159–71.
Ockendon, D. J. 1977. Rare self-incompatibility alleles in a purple cultivar of Brussels sprouts. *Heredity* **39**, 149–52.
Ockendon, D. J. 1980. Distribution of S-alleles and breeding structure of cape broccoli (*Brassica oleracea* var. *'italica'*). *Theoret. Appl. Genet.* **58**, 11–15.
Ockendon, D. J. and L. Currah 1979. The effect of protandry on the amount of outcrossing in the onion (*Allium cepa* L.). *Hort. Res.* **19**, 55–61.
Ockendon, D. J. and S. M. Walters 1970. Studies in *Potentilla anserina* L. *Watsonia* **8**, 135–44.
O'Donnell, S. and M. J. Lawrence 1984. The population genetics of the self-incompatibility polymorphism in *Papaver rhoeas*. IV. The estimation of the number of alleles in a population. *Heredity* **53**, 495–508.
Øllgaard, H. 1978. New species of *Taraxacum* from Denmark. *Bot. Not.* **131**, 487–521.
Ono, T. 1935. Chromosomen und sexualität von *Rumex acetosa*. *Tohoku Imp. Univ. Sci. Rep.* ser 4, **10**, 41–210.
Opler, P. A. and K. S. Bawa 1978. Sex ratios in tropical forest trees. *Evolution* **32**, 812–21.
Opler, P. A., H. G. Baker and G. W. Frankie 1975. Reproductive biology of some Costa Rican *Cordia* species (Boraginaceae) [*sic*, usually placed in Ehretiaceae – AJR]. *Biotropica* **7**, 234–47.
Ornduff, R. 1969. Reproductive biology in relation to systematics. *Taxon* **18**, 121.
Ornduff, R. 1970. Incompatibility and the pollen economy of *Jepsonia parryi*. *Am. J. Bot.* **57**, 1036–41.
Ornduff, R. 1971. The reproductive system of *Jepsonia heterandra*. *Evolution* **25**, 300–11.

Ornduff, R. 1975a. Complementary roles of halictids and syrphids in the pollination of *Jepsonia heterandra* (Saxifragaceae). *Evolution* **29**, 371–3.
Ornduff, R. 1975b. Heterostyly and pollen flow in *Hypericum aegypticum* (Guttiferae). *Bot. J. Linn. Soc.* **71**, 51–7.
Ornduff, R. 1976. The reproductive system of *Amsinckia grandiflora*, a distylous species. *Syst. Bot.* **1**, 57–66.
Ornduff, R. 1979a. The genetics of heterostyly in *Hypericum aegypticum*. *Heredity* **42**, 271–2.
Ornduff, R. 1979b. Pollen flow in *Primula vulgaris*. *Bot. J. Linn. Soc.* **78**, 1–10.
Ornduff, R. 1980a. Pollen flow in *Primula veris* (Primulaceae). *Pl. Syst. Evol.* **135**, 89–94.
Ornduff, R. 1980b. Heterostyly, population composition, and pollen flow in *Hedyotis caerulea*. *Am. J. Bot.* **67**, 95–103.

Palmer, R. G. 1971. Cytological studies of ameiotic and normal maize with references to premeiotic pairing. *Chromosoma* **35**, 233–46.
Pandey, K. K. 1956. Mutations of self-incompatibility alleles in *Trifolium pratense* and *T. repens*. *Genetics* **41**, 353–66.
Pandey, K. K. 1957. Genetics of incompatibility of *Physalis ixocarpa* Brot. A new system. *Am. J. Bot.* **44**, 879–87.
Pandey, K. K. 1959. Mutations of the self-incompatibility gene (S) and pseudo-compatibility in angiosperms. *Lloydia* **22**, 222–34.
Pandey, K. K. 1962. Interspecific incompatibility in *Solanum* species. *Am. J. Bot.* **49**, 874–82.
Pandey, K. K. 1967. Elements of the S-gene complex. II. Mutation and complementation at the $S_1$ locus in *Nicotiana alata*. *Heredity* **22**, 255–83.
Pandey, K. K. 1970. Time and site of the *S*-gene action, breeding systems and relationships in incompatibility. *Euphytica* **19**, 364–72.
Pandey, K. K. 1973. Phases in *S*-gene expression and *S*-allele interaction in the control of interspecific incompatibility. *Heredity* **31**, 381–400.
Pandey, K. K. 1979. Overcoming incompatibility and promoting genetic recombination in flowering plants. *NZ J. Bot.* **17**, 645–64.
Pandey, K. K. and J. H. Troughton 1974. Scanning electron microscopic observations of pollen grains and stigma in the self-incompatible heteromorphic species *Primula malacoides* Franch. and *Forysythia* × *intermedia* Zab. and genetics of sporopollenin deposition. *Euphytica* **23**, 337–44.
Paton, D. C. 1982. The influence of honeyeaters on flowering strategies of Australian plants. In *Pollination and evolution*, J. A. Armstrong, J. M. Powell and A. J. Richards (eds), 95–108. Sydney: Royal Botanic Garden.
Percival, M. S. 1961. Types of nectar in angiosperms. *New Phytol.* **60**, 235–81.
Percival, M. S. 1965. *Floral Biology*. Oxford: Pergamon.
Perring, F. H. and P. D. Sell 1968. *Critical supplement to the atlas of the British Flora*. London: Nelson.
Perttula, U. 1941. Untersuchungen über die generative und vegetative vermehrung der Blütenpflanzen in der wald-, hain-, wiesen, und hainfelsen vegetation. *Ann. Acad. Sci. Fenn.* ser. A. **58**, 1–388.
Philipp, M. 1980. Reproductive biology of *Stellaria longipes* Goldie as revealed by a cultivation experiment. *New Phytol.* **85**, 557–69.
Philipp, M. and O. Schou 1981. An unusual heteromorphic incompatibility system. Distyly, self-incompatibility, pollen load and fecundity in *Anchusa officinalis* (Boraginaceae). *New Phytol.* **89**, 693–703.
Philipson, M. N. 1978. Apomixis in *Cortaderia jubata* (Gramineae). *NZ J. Bot.* **16**, 45–59.

Phillips, M. A. and A. H. D. Brown 1977. Mating system and hybridity in *Eucalyptus pauciflora*. *Aust. J. Biol. Sci.* **30**, 337–44.

Pijl, L. van der 1978. Reproductive integration and sexual disharmony in floral functions. In *The pollination of flowers by insects*, A. J. Richards (ed.), 79–88. London: Academic Press.

Piper, J. G., B. Charlesworth and D. Charlesworth 1984. A high rate of self-fertilization and increased seed fertility of homostyle primroses. *Nature* **310**, 50–1.

Placke, A. 1958. Effect of gibberellic acid on corolla size. *Nature* **182**, 610.

Plitmann, U. and D. A. Levin 1983. Pollen–pistil relationships in the Polemoniaceae. *Evolution* **37**, 957–67.

Policansky, D. 1981. Sex choice and the size advantage in jack-in-the-pulpit (*Arisaema triphyllum*). *Proc. Natl. Acad. Sci., USA* **78**, 1306–1308.

Pope, O., D. M. Simpson and E. N. Duncan 1944. Effect of corn barriers on natural crossing in cotton. *J. Agric. Res.* **68**, 347–61.

Prell, H. 1921. Das Problem der Unfruchtbarkeit. *Naturw. Wochschr. N.F.* **20**, 440–6.

Prentice, H. C. 1984. The sex ratio in a dioecious endemic plant, *Silene diclinis*. *Genetica* **64**, 129–33.

Price, S. D. and S. C. H. Barrett 1984. The function and adaptive significance of tristyly in *Pontederia cordata* L. (Pontederiaceae). *Biol. J. Linn. Soc.* **21**, 315–29.

Primack, R. B. and D. G. Lloyd 1980. Andromonoecy in the New Zealand montane shrub Manuka, *Leptospermum scoparium* (Myrtaceae). *Am. J. Bot.* **67**, 361–8.

Primack, R. B. and J. A. Silander 1975. Measuring the relative importance of different pollinators to plants. *Nature* **255**, 143–4.

Proctor, M. C. F. 1978. Insect pollination syndromes in an evolutionary and ecosystemic context. In *The pollination of flowers by insects*, A. J. Richards (ed.), 105–16. London: Academic Press.

Proctor, M. C. F. and P. F. Yeo 1973. *The pollination of flowers*. London: Collins New Naturalist 54.

Punnett, R. C. 1950. The early days of genetics. *Heredity* **4**, 1–10.

Putwain, P. D. and J. L. Harper 1972. Studies in the dynamics of plant populations. V. Mechanisms governing the sex ratio in *Rumex acetosa* and *R. acetosella*. *J. Ecol.* **60**, 113–29.

Pyke, G. H. 1978a. Optimal foraging: Movement patterns of bumblebees between inflorescences. *Theor. Pop. Biol.* **13**, 72–98.

Pyke, G. H. 1978b. Optimal foraging in bumblebees and coevolution with their plants. *Oecologia* **36**, 281–96.

Pyke, G. H. 1978c. Optimal foraging in hummingbirds: Testing the marginal value theorem. *Am. Zool.* **18**, 627–40.

Pyke, G. H. 1979. Optimal foraging in bumblebees: Rule of movement between flowers within inflorescences. *Anim. Behav.* **27**, 1167–81.

Pyke, G. H. 1980a. Optimal foraging in bumblebees: Calculation of net rate of energy intake and optimal patch choice. *Theor. Pop. Biol.* **17**, 232–46.

Pyke, G. H. 1980b. Optimal foraging in nectar-feeding birds and coevolution with their plants. In *Foraging behaviour*, A. C. Kamil and T. D. Sargent (eds). New York: Garland Press.

Pyke, G. H. 1981. Optimal foraging in hummingbirds: A test of optimal foraging theory. *Anim. Behav.* **29**, 889–96.

Pyke, G. H. 1982a. Animal movements: An optimal foraging approach. In *The ecology of animal movement*, I. R. Swingland and P. J. Greenwood (eds). Oxford: Oxford University Press.

Pyke, G. H. 1982b. Evolution of inflorescence size and height in Waratahs (*Telopea speciosissima*): the difficulties of interpreting correlations between plant traits and fruit set. In *Pollination and evolution*, J. A. Armstrong, J. M. Powell and A. J. Richards (eds), 91–4. Sydney: Royal Botanic Gardens.

Pyke, G. H., H. R. Pulliam and E. L. Charnov 1977. Optimal foraging: a selective review of theory and tests. *Q. Rev. Biol.* **52**, 137–54.

Ramirez, B. W. 1969. Fig wasps: mechanism of pollen transfer. *Science* **163**, 580–1.

Rangaswamy, N. S. and K. R. Shivanna 1972. Overcoming self-incompatibility in *Petunia axillaris*. III. Two-site pollinations *in vitro*. *Phytomorphology* **21**, 284–9.

Rhoades, M. M. 1956. Genetic control of chromosome behaviour. *Maize Genet. Coop. News Lett.* **30**, 38–42.

Richards, A. J. 1970a. Eutriploid facultative agamospermy in *Taraxacum*. *New Phytol.* **69**, 761–74.

Richards, A. J. 1970b. Hybridisation in *Taraxacum*. *New Phytol.* **69**, 1103–21.

Richards, A. J. 1970c. Observations on *Taraxacum* sect. Erythrosperma in Slovakia. *Act. F.R.N. Univ. Comen.*, Bot. **18**, 81–120.

Richards, A. J. 1970d. Evolution of Alpines. *Bull. Alpine Garden Soc.* **38**, 160–9.

Richards, A. J. 1972. The *Taraxacum* flora of the British Isles. *Watsonia* **9** (suppl.), 1–141.

Richards, A. J. 1973. The origin of *Taraxacum* agamospecies. *Bot. J. Linn. Soc.* **66**, 189–211.

Richards, A. J. 1975a. Notes on the sex and age of *Potentilla fruticosa* L. in Upper Teesdale. *Trans. Nat. Hist. Soc. Northumbria* **42**, 85–97.

Richards, A. J. 1975b. The inheritance and behaviour of the rayed gene complex in *Senecio vulgaris*. *Heredity* **34**, 95–104.

Richards, A. J. 1975c. *Sorbus*. In *Hybridisation and the flora of the British Isles*, C. A. Stace (ed.), 233–8. London: Academic Press.

Richards, A. J. 1977. An account of *Primula* section *Petiolares* in cultivation. *J. Scot. Rock Gdn. Club* **15** (1977), 177–214.

Richards, A. J. 1981. An attempt to introduce New Zealand alpine plants. *Bull. Alpine Garden Soc.* **49**, 111–26.

Richards, A. J. 1982. The influence of minor structural changes in the flower on breeding systems and speciation in *Epipactis* Zinn. (Orchidaceae). In *Pollination and evolution*, J. A. Armstrong, J. M. Powell and A. J. Richards (eds), 47–53. Sydney: Royal Botanic Gardens.

Richards, A. J. 1984. The sex life of primroses. *Nature* **310**, 12–13.

Richards, A. J. and J. Blakemore 1975. Factors affecting the germination of turions in *Hydrocharis morsus-ranae* L. *Watsonia* **10**, 273–5.

Richards, A. J. and H. Ibrahim 1978. Estimation of neighbourhood size in two populations of *Primula veris*. In *The pollination of flowers by insects*, A. J. Richards (ed.), 165–74. London: Academic Press.

Richards, A. J. and Ibrahim 1982. The breeding system in *Primula veris* L. II. Pollen tube growth and seed set. *New Phytol.* **90**, 305–14.

Richards, A. J. and P. D. Sell 1976. *Taraxacum* in *Flora Europaea*, IV, T. G. Tutin *et al.* (eds), 332–43, 499–503. Cambridge: Cambridge University Press.

Richards, R. A. and N. Thurling 1973. The genetics of self-incompatibility in *Brassica campestris* L. ssp. *oleifera* Metzg. *Genetica* **44**, 428–38, 439–53.

Rick, C. M. and G. C. Hanna 1943. Determination of sex in *Asparagus officinalis* L. *Am. J. Bot.* **30**, 711–14.

Rick, C. M., M. Holle and R. W. Thorp 1978. Rates of cross pollination in *Lycopersicon pimpinellifolium*: impact of genetic variation in floral characters. *Pl. Syst. Evol.* **129**, 31–44.

Riley, H. P. 1936. The genetics and physiology of self-sterility in the genus *Capsella*. *Genetics* **21**, 24–39.

Riley, R. 1956. The influence of the breeding system on the genecology of *Thlaspi alpestre* L. *New Phytol.* **55**, 319–30.

Roberts, R. H. 1945. *Proc. Am. Soc. Hort. Sci.* **46**, 87.

Roggen, H. P., and A. J. van Dijk 1972. Breaking incompatibility in *Brassica oleracea* L. by steel-brush pollination. *Euphytica* **21**, 48–51.

Roggen, H. P., A. J. van Dijk and C. Dorsman 1972. 'Electric aided' pollination: A method of breaking incompatibility in *Brassica oleracea. Euphytica* **21**, 181–4.

Rollins, R. C. 1945. Evidence for genetic variation among apomictically produced plants of several $F_1$ progenies of guayale (*Parthenium argentatum*) and mariola (*P. incanum*). *Am. J. Bot.* **82**, 554–60.

Rosov, S. A. and N. D. Screbtsova 1958. Honeybees and selective fertilization of plants. XVII. *International Beekeeping Congress* **2**, 494–501.

Ross, M. D. 1973. The inheritance of self-incompatibility in *Plantago lanceolata. Heredity* **30**, 169–76.

Ross, M. D. 1978. The evolution of gynodioecy and subdioecy. *Evolution* **32**, 174–88.

Ross, M. D. 1982. Five evolutionary pathways to subdioecy. *Am. Nat.* **119**, 297–318.

Rousi, A. 1965. Biosystematic studies on the species aggregate *Potentilla anserina* L. *Suom. El. Kas. Seur. V. El. Julk.* **2**, 47–112.

Royama, T. 1971. Evolutionary significance of predators' response to local differences in prey density: a theoretical study. In *Dynamics of populations*, P. J. den Boer and G. R. Gradwell (eds). Wageningen: PUDOC.

Russell, S. D. 1981. Structure and quantitative cytology of male gametes of *Plumbago zeylanica* L. Abstracts, *XIII International Botanical Congress, Sydney*, **61**.

Salisbury, E. J. 1942. *The reproductive capacity of plants.* London: Bell.

Sampson, D. R. 1967. Frequency and distribution of self-incompatibility alleles in *Raphanus raphanistrum. Genetics* **56**, 241–51.

Sansome, F. W. 1938. Sex determination in *Silene otites* and related species. *J. Genet.* **35**, 387–96.

Saran, S. and Wet, J. M. J. de 1970. The mode of reproduction in *Dicanthium intermedium (sic)* (Gramineae). *Bull. Torrey Bot. Club* **97**, 6–13.

Sarukhan, J. 1977. On selective pressures and energy allocation in populations of *Ranunculus repens* L. *Ann. Missouri Bot. Gdn.* **63**, 290–308.

Savina, G. I. 1974. Fertilization in Orchidaceae. In *Fertilization in higher plants*, H. F. Linskens (ed.), 197–204. Amsterdam: North Holland.

Schaal, B. A. 1980. Measurement of gene flow in *Lupinus texensis. Nature* **284**, 450–1.

Schaal, B. A. and D. A. Levin 1978. Morphological differentiation and neighbourhood size in *Liatris cylindracea. Am. J. Bot.* **65**, 923–8.

Schaffner, J. H. 1921. Influence of environment on sexual expression in hemp. *Bot. Gaz.* **71**, 197–218.

Schaffner, J. H. 1923. The influence of relative length of daylight on the reversal of sex in hemp. *Ecology* **4**, 323–34.

Schick, J. M., R. J. Hoffmann and A. N. Lamb 1979. Asexual reproduction, population structure and genotype environment interactions in sea anemones. *Am. Zool.* **19**, 699–718.

Schmitt, J. 1980. Pollinator foraging behaviour and gene dispersal in *Senecio* (Compositae). *Evolution* **34**, 934–42.

Schoen, D. J. 1982a. The breeding system of *Gilia achilleifolia*: variation in floral characteristics and outcrossing rate. *Evolution* **36**, 352–60.

Schoen, D. J. 1982b. Genetic variation and the breeding system of *Gilia achilleifolia*. *Evolution* **36**, 361–70.

Schoener, T. W. 1969. Models of optimal size for solitary predators. *Am. Nat.* **103**, 277–313.

Schou, O. 1983. The distyly in *Primula elatior* (L.) Hill (Primulaceae), with a study of flowering phenology and pollen flow. *Bot. J. Linn. Soc.* **86**, 261–74.

Schou, O. 1984. The dry and wet stigmas of *Primula obconica*: Ultrastructural and cytochemical dimorphisms. *Protoplasma* **121**, 99–113.

Schou, O. and M. Philipp 1983. An unusual heteromorphic incompatibility system. 3. On the genetic control of distyly and self-incompatibility in *Anchusa officinalis* L. (Boraginaceae). *Theoret. Appl. Genet.* **68**, 139–44.

Schwarz, O. 1968. Beiträge zur Kenntnis der Gattung *Primula*. *Wiss. Z. F.-S. Univ. Jena, Mat.-Nat. R.* **17**, 307–32.

Sedgely, M. 1974. Assessment of serological techniques for *S*-allele identification in *Brassica oleracea*. *Euphytica* **23**, 543–51.

Selander, R. K. 1976. Genic variation in natural populations. In *Molecular Evolution*, F. J. Ayala (ed.), 21–45. Sunderland, Mass: Sinauer.

Sharma, S. and K. R. Shivanna 1982. Effects of pistil extracts on *in vitro* responses of compatible and incompatible pollen in *Petunia hybrida* Vilm. *Ind. J. Exp. Biol.* **20**, 255–6.

Shivanna, K. R. and J. Heslop-Harrison 1981. Membrane state and pollen viability. *Ann. Bot.* **47**, 759–70.

Shivanna, K. R. and N. S. Rangaswamy 1969. Overcoming self-incompatibility in *Petunia axillaris*. I. Delayed pollination, pollination with stored pollen, and bud pollination. *Phytomorphology* **19**, 372–80.

Shivanna, K. R., J. Heslop-Harrison and Y. Heslop-Harrison 1981. Heterostyly in *Primula*. 2. Sites of pollen inhibition, and effects of pistil constituents on compatible and incompatible pollen-tube growth. *Protoplasma* **95**, 229–54.

Shivanna, K. R., J. Heslop-Harrison and Y. Heslop-Harrison 1983. Heterostyly in *Primula*. 3. Pollen water economy: a factor in the intramorph incompatibility response. *Protoplasma* **117**, 175–84.

Shull, G. H. 1929. Species hybridizations among old and new species of shepherd's purse. *Int. Congr. Pl. Sci.* **I**, 837–88.

Simmonds, N. W. 1971. The breeding system of *Chenopodium quinoa*. I. Male sterility. *Heredity* **27**, 73–82.

Simmonds, N. W. 1976. *Evolution of Crop Plants*. London: Longman.

Simpson, D. A. 1984. A short history of the introduction and spread of *Elodea* Michx. in the British Isles. *Watsonia* **15**, 1–9.

Sirks, M. J. 1927. The genotypical problems of self and cross-incompatibility. *Mem. Hort. Soc. New York* **3**, 325–43.

Smith, B. W. 1963. The mechanism of sex determination in *Rumex hastatulus*. *Genetics* **48**, 1265–88.

Snaydon, R. W. 1970. Rapid population differentiation in a mosaic environment. 1. The response of *Anthoxanthum odoratum* populations to soils. *Evolution* **24**, 257–69.

Soane, I. D. and A. R. Watkinson 1979. Clonal variation in populations of *Ranunculus repens*. *New Phytol.* **82**, 557–73.

Soest, J. L. van 1963. *Taraxacum* species from India, Pakistan and neighbouring countries. *Wentia* **10**, 1–91.

Solbrig, O. T. and R. C. Rollins 1977. The evolution of autogamy in species of the mustard genus *Leavenworthia*. *Evolution* **31**, 265–81.

Solbrig, O. T. and B. B. Simpson 1974. Components of regulation of a population of dandelions in Michigan. *J. Ecol.* **62**, 473–86.

Solbrig, O. T., S. K. Jain, G. B. Johnson and P. H. Raven 1979. *Topics in plant population biology*. New York: Columbia University Press.

Solnetzeva, M. P. 1974. Disturbances in the progress of fertilization in angiosperms under hemigamy. In *Fertilization in higher plants*, H. F. Linskens (ed.), 311–24. Amsterdam: North Holland.

Sørensen, T. 1954. Adaptations of small plants to deficient nutrition and a short growing season. Illustrated by cultivation experiments with *Capsella bursa-pastoris* (L.) Med. *Bot. Tidsskr.* **51**, 339–61.

Sørensen, T. 1958. Sexual chromosome-aberrants in triploid apomictic *Taraxaca. Bot. Tidskr.* **54**, 1–22.

Sørensen, T. and G. Gudjónsson 1946. Spontaneous chromosome-aberrants in apomictic *Taraxaca. K. Danske Vid. Selsk. Biol. Medd.* **4**, 3–48.

Soyrinki, N. 1938. Studien über die generative und vegetative vermehrung der Samenpflanzen in der alpinen vegetation Petsamo-Lapplands. *Ann. Bot. Soc. Zool-Bot. Fenn. Vanamo* **11**, 1–311.

Spoor, W. 1976. Self-incompatibility in *Lolium perenne* L. *Heredity* **37**, 417–21.

Sprengel, C. K. 1793. *Das Entdecke Geheimnis der Natur im Bau und in der Befruchtung der Blumen*. Berlin.

Stace, C. A. 1975. *Hybridisation and the Flora of the British Isles*. London: Academic Press.

Stanley, R. G. and F. A. Loewus 1964. Boron and myo-inositol in pollen pectin biosynthesis. In *Pollen Physiology* and *Germination*, H. F. Linskens (ed.), 128–36. Amsterdam: North Holland.

Staudt, G. 1952. Genetische Untersuchungen an *Fragaria orientalis* Los. und ihre Bedeutung für Arbildung und Geschlechts-Differenzierung in der Gattung *Fragaria* L. *Z. Indukt. Abst. Ver.* **84**, 361–416.

Stebbins, G. L. 1950. *Variation and evolution in plants*. New York: Columbia University Press.

Stebbins, G. L. and L. Ferlan 1956. Population variability, hybridization and introgression in some species of *Ophrys. Evolution* **10**, 32–46.

Stelleman, P. 1978. The possible role of insect visits in pollination of reputedly anemophilous plants, exemplified by *Plantago lanceolata*, and syrphid flies. In *The pollination of flowers by insects*, A. J. Richards (ed.), 41–6. London: Academic Press.

Stelleman, P. 1979. The significance of biotic pollination in a nominally anemophilous plant: *Plantago lanceolata. Proc. Kon. Nederl. Akad. Wet.* ser. C **87**, 95–119.

Stephens, S. G. and M. D. Finker 1953. Natural crossing in cotton. *Econ. Bot.* **7**, 257–69.

Stevens, D. P. 1985. Studies in *Saxifraga granulata* L. Ph.D. thesis, University of Newcastle upon Tyne.

Stevens, D. P. and A. J. Richards 1985. Gynodiecy in *Saxifraga granulata. Pl. Syst. Evol.* **151**, 43–54.

Stevens, V. A. M. and B. G. Murray 1981. Studies on heteromorphic self-incompatibility systems: The cytochemistry and ultrastructure of the tapetum of *Primula obconica. J. Cell Sci.* **50**, 419–31.

Stevens, V. A. M. and B. G. Murray 1982. Studies on heteromorphic self-incompatibility systems: physiological aspects of the incompatibility system of *Primula obconica. Theoret. Appl. Genet.* **61**, 245–56.

Stiles, F. G. 1975. Ecology, flowering phenology and hummingbird pollination of some Costa Rican *Heliconia* species. *Ecoloygy* **56**, 285–301.

Stoddart, J. A. 1983. The accumulation of genetic variation in a parthenogenetic snail. *Evolution* **37**, 546–54.

Storey, W. B. 1975. In *Advances in fruit breeding*, J. Janick and J. N. Moore (eds), 568–89. West Lafayette: Purdue University Press.

Stout, A. B. 1916. Self- and cross-pollinations in *Cichorium intybus* with reference to sterility. *Mem. New York Bot. Gdn* **6**, 333–454.

Stout, A. B. 1917. Fertility in *Cichorium intybus*. The sporadic occurrence of self-fertile plants among the progeny of self-sterile plants. *Am. J. Bot.* **4**, 375–95.

Stout, A. B. 1918a. Experimental studies of self-incompatibilities in fertilization. *Proc. Soc. Exp. Biol. Med.* **15**, 51–4.

Stout, A. B. 1918b. Fertility in *Cichorium intybus*: Self-compatibility and self-incompatibility among the offspring of self-fertile lines of descent. *J. Genetics* **8**, 71–103.

Stout, A. B. and C. Chandler 1942. Hereditary transmissions of induced tetraploidy and compatibility in fertilisation. *Science* **96**, 257.

Stoutamire, W. P. 1983. Wasp-pollinated species of *Caladenia* (Orchidaceae) in south-western Australia. *Aust. J. Bot.* **31**, 383–94.

Strenseth, N. C., L. R. Kirkendall and N. Moran 1985. On the evolution of pseudogamy. *Evolution* **39**, 294–307.

Strid, A. 1970. Studies in the Aegean Flora. XVI. Biosystematics of the *Nigella arvensis* complex. *Opera Botanica* **28**, 1–169.

Synge, A. D. 1947. Pollen collection by honeybees (*Apis mellifera*). *J. Anim. Ecol.* **16**, 122–38.

Taliaferro, C. M. and E. C. Bashaw 1966. Inheritance and control of obligate apomixis in breeding buffelgrass, *Pennisetum ciliare*. *Crop. Sci.* **6**, 473–6.

Taylor, P. D. 1984. Evolutionarily stable reproductive allocations in heterostylous plants. *Evolution* **38**, 408–16.

Ter-Avanesian, D. V. 1978. The effect of varying the number of pollen grains used in fertilisation. *Theoret. Appl. Genet.* **52**, 77–9.

Thomas, P. T. 1940. Reproductive versatility in *Rubus*. II. The chromosome and development. *J. Genet.* **40**, 119–28.

Thomson, J. D. 1982. Patterns of visitation by animal pollinators. *Oikos* **39**, 241–50.

Thomson, J. D. and R. C. Plowright 1980. Pollen carryover, nectar rewards, and pollinator behaviour with special reference to *Diervilla lonicera*. *Oecologia* **46**, 68–74.

Trela, Z. 1963. Embryological studies in *Anemone nemorosa* L. *Act. Biol. Crac.* **6**, 1–14.

Turkington, R. and J. L. Harper 1979. The growth, distribution and relationships of *Trifolium repens* in a permanent pasture. IV. Fine-scale biotic differentiation. *J. Ecol.* **67**, 245–54.

Uno, G. E. 1982. Comparative reproductive biology of hermaphroditic and male-sterile *Iris douglasiana* Herb. (Iridaceae). *Am. J. Bot.* **69**, 818–23.

Usberti, J. A. and S. K. Jain 1978. Variation in *Panicum maximum*; a comparison of sexual and asexual populations. *Bot. Gaz.* **139**, 112–16.

Vaarama, A. and O. Jääskeläinen 1973. Studies on gynodioecism in the Finnish populations of *Geranium sylvaticum* L. *Ann. Acad. Sci. Fenn.* A **108**, 1–39.

Valdeyron, G., B. Dommée and P. Vernet 1977. Self-fertilisation in male-fertile plants of a gynodioecious species: *Thymus vulgaris* L. *Heredity* **39**, 243–9.

Vansell, G. H., W. G. Watkins and R. K. Bishop 1942. Orange nectar and pollen in relation to bee activity. *J. Econ. Entomol.* **35**, 321–3.

Vasek, F. C. 1965. Outcrossing in natural populations. II. *Clarkia unguiculata*. *Evolution* **19**, 152–6.

Vasek, F. C. and J. Harding 1976. Outcrossing in natural populations. V. Analysis of outcrossing, inbreeding and selection in *Clarkia exilis* and *Clarkia tremblorien-sis*. *Evolution* **80**, 403–11.

Vepsalainen, K. and O. Jarvinen 1979. Apomictic parthenogenesis and the pattern of the environment. *Am. Zool.* **19**, 739–51.

Vogel, S. 1962. Duftdrüsen im Dienste der Bestäubung: über Bau und Funktion der Osmophoren. *Abh. Math.-Naturw. Kl. Akad. Wiss. Mainz.* **10**, 599–763.

Vogel, S. 1966. Scent organs of orchid flowers and their relation to insect pollination. *Proc. Fifth World Orchid Conference*, 253–59. Long Beach, California.

Vogel, S. 1978. Evolutionary shifts from reward to deception in pollen flowers. In *The pollination of flowers by insects*, A. J. Richards (ed.), 89–96. London: Academic Press.

Vries, H. de 1907. *Plant breeding*. Chicago: Open Court Publ. Co.

Waddington, C. H. 1957. *The strategy of genes*. London: Allen and Unwin.

Waddington, K. D. 1979. Divergence in inflorescence height: An evolutionary response to pollinator fidelity. *Oecologia* **40**, 43–50.

Waddington, K. D. 1981. Factors influencing pollen flow in bumblebee-pollinated *Delphinium virescens*. *Oikos* **37**, 153–9.

Waddington, K. D. 1983. Foraging behaviour of pollinators. In *Pollination Biology*, L. Real (ed.), 213–39. New York: Academic Press.

Waddington, K. D. and B. Heinrich 1981. Patterns of movement and floral choice by foraging bees. In *Foraging behaviour: ecological, ethological and psychological approaches*, A. Kamil and T. Sargent (eds), 215–30. New York: Garland STPM Press.

Waddington, K. D., T. Allen and B. Heinrich 1981. Floral preferences of bumblebees (*Bombus edwardsii*) in relation to intermittent versus continuous rewards. *Anim. Behav.* **29**, 779–84.

Wade, K. M. 1981a. Experimental studies on the distribution of the sexes of *Mercurialis perennis* L. II. Transplanted populations under different canopies in the field. *New Phytol.* **87**, 439–46.

Wade, K. M. 1981b. Experimental studies on the distribution of the sexes of *Mercurialis perennis* L. III. Transplanted populations under light screens. *New Phytol.* **87**, 447–55.

Wade, K. M., R. A. Armstrong and S. R. J. Woodell 1981. Experimental studies on the distribution of the sexes of *Mercurialis perennis* L. 1. Field observations and canopy removal experiments. *New Phytol.* **87**, 431–8.

Wahlund, S. 1928. Zusammensetzung von Populationen und Korrelations-erscheinungen vom Standpunkt der Vererbungslehre aus betrachtet. *Hereditas* **11**, 65–106.

Waller, D. M. 1984. Differences in fitness between seedlings derived from cleistogamous and chasmogamous flowers in *Impatiens capensis*. *Evolution* **38**, 427–40.

Warmke, H. E. 1946. Sex determination and sex balance in *Melandrium album*. *Am. J. Bot.* **33**, 648–60.

Waser, N. M. and M. V. Price 1981. Pollinator choice, and stabilizing selection for flower color in *Delphinium nelsonii*. *Evolution* **35**, 376–90.

Waser, N. M. and M. V. Price 1982. A comparison of pollen and fluorescent dye carryover by natural pollinators of *Ipomopsis aggregata* (Polemoniaceae). *Ecology* **63**, 1168–72.

Waser, N. M. and M. V. Price 1983. Optimal and actual outcrossing in plants, and the nature of plant-pollinator interaction. In *Handbook of experimental pollination biology*, C. E. Jones and R. J. Little (eds), 341–59. New York: Van Nostrand.

Webb, C. J. 1979a. Breeding systems and the evolution of dioecy in New Zealand apioid Umbelliferae. *Evolution* **33**, 662–72.

Webb, C. J. 1979b. Breeding system and seed set in *Euonymus europaeus* (Celastraceae). *Pl. Syst. Evol.* **132**, 299–303.

Webb, C. J. 1981a. Test of a model predicting equilibrium frequencies of females in populations of gynodioecious Angiosperms. *Heredity* **46**, 397–405.

Webb, C. J. 1981b. Andromonoecism, protandry, and sexual selection in Umbelliferae. *NZ J. Bot.* **19**, 335–8.

Webb, C. J. and D. G. Lloyd 1980. Sex ratios in New Zealand apioid Umbelliferae. *NZ J. Bot.* **18**, 121–6.

Weller, S. G. 1976. The genetic control of tristyly in *Oxalis* section Ionoxalis. *Heredity* **37**, 387–92.

Weller, S. G. 1980. Pollen flow and fecundity in populations of *Lithospermum carolinense*. *Am. J. Bot.* **67**, 1334–41.

Wendelbo, P. 1959. *Taraxacum gotlandicum*, a pre-boreal relic in the Norwegian Flora? *Nytt. Mag. Bot.* **7**, 161–7.

Wendelbo, P. 1961. An account of *Primula* subgenus *Sphondylia* with a review of the subdivisions of the genus. *Arbok Univ. Bergen, Mat-Nat.* **11**, 33–43.

Westergaard, M. 1940. Studies on polyploidy and sex determination in polyploid forms of *Melandrium album*. *Dansk. Bot. Arkiv.* **10**(5), 1–131.

Westergaard, M. 1946. Aberrant Y chromosomes and sex expression in *Melandrium album*. *Hereditas* 32, 419–43.

Westergaard, M. 1948. The relation between chromosome constitution and sex in the offspring of triploid *Melandrium*. *Hereditas* **34**, 257–79.

Westergaard, M. 1953. Über den mechanismus der Geschlechtsbestimmung bei *Melandrium album*. *Die Naturwiss.* **9**, 253–60.

Westergaard, M. 1958. The mechanism of sex determination in dioecious flowering plants. *Adv. Gen.* **9**, 217–81.

Wet, J. M. de and H. T. Stalker 1974. Gametophytic apomixis and evolution in plants. *Taxon* **23**, 689–97.

White, M. J. D. 1970. Heterozygosity and genetic polymorphism in parthenogenetic animals. In *Essays in evolution and genetics in honor of Theodosius Dobzhansky*, M. K. Hecht and W. C. Steere (eds), 237–262. New York: Appleton-Century-Croft.

Whitehouse, H. L. K. 1950. Multiple-allelomorph incompatibility of pollen and style in the evolution of the angiosperms. *Ann. Bot. n.s.* **14**, 198–216.

Willemse, M. T. M. 1974. Megagametogenesis and formation of neocytoplasm in *Pinus sylvestris* L. In *Fertilization in higher plants*, H. F. Linskens (ed.), 97–104. Amsterdam: North Holland.

Williams, G. C. 1975. *Sex and evolution*. Princeton, NJ: Princeton University Press.

Williams, N. H. and C. H. Dodson 1972. Selective attraction of male euglossid bees to orchid floral fragrancies and its importance in long distance pollen flow. *Evolution* **26**, 84–95.

Williams, W. 1960. Relative variability of inbred lines and $F_1$ hybrids in *Lycopersicon esculentum*. *Genetics* **45**, 1457–65.

Williams, W. 1964. *Genetic principles and plant breeding*. Oxford: Blackwell.

Winge, O. 1931. X- and Y-linked inheritance in *Melandrium*. *Hereditas* **15**, 127–65.

Wolf, L. L. 1975. Energy intake and expenditures in a nectar-feeding sunbird. *Ecology* **56**, 92–104.

Wolf, L. L. and F. R. Hainsworth 1971. Time and energy budgets of territorial hummingbirds. *Ecology* **52**, 980–8.

Wolf, L. L., F. R. Hainsworth and F. G. Stiles 1972. Energetics of foraging: rate and efficiency of nectar extraction by hummingbirds. *Science* **176**, 1351–2.

Wolfe, H. S. *et al.* 1934. *Florida Univ. Agr. Expt. Sta. (Gainesville) Bull.* **272**.

Woodell, S. R. J. 1960a. Studies in British Primulas. VII. Development of seed from reciprocal crosses between *P. vulgaris* Huds and *P. veris* L. *New Phytol.* **59**, 302–13.

Woodell, S. R. J. 1960b. Studies in British Primulas. VIII. Development of seed from reciprocal crosses between *P. vulgaris* Huds and *P. elatior* (L.) Hill. and between *P. veris* L. and *P. elatior* (L.) Hill. *New Phytol.* **59**, 314.

Woodell, S. R. J. 1978. Directionality in bumblebees in relation to environmental factors. In *The pollination of flowers by insects*, A. J. Richards (ed.), 31–9. London: Academic Press.

Woodruff, R. C. and J. N. Thompson 1980. Hybrid release of mutator activity and the genetic structure of natural populations. *Evol. Biol.* **12**, 129–62.

Wright, J. W. 1953. Pollen-dispersion studies: some practical applications. *J. Forestry* **51**, 114–18.

Wright, S. 1938. Size of population and breeding structure in relation to evolution. *Science* **87**, 430–1.

Wright, S. 1939. The distribution of self-sterility alleles in populations. *Genetics* **24**, 538–52.

Wright, S. 1940. Breeding structure of populations in relation to speciation. *Am. Nat.* **74**, 232–48.

Wright, S. 1943a. Isolation by distance. *Genetics* **28**, 114–38.

Wright, S. 1943b. An analysis of local variability of flower color in *Linanthus parryae*. *Genetics* **28**, 139–56.

Wright, S. 1946. Isolation by distance under diverse systems of mating. *Genetics* **31**, 39–59.

Wright, S. 1951. The genetical structure of populations. *Ann. Eugen.* **15**, 323–54.

Wright, S. 1978. *Evolution and the genetics of populations.* Vol. 4. *Variability within and among natural populations.* Chicago: University of Chicago Press.

Wu, L., A. D. Bradshaw and D. A. Thurman 1975. The potential for evolution of heavy metal tolerance in plants. III. The rapid evolution of copper tolerance in *Agrostis stolonifera*. *Heredity* **34**, 165–87.

Wyatt, R. 1981. Components of reproductive output in five tropical legumes. *Bull. Torrey Bot. Club* **108**, 67–75.

Wyatt, R. 1984. The evolution of self-pollination in granite outcrop species of *Arenaria* (Caryophyllaceae). Morphological correlates. *Evolution* **38**, 804–16.

Yamamota, Y. 1938. Karyogenetische Untersuchungen der Gattung *Rumex*. VI. Geschechtsbestimmung bei eu- und aneuploiden Pflanzen von *Rumex acetosa* L. *Kyoto Imp. Univ., Mem. Coll. Agri.* **43**, 1–59.

Yampolsky, E. and H. Yampolsky 1922. Distribution of sex forms in the phanerogamic flora. *Bibl. Genet.* **3**, 1–62.

Yeo, P. F. 1975. Some aspects of heterostyly. *New Phytol.* **75**, 147–53.

Yudin, B. F. 1970. Capacity for parthenogenesis and effectiveness of selection on the basis of this character in diploid and autotetraploid maize. *Genetika* **6**, 13–22.

Zahn, K. H. 1921–3. *Hieracium* in *Pflanzenreich* IV 280; **75**, 1–288, **76**, 289–576, **77**, 577–864, **79**, 865–1146, **82**, 1147–1705, H. G. A. Engler (ed.).

Zimmerman, M. 1979. Optimal foraging: A case for random movement. *Oecologia* **43**, 261–7.

Zimmerman, M. 1982. The effect of nectar production on neighbourhood size. *Oecologia* **52**, 104–8.
Zuk, J. 1963. An investigation on polyploidy and sex determination within the genus *Rumex*. *Act. Soc. Bot. Pol.* **23** (1), 5–67.

# Index